MODERN ANALYSIS OF AUTOMORPHIC FORMS BY EXAMPLE, VOLUME 1

This is Volume 1 of a two-volume book that provides a self-contained introduction to the theory and application of automorphic forms, using examples to illustrate several critical analytical concepts surrounding and supporting the theory of automorphic forms. The two-volume book treats three instances, starting with some small unimodular examples, followed by adelic GL2, and finally GLn. Volume 1 features critical results, which are proven carefully and in detail, including discrete decomposition of cuspforms, meromorphic continuation of Eisenstein series, spectral decomposition of pseudo-Eisenstein series, and automorphic Plancherel theorem. Volume 2 features automorphic Green's functions, metrics and topologies on natural function spaces, unbounded operators, vector-valued integrals, vector-valued holomorphic functions, and asymptotics. With numerous proofs and extensive examples, this classroom-tested introductory text is meant for a second-year or advanced graduate course in automorphic forms, and also as a resource for researchers working in automorphic forms, analytic number theory, and related fields.

Paul Garrett is Professor of Mathematics at the University of Minnesota–Minneapolis. His research focuses on analytical issues in the theory of automorphic forms. He has published numerous journal articles as well as five books.

CAMBRIDGE STUDIES IN ADVANCED MATHEMATICS

All the titles listed below can be obtained from good booksellers or from Cambridge University Press. For a complete series listing visit: www.cambridge.org/mathematics.

Already published

138 C. Muscalu & W. Schlag *Classical and multilinear harmonic analysis, II*
139 B. Helffer *Spectral theory and its applications*
140 R. Pemantle & M. C. Wilson *Analytic combinatorics in several variables*
141 B. Branner & N. Fagella *Quasiconformal surgery in holomorphic dynamics*
142 R. M. Dudley *Uniform central limit theorems (2nd Edition)*
143 T. Leinster *Basic category theory*
144 I. Arzhantsev, U. Derenthal, J. Hausen & A. Laface *Cox rings*
145 M. Viana *Lectures on Lyapunov exponents*
146 J.-H. Evertse & K. Győry *Unit equations in Diophantine number theory*
147 A. Prasad *Representation theory*
148 S. R. Garcia, J. Mashreghi & W. T. Ross *Introduction to model spaces and their operators*
149 C. Godsil & K. Meagher *Erdős–Ko–Rado theorems: Algebraic approaches*
150 P. Mattila *Fourier analysis and Hausdorff dimension*
151 M. Viana & K. Oliveira *Foundations of ergodic theory*
152 V. I. Paulsen & M. Raghupathi *An introduction to the theory of reproducing kernel Hilbert spaces*
153 R. Beals & R. Wong *Special functions and orthogonal polynomials*
154 V. Jurdjevic *Optimal control and geometry: Integrable systems*
155 G. Pisier *Martingales in Banach spaces*
156 C. T. C. Wall *Differential topology*
157 J. C. Robinson, J. L. Rodrigo & W. Sadowski *The three-dimensional Navier–Stokes equations*
158 D. Huybrechts *Lectures on K3 surfaces*
159 H. Matsumoto & S. Taniguchi *Stochastic analysis*
160 A. Borodin & G. Olshanski *Representations of the infinite symmetric group*
161 P. Webb *Finite group representations for the pure mathematician*
162 C. J. Bishop & Y. Peres *Fractals in probability and analysis*
163 A. Bovier *Gaussian processes on trees*
164 P. Schneider *Galois representations and* (φ, Γ)*-modules*
165 P. Gille & T. Szamuely *Central simple algebras and Galois cohomology* (2nd Edition)
166 D. Li & H. Queffelec *Introduction to Banach spaces, I*
167 D. Li & H. Queffelec *Introduction to Banach spaces, II*
168 J. Carlson, S. Müller-Stach & C. Peters *Period mappings and period domains* (2nd Edition)
169 J. M. Landsberg *Geometry and complexity theory*
170 J. S. Milne *Algebraic groups*
171 J. Gough & J. Kupsch *Quantum fields and processes*
172 T. Ceccherini-Silberstein, F. Scarabotti & F. Tolli *Discrete harmonic analysis*
173 P. Garrett *Modern analysis of automorphic forms by example, I*
174 P. Garrett *Modern analysis of automorphic forms by example, II*
175 G. Navarro *Character theory and the McKay conjecture*
176 P. Fleig, H. P. A. Gustafsson, A. Kleinschmidt & D. Persson *Eisenstein series and automorphic representations*
177 E. Peterson *Formal geometry and bordism operators*
178 A. Ogus *Lectures on Logarithmic Algebraic Geometry*

Modern Analysis of Automorphic Forms by Example, Volume 1

PAUL GARRETT
University of Minnesota–Minneapolis

CAMBRIDGE
UNIVERSITY PRESS

CAMBRIDGE
UNIVERSITY PRESS

University Printing House, Cambridge CB2 8BS, United Kingdom

One Liberty Plaza, 20th Floor, New York, NY 10006, USA

477 Williamstown Road, Port Melbourne, VIC 3207, Australia

314–321, 3rd Floor, Plot 3, Splendor Forum, Jasola District Centre, New Delhi - 110025, India

79 Anson Road, #06-04/06, Singapore 079906

Cambridge University Press is part of the University of Cambridge.

It furthers the University's mission by disseminating knowledge in the pursuit of education, learning, and research at the highest international levels of excellence.

www.cambridge.org
Information on this title: www.cambridge.org/9781107154001
DOI: 10.1017/9781316650332

First published 2018

A catalogue record for this publication is available from the British Library.

ISBN 978-1-107-15400-1 Hardback

Contents

Introduction and Historical Notes

The aim of this book is to offer persuasive proof of several important analytical results about automorphic forms, among them spectral decompositions of spaces of automorphic forms, discrete decompositions of spaces of cuspforms, meromorphic continuation of Eisenstein series, spectral synthesis of automorphic forms, a Plancherel theorem, and various notions of convergence of spectral expansions. Rather than assuming prior knowledge of the necessary analysis or giving extensive external references, this text provides customized discussions of that background, especially of ideas from 20th-century analysis that are often neglected in the contemporary standard curriculum. Similarly, I avoid assumptions of background that would certainly be useful in studying automorphic forms but that beginners cannot be expected to have. Therefore, I have kept external references to a minimum, treating the modern analysis and other background as a significant part of the discussion.

Not only for reasons of space, the treatment of automorphic forms is deliberately neither systematic nor complete, but instead provides three families of examples, in all cases aiming to illustrate aspects beyond the introductory case of $SL_2(\mathbb{Z})$ and its congruence subgroups.

The first three chapters set up the three families of examples, proving essential preparatory results and many of the basic facts about automorphic forms, while merely stating results whose proofs are more sophisticated or difficult. The proofs of the more difficult results occupy the remainder of the book, as in many cases the arguments require various ideas not visible in the statements.

The first family of examples is introduced in Chapter 1, consisting of *waveforms* on quotients having dimensions 2, 3, 4, 5 with a single *cusp*, which is just a *point*. In the two-dimensional case, the space on which the functions live is the usual quotient $SL_2(\mathbb{Z})\backslash\mathfrak{H}$ of the complex upper half-plane $\mathfrak{H}$. The three-dimensional case is related to $SL_2(\mathbb{Z}[i])$, and the four-dimensional and five-dimensional cases are similarly explicitly described. Basic discussion of the physical spaces themselves involves explication of the groups acting on them, and decompositions of these groups in terms of subgroups, as well as the expression of the physical spaces as G/K for K a maximal compact subgroup of G. There are natural invariant measures and integrals on G/K and on $\Gamma\backslash G/K$ whose salient properties can be described quickly, with proofs deferred to a later point. Similarly, a natural Laplace-Beltrami operator Δ on G/K and $\Gamma\backslash G/K$ can be described easily, but with proofs deferred. The first serious result specific to automorphic forms is about *reduction theory*, that is, determination of a nice set in G/K that *surjects* to the quotient $\Gamma\backslash G/K$, for specific *discrete* subgroups Γ of G. The four examples in this simplest scenario all admit very simple sets of representatives, called *Siegel sets*, in every case a product of a ray and a box, with Fourier expansions possible along the box-coordinate, consonant with a decomposition of part of the group G (Iwasawa decomposition). This greatly simplifies both statements and proofs of fundamental theorems.

In the simplest family of examples, the space of *cuspforms* consists of those functions on the quotient $\Gamma\backslash G/K$ with 0th Fourier coefficient identically 0. The basic theorem, quite nontrivial to prove, is that the space of cuspforms in $L^2(\Gamma\backslash G/K)$ has a basis consisting of eigenfunctions for the invariant Laplacian Δ. This result is one form of the *discrete decomposition of cuspforms*. We delay its proof, which uses many ideas not apparent in the statement of the theorem. The orthogonal complement to cuspforms in $L^2(\Gamma\backslash G/K)$ is readily characterized as the space of *pseudo-Eisenstein series*, parametrized here by test functions on $(0, +\infty)$. However, these simple, explicit automorphic forms are never eigenfunctions for Δ. Rather, via Euclidean Fourier-Mellin inversion, they are expressible as *integrals* of (genuine) Eisenstein series, the latter eigenfunctions for Δ, but unfortunately not in $L^2(\Gamma\backslash G/K)$. Further, it turns out that the best expression of pseudo-Eisenstein series in terms of genuine Eisenstein series E_s involves the latter with complex parameter outside the region of convergence of the defining series. Thus arises the need to *meromorphically continue* the Eisenstein series in that complex parameter. Genuine proof of meromorphic continuation, with control over the behavior of the meromorphically continued function, is another basic but nontrivial result, whose proof is delayed. Granting those postponed proofs, a *Plancherel theorem* for the space of pseudo-Eisenstein series follows from their expansion in terms of genuine Eisenstein series, together with attention to integrals as *vector-valued* (rather than merely numerical), with the important corollary that such integrals commute with continuous operators on the vector space. This and other aspects of vector-valued integrals are treated at length in an appendix. Then we obtain the Plancherel theorem for the whole space of L^2 waveforms. Even for the simplest examples, these few issues illustrate the goals of this book: discrete decomposition of spaces of cuspforms, meromorphic continuation of Eisenstein series, and a Plancherel theorem.

In Chapter 2 is the second family of examples, *adele groups* GL_2 over number fields. These examples subsume classical examples of quotient $\Gamma_0(N)\backslash\mathfrak{H}$ with several cusps, reconstituting things so that operationally there is a *single* cusp. Also, examples of Hilbert modular groups and Hilbert modular forms are subsumed by rewriting things so that the vagaries of class numbers and unit groups become irrelevant. Assuming some basic algebraic number theory, we prove p-adic analogues of the group decomposition results proven earlier in Chapter 1 for the purely archimedean examples. Integral operators made from C_c^o functions on the p-adic factor groups, known as *Hecke operators*, are reasonable p-adic analogues of the archimedean factors' Δ, although the same integral operators do make the same sense on archimedean factors. Again, the first serious result for these examples is that of *reduction theory*, namely, that there is a single nice set, an adelic form of a *Siegel set*, again nearly the product of a ray and a box, that surjects to the quotient $Z^+GL_2(k)\backslash GL_2(\mathbb{A})$, where Z^+ is itself a ray in the center of the group. The first serious analytical result is again about *discrete decomposition of spaces of cuspforms*, where now relevant operators are both the invariant Laplacians and the Hecke operators. Again, the deferred proof is much more substantial than the statement and needs ideas not visible in the assertion itself. The orthogonal complement to cuspforms is again describable as the L^2 span of *pseudo-Eisenstein series*, now with a discrete parameter, a Hecke character (grossencharacter) of the ground field, in addition to the test function on $(0, +\infty)$. The pseudo-Eisenstein series are never eigenfunctions for invariant Laplacians or for Hecke operators. Within each family, indexed by Hecke characters, every pseudo-Eisenstein series again decomposes via Euclidean Fourier-Mellin inversion as an integral of (genuine) Eisenstein series with the same discrete parameter. The genuine Eisenstein series are eigenfunctions for invariant Laplacians and are eigenfunctions for Hecke operators at almost all finite places, but are not square-integrable. Again, the best assertion of spectral decomposition requires a meromorphic continuation of the genuine Eisenstein series in the continuous parameter. Then a Plancherel theorem for pseudo-Eisenstein series for each discrete parameter value follows from the integral representation in terms of genuine Eisenstein series and general properties of vector-valued integrals. These are assembled into a Plancherel theorem for all L^2 automorphic forms. An appendix computes *periods* of Eisenstein series along copies of $GL_1(\widetilde{k})$ of quadratic field extensions $\widetilde{k}$ of the ground field.

Chapter 3 treats the most complicated of the three families of examples, including automorphic forms for $SL_n(\mathbb{Z})$, both purely archimedean and adelic.

Again, some relatively elementary set-up regarding group decompositions is necessary and carried out immediately. Identification of invariant differential operators and Hecke operators at finite places is generally similar to that for the previous example GL_2. A significant change is the proliferation of types of *parabolic subgroups* (essentially, subgroups conjugate to subgroups containing upper-triangular matrices). This somewhat complicates the notion of *cuspform*, although the general idea, that zeroth Fourier coefficients vanish, is still correct, if suitably interpreted. Again, the space of square-integrable cuspforms decomposes discretely, although the complexity of the proof for these examples increases significantly and is again delayed. The increased complication of parabolic subgroups also complicates the description of the orthogonal complement to cuspforms in terms of pseudo-Eisenstein series. For purposes of spectral decomposition, the discrete parameters now become more complicated than the GL_2 situation: *cuspforms* on the Levi components (diagonal blocks) in the parabolics generalize the role of Hecke characters. Further, the continuous complex parametrizations need to be over larger-dimensional Euclidean spaces. Thus, I restrict attention to the two extreme cases: minimal parabolics (also called *Borel subgroups*), consisting exactly of upper-triangular matrices, and *maximal proper* parabolics, which have exactly two diagonal blocks. The minimal parabolics use no cuspidal data but for $SL_n(\mathbb{Z})$ have an $(n-1)$-dimensional complex parameter. The maximal proper parabolics have just a one-dimensional complex parameter but typically need two cuspforms on smaller groups, one on each of the two diagonal blocks. The general qualitative result that the L^2 orthogonal complement to cuspforms is spanned by pseudo-Eisenstein series of various types does still hold, and the various types of pseudo-Eisenstein series are integrals of genuine Eisenstein series with the same discrete parameters. Again, the best description of these integrals requires the meromorphic continuation of the Eisenstein series. For nonmaximal parabolics, Bochner's lemma (recalled and proven in an appendix) reduces the problem of meromorphic continuation to the maximal proper parabolic case, with cuspidal data on the Levi components. Elementary devices such as Poisson summation, which suffice for meromorphic continuation for GL_2, as seen in the appendix to Chapter 2, are inadequate to prove meromorphic continuation involving the nonelementary cuspidal data. I defer the proof. Plancherel theorems for the spectral fragments follow from the integral representations in terms of genuine Eisenstein series, together with properties of vector-valued integrals.

The rest of the book gives proofs of those foundational analytical results, discreteness of cuspforms and meromorphic continuation of Eisenstein series, at various levels of complication and by various devices. Perhaps surprisingly, the required analytical underpinnings are considerably more substantial than an unsuspecting or innocent bystander might imagine. Further, not everyone interested in the *truth* of foundational analytical facts about automorphic forms will necessarily care about their *proofs*, especially upon discovery that that burden is greater than anticipated. These obvious points reasonably explain the compromises made in many sources. Nevertheless, rather than either gloss over the analytical issues, refer to encyclopedic treatments of modern analysis on a scope quite unnecessary for our immediate interests, or give suggestive but misleading neoclassical heuristics masquerading as adequate arguments for what is truly needed, the remaining bulk of the book aims to discuss analytical issues at a technical level truly sufficient to convert appealing heuristics to persuasive, genuine proofs. For that matter, one's own lack of interest in the proofs might provide all the more interest in knowing that things widely believed are in fact provable by standard methods.

Chapter 4 explains enough Lie theory to understand the invariant differential operators on the ambient archimedean groups G, both in the simplest small examples and, more generally, determining the invariant Laplace-Beltrami operators explicitly in coordinates on the four simplest examples.

Chapter 5 explains how to integrate on quotients, without concern for explicit sets of representatives. Although in very simple situations, such as quotients $\mathbb{R}/\mathbb{Z}$ (the circle), it is easy to manipulate sets of representatives (the interval $[0, 1]$ for the circle), this eventually becomes infeasible, despite the traditional example of the explicit fundamental domain for $SL_2(\mathbb{Z})$ acting on the upper half-plane $\mathfrak{H}$. That is, much of the picturesque detail is actually inessential, which is fortunate because that level of detail is also unsustainable in all but the simplest examples.

Chapter 6 introduces natural actions of groups on spaces of functions on *physical* spaces on which the groups act. In some contexts, one might make a more elaborate *representation theory* formalism here, but it is possible to reap many of the benefits of the ideas of representation theory without the usual superstructure. That is, the *idea* of a linear action of a topological group on a topological vector space of functions on a physical space is the beneficial notion, with or without *classification*. It is true that at certain technical moments, classification results are crucial, so although we do not prove either the Borel-Casselman-Matsumoto classification in the p-adic case [Borel 1976], [Matsumoto 1977], [Casselman 1980], nor the subrepresentation theorem [Casselman 1978/1980], [Casselman-Miličić 1982] in the archimedean case, ideally the roles of these results are made clear. Classification results per se, although difficult and interesting problems, do not necessarily affect the foundational analytic aspects of automorphic forms.

Chapter 7 proves the discreteness of spaces of cuspforms, in various senses, in examples of varying complexity. Here, it becomes apparent that genuine proofs, as opposed to heuristics, require some sophistication concerning topologies on natural function spaces, beyond the typical Hilbert, Banach, and Fréchet spaces. Here again, there is a forward reference to the extended appendix on function spaces and classes of topological vector spaces necessary for practical analysis. Further, even less immediately apparent, but in fact already needed in the discussion of decomposition of pseudo-Eisenstein series in terms of genuine Eisenstein series, we need a coherent and effective theory of *vector-valued integrals*, a complete, succinct form given in the corresponding appendix, following Gelfand and Pettis, making explicit the most important corollaries on uniqueness of invariant functions, differentiation under integral signs with respect to parameters, and related.

Chapter 8 fills an unobvious need, proving that automorphic forms that are of *moderate growth* and are eigenfunctions, for Laplacians have asymptotics given by their *constant terms*. In the smaller examples, it is easy to make this precise. For SL_n with $n \geq 3$, some effort is required for an accurate statement. As corollaries, L^2 cuspforms that are eigenfunctions are of rapid decay, and Eisenstein series have relatively simple asymptotics given by their constant terms. Thus, we discover again the need to prove that Eisenstein series have vector-valued meromorphic continuations, specifically as moderate-growth functions.

Chapter 9 carefully develops ideas concerning unbounded symmetric operators on Hilbert spaces, thinking especially of operators related to Laplacians Δ, and especially those such that $(\Delta - \lambda)^{-1}$ is a compact-operator-valued meromorphic function of $\lambda \in \mathbb{C}$. On one hand, even a naive conception of the general behavior of Laplacians is fairly accurate, but this is due to a subtle fact that needs proof, namely, the *essential self-adjointness* of Laplacians on natural spaces such as $\mathbb{R}^n$, multi-toruses $\mathbb{T}^n$, spaces G/K, and even spaces $\Gamma\backslash G/K$. This has a precise sense: the (invariant) Laplacian restricted to *test functions* has a *unique* self-adjoint extension, which then is necessarily its *graph-closure*. Thus, the naive presumption, implicit or explicit, that the graph closure is a (maximal) self-adjoint extension is *correct*. On the other hand, the proof of meromorphic continuation of Eisenstein series in [Colin de Verdière 1981, 1982/1983] makes essential use of some quite counterintuitive features of (Friedrichs's) self-adjoint extensions of *restrictions* of self-adjoint operators, which therefore merit careful attention. In this context, the basic examples are the usual Sobolev spaces on $\mathbb{T}$ or $\mathbb{R}$ and the quantum harmonic oscillator $-\Delta + x^2$ on $\mathbb{R}$. An appendix recalls the proof of the spectral theorem for compact, self-adjoint operators.

Chapter 10 extends the idea from [Lax-Phillips 1976] to prove that larger spaces than spaces of cuspforms decompose *discretely* under the action of self-adjoint extensions $\widetilde{\Delta}_a$ of suitable restrictions Δ_a of Laplacians. Namely, the space of *pseudo-cuspforms* L^2_a at cutoff height a is specified, not by requiring constant terms to vanish *entirely*, but by requiring that all constant terms vanish above height a. The discrete decomposition is proven, as expected, by showing that the resolvent $(\widetilde{\Delta}_a - \lambda)^{-1}$ is a meromorphic compact-operator-valued function of λ, and invoking the spectral theorem for self-adjoint compact operators. The compactness of the resolvent is a Rellich-type compactness result, proven by observing that $(\widetilde{\Delta}_a - \lambda)^{-1}$ maps L^2_a to a Sobolev-type space $\mathfrak{B}^1_a$ with a finer topology on $\mathfrak{B}^1_a$ than the subspace topology and that the inclusion $\mathfrak{B}^1_a \to L^2_a$ is compact.

Chapter 11 uses the discretization results of Chapter 10 to prove meromorphic continuations and functional equations of a variety of Eisenstein series, following [Colin de Verdière 1981, 1982/1983]'s application of the discreteness result in [Lax-Phillips 1976]. This is carried out first for the four simple examples, then for maximal proper parabolic Eisenstein series for $SL_n(\mathbb{Z})$, with cuspidal data. In both the simplest cases and the higher-rank examples, we identify the *exotic eigenfunctions* as being certain truncated Eisenstein series.

Chapter 12 uses several of the analytical ideas and methods of the previous chapters to reconsider automorphic Green's functions, and solutions to other differential equations in automorphic forms, by spectral methods. We prove a *pretrace formula* in the simplest example, as an application of a comparably simple instance of a *subquotient theorem*, which follows from asymptotics of solutions of second-order ordinary differential equations, recalled in a later appendix. We recast the pretrace formula as a demonstration that an automorphic Dirac δ-function lies in the expected *global automorphic Sobolev space*. The same argument gives a corresponding result for any compact automorphic *period*. Subquotient/subrepresentation theorems for groups such as $G = SO(n, 1)$ (rank-one groups with abelian unipotent radicals) appeared in [Casselman-Osborne 1975], [Casselman-Osborne 1978]. For higher-rank groups $SL_n(\mathbb{Z})$, the corresponding subrepresentation theorem is [Casselman 1978/1980], [Casselman-Miličić 1982]. Granting that, we obtain a corresponding pretrace formula for a class of compactly supported automorphic distributions, showing that these distributions lie in the expected global automorphic Sobolev spaces.

Chapter 13 is an extensive appendix with many examples of natural spaces of functions and appropriate topologies on them. One point is that too-limited types of topological vector spaces are inadequate to discuss natural function spaces arising in practice. We include essential standard arguments characterizing locally convex topologies in terms of families of seminorms. We prove the *quasi-completeness* of all natural function spaces, weak duals, and spaces of maps between them. Notably, this includes spaces of distributions.

Chapter 14 proves existence of Gelfand-Pettis vector-valued integrals of compactly supported continuous functions taking values in locally convex, quasi-complete topological vector space. Conveniently, the previous chapter showed that all function spaces of practical interest meet these requirements. The fundamental property of Gelfand-Pettis integrals is that for V-valued f, $T : V \to W$ continuous linear,

$$T\Big(\int f\Big) = \int T \circ f$$

at least for f continuous, compactly supported, V-valued, where V is quasi-complete and locally convex. That is, continuous linear operators pass inside the integral. In suitably topologized natural function spaces, this situation includes differentiation with respect to a parameter. In this situation, as corollaries we can easily prove uniqueness of invariant distributions, density of smooth vectors, and similar.

Chapter 15 carefully discusses holomorphic V-valued functions, using the Gelfand-Pettis integrals as well as a variant of the Banach-Steinhaus theorem. That is, weak holomorphy implies (strong) holomorphy, and the expected Cauchy integral formulas and Cauchy-Goursat theory apply almost *verbatim* in the vector-valued situation. Similarly, we prove that for f a V-valued function on an interval $[a, b]$, $\lambda \circ f$ being C^k for all $\lambda \in V^*$ implies that f itself is C^{k-1} as a V-valued function.

Chapter 16 reviews basic results on asymptotic expansions of integrals and of solutions to second-order ordinary differential equations. The methods are deliberately general, rather than invoking specific features of special functions, to illustrate methods that are applicable more broadly. The simple subrepresentation theorem in Chapter 12 makes essential use of asymptotic expansions.

Our coverage of modern analysis does not aim to be either systematic or complete but well-grounded and adequate for the aforementioned issues concerning automorphic forms. In particular, several otherwise-apocryphal results are treated carefully. We want a sufficient viewpoint so that attractive heuristics, for example, from physics, can become succinct, genuine proofs. Similarly, we do *not* presume familiarity with Lie theory, nor algebraic groups, nor representation theory, nor algebraic geometry, and certainly not with classification of representations of Lie groups or p-adic groups. All these are indeed useful, in the long run, but it is unreasonable to demand mastery of these before thinking about analytical issues concerning automorphic forms. Thus, we directly develop some essential ideas in these supporting topics, sufficient for immediate purposes here. [Lang 1975] and [Iwaniec 2002] are examples of the self-supporting exposition intended here.

Naturally, any novelty here is mostly in the presentation, rather than in the facts themselves, most of which have been known for several decades. Sources and origins can be most clearly described in a historical context, as follows.

The reduction theory in [1.5] is merely an imitation of the very classical treatment for $SL_2(\mathbb{Z})$, including some modern ideas, as in [Borel 1997]. The subtler versions in [2.2] and [3.3] are expanded versions of the first part of [Godement 1962–1964], a more adele-oriented reduction theory than [Borel 1965/1966b], [Borel 1969], and [Borel-HarishChandra 1962]. Proofs [1.9.1], [2.8.6], [3.10.1-2], [3.11.1] of convergence of Eisenstein series are due to Godement use similar ideas, reproduced for real Lie groups in [Borel 1965/1966a]. Convergence arguments on larger groups go back at least to [Braun 1939]'s treatment of convergence of Siegel Eisenstein series. Holomorphic Hilbert-Blumenthal modular forms were studied by [Blumenthal 1903/1904]. What would now be called degenerate Eisenstein series for GL_n appeared in [Epstein 1903/1907]. [Picard 1882, 1883, 1884] was one of the earliest investigations beyond the elliptic modular case. Our notion of *truncation* is from [Arthur 1978] and [Arthur 1980].

Eigenfunction expansions and various notions of convergence are a pervasive theme here and have a long history. The idea that periodic functions should be expressible in terms of sines and cosines is at latest from [Fourier 1822], including what we now call the Dirichlet kernel, although [Dirichlet 1829] came later. Somewhat more generally, eigenfunction expansions for Sturm-Liouville problems appeared in [Sturm 1836] and [Sturm 1833a,b,1836a,b] but were not made rigorous until [Bôcher 1898/1899] and [Steklov 1898] (see [Lützen 1984]). Refinements of the spectral theory of ordinary differential equations continued in [Weyl 1910], [Kodaira 1949], and others, addressing issues of non-compactness and unboundedness echoing complications in the behavior of Fourier transform and Fourier inversion on the line [Bochner 1932], [Wiener 1933]. Spectral theory and eigenfunction expansions for integral equations, which we would now call compact operators [9.A], were recognized as more tractable than direct treatment of differential operators soon after 1900: [Schmidt 1907], [Myller-Lebedev 1907], [Riesz 1907], [Hilbert 1909], [Riesz 1910], [Hilbert 1912]. Expansions in spherical harmonics were used in the 18th century by S.P. Laplace and J.-L. Lagrange, and eventually subsumed in the representation theory of compact Lie groups [Weyl 1925/1926], and in eigenfunction expansions on Riemannian manifolds and Lie groups, as in [Minakshisundaram-Pleijel 1949], [Povzner 1953], [Avakumović 1956], [Berezin 1956], and many others.

Spectral decomposition and synthesis of various types of automorphic forms is more recent, beginning with [Maaß 1949], [Selberg 1956], and [Roelcke 1956a, 1956b]. The spectral decomposition for automorphic forms on general reductive groups is more complicated than might have been anticipated by the earliest pioneers. Subtleties are already manifest in [Gelfand-Fomin 1952], and then in [Gelfand-Graev 1959], [Harish-Chandra 1959], [Gelfand-PiatetskiShapiro 1963], [Godement 1966b], [Harish-Chandra 1968], [Langlands 1966], [Langlands 1967/1976], [Arthur 1978], [Arthur 1980], [Jacquet 1982/1983], [Moeglin-Waldspurger 1989], [Moeglin-Waldspurger 1995], [Casselman 2005], [Shahidi 2010]. Despite various formalizations, spectral synthesis of automorphic forms seems most clearly understood in fairly limited scenarios: [Godement 1966a], [Faddeev 1967], [Venkov 1971], [Faddeev-Pavlov 1972], [Arthur 1978], [Venkov 1979], [Arthur 1980], [Cogdell-PiatetskiShapiro 1990], largely due to issues of convergence, often leaving discussions in an ambiguous realm of (nevertheless interesting) heuristics.

Regarding meromorphic continuation of Eisenstein series: our proof [2.B] for the case [2.9] of GL_2 is an adaptation of the Poisson summation argument from [Godement 1966a]. The essential idea already occurred in [Rankin 1939] and [Selberg 1940]. [Elstrodt-Grunewald-Mennicke 1985] treated examples including our example $SL_2(\mathbb{Z}[i])$, and in that context [Elstrodt-Grunewald-Mennicke 1987] treats special cases of the *period* computation of [2.C]. For Eisenstein series in rank one groups, compare also [Cohen-Sarnak 1980], which treats a somewhat larger family including our simplest four examples, and then [Müller 1996]. The minimal-parabolic example in [3.12] using Bochner's lemma [3.A] essentially comes from an appendix in [Langlands 1967/1976]. The arguments for the broader class of examples in Chapter 11 are adaptations of [Colin de Verdière 1981, 1982/1983], using discretization effects of pseudo-Laplacians from Chapter 10, which adapts the idea of [Lax-Phillips 1976]. Certainly one should compare the arguments in [Harish-Chandra 1968], [Langlands 1967/1976], [Wong 1990], and [Moeglin-Waldspurger 1995], the last of which gives a version of Colin de Verdière's idea due to H. Jacquet.

The discussion of group actions on function spaces in Chapter 6 is mostly very standard. Apparently the first occurrence of the Gelfand-Kazhdan criterion idea is in [Gelfand 1950]. An extension of that idea appeared in [Gelfand-Kazhdan 1975].

The arguments for discrete decomposition of cuspforms in Chapter 11 are adaptations of [Godement 1966b]. The discrete decomposition examples for larger spaces of pseudo-cuspforms in Chapter 10 use the idea of [Lax-Phillips 1976]. The idea of this decomposition perhaps goes back to [Gelfand-Fomin 1952] and, as with many of these ideas, was elaborated on in the iconic sources [Gelfand-Graev 1959], [Harish-Chandra 1959], [Gelfand-PiatetskiShapiro 1963], [Godement 1966b], [Harish-Chandra 1968], [Langlands 1967/1976], and [Moeglin-Waldspurger 1989].

Difficulties with pointwise convergence of Fourier series of continuous functions, and problems in other otherwise-natural Banach spaces of functions, were well appreciated in the late 19th century. There was a precedent for constructs avoiding strictly pointwise conceptions of functions in the very early 20th century, when B. Levi, G. Fubini, and D. Hilbert used Hilbert space constructs to legitimize Dirichlet's minimization principle, in essence that a nonempty closed convex set should have a (unique) point nearest a given point not in that set. The too-general form of this principle is false, in that both existence and uniqueness easily fail in Banach spaces, in natural examples, but the principle is correct in Hilbert spaces. Thus, natural Banach spaces of pointwise-valued functions, such as continuous functions on a compact set with sup norm, do not support this minimization principle. Instead, Hilbert-space versions of continuity and differentiability are needed, as in [Levi 1906]. This idea was systematically developed by [Sobolev 1937, 1938, 1950]. We recall the L^2 Sobolev spaces for circles in [9.5] and for lines in [9.7] and develop various (global) automorphic versions of Sobolev spaces in Chapters 10, 11, and 12.

For applications to analytic number theory, automorphic forms are often constructed by *winding up* various simpler functions containing parameters, forming *Poincaré series* [Cogdell-PiatetskiShapiro 1990] and [Cogdell-PiatetskiShapiro-Sarnak 1991]. Spectral expansions are the standard device for demonstration of meromorphic continuation in the parameters, if it exists at all, which is a nontrivial issue [Estermann 1928], [Kurokawa 1985a,b]. For the example of automorphic Green's functions, namely, solutions to equations $(\Delta - s(s-1))u = \delta_w^{\text{afc}}$ with invariant Laplacian Δ on $\mathfrak{H}$ and automorphic Dirac δ on the right, [Huber 1955] had considered such matters in the context of lattice-point problems in hyperbolic spaces, and, independently, [Selberg 1954] had addressed this issue in lectures in Göttingen. [Neunhöffer 1973] carefully considers the convergence and meromorphic continuation of a solution of that equation formed by winding up. See also [Elstrodt 1973]. The complications or failures of pointwise convergence of the spectral synthesis expressions can often be avoided entirely by considering convergence in suitable global automorphic Sobolev spaces described in Chapter 12. See [DeCelles 2012] and [DeCelles 2016] for developments in this spirit.

Because of the naturality of the issue and to exploit interesting idiosyncrasies, we pay considerable attention to invariant Laplace-Beltrami operators and their eigenfunctions. To have genuine proofs, rather than heuristics, Chapter 9 attends to rigorous notions of unbounded operators on Hilbert spaces [vonNeumann 1929], with motivation toward [vonNeumann 1931], [Stone 1929, 1932], [Friedrichs 1934/1935], [Krein 1945], [Krein 1947]. In fact, [Friedrichs 1934/1935]'s special construction [9.2] has several useful idiosyncracies, exploited in Chapters 10 and 11. Incidentally, the apparent fact that the typically naive treatment of many natural Laplace-Beltrami operators without boundary conditions does not lead to serious mistakes is a corollary of their *essential self-adjointness* [9.9], [9.10]. That is, in many situations, the naive form of the operator admits a *unique* self-adjoint extension, and this extension is the *graph closure* of the original. Thus, in such situations, a naive treatment is provably reasonable. However, the Lax-Phillips discretization device, and Colin de Verdière's use of it to prove meromorphic continuation of Eisenstein series and also to convert certain inhomogeneous differential equations to homogeneous ones, illustrate the point that *restrictions* of essentially self-adjoint operators need not remain essentially self-adjoint. With hindsight, this possibility is already apparent in the context of Sturm-Liouville problems [9.3].

The *global automorphic Sobolev spaces* of Chapter 12 already enter in important auxiliary roles as the spaces $\mathfrak{B}^1$, $\mathfrak{B}^1_a$ in Chapter 10's proofs of discrete decomposition of spaces of pseudo-cuspforms, and $\mathfrak{E}^1$ and $\mathfrak{E}^1_a$ in [11.7-11.11] proving meromorphic continuation of Eisenstein series. The basic estimate called a *pretrace formula* occurred as a precursor to *trace formulas*, as in [Selberg 1954], [Selberg 1956], [Hejhal 1976/1983], and [Iwaniec 2002]. The notion of global automorphic Sobolev spaces provides a reasonable context for discussion of automorphic Green's functions, other automorphic distributions, and solutions of partial differential equations in automorphic forms. The heuristics for Green's functions [Green 1828], [Green 1837] had repeatedly shown their utility in the 19th century. Differential equations $(-\Delta - \lambda)u = \delta$ related to Green's functions had been used by physicists [Dirac 1928a/1928b, 1930], [Thomas 1935], [Bethe-Peierls 1935], with excellent corroboration by physical experiments and are nowadays known as *solvable models*. At the time, and currently, in physics contexts they are rewritten as $((-\Delta + \delta) - \lambda)u = 0$, viewing $-\Delta + \delta$ as a perturbation of $-\Delta$ by a *singular potential* δ, a mathematical idealization of a very short-range force. This was treated rigorously in [Berezin-Faddeev 1961]. The necessary systematic estimates on eigenvalues of integral operators use a *subquotient theorem*, which we prove for the four simple examples, as in that case the issue is about asymptotics of solutions of second-order differential equations, classically understood as recalled in an appendix (Chapter 16). The general result is the *subrepresentation theorem* from [Casselman 1978/1980], [Casselman-Miličić 1982], improving the *subquotient theorem* of [HarishChandra 1954]. In [Varadarajan 1989] there are related computations for $SL_2(\mathbb{R})$.

In the discussion of natural function spaces in Chapter 13, in preparation for the vector-valued integrals of the following chapter, the notion of *quasi-completeness* proves to be the correct general version of completeness. The *incompleteness* of weak duals has been known at least since [Grothendieck 1950], which gives a systematic analysis of completeness of various types of duals. This larger issue is systematically discussed in [Schaefer-Wolff 1966/1999], pp. 147–148 and following. The significance of the compactness of the closure of the convex hull of a compact set appears, for example, in the discussion of vector-valued integrals in [Rudin 1991], although the latter does not make clear that this condition is fulfilled in more than Fréchet spaces and does not mention quasi-completeness. To apply these ideas to distributions, one might cast about for means to prove the compactness condition, eventually hitting on the hypothesis of quasi-completeness in conjunction with ideas from the proof of the Banach-Alaoglu theorem. Indeed, in [Bourbaki 1987] it is shown (by apparently different methods) that quasi-completeness implies this compactness condition. The fact that a bounded subset of a countable strict inductive limit of closed subspaces must actually be a bounded subset of one of the subspaces, easy to prove once conceived, is attributed to Dieudonne and Schwartz in [Horvath 1966]. See also [Bourbaki 1987], III.5 for this result. Pathological behavior of uncountable colimits was evidently first exposed in [Douady 1963].

In Chapter 14, rather than *constructing* vector-valued integrals as limits following [Bochner 1935], [Birkhoff 1935], et alia, we use the [Gelfand 1936]-[Pettis 1938] *characterization* of integrals, which has good functorial properties and gives a forceful reason for *uniqueness*. The issue is *existence*. Density of smooth vectors follows [Gårding 1947]. Another of application of holomorphic and meromorphic vector-valued functions is to *generalized functions*, as in [Gelfand-Shilov 1964], studying *holomorphically parametrized families* of distributions. A hint appears in the discussion of holomorphic vector-valued functions in [Rudin 1991]. A variety of developmental episodes and results in the Banach-space-valued case is surveyed in [Hildebrandt 1953]. Proofs and application of many of these results are given in [Hille-Phillips 1957]. (The first edition, authored by Hille alone, is sparser in this regard.) See also [Brooks 1969] to understand the viewpoint of those times.

Ideas about vector-valued holomorphic and differentiable functions, in Chapter 15, appeared in [Schwartz 1950/1951], [Schwartz 1952], [Schwartz 1953/1954], and in [Grothendieck 1953a,1953b].

The asymptotic expansion results of Chapter 16 are standard. [Blaustein-Handelsman 1975] is a standard source for asymptotics of integrals. *Watson's lemma* and *Laplace's method* for integrals have been used and rediscovered repeatedly. Watson's lemma dates from at latest [Watson 1918], and Laplace's method at latest from [Laplace 1774]. [Olver 1954] notes that Carlini [Green 1837] and [Liouville 1837] investigated relatively simple cases of asymptotics at irregular singular points of ordinary differential equations, without complete rigor. According to [Erdélyi 1956] p. 64, there are roughly two proofs that the standard argument produces genuine asymptotic expansions for solutions of the differential equation. Poincaré's approach, elaborated by J. Horn, expresses solutions as Laplace transforms and invokes Watson's lemma to obtain asymptotics. G.D. Birkhoff and his students constructed auxiliary differential equations from partial sums of the asymptotic expansion, and compared these auxiliary equations to the original [Birkhoff 1908], [Birkhoff 1909], [Birkhoff 1913]. Volterra integral operators are important in both approaches, insofar as asymptotic expansions behave better under *integration* than under *differentiation*. Our version of the Birkhoff argument is largely adapted from [Erdélyi 1956].

Many parts of this exposition are adapted and expanded from [Garrett vignettes], [Garrett mfms-notes], [Garrett fun-notes], and [Garrett alg-noth-notes]. As is surely usual in book writing, many of the issues here had plagued me for decades.

1

Four Small Examples

We recall basic notions related to automorphic forms on some simple arithmetic quotients, including the archetypical quotient $SL_2(\mathbb{Z})\backslash\mathfrak{H}$ of the complex upper half-plane $\mathfrak{H}$ and the related $SL_2(\mathbb{Z})\backslash SL_2(\mathbb{R})$. To put this in a somewhat larger context,[1] we consider parallel examples $\Gamma\backslash X$ and $\Gamma\backslash G$ for a few other groups G, discrete subgroups Γ, and spaces $X \approx G/K$ for compact subgroups K of G. The

[1] In slightly more sophisticated terms inessential to this discussion: the four examples G immediately considered are *real-rank one* semi-simple Lie groups, and the discrete subgroups Γ are *unicuspidal* in the sense that $\Gamma\backslash G/K$ is reasonably compactified by adding just a single *cusp*, where K is a (maximal) compact subgroup of G. That is, the *reduction theory* of $\Gamma\backslash G$ is especially simple in these four cases. Examples with larger real rank, such as GL_n with $n \geq 3$, are considered later.

other three examples share several of the features of $G = SL_2(\mathbb{R})$, $\Gamma = SL_2(\mathbb{Z})$, and $X = \mathfrak{H} \approx G/K$ with $K = SO_2(\mathbb{R})$, allowing simultaneous treatment.

For many reasons, even if we are only interested in harmonic analysis on quotients $\Gamma\backslash X$, it is necessary to consider spaces of functions on the *overlying* spaces $\Gamma\backslash G$, on which G acts by right translations, with a corresponding translation action on functions.

Some basic discussions *not* specific to the four examples are postponed, such as determination of invariant Laplacians in coordinates, self-adjointness properties of invariant Laplacians, proof of the formula for the left G-invariant measure on $X = G/K$, unwinding properties of integrals and sums, continuity of the action of G on *test functions* on $\Gamma\backslash G$, density of test functions in $L^2(\Gamma\backslash X)$, vector-valued integrals, holomorphic vector-valued functions, and other generalities.

We also postpone the relatively *specific* proofs of the major theorems *stated* in the final sections of this chapter, concerning the spectral decomposition of automorphic forms, meromorphic continuation of Eisenstein series, and the theory of the constant term. Those proofs make pointed use of finer details from the more sophisticated analysis.

1.1 Groups $G = SL_2(\mathbb{R}), SL_2(\mathbb{C}), Sp^*_{1,1}$, and $SL_2(\mathbb{H})$

These four groups share some convenient simplifying features, which we will exploit. The first two examples G are easy to describe:

$$G = \begin{cases} SL_2(\mathbb{R}) = \text{a } \textit{special linear} \text{ group over } \mathbb{R} \\ \qquad\qquad = \text{two-by-two determinant-1 real matrices} \\ SL_2(\mathbb{C}) = \text{a } \textit{special linear} \text{ group over } \mathbb{C} \\ \qquad\qquad = \text{two-by-two determinant-1 complex matrices} \end{cases}$$

We will have occasion to use the *general linear* groups $GL_2(R)$ of 2-by-2 invertible matrices with entries in a ring R. Our other two example groups are conveniently described in terms of the Hamiltonian *quaternions* $\mathbb{H} = \mathbb{R} + \mathbb{R}i + \mathbb{R}j + \mathbb{R}k$, with the usual relations

$$i^2 = j^2 = k^2 = -1 \qquad ij = -ji = k \qquad jk = -kj = i \qquad ki = -ik = j$$

The quaternion conjugation is $\overline{\alpha} = \overline{a + bi + cj + dk} = a - bi - cj - dk$ for $\alpha = a + bi + cj + dk$, the *norm* is $N\alpha = \alpha \cdot \overline{\alpha}$, and $|\alpha| = (N\alpha)^{\frac{1}{2}}$. $\mathbb{H}$ can be modeled in two-by-two complex matrices by

$$\rho(a + bi + cj + dk) = \begin{pmatrix} a + bi & c + di \\ -c + di & a - bi \end{pmatrix}$$

with $\det \rho(\alpha) = N\alpha$. For a quaternion matrix g, let g^* be the transpose of the entry-wise conjugate:

$$\begin{pmatrix} \alpha & \beta \\ \gamma & \delta \end{pmatrix}^* = \begin{pmatrix} \overline{\alpha} & \overline{\gamma} \\ \overline{\beta} & \overline{\delta} \end{pmatrix} \qquad (\text{for } \alpha, \beta, \gamma, \delta \in \mathbb{H})$$

The third example group is a kind of *symplectic group*: letting $S = \begin{pmatrix} 0 & 1 \\ 1 & 0 \end{pmatrix}$, this group is

$$G = Sp^*_{1,1} = \{g \in GL_2(\mathbb{H}) : g^* S g = S\}$$

The fourth example is a *special linear* group $G = SL_2(\mathbb{H})$. In the latter, SL_2 is more convenient than GL_2, having a smaller center. However, because $\mathbb{H}$ is not commutative, the notion of *determinant* is problemmatic. One way to skirt the issue is to imbed $r : GL_2(\mathbb{H}) \to GL_4(\mathbb{C})$: with quaternions $\alpha, \beta, \gamma, \delta$,

$$r\begin{pmatrix} \alpha & \beta \\ \gamma & \delta \end{pmatrix} = \begin{pmatrix} \rho(\alpha) & \rho(\beta) \\ \rho(\gamma) & \rho(\delta) \end{pmatrix}$$

identified with a 4-by-4 complex matrix, using the map ρ of $\mathbb{H}$ to 2-by-2 complex matrices, and require that the *image* in $GL_4(\mathbb{C})$ be in the subgroup $SL_4(\mathbb{C})$ where determinant is 1:

$$SL_2(\mathbb{H}) = \{g \in GL_2(\mathbb{H}) : r(g) \in SL_4(\mathbb{C})\}$$

Standard subgroups of any of these groups G are

$$P = \{\begin{pmatrix} * & * \\ 0 & * \end{pmatrix}\} \quad N = \{\begin{pmatrix} 1 & * \\ 0 & 1 \end{pmatrix}\} \quad M = \{\begin{pmatrix} * & 0 \\ 0 & * \end{pmatrix}\}$$

$$A^+ = \{\begin{pmatrix} t & 0 \\ 0 & t^{-1} \end{pmatrix} : t > 0\}$$

The *Levi-Malcev decomposition* $P = NM$ is elementary to check. By direct computation from the defining relations of the groups, one finds

$$M = \begin{cases} \{\begin{pmatrix} m & 0 \\ 0 & m^{-1} \end{pmatrix} : m \in \mathbb{R}^\times\} & (\text{for } G = SL_2(\mathbb{R})) \\ \{\begin{pmatrix} m & 0 \\ 0 & m^{-1} \end{pmatrix} : m \in \mathbb{C}^\times\} & (\text{for } G = SL_2(\mathbb{C})) \\ \{\begin{pmatrix} m & 0 \\ 0 & \overline{m}^{-1} \end{pmatrix} : m \in \mathbb{H}^\times\} & (\text{for } G = Sp^*_{1,1}) \\ \{\begin{pmatrix} a & 0 \\ 0 & d \end{pmatrix} : N(ad) = 1, \ a, d \in \mathbb{H}^\times\} & (\text{for } G = SL_2(\mathbb{H})) \end{cases}$$

and

$$N = \begin{cases} \{\begin{pmatrix} 1 & x \\ 0 & 1 \end{pmatrix} : x \in \mathbb{R}\} & (\text{for } G = SL_2(\mathbb{R})) \\ \{\begin{pmatrix} 1 & x \\ 0 & 1 \end{pmatrix} : x \in \mathbb{C}\} & (\text{for } G = SL_2(\mathbb{C})) \\ \{\begin{pmatrix} 1 & x \\ 0 & 1 \end{pmatrix} : x \in \mathbb{H},\ x + \overline{x} = 0\} & (\text{for } G = Sp^*_{1,1}) \\ \{\begin{pmatrix} 1 & x \\ 0 & 1 \end{pmatrix} : x \in \mathbb{H}\} & (\text{for } G = SL_2(\mathbb{H})) \end{cases}$$

The subgroup P is the *standard (proper) parabolic*, N is its *unipotent radical*, M is the standard *Levi-Malcev component*, and A^+ is the *standard split component*. We will use these (standard) names without elaborating on their history or their connotations.

In these examples, the (spherical) Bruhat decomposition is

$$G = \bigsqcup_{w=1,w_o} PwP = P \sqcup Pw_oP = P \sqcup Pw_oN$$

where $w_o = \begin{pmatrix} 0 & -1 \\ 1 & 0 \end{pmatrix}$, with the last equality following because w_o normalizes M:

$$Pw_oP = Pw_oMN = P(w_oMw_o^{-1})w_oN = Pw_oN$$

The element w_o is the *long Weyl element*. The *small (Bruhat) cell* is P itself, and the *big (Bruhat) cell* is Pw_oP. The (spherical, geometric) Weyl group is $\{1, w_o\}$. It is a group modulo the center of G. The proof of the Bruhat decomposition is straightforward: $g = \begin{pmatrix} a & b \\ c & d \end{pmatrix} \in P$ if and only if $c = 0$. Otherwise, $c \neq 0$, and we try to find $p \in P$ and $n \in N$ such that $g = pw_on$. To simplify, since $c \neq 0$, it is invertible, so, in a form applicable to all four cases, we can left multiply by $\begin{pmatrix} c & 0 \\ 0 & c^{-1} \end{pmatrix} \in M$ to make $c = 1$ without loss of generality. Then try to solve

$$\begin{pmatrix} a & b \\ 1 & d \end{pmatrix} = g = pw_on = \begin{pmatrix} p_{11} & p_{12} \\ 0 & 1 \end{pmatrix} \begin{pmatrix} 0 & -1 \\ 1 & 0 \end{pmatrix} \begin{pmatrix} 1 & n_{12} \\ 0 & 1 \end{pmatrix}$$
$$= \begin{pmatrix} p_{12} & p_{12}n_{12} - p_{11} \\ 1 & n_{12} \end{pmatrix}$$

From the lower right entry, apparently $n_{12} = d$. For the case $G = Sp^*_{1,1}$ the additional condition must be checked, as follows. Observe that inverting $g^*Sg = S$ gives $g^{-1}S^{-1}(g^*)^{-1} = S^{-1}$, and then $S = gSg^*$. In particular, this gives a relation between the c, d entries of g:

$$\begin{pmatrix} 0 & 1 \\ 1 & 0 \end{pmatrix} = S = gSg^* = \begin{pmatrix} * & * \\ c & d \end{pmatrix} \begin{pmatrix} 0 & 1 \\ 1 & 0 \end{pmatrix} \begin{pmatrix} * & \overline{c} \\ * & \overline{d} \end{pmatrix}$$
$$= \begin{pmatrix} * & * \\ * & c\overline{d} + d\overline{c} \end{pmatrix}$$

For $c = 1$, this gives $d + \overline{d} = 0$, which is the condition for $\begin{pmatrix} 1 & d \\ 0 & 1 \end{pmatrix} \in N$ in that case. Thus, in all cases, right multiplying g by $\begin{pmatrix} 1 & -d \\ 0 & 1 \end{pmatrix} \in N$ makes $d = 0$, without loss of generality. Thus, it suffices to solve

$$\begin{pmatrix} a & b \\ 1 & 0 \end{pmatrix} = g = pw_o = \begin{pmatrix} p_{11} & p_{12} \\ 0 & 1 \end{pmatrix} \begin{pmatrix} 0 & -1 \\ 1 & 0 \end{pmatrix} = \begin{pmatrix} p_{12} & -p_{11} \\ 1 & 0 \end{pmatrix}$$

That is,

$$gw_o^{-1} = \begin{pmatrix} -b & a \\ 0 & 1 \end{pmatrix} = p$$

Since $g \in G$, the entries a, b satisfy whatever relations G requires, and $p \in G$. This proves the Bruhat decomposition.

1.2 Compact Subgroups $K \subset G$, Cartan Decompositions

We describe the standard maximal[2] compact subgroups $K \subset G$ for the four examples G. With $\mathbb{H}^1$ the quaternions of norm 1, in a notation consistent with that for $Sp^*_{1,1}$, write

$$Sp^*_1 = \{g \in GL_1(\mathbb{H}) : g^*g = 1\} = \{g \in \mathbb{H}^\times : \overline{g}g = 1\} = \mathbb{H}^1$$

Letting 1_2 be the two-by-two identity matrix, the four maximal compact subgroups are

$$K = \begin{cases} SO_2(\mathbb{R}) & = \{g \in SL_2(\mathbb{R}) : g^\top g = 1_2\} \quad (\text{for } G = SL_2(\mathbb{R})) \\ SU_2 & = \{g \in SL_2(\mathbb{C}) : g^* g = 1_2\} \quad (\text{for } G = SL_2(\mathbb{C})) \\ Sp^*_1 \times Sp^*_1 & = \mathbb{H}^1 \times \mathbb{H}^1 \quad (\text{for } G = Sp^*_{1,1}) \\ Sp^*_2 & = \{g \in GL_2(\mathbb{H}) : g^* g = 1_2\} \quad (\text{for } G = SL_2(\mathbb{H})) \end{cases}$$

[2] The maximality of each of these subgroups K among all compact subgroups in the corresponding G is *not* obvious but is not used in the sequel.

In all four cases, the indicated groups *are* compact. Verification of the compactness of the first three is straightforward because their defining equations present them as spheres or products of spheres. Verification that Sp_2^* is compact and is a subgroup of $SL_2(\mathbb{H})$ merits discussion. For the fourth, observe that the defining condition

$$\begin{pmatrix} 1 & 0 \\ 0 & 1 \end{pmatrix} = \begin{pmatrix} a & b \\ c & d \end{pmatrix}^* \begin{pmatrix} a & b \\ c & d \end{pmatrix} = \begin{pmatrix} |a|^2 + |c|^2 & \overline{a}b + \overline{c}d \\ \overline{b}a + \overline{d}c & |b|^2 + |d|^2 \end{pmatrix}$$

makes Sp_2^* a closed subset of a product of two seven-spheres, $|a|^2 + |c|^2 = 1$ and $|b|^2 + |d|^2 = 1$, thus, compact. Further, Sp_2^* lies inside $SL_2(\mathbb{H})$ rather than merely $GL_2(\mathbb{H})$. For the moment, we will prove a slightly weaker property, that the relevant determinant is ± 1. Use the feature

$$\rho(\overline{\alpha}) = \varepsilon \rho(\alpha)^\top \varepsilon^{-1} \qquad \left(\text{where } \varepsilon = \begin{pmatrix} 0 & -1 \\ 1 & 0 \end{pmatrix}, \text{ for } \alpha \in \mathbb{H}\right)$$

of the imbedding ρ of $\mathbb{H}$ in 2-by-2 complex matrices, and again let

$$r\begin{pmatrix} a & b \\ c & d \end{pmatrix} = \begin{pmatrix} \rho(a) & \rho(b) \\ \rho(c) & \rho(d) \end{pmatrix} \qquad (\text{for } a, b, c, d \in \mathbb{H})$$

viewed as mapping to 4-by-4 complex matrices. Then $r(g^*) = J \cdot r(g) \cdot J^{-1}$, where

$$J = \begin{pmatrix} \varepsilon & \\ & \varepsilon \end{pmatrix} = \begin{pmatrix} & -1 & & \\ 1 & & & \\ & & & -1 \\ & & 1 & \end{pmatrix}$$

and $g \in GL_2(\mathbb{H})$. Thus, for $g^* g = 1_2 \in GL_2(\mathbb{H})$,

$$1_4 = r(1_2) = r(g^*g) = r(g^*) \cdot r(g) = J \cdot r(g)^\top \cdot J^{-1} \cdot r(g)$$

In other words, $r(g)^\top J r(g) = J$.[3] Taking determinants shows $\det r(g)^2 = 1$, so $\det r(g) = \pm 1$. Thus, g in the *connected component* of Sp_2^* containing 1 has $\det r(g) = 1$.

The copy K of $Sp_1^* \times Sp_1^*$ inside $Sp_{1,1}^*$ is not immediately visible in these coordinates, which were chosen to make the *parabolic* P visible. That is, defining $Sp_{1,1}^*$ as the isometry group of the quaternion Hermitian form S obscures the nature of the (maximal) compact K. Changing coordinates by replacing S by

$$S' = \tfrac{1}{2}\begin{pmatrix} 1 & -1 \\ 1 & 1 \end{pmatrix} S \begin{pmatrix} 1 & -1 \\ 1 & 1 \end{pmatrix}^\top = \tfrac{1}{2}\begin{pmatrix} 1 & -1 \\ 1 & 1 \end{pmatrix}\begin{pmatrix} 0 & 1 \\ 1 & 0 \end{pmatrix}\begin{pmatrix} 1 & 1 \\ -1 & 1 \end{pmatrix}$$
$$= \begin{pmatrix} -1 & 0 \\ 0 & 1 \end{pmatrix}$$

[3] Thus, $r(g)$ is inside a *symplectic group* denoted $Sp_4(\mathbb{C})$ or $Sp_2(\mathbb{C})$, depending on convention.

gives

$$\begin{pmatrix} 1 & -1 \\ 1 & 1 \end{pmatrix} Sp^*_{1,1} \begin{pmatrix} 1 & -1 \\ 1 & 1 \end{pmatrix}^{-1} = \{g \in GL_2(\mathbb{H}) : g^* S' g = S'\}$$

and makes the two copies of Sp^*_1 visible on the diagonal:

$$\left\{ k = \begin{pmatrix} * & 0 \\ 0 & * \end{pmatrix} : k^* S' k = S' \right\} = \left\{ k = \begin{pmatrix} \mu & 0 \\ 0 & \nu \end{pmatrix} : \mu, \nu \in \mathbb{H}^1 \right\}$$

That is,

$$K = \begin{pmatrix} 1 & -1 \\ 1 & 1 \end{pmatrix}^{-1} \cdot \left\{ \begin{pmatrix} \mu & 0 \\ 0 & \nu \end{pmatrix} : \mu, \nu \in \mathbb{H}^1 \right\} \cdot \begin{pmatrix} 1 & -1 \\ 1 & 1 \end{pmatrix}$$
$$= \left\{ \begin{pmatrix} \frac{\mu+\nu}{2} & \frac{-\mu+\nu}{2} \\ \frac{-\mu+\nu}{2} & \frac{\mu+\nu}{2} \end{pmatrix} : \mu, \nu \in \mathbb{H}^1 \right\}$$

[1.2.1] Claim: $K \cap P = K \cap M$, and

$$K \cap M = \begin{cases} \pm 1_2 & (\text{for } G = SL_2(\mathbb{R})) \\ \{\begin{pmatrix} \mu & 0 \\ 0 & \mu^{-1} \end{pmatrix} : \mu \in \mathbb{C}^\times,\ |\mu| = 1\} & (\text{for } G = SL_2(\mathbb{C})) \\ \{\begin{pmatrix} \mu & 0 \\ 0 & \mu \end{pmatrix} : \mu \in \mathbb{H}^1\} & (\text{for } G = Sp^*_{1,1}) \\ \{\begin{pmatrix} \mu & 0 \\ 0 & \nu \end{pmatrix} : \mu,\ \nu \in \mathbb{H}^1\} & (\text{for } G = SL_2(\mathbb{H})) \end{cases}$$

Proof: In all but the third case, this follows from the description of K. For example, for $G = SL_2(\mathbb{R})$ and $K = SO_2(\mathbb{R})$, take $p = \begin{pmatrix} a & b \\ 0 & a^{-1} \end{pmatrix} \in P$ and examine the relation $p^\top p = 1_2$ for p to be in K:

$$\begin{pmatrix} 1 & 0 \\ 0 & 1 \end{pmatrix} = p^\top p = \begin{pmatrix} a & 0 \\ b & a^{-1} \end{pmatrix} \begin{pmatrix} a & b \\ 0 & a^{-1} \end{pmatrix}$$
$$= \begin{pmatrix} a^2 & (a+a^{-1})b \\ (a+a^{-1})b & b^2 + a^{-2} \end{pmatrix}$$

From the upper-left entry, $a = \pm 1$. From the off-diagonal entries, $b = 0$. The arguments for $SL_2(\mathbb{C})$ and $SL_2(\mathbb{H})$ are similar. For $Sp^*_{1,1}$, comparison to the coordinates that diagonalize $K \approx Sp_1^* \times Sp_1^*$ gives

$$\{\begin{pmatrix} k_1 & 0 \\ 0 & k_2 \end{pmatrix} : k_1, k_2 \in \mathbb{H}^1\} = \begin{pmatrix} 1 & -1 \\ 1 & 1 \end{pmatrix} K \begin{pmatrix} 1 & -1 \\ 1 & 1 \end{pmatrix}^{-1}$$

$$\ni \begin{pmatrix} 1 & -1 \\ 1 & 1 \end{pmatrix} \begin{pmatrix} a & b \\ 0 & (a^*)^{-1} \end{pmatrix} \begin{pmatrix} 1 & -1 \\ 1 & 1 \end{pmatrix}^{-1}$$

$$= \frac{1}{2} \begin{pmatrix} a+(a^*)^{-1}-b & a-(a^*)^{-1}+b \\ a-(a^*)^{-1}-b & a+(a^*)^{-1}+b \end{pmatrix}$$

For example, adding the elements of the bottom row gives $a = k_2 \in \mathbb{H}^1$ and also $(a^*)^{-1} = a$. From either off-diagonal entry, $b = 0$. ///

In all four cases, the same discussion gives $M = A^+ \cdot (P \cap K) = A^+ \cdot (M \cap K)$.

The following will be essential in [7.1]:

[1.2.2] Claim: *(Cartan decomposition)* $G = KA^+K$.

Proof: First, treat $G = SL_2(\mathbb{R})$. Prove that every $g \in G$ can be written as $g = sk$ with $s^\top = s$ and s positive-definite. To find such s, assume for the moment that it exists, and consider

$$g \cdot g^\top = (sk) \cdot (sk)^\top = sk \cdot k^{-1}s = s^2$$

Certainly $gg^\top$ is symmetric and positive-definite, so having a positive-definite symmetric square root of positive-definite symmetric t would produce s. Such t gives a positive, symmetric operator on $\mathbb{R}^2$, which by the spectral theorem has an orthonormal basis of eigenvalues. That is, there is $h \in K$ such that $hth^\top = \delta$ is diagonal, necessarily with positive diagonal entries. With $\delta^{\frac{1}{2}}$ being the positive diagonal square root of δ,

$$(h^\top \delta^{\frac{1}{2}} h)^2 = h^\top \delta^{\frac{1}{2}} h \cdot h^\top \delta^{\frac{1}{2}} h = h^\top \delta^{\frac{1}{2}} \cdot \delta^{\frac{1}{2}} h = h^\top \delta h = t$$

Thus, take $s = h^\top \delta^{\frac{1}{2}} h$, and every $g \in G$ can be written as $g = ks$. Indeed, we have more:

$$g = ks = k \cdot h^\top \delta^{\frac{1}{2}} h = (k \cdot h^\top) \cdot \delta^{\frac{1}{2}} \cdot h \in K \cdot A^+ \cdot K$$

giving the claim in this case. The cases of $G = SL_2(\mathbb{C})$ is similar, using $g = sk$ with $s = s^*$ Hermitian positive-definite and $k^* = k^{-1} \in K$, invoking the spectral theorem for Hermitian positive-definite operators. The same argument succeeds for $G = SL_2(\mathbb{H})$ with quaternion conjugation replacing complex, with a suitably adapted spectral theorem for $s \in GL_2(\mathbb{H})$ with $s^* = s$ and x^*sc real and positive for all nonzero 2-by-1 quaternion matrices x.[4]

The case of $G = Sp^*_{1,1}$ essentially reduces to the case of $SL_2(\mathbb{H})$, as follows. Since $g^*Sg = S$, $SgS^{-1} = (g^*)^{-1}$. Anticipating the Cartan decomposition $g = sk$, from $gg^* = ss^* = s^2$, by the quaternionic version of the spectral theorem, there is $k \in Sp^*_2$ such that $k^{-1}gg^*k = \Lambda$ with Λ positive real diagonal. We want to adjust k to be in $Sp^*_{1,1} \cap Sp^*_2$, while preserving the property $k^{-1}gg^*k = \Lambda$. Unless gg^* is *scalar*, the diagonal entries are distinct. By $SgS^{-1} = (g^*)^{-1}$ and $Sg^*S^{-1} = g^{-1}$ for $g \in G$,

$$\begin{aligned}\Lambda^{-1} &= (\Lambda^*)^{-1} = S\Lambda S^{-1} = S(k^{-1}gg^*k)S^{-1}\\ &= (SkS^{-1})^{-1} \cdot Sgg^*S^{-1} \cdot SkS^{-1} = (SkS^{-1})^{-1} \cdot (gg^*)^{-1} \cdot SkS^{-1}\end{aligned}$$

Inverting gives $\Lambda = (SkS^{-1})^{-1} \cdot gg^* \cdot SkS^{-1}$. Also $\Lambda = k^{-1}gg^*k$, so

$$(SkS^{-1}) \cdot \Lambda \cdot (SkS^{-1})^{-1} = gg^* = k \cdot \Lambda \cdot k^{-1}$$

That is, $k^{-1} \cdot SkS^{-1}$ commutes with Λ, and $\delta = k^{-1} \cdot SkS^{-1}$ is at worst diagonal:

$$SkS^{-1} = k \cdot \delta = k \cdot \begin{pmatrix} a & 0 \\ 0 & d \end{pmatrix}$$

Since $\delta \in Sp^*_2$, $a \cdot \bar{a} = 1$ and $d \cdot \bar{d} = 1$. To preserve $k^{-1}gg^*k = \Lambda$, to adjust k to be in $K = Sp^*_2 \cap Sp^*_{1,1}$, adjust k by diagonal matrices ε in Sp^*_2. The condition for $k\varepsilon$ to be in K is

$$(k \cdot \varepsilon) = ((k\varepsilon)^*)^{-1} = S(k\varepsilon)S^{-1} = SkS^{-1} \cdot S\varepsilon S^{-1} = k \cdot \delta \cdot S\varepsilon S^{-1}$$

so take $\varepsilon = S^{-1}\delta S$. The rest of the argument runs as in the first three cases. ///

1.3 Iwasawa Decomposition $G = PK = NA^+K$

The subgroups P and K are *not* normal in G, so the *Iwasawa decompositions* $G = PK = \{pk : p \in P,\ k \in K\}$ do *not* express G as a product group. Nevertheless, these decompositions are essential.

[1.3.1] Claim: (*Iwasawa decomposition*) $G = PK = NA^+K$. In particular, the map $N \times A^+ \times K \longrightarrow G$ by $n \times a \times k \longrightarrow nak$ is an *injective* set map (and is a diffeomorphism).

[4] In all three of these cases, a Rayleigh-Ritz approach gives a sufficient spectral theorem, as follows. Let F be $\mathbb{R}$, $\mathbb{C}$, or $\mathbb{H}$. Let $\langle x, y\rangle = y^*x$ for 2-by-1 matrices x, y over F. Let $T : F^2 \to F^2$ be right F-linear and positive Hermitian in the sense that $\langle Tx, x\rangle$ is positive, real for $x \neq 0$. Then x with $\langle x, x\rangle = 1$ maximizing $\langle Tx, x\rangle$ is an eigenvector for T. For nonscalar T, the unit vector y minimizing $\langle Ty, y\rangle$ is an eigenvector for T orthogonal to x. Letting k be the matrix with columns x, y, the conjugated matrix $k^{-1}Tk$ is diagonal.

Proof: For $g = \begin{pmatrix} a & b \\ c & d \end{pmatrix} \in G$, in the easy case that $c = 0$, then $g \in P$. In all cases, once we have $g = nm \in P$, we can adjust g on the right by $M \cap K$ to put the Levi component m into A^+.

One approach is to think of right multiplication by K as *rotating* the lower row $(c\ d)$ of $g \in G$ to put it into the form $(0\ *)$ of the lower row of an element of P. For $g = \begin{pmatrix} a & b \\ c & d \end{pmatrix} \in G = SL_2(\mathbb{R})$: right multiplication by the explicit element

$$k = \begin{pmatrix} \frac{d}{\sqrt{c^2+d^2}} & \frac{c}{\sqrt{c^2+d^2}} \\ \frac{-c}{\sqrt{c^2+d^2}} & \frac{d}{\sqrt{c^2+d^2}} \end{pmatrix} \in K = SO_2(\mathbb{R})$$

puts $gk \in P$:

$$\begin{pmatrix} a & b \\ c & d \end{pmatrix} \cdot \begin{pmatrix} \frac{d}{\sqrt{c^2+d^2}} & \frac{c}{\sqrt{c^2+d^2}} \\ \frac{-c}{\sqrt{c^2+d^2}} & \frac{d}{\sqrt{c^2+d^2}} \end{pmatrix} = \begin{pmatrix} * & * \\ 0 & * \end{pmatrix}$$

Similarly, for $g = \begin{pmatrix} a & b \\ c & d \end{pmatrix} \in G = SL_2(\mathbb{C})$, right multiplication by

$$k = \begin{pmatrix} \frac{d}{\sqrt{|c|^2+|d|^2}} & \frac{\overline{c}}{\sqrt{|c|^2+|d|^2}} \\ \frac{-c}{\sqrt{|c|^2+|d|^2}} & \frac{\overline{d}}{\sqrt{|c|^2+|d|^2}} \end{pmatrix} \in K = SU(2)$$

gives $gk \in P$. Likewise, for $G = SL_2(\mathbb{H})$, nearly the same explicit expression as for $SL_2(\mathbb{C})$ succeeds, with complex conjugation replaced by quaternion conjugation, accommodating the noncommutativity:[5]

$$\begin{pmatrix} a & b \\ c & d \end{pmatrix} \cdot \begin{pmatrix} \frac{c^{-1}d}{\sqrt{1+|c^{-1}d|^2}} & \frac{1}{\sqrt{1+|c^{-1}d|^2}} \\ \frac{-1}{\sqrt{1+|c^{-1}d|^2}} & \frac{\overline{c^{-1}d}}{\sqrt{1+|c^{-1}d|^2}} \end{pmatrix} = \begin{pmatrix} * & * \\ 0 & * \end{pmatrix} \in P$$

For $g = \begin{pmatrix} a & b \\ c & d \end{pmatrix} \in G = Sp^*_{1,1}$, we hope that a matrix k of a similar form lies in $K \approx Sp^*_1 \times Sp^*_1$, and then $gk \in P$. To be sure that the defining relation for

[5] This explicit element lies in the connected component of Sp^*_2 containing 1, so this argument for the Iwasawa decomposition is complete whether or not we have verified that $Sp^*_2 \subset SL_2(\mathbb{H})$.

$Sp^*_{1,1}$ is fulfilled, use the more explicit coordinates

$$K = \left\{ \begin{pmatrix} \frac{\mu+\nu}{2} & \frac{-\mu+\nu}{2} \\ \frac{-\mu+\nu}{2} & \frac{\mu+\nu}{2} \end{pmatrix} : \mu, \nu \in \mathbb{H}^1 \right\}$$

To reduce the issue to more manageable pieces, left multiply $g = \begin{pmatrix} a & b \\ c & d \end{pmatrix}$ by $\begin{pmatrix} c^* & 0 \\ 0 & c^{-1} \end{pmatrix}$ to make $c = 1$. As earlier, $g^*Sg = S$ implies $gSg^* = S$, so that $c\overline{d} + d\overline{c} = 0$, and with $c = 1$ we have $d + \overline{d} = 0$. Also, $|1+d|^2 = 1 + |d|^2$. Thus, with $\mu = \frac{d+1}{|d+1|}$ and $\nu = \frac{d-1}{|d-1|}$, K contains

$$\begin{pmatrix} \frac{\mu+\nu}{2} & \frac{-\mu+\nu}{2} \\ \frac{-\mu+\nu}{2} & \frac{\mu+\nu}{2} \end{pmatrix} = \begin{pmatrix} \frac{d}{\sqrt{1+|d|^2}} & \frac{-1}{\sqrt{1+|d|^2}} \\ \frac{-1}{\sqrt{1+|d|^2}} & \frac{d}{\sqrt{1+|d|^2}} \end{pmatrix}$$

Then $gk \in P$, giving the Iwasawa decomposition in this case. In all cases, the fact that $N \cap A^+ = \{1\}$ and $NA^+ \cap K = \{1\}$ proves the injectivity of the multiplication $n \times a \times k \to nak$. ///

The following assertion is a generalization of the standard fact that, with $z = x + iy \in \mathfrak{H}$,

$$\mathrm{Im}\,(gz) = \frac{y}{|cz+d|^2} \qquad \left(\text{for } g = \begin{pmatrix} a & b \\ c & d \end{pmatrix} \in SL_2(\mathbb{R})\right)$$

This is the foundation for *reduction theory* for these examples, that is, for determination of the behavior of images $\gamma \cdot gK$ as γ varies in Γ, as below. Let

$$a_y = \begin{pmatrix} \sqrt{y} & 0 \\ 0 & 1/\sqrt{y} \end{pmatrix} \qquad (\text{with } y > 0)$$

[1.3.2] Claim: For Iwasawa decomposition $g = na_yk$ with $n \in N$, $y > 0$, and $k \in K$, say that y is the *height* of g. In all four cases, with $n_x \in N$ and $y > 0$,

$$\text{height}(g \cdot n_x a_y) = \frac{y}{|cy|^2 + |cx+d|^2} \qquad \left(\text{for } g = \begin{pmatrix} a & b \\ c & d \end{pmatrix} \in G\right)$$

Proof: This is a direct computation.

$$gn_xa_y = \begin{pmatrix} a & b \\ c & d \end{pmatrix} \begin{pmatrix} 1 & x \\ 0 & 1 \end{pmatrix} \begin{pmatrix} \sqrt{y} & 0 \\ 0 & 1/\sqrt{y} \end{pmatrix} = \begin{pmatrix} a\sqrt{y} & \frac{ax+b}{\sqrt{y}} \\ c\sqrt{y} & \frac{cx+d}{\sqrt{y}} \end{pmatrix}$$
$$= \begin{pmatrix} * & 0 \\ 0 & c\sqrt{y} \end{pmatrix} \begin{pmatrix} * & * \\ 1 & \frac{x+c^{-1}d}{y} \end{pmatrix}$$

For $G = SL_2(\mathbb{R})$, $SL_2(\mathbb{C})$, and $SL_2(\mathbb{H})$ with respect compact subgroups K, for D in $\mathbb{R}$, $\mathbb{C}$, $\mathbb{H}$, respectively,

$$k = \begin{pmatrix} \frac{D}{\sqrt{1+|D|^2}} & \frac{1}{\sqrt{1+|D|^2}} \\ \frac{-1}{\sqrt{1+|D|^2}} & \frac{\overline{D}}{\sqrt{1+|D|^2}} \end{pmatrix} \in K$$

In those three cases, letting $D = \frac{x+c^{-1}d}{y}$,

$$\begin{aligned} gn_x a_y \cdot k &= \begin{pmatrix} * & 0 \\ 0 & c\sqrt{y} \end{pmatrix} \begin{pmatrix} * & * \\ 1 & \frac{x+c^{-1}d}{y} \end{pmatrix} \begin{pmatrix} \frac{D}{\sqrt{1+|D|^2}} & \frac{1}{\sqrt{1+|D|^2}} \\ \frac{-1}{\sqrt{1+|D|^2}} & \frac{\overline{D}}{\sqrt{1+|D|^2}} \end{pmatrix} \\ &= \begin{pmatrix} * & 0 \\ 0 & c\sqrt{y} \end{pmatrix} \begin{pmatrix} * & * \\ 0 & \frac{1+|D|^2}{\sqrt{1+|D|^2}} \end{pmatrix} \\ &= \begin{pmatrix} * & * \\ 0 & |c|\sqrt{y}\sqrt{1+|D|^2} \end{pmatrix} \cdot \begin{pmatrix} * & 0 \\ 0 & \frac{c}{|c|} \end{pmatrix} \end{aligned}$$

noting that $\begin{pmatrix} * & 0 \\ 0 & \frac{c}{|c|} \end{pmatrix} \in K$. Simplifying,

$$|c|\sqrt{y}\sqrt{1+|D|^2} = |c|\sqrt{y}\sqrt{1+\left|\frac{x+c^{-1}d}{y}\right|^2} = \sqrt{\frac{|cy|^2+|cx+d|^2}{y}}$$

Thus, in these three cases,

$$gn_x a_y \in N \begin{pmatrix} \sqrt{y'} & 0 \\ 0 & 1/\sqrt{y'} \end{pmatrix} K \quad \text{with} \quad y' = \frac{y}{|cy|^2+|cx+d|^2}$$

For $G = Sp^*_{1,1}$, the explicit element of K is slightly different

$$k = \begin{pmatrix} \frac{D}{\sqrt{1+|D|^2}} & \frac{-1}{\sqrt{1+|D|^2}} \\ \frac{-1}{\sqrt{1+|D|^2}} & \frac{D}{\sqrt{1+|D|^2}} \end{pmatrix} \in K$$

but the conclusion will be the same: with $D = \frac{x+c^{-1}d}{y}$

$$\begin{aligned} gn_x a_y \cdot k &= \begin{pmatrix} * & 0 \\ 0 & c\sqrt{y} \end{pmatrix} \begin{pmatrix} * & * \\ 1 & \frac{x+c^{-1}d}{y} \end{pmatrix} \begin{pmatrix} \frac{D}{\sqrt{1+|D|^2}} & \frac{-1}{\sqrt{1+|D|^2}} \\ \frac{-1}{\sqrt{1+|D|^2}} & \frac{D}{\sqrt{1+|D|^2}} \end{pmatrix} \\ &= \begin{pmatrix} * & 0 \\ 0 & c\sqrt{y} \end{pmatrix} \begin{pmatrix} * & * \\ 0 & \frac{-1+D^2}{\sqrt{1+|D|^2}} \end{pmatrix} \end{aligned}$$

For $Sp^*_{1,1}$, as in earlier computations, the relation $h^*Sh = S$ gives $hSh^* = S$, so for $h = \begin{pmatrix} * & * \\ 1 & D \end{pmatrix}$ we find $D + \overline{D} = 0$. That is, D is purely imaginary, so

$D^2 = -|D|^2$, and

$$g n_x a_y \cdot k = \begin{pmatrix} * & * \\ 0 & -c\sqrt{y}\sqrt{1+|D|^2} \end{pmatrix} = \begin{pmatrix} * & * \\ 0 & |c|\sqrt{y}\sqrt{1+|D|^2} \end{pmatrix} \cdot \begin{pmatrix} * & 0 \\ 0 & \frac{-c}{|c|} \end{pmatrix}$$

The remainder of the computation is identical to the other three cases. ///

1.4 Some Convenient Euclidean Rings

We recall proofs that, just as the ordinary integers are Euclidean, the Gaussian integers $\mathbb{Z}[i]$ and Hurwitz quaternion integers are *Euclidean*. This will greatly simplify the geometry of quotients $\Gamma\backslash X$ in [1.5.1] by assuring that there is just a single *cusp*.

Recall the simplest version of *Euclidean-ness* for a ring R with 1: there is a function $\|\cdot\| : R \to \mathbb{Z}$ such that $\|r\| \geq 0$ and $\|r\| = 0$ implies $r = 0$, such that $\|rr'\| = \|r\| \cdot \|r'\|$, and, for every $a \in R$ and every $0 \neq d \in R$, there is $q \in R$ such that $\|a - qd\| < \|d\|$.

Since $\|1\| = \|1^2\| = \|1\| \cdot \|1\|$ and $0 < \|1\|$, necessarily $\|1\| = 1$. *Units* $r \in R^\times$ have $\|r\| = 1$, because $rs = 1$ gives $\|r\| \cdot \|s\| = \|rs\| = \|1\| = 1$, and $\|\cdot\|$ takes nonnegative integer values.

Euclidean-ness implies that every left ideal is principal: let d be an element having the smallest norm in a given nonzero left ideal I. For any $a \in I$, there is $q \in R$ such that $\|a - qd\| < \|d\|$. Thus, $\|a - qd\| = 0$, and $a = qd$.

To show that $R = \mathbb{Z}[i]$ is Euclidean with respect to the square of the usual complex absolute value $\|\cdot\| = |\cdot|^2$, for $a \in \mathbb{Z}[i]$ and given $0 \neq d \in \mathbb{Z}[i]$, we need to find $q \in \mathbb{Z}[i]$ such that $\|a - dq\| < \|d\|$. The requirement $\|a - qd\| < \|d\|$ is equivalent to $\|a/d - q\| < 1$. Thus, given $a/d \in \mathbb{Q}(i)$, we want $q \in \mathbb{Z}[i]$ within distance-squared 1. With $a/d = u + iv$ with $u, v \in \mathbb{Q}$, taking $u', v' \in \mathbb{Z}$ such that $|u - u'| \leq \frac{1}{2}$ and $|v - v'| \leq \frac{1}{2}$ gives the desired $\|a/d - (u' + iv')\| \leq (\frac{1}{2})^2 + (\frac{1}{2})^2 < 1$.

In the rational quaternions $\mathbb{H}_\mathbb{Q} = \mathbb{Q} + \mathbb{Q}i + \mathbb{Q}j + \mathbb{Q}k$, the natural choice $\mathbb{Z} + \mathbb{Z}i + \mathbb{Z}j + \mathbb{Z}k$ for *integers* is not optimal. Instead, we use the slightly larger ring of *Hurwitz integers*:

$$\mathfrak{o} = (\mathbb{Z} + \mathbb{Z}i + \mathbb{Z}j + \mathbb{Z}k) + \mathbb{Z} \cdot \frac{1 + i + j + k}{2}$$

We prove that the Hurwitz integers are Euclidean, using the square of the quaternion norm: $\| \cdot \| = | \cdot |^2$. To see that the norm-squared takes integer values on $\mathfrak{o}$, the only possible difficulty might be a denominator of 4, which does not occur because, for all $a, b, c, d \in \mathbb{Z}$,

$$(2a+1)^2 + (2b+1)^2 + (2c+1)^2 + (2d+1)^2 = 0 \bmod 4$$

Given $a \in \mathfrak{o}$ and $0 \neq d \in \mathfrak{o}$, to show that there is $q \in \mathfrak{o}$ such that $\|a - qd\| < \|d\|$ is equivalent to $\|ad^{-1} - q\| < 1$. For $ad^{-1} = x + yi + zj + wk$ with $x, y, z, w \in \mathbb{Q}$, there are $x', y', z', w' \in \mathbb{Z}$ differing by at most $\frac{1}{2}$ in absolute value from the respective x, y, z, w. However, the resulting estimate

$$\begin{aligned} &\|(x + yi + zj + wk) - (x' + y'i + z'j + w'k)\| \\ &\quad \le (\tfrac{1}{2})^2 + (\tfrac{1}{2})^2 + (\tfrac{1}{2})^2 + (\tfrac{1}{2})^2 = 1 \end{aligned}$$

is insufficient. Nevertheless, being slightly more precise, if $|x - x'| < \frac{1}{2}$ or $|y - y'| < \frac{1}{2}$ or $|z - z'| < \frac{1}{2}$ or $|w - w'| < \frac{1}{2}$, then we *do* have the desired

$$\|(x + yi + zj + wk) - (x' + y'i + z'j + w'k)\| < 1$$

That is, the only case of failure is $|x - x'| = |y - y'| = |z - z'| = |w - w'| = \frac{1}{2}$. Subtracting 1 from x, y, z, w if necessary, without loss of generality $x - x' = y - y' = z - z' = w - w' = \frac{1}{2}$. In that case,

$$(x + yi + zj + wk) - \Big((x' + y'i + z'j + w'k) + \frac{1 + i + j + k}{2}\Big) = 0$$

proving that the Hurwitz integers are Euclidean. A qualitative version of the Euclidean-ness of $\mathfrak{o}$ will be useful in one of the proofs of *unicuspidality*: for $\alpha = x + yi + zj + wk$ with $x, y, z, w \in \mathbb{Q}$, there is $x' + y'i + z'j + w'k \in \mathfrak{o}$ such that

$$|(x + yi + zj + wk) - (x' + y'i + z'j + w'k)| \le \frac{\sqrt{13}}{4}$$

Adjust α by an element of $\mathbb{Z} + \mathbb{Z}i + \mathbb{Z}j + \mathbb{Z}k$ so that, without loss of generality, all coefficients are of absolute value at most $\frac{1}{2}$. If any one coefficient is smaller than $\frac{1}{4}$, then $|\alpha|^2 \le (\frac{1}{2})^2 + (\frac{1}{2})^2 + (\frac{1}{2})^2 + (\frac{1}{4})^2 = \frac{13}{16}$ as desired. When *all* coefficients are between $\frac{1}{4}$ and $\frac{1}{2}$ in absolute value, make them all of the same *sign* by adding or subtracting 1 to either one or two, paying the price that those one or two are of absolute value between $\frac{1}{2}$ and $\frac{3}{4}$, while the others are still of absolute value between $\frac{1}{4}$ and $\frac{1}{2}$. Adding or subtracting $(1 + i + j + k)/2$ depending on sign, all coefficients are between $-\frac{1}{4}$ and $\frac{1}{4}$, and the quaternion norm of the result is at most $\frac{1}{2} \le \frac{\sqrt{13}}{4}$.

1.5 Discrete Subgroups $\Gamma \subset G$, Reduction Theory

We specify discrete[6] subgroups Γ of each of the examples G, so that $\Gamma\backslash G/K$ *has just one cusp*, in a sense made precise subsequently. *Reduction theory* is the exhibition of a simple approximate collection of representatives for the quotient $\Gamma\backslash G/K$ sufficient to understand the most basic geometric features of that quotient.[7] The simple outcome in the present examples, *unicuspidality*, simplifies *meromorphic continuation of Eisenstein series* and simplifies the form of the *spectral decomposition* of the space of square-integrable automorphic forms on $\Gamma\backslash G/K$. The four cases are[8]

$$\Gamma = \begin{cases} SL_2(\mathbb{Z}) & (\text{for } G = SL_2(\mathbb{R})) \quad (\text{elliptic modular group}) \\ SL_2(\mathbb{Z}[i]) & (\text{for } G = SL_2(\mathbb{C})) \quad (\text{a Bianchi modular group}) \\ Sp^*_{1,1}(\mathfrak{o}) & (\text{for } G = Sp^*_{1,1}) \\ SL_2(\mathfrak{o}) & (\text{for } G = SL_2(\mathbb{H})) \end{cases}$$

where $Sp^*_{1,1}(\mathfrak{o})$ and $SL_2(\mathfrak{o})$ denote the elements of $Sp^*_{1,1}$ and $SL_2(\mathbb{H})$ with entries in the ring of Hurwitz integers $\mathfrak{o}$.

In all examples, $\Gamma \cap P = (\Gamma \cap M) \cdot (\Gamma \cap N)$. We have

$$\Gamma \cap N = \{\begin{pmatrix} 1 & x \\ 0 & 1 \end{pmatrix}\} \text{ where } \begin{cases} x \in \mathbb{Z} & (\text{for } \Gamma = SL_2(\mathbb{Z})) \\ x \in \mathbb{Z}[i] & (\text{for } \Gamma = SL_2(\mathbb{Z}[i])) \\ x \in \mathbb{Z}i + \mathbb{Z}j + \mathbb{Z}k & (\text{for } \Gamma = Sp^*_{1,1}(\mathfrak{o})) \\ x \in \mathfrak{o} & (\text{for } \Gamma = SL_2(\mathbb{H})) \end{cases}$$

6 As usual, a subset D of a topological space X is *discrete* when every point $x \in D$ has a neighborhood U such that $U \cap D = \{x\}$. The topologies on our groups G are the subspace topologies from the ambient real vector spaces of 2-by-2 real, complex, or quaternion matrices.

7 In some contexts, the goal of determination of an *exact, explicit* collection of representatives in G/K for the quotient $\Gamma\backslash G/K$ is given high priority. A precise collection of representative is often called a *fundamental domain*. However, in general, determination of an explicit fundamental domain is infeasible. Fortunately, it is also inessential.

8 The elliptic modular group has its origins in dim antiquity. [Picard 1883] and [Picard 1884] looked at similar subgroups of small noncompact unitary groups. [Bianchi 1892] looked at a family of discrete subgroups of $SL_2(\mathbb{C})$, such as $SL_2(\mathbb{Z}[i])$. W. de Sitter proposed a model of space-time in which the cosmological constant dominates and matter is negligible, with symmetry group $SO(4,1)$, and $Sp^*_{1,1}$ is a twofold cover of $SO(4,1)$. No automorphic forms directly entered his work, but his attention to specific groups, as in the more theoretical work of [Bargmann 1947] and [Wigner 1939], provided examples that were eventually appreciated for their illustration of phenomena with mathematical significance beyond physics itself. [Hurwitz 1898] studied the quaternion integers $\mathfrak{o}$ which bear his name. See also [Hurwitz 1919] and [Conway-Smith 2003].

As in the discussion of Euclidean-ness, the quotients $(\Gamma \cap N)\backslash N$ have (redundant) representatives $(\Gamma \cap N)\backslash N = \{\begin{pmatrix} 1 & x \\ 0 & 1 \end{pmatrix}\}$, where

$$\begin{cases} x \in \mathbb{R},\ |x| \le \frac{1}{2} & (\text{for } \Gamma = SL_2(\mathbb{Z})) \\ x \in \mathbb{C},\ |x| \le \frac{1}{\sqrt{2}} & (\text{for } \Gamma = SL_2(\mathbb{Z}[i])) \\ x = ai + bj + ck,\ |x| \le \frac{\sqrt{3}}{2} & (\text{for } \Gamma = Sp^*_{1,1}(\mathfrak{o})) \\ x \in \mathbb{H},\ |x| \le \frac{\sqrt{13}}{4} & (\text{for } \Gamma = SL_2(\mathbb{H})) \end{cases}$$

In particular, $(\Gamma \cap N)\backslash N$ is *compact*. We have

$$\Gamma \cap M = \{\begin{pmatrix} a & 0 \\ 0 & d \end{pmatrix}\} \text{ where } \begin{cases} a = d^{-1} \in \mathbb{Z}^\times & (\text{for } \Gamma = SL_2(\mathbb{Z})) \\ a = d^{-1} \in \mathbb{Z}[i]^\times & (\text{for } \Gamma = SL_2(\mathbb{Z}[i])) \\ a = (d^*)^{-1} \in \mathfrak{o}^\times & (\text{for } \Gamma = Sp^*_{1,1}(\mathfrak{o})) \\ a, d \in \mathfrak{o}^\times & (\text{for } \Gamma = SL_2(\mathbb{H})) \end{cases}$$

The groups of units $\mathbb{Z}^\times$ and $\mathbb{Z}[i]^\times$ are well known and finite. The group $\mathfrak{o}^\times$ is also *finite* but less trivial. As noted earlier, $\alpha \in \mathfrak{o}^\times$ implies $|\alpha| = 1$. Certainly $\mathfrak{o} \subset \frac{1}{2} \cdot (\mathbb{Z} + \mathbb{Z}i + \mathbb{Z}j + \mathbb{Z}k)$ and

$$\frac{a^2 + b^2 + c^2 + d^2}{4} = \left|\frac{a + bi + cj + dk}{2}\right|^2 \le 1$$

implies $|a| \le 2$, $|b| \le 2$, $|c| \le 2$, and $|d| \le 2$, giving a crude bound on the number of possibilities for integers a, b, c, d.

For compact $C \subset N$, a *standard Siegel set* is a subset of G of the form

$$\mathfrak{S}_{t,C} = \{na_y k : n \in C,\ k \in K,\ y \ge t\}$$

This is essentially a half-infinite rectangle right-multiplied by K. On other occasions, a Siegel set is construed as a subset of the quotient $(\Gamma \cap N)\backslash G$ or as a subset of G/K. These distinctions are inessential. Let $\Gamma_\infty = P \cap \Gamma$. Left multiplication by N does not change heights on G. Since $\Gamma_\infty \subset N \times (M \cap \Gamma)$ and $M \cap \Gamma \subset M \cap K$, left multiplication by Γ_∞ does not change heights. Siegel sets are a simple type of set among that, as it turns out, we can find approximate sets of representatives for the quotient $\Gamma\backslash G/K$. That is, *reduction theory* for these examples is relatively simple:

[1.5.1] Theorem: For all four examples, $\Gamma\backslash G$ is *unicuspidal*, in the sense that there is $t > 0$ and compact $C \subset N$ such that a *single Siegel set covers G:*

$$\bigcup_{\gamma \in \Gamma} \gamma \cdot \mathfrak{S}_{t,C} = G$$

Proof: For fixed x, y, the function $q(c, d) = q_{x,y}(c, d) = |cx + d|^2 + |cy|^2$ is a homogeneous real-valued quadratic polynomial function on $\mathbb{R} \oplus \mathbb{R}$, $\mathbb{C} \oplus \mathbb{C}$, or $\mathbb{H} \oplus \mathbb{H}$, in the respective cases, with $y > 0$ and appropriate x. It is *positive definite*: $q(c, d) = 0$ implies $c = 0 = d$. Thus, $q(c, d)$ is *comparable* to $|c|^2 + |d|^2$: there are positive constants A, B depending on x, y such that

$$A \cdot (|c|^2 + |d|^2) \leq q(c, d) \leq B \cdot (|c|^2 + |d|^2)$$

The number of points (c, d) in a lattice $\mathbb{Z} \oplus \mathbb{Z}$, $\mathbb{Z}[i] \oplus \mathbb{Z}[i]$, or $\mathfrak{o} \oplus \mathfrak{o}$ inside a ball of finite radius is *finite*. In particular, in the orbit $\Gamma \cdot n_x a_y$ there are only *finitely many* values height$(\gamma \cdot n_x a_y)$ above a given bound $t > 0$. In particular, the supremum of these heights is *attained*. Thus, every Γ-orbit contains (at least one) $n_x a_y$ of maximum height, and $|cx + d|^2 + |cy|^2 \geq 1$ for all lower rows $(c\ d)$ of $\gamma \in \Gamma$. In particular, with $c = 1$ and $d = 0$, $|x|^2 + |y|^2 \geq 1$.

Thus, given $n_x a_y K \in G/K$, adjust on the left by $\gamma \in \Gamma$ so that $\gamma n_x a_y K$ is (one of) the highest in its orbit on G/K. In particular, this makes $|x|^2 + |y|^2 \geq 1$. From the specific estimates on parameters ξ for representatives n_ξ of $(\Gamma \cap N)\backslash N$, in all cases there is $0 < t < 1$ such that $|\xi| \leq 1 - t$ for all representatives. Thus, if $|x| > 1 - t$, further adjust on the left by $\gamma \in \Gamma \cap N$ so that $|x| \leq 1 - t$, *without altering the height*. Thus, the new $n_x a_y$ is still among the highest in its orbit, and $|x|^2 + |y|^2 \geq 1$ still holds. Thus

$$|y|^2 \geq 1 - |x|^2 \geq 1 - (1 - t)^2 = t(2 - t) \geq t$$

Therefore, every Γ-orbit has a representative in the Siegel set $\mathfrak{S}_{t,C}$, where $C = \{n_x \in N : |x| \leq 1 - t\}$. ///

The following part of *reduction theory* is more technical, but essential.

[1.5.2] Theorem: For given $t, t' > 0$ and compact subsets C, C' of N, there are only finitely many $\gamma \in \Gamma$ such that $\mathfrak{S}_{t,C} \cap \gamma \cdot \mathfrak{S}_{t',C'} \neq \phi$. Further, given $t > 0$, for sufficiently large $t' > 0$, $\mathfrak{S}_{t,C} \cap \gamma \mathfrak{S}_{t',C'} \neq \phi$ implies $\gamma \in \Gamma_\infty$.

Proof: Continue in the context of the proof of the previous theorem. Given $t > 0$, take $t' > 1/t$. For $\gamma = \begin{pmatrix} a & b \\ c & d \end{pmatrix} \notin \Gamma_\infty$, $c \neq 0$, so $|c| \geq 1$. For $y \geq t'$ and arbitrary x,

$$\text{height}(\gamma \cdot n_x a_y) = \frac{y}{|cx + d|^2 + |cy|^2} \leq \frac{y}{|c|^2 \cdot y^2} \leq \frac{1}{1 \cdot y} \leq \frac{1}{t'} < t$$

Thus, $\mathfrak{S}_{t,C} \cap \gamma \cdot \mathfrak{S}_{t',C'} \neq \phi$ implies $\gamma \in \Gamma_\infty$ for such t, t'.

For arbitrary $0 < t \le t'$, to show *finiteness* of the set of γ so that $\mathfrak{S}_{t,C} \cap \gamma\mathfrak{S}_{t',C'} \neq \phi$, take t'' strictly larger than t, t', $1/t$, and $1/t'$. The two sets

$$\Omega = \{n_x a_y k : n \in C,\ k \in K,\ t \le y \le t''\}$$
$$\Omega' = \{n_x a_y k : n \in C',\ k \in K,\ t' \le y \le t''\}$$

are *compact*, and $\mathfrak{S}_{t,C} = \mathfrak{S}_{t'',C} \cup \Omega$ and $\mathfrak{S}_{t',C'} = \mathfrak{S}_{t'',C'} \cup \Omega'$. For the asserted finiteness, it suffices to treat the pieces separately.

By the previous paragraph, $\mathfrak{S}_{t,C} = \mathfrak{S}_{t'',C} \cup \Omega$ meets $\gamma\mathfrak{S}_{t'',C'}$ only for $\gamma \in \Gamma_\infty$. Since $\Gamma_\infty = (\Gamma \cap N) \cdot (\Gamma \cap M)$ and $\Gamma \cap M$ is *finite*, it suffices to consider $\gamma = n_x \in \Gamma \cap N$. In that case, $\gamma\mathfrak{S}_{t'',C'} = \mathfrak{S}_{t'',C'+x}$. By the Iwasawa decomposition, $\mathfrak{S}_{t,C} \cap \mathfrak{S}_{t'',C'+x} \neq \phi$ if and only if $C \cap (C' + x) \neq \phi$. Equivalently, $x \in C - C'$ and x is in a lattice $\mathbb{Z}$, $\mathbb{Z}[i]$, $\mathbb{Z}i + \mathbb{Z}j + \mathbb{Z}k$, or $\mathfrak{o}$, respectively. The set $C - C'$ of element-wise differences is *compact*. Either by the lemma below or by more elementary considerations, the set of such x is *finite*.

Similarly $\mathfrak{S}_{t'',C}$ meets $\gamma\Omega'$ only for $\gamma \in \Gamma_\infty$, and there are only finitely many possibilities.

The interaction of Ω and Ω' is subtler. For γ such that $\Omega \cap \gamma\Omega' \neq \phi$, there are $\omega \in \Omega$ and $\omega' \in \Omega'$ such that $\omega = \gamma\omega'$. That is,

$$\gamma = \omega(\omega')^{-1} \cap \Gamma \subset \Omega \cdot \Omega'^{-1} \cap \Gamma$$

where Ω'^{-1} is element-wise inversion. Inversion and multiplication are continuous maps[9] $G \to G$ and $G \times G \to G$, so they map compacts to compacts, so $\Omega\Omega'^{-1}$ is compact. By the following lemma, such a set is finite. ///

[1.5.3] Lemma: The intersection $\Omega \cap \Gamma$ of a compact subset Ω of a topological group[10] G and a discrete sub*group* $\Gamma \subset G$ is *finite*.

Proof: First, we prove that Γ is *closed*, which would not necessarily hold for a discrete sub*set*. For us, a *topological group* is locally compact, Hausdorff, and countably based. Let U be a neighborhood of 1 such that $U \cap \Gamma = \{1\}$. By continuity of inversion and multiplication, there is a neighborhood U_1 of 1 such that $U_1^{-1} \cdot U_1 \subset U$. For $g \notin \Gamma$ but g in the closure of Γ in G, the neighborhood

9 The *multiplication* of real, complex, or quaternion matrices is polynomial in the entries and so is continuous. The continuity of *inversion* can be seen via the explicit formula in terms of determinants of minors over a field k, for example: for $g \in SL_n(k)$, letting A_{ij} be the $(n-1)$-by-$(n-1)$ matrix obtained by deleting the ith row and j^{th} column, $(-1)^{i+j} \det A_{ij}$ is the ij^{th} entry of g^{-1}.

10 As usual, a *topological group* is a locally compact, Hausdorff topological space G with a countable basis, and so that the inverse map $g \to g^{-1}$ and multiplication $g_1 \times g_2 \to g_1 g_2$ are continuous.

gU_1 of g contains infinitely many elements of Γ. For $\gamma \neq \delta$ two such,

$$1 \neq \gamma^{-1} \cdot \delta \in (gU_1)^{-1} \cdot (gU_1) = U_1^{-1} \cdot U_1 \subset U$$

contradiction. Thus, Γ is closed in G.

In a Hausdorff space G, a compact subset C is *closed*, so $C \cap \Gamma$ is closed. A closed subset of a compact set is compact. Thus, $C \cap \Gamma$ is compact, and it is (still) discrete. Discrete compact sets are *finite*, proven as follows. For each $\gamma \in C \cap \Gamma$, let N_γ be a neighborhood of γ in G containing no other element of Γ. The open cover $\{N_\gamma : \gamma \in C \cap \Gamma\}$ of $C \cap \Gamma$ has a finite subcover $N_{\gamma_1} \cup \ldots \cup N_{\gamma_n}$. Since $N_{\gamma_i} \cap \Gamma = \gamma_i$, necessarily $C \cap \Gamma$ is finite. ///

1.6 Invariant Measures, Invariant Laplacians

Proofs of the assertions in this section require substantial preparation, and succeed for very general reasons, so are postponed to [5.2] and [4.2]. In all four examples, the subgroup NA^+ of P is *transitive* on $X = G/K$, by the Iwasawa decomposition $G = NA^+K$, giving a bijection

$$X = G/K = (NA^+K)/K \approx (NA^+)/(NA^+ \cap K) = NA^+$$

In coordinate-independent formulations, notation $x \in X$ is reasonable, despite the fact that, somewhat incompatibly, when convenient we will use coordinates

$$(x, y) \longrightarrow n_x a_y = \begin{pmatrix} 1 & x \\ 0 & 1 \end{pmatrix} \begin{pmatrix} \sqrt{y} & 0 \\ 0 & 1/\sqrt{y} \end{pmatrix}$$

on $NA^+ \approx G/K = X$, as earlier, with $y > 0$, and x in $\mathbb{R}$, $\mathbb{C} \approx \mathbb{R}^2$, $\mathbb{R}i + \mathbb{R}j + \mathbb{R}k \approx \mathbb{R}^3$, or $\mathbb{H} \approx \mathbb{R}^4$, respectively. These coordinates $(x, y) \in \mathbb{R}^{\ell-1} \times (0, +\infty)$ are standard coordinates on *real hyperbolic ℓ-space*, with $\ell = 2, 3, 4, 5$.

Although we will eventually need the *right* translation action of G on G and on functions on G, for the moment we are considering the quotient $X = G/K$. Since K is not a normal subgroup, there is no sensible right translation action of G on $X = G/K$, only the *left* translation action $g \cdot (g_o K) = (g g_o)K$.

The group G acts on the collection $C_c^o(X)$ of continuous, compactly supported functions on $X = G/K$ by left translation

$$L_g f(x) = f(g^{-1}x)$$

with the inverse inserted to have the associativity

$$L_{g_1} L_{g_2} f = L_{g_1 g_2} f$$

A *G-invariant measure/integral* μ on the quotient $X = G/K$ is characterized by the property

$$\int_X L_g f \, d\mu = \int_X f \, d\mu \qquad (\text{for all } g \in G, \ f \in C_c^o(X))$$

In [14.4] we will see that such an invariant measure/integral is *unique* up to scalar multiples and is given in the $x, y \to n_x a_y$ coordinates by

$$f \longrightarrow \int_0^\infty \int_{\mathbb{R}^{\ell-1}} f(x,y) \, \frac{dx\,dy}{|y|^\ell} \qquad (\ell = 2, 3, 4, 5, \text{ respectively})$$

[1.6.1] Corollary: *(of reduction theory)* The invariant volume of $\Gamma\backslash X = \Gamma\backslash G/K$ is *finite.*

Proof: Since there is a Siegel set $\mathfrak{S}_{t,C}$ that surjects to $\Gamma\backslash X$ for some compact $C \subset N$ and some $t > 0$, it suffices to show that the invariant measure of a Siegel set is finite. In the $x, y \to n_x a_y$ coordinates,

$$\int_{\mathfrak{S}_{t,C}} 1 \, d\mu = \int_t^\infty \int_{\{x: n_x \in C\}} 1 \, \frac{dx\,dy}{y^\ell} = (N\text{-volume of } C) \cdot \int_t^\infty \frac{dy}{y^\ell}$$

where $\ell = 2, 3, 4, 5$ in the respective examples. Each of these integrals is finite. ///

Test functions $C_c^\infty(G/K)$ should be compactly supported, infinitely differentiable functions on G/K. However, while these groups G are smooth manifolds, it is less clear whether G/K is a smooth manifold. This potential issue is rendered irrelevant by taking

$$C_c^\infty(G/K) = \{\text{right } K\text{-invariant test functions on } G\} = C_c^\infty(G)^K$$

where right K-invariance means $f(gk) = f(g)$ for all $g \in G$ and $k \in K$. The *invariance* of a G-invariant Laplacian Δ on the quotient $X = G/K$ is the property

$$\Delta(L_g f) = L_g(\Delta f) \qquad (\text{for all } g \in G, \ f \in C_c^\infty(X))$$

In [4.2] we will see that such a Laplacian is essentially canonical, and in the x, y coordinates is, *up to constants which we might want to adjust later for notational convenience,*

$$\Delta = \begin{cases} y^2\Big(\frac{\partial^2}{\partial x^2} + \frac{\partial^2}{\partial y^2}\Big) & (G = SL_2(\mathbb{R})) \\[2mm] y^2\Big(\frac{\partial^2}{\partial x_1^2} + \frac{\partial^2}{\partial x_2^2} + \frac{\partial^2}{\partial y^2}\Big) - y\frac{\partial}{\partial y} & (G = SL_2(\mathbb{C})) \\[2mm] y^2\Big(\frac{\partial^2}{\partial x_1^2} + \frac{\partial^2}{\partial x_2^2} + \frac{\partial^2}{\partial x_3^2} + \frac{\partial^2}{\partial y^2}\Big) - 2y\frac{\partial}{\partial y} & (G = Sp^*_{1,1}) \\[2mm] y^2\Big(\frac{\partial^2}{\partial x_1^2} + \frac{\partial^2}{\partial x_2^2} + \frac{\partial^2}{\partial x_3^2} + \frac{\partial^2}{\partial x_4^2} + \frac{\partial^2}{\partial y^2}\Big) - 3y\frac{\partial}{\partial y} & (G = SL_2(\mathbb{H})) \end{cases}$$

where

$$\begin{cases} x \in \mathbb{R} \text{ for } G = SL_2(\mathbb{R}) \\ x = x_1 + ix_2 \in \mathbb{C} \text{ for } G = SL_2(\mathbb{C}) \\ x = x_1 i + x_2 j + x_3 k \in \mathbb{H} \text{ for } G = Sp^*_{1,1}, \\ x = x_1 + x_2 i + x_3 j + x_4 k \text{ for } G = SL_2(\mathbb{H}) \end{cases}$$

In [6.6] we will see the *symmetry* property of Δ:

$$\int_X \Delta f \cdot \overline{F}\, d\mu = \int_X f \cdot \overline{\Delta F}\, d\mu \qquad (\text{for } f, F \in C_c^\infty(X))$$

Also, the *negative-semi-definite* property

$$\int_X \Delta f \cdot \overline{f}\, d\mu \le 0 \qquad (\text{for } f \in C_c^\infty(X))$$

For $G = SL_2(\mathbb{R})$, where $\ell = 2$, by chance the first-order term in y in Δ disappears, the powers of y in Δ and μ cancel, and the symmetry property reduces to symmetry of the Euclidean Laplacian, just integration by parts. Although attractive, this coincidence is inessential.

The left G-invariance of μ and Δ ensure that they *descend* to the quotient $\Gamma\backslash X \approx \Gamma\backslash G/K$. We will use the same symbols for the versions on $\Gamma\backslash X$. As we see in [5.2], uniqueness of invariant measure/integral entails *unwinding* identities such as

$$\int_{\Gamma\backslash X} \Big(\sum_{\gamma\in\Gamma} L_\gamma f\Big)\, d\mu = \int_{\Gamma\backslash X} \Big(\sum_{\gamma\in\Gamma} f(\gamma^{-1}x)\Big)\, d\mu(x) = \int_X f\, d\mu$$

for $f \in C_c^o(X)$. For the Laplacian,

$$\Delta f \circ q = \Delta(f \circ q) \qquad (\text{for } f \in C_c^\infty(\Gamma\backslash X))$$

The symmetry and negative semi-definiteness of Δ descend to functions on the quotient, $C^\infty(\Gamma\backslash X) = C^\infty(\Gamma\backslash G)^K$, where we take advantage of the fact that $\Gamma\backslash G$ is a smooth manifold. Again, see [6.6] for proofs.

As expected, with invariant measure μ descended to $\Gamma\backslash X$, the usual Hermitian inner product is[11]

$$\langle f, F\rangle_{\Gamma\backslash X} = \int_{\Gamma\backslash X} f \cdot \overline{F}\, d\mu$$

[11] While integrals of Γ-invariant functions on $\mathfrak{H}$ on the *quotient* $\Gamma\backslash X$ can be understood in an elementary way as integrals over explicit fundamental domains, such a viewpoint impedes understanding of integration by parts on $C_c^\infty(\Gamma\backslash X)$. It is better to use an *intrinsic* integral on the *quotient*, characterized by the earlier unwinding relation, as in [5.2].

with associated norm

$$|f|_{L^2(\Gamma\backslash X)} = \langle f, f\rangle_{\Gamma\backslash X}^{\frac{1}{2}}$$

As usual, in the characterization

$$L^2(\Gamma\backslash X) = \{\text{measurable } f : |f|_{L^2(\Gamma\backslash X)} < \infty\}$$

elements of $L^2(\Gamma\backslash X)$ are *equivalence classes* of *measurable* functions, with equivalence being equality almost everywhere. In [6.5] we will see that this characterization is equivalent to a characterization as L^2-completion of test functions $C_c^\infty(\Gamma\backslash X)$.

1.7 Discrete Decomposition of $L^2(\Gamma\backslash G/K)$ Cuspforms

The theorems stated below will be proven later, in [7.1–7.7], but we can set up precise statements.

In this section and much of the sequel, *waveform*, *automorphic form*, and *automorphic function* will be used roughly as synonyms, referring to $\mathbb{C}$-valued functions on $\Gamma\backslash X$, meeting further conditions depending on the situation. Such functions are identifiable with Γ-invariant functions on X, by composing with the quotient map $X \to \Gamma\backslash X$.

The *constant term* $c_P f$ of a waveform f on $\Gamma\backslash X$ is a function on $X = G/K$ defined by

$$(\text{constant term})f(x) = c_P f(x) = \int_{(N\cap\Gamma)\backslash N} f(n\cdot x)\, dn$$

Here the group N is abelian, isomorphic to $\mathbb{R}^{\ell-1}$ for $\ell = 2, 3, 4, 5$, and $N \cap \Gamma$ is a discrete subgroup with compact quotient $(N \cap \Gamma)\backslash N$. We give N the measure from the coordinate $x \to n_x$ with $x \in \mathbb{R}^{\ell-1}$, and as earlier and in [5.2] give the quotient the unique compatible measure for unwindings

$$\int_{(N\cap\Gamma)\backslash N} \Big(\sum_{\gamma\in N\cap\Gamma} \varphi(\gamma n)\Big)\, dn = \int_N \varphi(n)\, dn$$

for all $\varphi \in C_c^o(N)$. By changing variables, we see that, although the constant term has probably lost left Γ-invariance, $c_P f$ is a left N-invariant function on $X = G/K$:

$$\begin{aligned} c_P f(n'x) &= \int_{(N\cap\Gamma)\backslash N} f(n\cdot n'x)\, dn = \int_{(N\cap\Gamma)\backslash N} f((nn')\cdot x)\, dn \\ &= \int_{(N\cap\Gamma)\backslash N} f(n\cdot x)\, dn \qquad (\text{for } n' \in N) \end{aligned}$$

Thus, constant terms of functions f on $\Gamma\backslash G/K$ can be viewed as functions on the *ray*

$$N\backslash X = N\backslash G/K = N\backslash(NA^+K)/K) \approx A^+ \approx (0,\infty)$$

Similarly, because $\Gamma_\infty = P\cap\Gamma$ normalizes $N\cap\Gamma$, the constant term is left Γ_∞-invariant.[12] Altogether, c_Pf is left invariant by the group $N\Gamma_\infty$.

All this presumes that c_Pf has at least as much sense as a function with pointwise values as did f, but we need more than that. For example, unfortunately, it turns out that $f\in L^2(\Gamma\backslash G/K)$ does *not* imply that $c_Pf\in L^2(N\backslash G/K)$. More cautiously, suppose f is locally L^1, meaning that $|f|$ has finite integrals over compact subsets of $\Gamma\backslash G$. Fubini's theorem implies that a compactly supported integral of f in *one* of several variables is again locally L^1. This applies to $n\times y\to f(na_y)$. The nature of the constant term map is clarified in [1.8].

Cuspforms are waveforms f meeting the *Gelfand condition* $c_Pf=0$. In some contexts, the term *cuspform* further connotes Δ*-eigenfunctions* in $L^2(\Gamma\backslash G/K)$, but for present purposes, the latter usage is too restrictive. A genuine minor complication is that L^2 functions do not have good pointwise values, so vanishing of the constant term must mean *almost everywhere* for L^2 functions. Thus, it is often better to consider the constant-term map as a map on *distributions*, and the Gelfand condition as a distributional vanishing condition on distributions, as in [1.8]. As usual, put

$$L^2_o(\Gamma\backslash X) = \{L^2\text{-cuspforms}\} = \{f\in L^2(\Gamma\backslash G/K) : c_Pf=0\}$$

The first main theorem, proven in [7.1–7.7], is the *discrete decomposition of the space of cuspforms:* one version is

[1.7.1] Theorem: The space $L^2_o(\Gamma\backslash G/K)$ of square-integrable cuspforms is a *closed* subspace of $L^2(\Gamma\backslash G/K)$, and has an orthonormal basis of Δ-eigenfunctions. Each eigenspace is finite-dimensional, and the number of eigenvalues below a given bound is finite. *(Proof in [7.1–7.7].)*

The *closed-ness* of the space of L^2 cuspforms comes from recharacterization of it in terms of pseudo-Eisenstein series, in [1.8].

In contrast, the full space $L^2(\Gamma\backslash X)$ does *not* have a basis of Δ-eigenfunctions: as proven in [1.12], the orthogonal complement of cuspforms in $L^2(\Gamma\backslash X)$ mostly consists of *integrals* of *non-*L^2 eigenfunctions for Δ, the *Eisenstein series* E_s, introduced in [1.9].

[12] In the present examples, $\Gamma_\infty = P\cap\Gamma$ is only finite index larger than $N\cap\Gamma$, but in other examples this index can be infinite.

The operator Δ presents some technical issues. For example, while $L^2(\Gamma\backslash X)$ lies inside the collection of *distributions* on $\Gamma\backslash X$, and interpreting Δ distributionally would make it well defined on all of $L^2(\Gamma\backslash X)$, it would not *stabilize* $L^2(\Gamma\backslash X)$. This would seem to obstruct use of its symmetry or self-adjointness as an (unbounded) operator on a Hilbert space. On the other hand, indeed, no version of Δ can be defined on all of $L^2(\Gamma\backslash X)$ while retaining the symmetry $\langle \Delta f, F\rangle = \langle f, \Delta F\rangle$ for test functions f, F in $L^2(\Gamma\backslash X)$. This situation requires careful treatment of unbounded, densely defined operators on Hilbert spaces, as in [9.1–9.2].

1.8 Pseudo-Eisenstein Series

Returning to $L^2(\Gamma\backslash X)$, we want to express the orthogonal complement of cuspforms $L^2_o(\Gamma\backslash X)$ in terms of Δ-eigenfunctions, as discussed in [1.12] and [1.13]. To exhibit explicit L^2 functions demonstrably spanning the orthogonal complement to cuspforms, we intend to recast the Gelfand vanishing condition. First, for $f \in L^2(\Gamma\backslash X)$, the constant term $c_P f$ is a left $N\Gamma_\infty$-invariant function on G. It vanishes *as a distribution* if and only if

$$\int_{N\Gamma_\infty\backslash G} \varphi \cdot c_P f = 0 \qquad (\text{for all } \varphi \in C_c^\infty(N\Gamma_\infty\backslash G))$$

with right G-invariant measure on $N\Gamma_\infty\backslash G$ as in [5.2]. In fact, since f is right K-invariant, $c_P f$ is right K-invariant, so we need only test against $\varphi \in C_c^\infty(N\Gamma_\infty\backslash G)^K$. The isomorphisms

$$N\backslash X \approx N\backslash G/K \approx N\backslash(NA^+K)/K \approx A^+$$

identify $N\backslash X$ with the *ray* $A^+ \approx (0, +\infty)$, and *identify* right K-invariant functions φ on $N\backslash G$ with functions of $y = \text{height}(na_y k)$. As in the previous section, for f in L^2, since f is *locally* integrable its constant term is locally integrable, by Fubini's theorem. Thus, $c_P f$ can be integrated against test functions on $N\backslash G/K$.

Given φ in $C_c^\infty(N\Gamma_\infty\backslash G)^K$, the corresponding *pseudo-Eisenstein series* Ψ_φ should be a function in $C_c^\infty(\Gamma\backslash X)$ fitting into an *adjunction*:

$$\int_{N\Gamma_\infty\backslash G} \varphi \cdot c_P f = \int_{\Gamma\backslash G} \Psi_\varphi \cdot f \qquad (\text{for } f \in L^2(\Gamma\backslash X))$$

This adjunction will involve an *unwinding/winding-up*, so we might prefer that $c_P f$ be *continuous*, to easily invoke properties of vector-valued integrals [14.1]. For general reasons [6.1], $C_c^o(\Gamma\backslash G)$ is dense in $L^2(\Gamma\backslash G)$ in the L^2 topology, and for general reasons [5.1] the *left* action of $(N \cap \Gamma)\backslash N$ on the Fréchet space $C^o((N \cap \Gamma)\backslash G)$ is a continuous map $(N \cap \Gamma)\backslash N \times C^o((N \cap \Gamma)\backslash G) \to$

$C^o(N\backslash G)$, so $c_P f$ exists as a $C^o(N\backslash G)$-valued Gelfand-Pettis integral [14.1]. For $f \in C^o(\Gamma\backslash G)$, the integral of $c_P f$ against $\varphi \in C_c^\infty(N\backslash G/K)$ is the integral of a compactly supported, continuous function.

Direct computation yields a canonical *expression* for the desired Ψ_φ, using the left $N\Gamma_\infty$-invariance of φ and the left Γ-invariance of f, as follows. First, *unwinding* as in [5.2],

$$\int_{N\Gamma_\infty\backslash G} \varphi \cdot c_P f = \int_{N\Gamma_\infty\backslash G} \varphi(g) \Big(\int_{N\cap\Gamma\backslash N} f(ng)\, dn \Big)\, d\mu(g)$$
$$= \int_{\Gamma_\infty\backslash G} \varphi(g)\, f(g)\, d\mu(g)$$

Winding up, using the left Γ-invariance of f,

$$\int_{\Gamma_\infty\backslash G} f(g)\,\varphi(g)\, d\mu(g) = \int_{\Gamma\backslash G} \sum_{\gamma\in\Gamma_\infty\backslash\Gamma} f(\gamma\cdot g)\,\varphi(\gamma\cdot g)\, d\mu(g)$$
$$= \int_{\Gamma\backslash G} f(g) \Big(\sum_{\gamma\in\Gamma_\infty\backslash\Gamma} \varphi(\gamma g) \Big)\, d\mu(g)$$

The inner sum in the last integral is the pseudo-Eisenstein series[13] attached to φ:

$$\Psi_\varphi(g) = \sum_{\gamma\in\Gamma_\infty\backslash\Gamma} \varphi(\gamma g)$$

The convergence of the sum needs attention:

[1.8.1] Claim: The series for a pseudo-Eisenstein series Ψ_φ is *locally finite*, meaning that for g in a fixed compact in G, there are only finitely many nonzero summands in $\Psi_\varphi(g) = \sum_\gamma \varphi(\gamma g)$. Thus, $\Psi_\varphi \in C_c^\infty(\Gamma\backslash X)$.

Proof: Given $\varphi \in C_c^\infty(N\backslash G/K)$, let $C \subset G$ be compact so that $N \cdot C$ contains the support of φ. Fix compact $C_o \subset G$ in which $g \in G$ is constrained to lie. Then a summand $\varphi(\gamma g)$ is nonzero only if $\gamma g \in N \cdot C$, which holds only if

$$\gamma \in \Gamma_\infty \cdot C \cdot g^{-1}$$

so

$$\gamma \in \Gamma \cap \Gamma_\infty \cdot C \cdot C_o^{-1}$$

In the quotient $G \to \Gamma_\infty\backslash G$, the image of Γ is closed and discrete. The image of the compact set $N \cdot C \cdot C_o^{-1}$ under the continuous quotient map is *compact*,

[13] In 1966 [Godement 1966b] called these *incomplete theta series*. More recently [Moeglin-Waldspurger 1995] reinforced the precedent of calling them *pseudo-Eisenstein series*.

since $(\Gamma \cap N)\backslash N$ is compact and continuous images of compacts are compact. Thus, left modulo Γ_∞, that intersection is the intersection of a *closed* discrete set and a compact set, so *finite*. (Compare the [1.5.2] from reduction theory.) Therefore, the series is *locally finite* and defines a smooth function on $\Gamma\backslash G$. Summing over left translates certainly retains right K-invariance.

To show that Ψ_φ has compact support in $\Gamma\backslash G$, proceed similarly. That is, for a summand $\varphi(\gamma g)$ to be nonzero, it must be that $g \in \Gamma \cdot C$. The image $\Gamma\backslash(\Gamma \cdot C)$ is compact, being the continuous image of the compact set C under the continuous map $G \to \Gamma\backslash G$, proving the compact support. ///

[1.8.2] Corollary: *Square-integrable cuspforms* are the orthogonal complement in $L^2(\Gamma\backslash X)$ to the subspace of $L^2(\Gamma\backslash X)$ spanned by the pseudo-Eisenstein series Ψ_φ with $\varphi \in C_c^\infty(N\backslash X)$. The map $f \to c_P f$ is continuous from $L^2(\Gamma\backslash G/K)$ to *distributions* on $N\backslash G/K$.

Proof: Again, as previously, for general reasons [6.1] $C_c^o(\Gamma\backslash G/K)$ is dense in $L^2(\Gamma\backslash G/K)$, and the constant terms $c_P f$ are continuous for such f, so integrals against $\varphi \in C_c^\infty(N\backslash G/K)$ exist. Then the adjunction gives

$$\left| \int_{N\backslash G/K} c_P f \cdot \overline{\varphi} \right| = \left| \langle f, \Psi_\varphi \rangle \right| \leq |f|_{L^2} \cdot |\Psi_\varphi|_{L^2}$$

Thus, $f \to \int_{N\backslash G/K} c_P f \cdot \overline{\varphi}$ is a *continuous* linear functional on L^2. In particular, the kernels are closed, and the intersection of all these is the space of L^2 cuspforms. The inequality is exactly the continuity of $f \to c_P f$ with the weak dual topology [13.14] on distributions on $(0, \infty) \approx N\backslash G/K$. ///

Because Δ commutes with the group action, the effect of Δ on a pseudo-Eisenstein series is reflected entirely in its effect on the data: the sum is locally finite, so interchange of the operator and the sum is easy, giving

$$\begin{aligned} \Delta \Psi_\varphi &= \Delta \sum_{\gamma \in \Gamma_\infty \backslash \Gamma} \varphi \circ \gamma = \sum_{\gamma \in \Gamma_\infty \backslash \Gamma} \Delta(\varphi \circ \gamma) \\ &= \sum_{\gamma \in \Gamma_\infty \backslash \Gamma} (\Delta \varphi) \circ \gamma = \Psi_{\Delta \varphi} \end{aligned}$$

This correctly suggests that a suitable dense subspace of $L^2_o(\Gamma\backslash X)$ is indeed *stable* under Δ. However, at this point we do not have a good device to prove density of *smooth* cuspforms with *sufficient decay* to prove symmetry $\langle \Delta f, F \rangle = \langle f, \Delta F \rangle$. For that matter, there is no reason to expect *test functions* in $L^2_o(\Gamma\backslash X)$ to be dense because smooth-truncation to arrange compact support can succeed directly in $y \gg 1$ but disturbs the constant term as $y \to 0^+$. A convincing argument for smoothness of cuspforms and behavior of Δ on them can be given after the decomposition result of [7.1–7.7].

1.9 Eisenstein Series

We can attempt to make a pseudo-Eisenstein series Ψ_φ which is a Δ-eigenfunction, by using a function φ on $N\backslash X = N\backslash G/K$, which is a Δ-eigenfunction. Using the y-coordinate on $N\backslash G/K \approx A^+$, the differential equation is

$$\lambda \cdot \varphi = \Delta\varphi = \left(y^2 \frac{\partial^2}{\partial y^2} - (\ell - 2)y\frac{\partial}{\partial y}\right)\varphi = y^2\varphi'' - (\ell - 2)y\varphi'$$

The differential equation $y^2\varphi'' - (\ell - 2)y\varphi' - \lambda\varphi = 0$ is of *Euler type*, that is, will have solutions of the form y^α, with α determined by

$$\begin{aligned} 0 &= y^2 \cdot \alpha(\alpha - 1)y^{\alpha-2} - (\ell - 2)y \cdot \alpha y^{\alpha-1} - \lambda \cdot y^\alpha \\ &= y^\alpha \cdot (\alpha(\alpha - 1) - (\ell - 2)\alpha - \lambda) \end{aligned}$$

That is, for given λ, the corresponding exponents α are found by solving the *indicial equation* $\alpha(\alpha - 1) - (\ell - 2)\alpha - \lambda = 0$, so $\lambda = \alpha(\alpha - (\ell - 1))$. This computation suggests incorporating the factor $\ell - 1$ into the exponent. Thus, with a function η on $N\backslash G/K$ defined by, with $n \in N$, $k \in K$,

$$\eta(na_y k) = y^{\ell-1} \qquad \text{(with } \ell = 2, 3, 4, 5\text{, respectively)}$$

we have

$$\Delta\eta^s = (\ell - 1)^2 \cdot s(s - 1) \cdot \eta^s$$

Unfortunately, η^s is not in $C_c^\infty(N\backslash G/K)$, although it is smooth. The *genuine* Eisenstein series E_s on $\Gamma\backslash X$ is[14]

$$E_s(g) = \sum_{\gamma \in \Gamma_\infty\backslash\Gamma} \eta(\gamma \cdot g)^s$$

The following claim has a much more elementary proof in the two simplest cases $\Gamma = SL_2(\mathbb{Z}), SL_2(\mathbb{Z}[i])$, but something more is required for $\Gamma = Sp^*_{1,1}$ and $SL_2(\mathbb{H})$. We give an argument that applies uniformly to all four:

[14] There is no universal choice of normalization. Here, the choice is made so that the critical strip is $0 \le \mathrm{Re}(s) \le 1$, the rightmost pole is at $s = 1$, and the functional equation relates E_s and E_{1-s}. In more general contexts, other considerations dominate.

[1.9.1] Claim: For $\mathrm{Re}(s) > 1$, the series expression for E_s converges absolutely and uniformly on compacts, to a *continuous* function on $\Gamma\backslash X$, of *moderate growth*

$$|E_s(g)| \ll_{t,C} \eta^{\mathrm{Re}(s)} \qquad (\text{on } \mathfrak{S}_{t,C})$$

with implied constant depending on t, C.

Proof: It suffices to consider $s = \sigma$ real. From [1.3.2], in $x, y \to n_x a_y$ coordinates,

$$\eta\left(\begin{pmatrix} * & * \\ c & d \end{pmatrix} n_x a_y\right) = \left(\frac{y}{|cx+d|^2 + |cy|^2}\right)^{\ell-1}$$

and $(c\ d) \to |cx+d|^2 + |cy|^2$ is a positive-definite quadratic function on the real vector space in which $(c\ d)$ lies. The coefficients of the quadratic function depend continuously on x, y, so for x, y in a fixed compact there are *uniform* constants A, B such that

$$A \cdot (|c|^2 + |d|^2) \leq |cx+d|^2 + |cy|^2 \leq B \cdot (|c|^2 + |d|^2)$$

In particular, for another pair x', y' in the same compact,

$$\begin{aligned} |cx'+d|^2 + |cy'|^2 \leq B \cdot (|c|^2 + |d|^2) &= \frac{B}{A} \cdot \left(A \cdot (|c|^2 + |d|^2)\right) \\ &\leq \frac{B}{A} \cdot \left(|cx+d|^2 + |cy|^2\right) \end{aligned}$$

Thus, convergence of the series is equivalent to convergence of an averaged form, namely,

$$\int_C \sum_{\Gamma_\infty\backslash\Gamma} \eta(\gamma n_x a_y)^\sigma \, \frac{dx\, dy}{y^\ell}$$

Similarly, because the inf of lengths of nonzero vectors in a lattice in a real vector space is *positive*, there is a *uniform* nonzero lower bound for $|cx+d|^2 + |cy|^2$ for $n_x a_y \in C$ and $(c\ d)$ a lower row in Γ. That is, the sup of $\eta(\gamma g)$ over $\gamma \in \Gamma$ and $g \in C$ is finite and is attained. Let the sup be $\mu^{\ell-1}$ for $\mu > 0$. Then $\Gamma \cdot C$ is contained in

$$Y = \{n_x a_y \in X : y^{\ell-1} = \eta(n_x a_y) \leq \mu^{\ell-1}\} = \{n_x a_y \in X : y \leq \mu\}$$

By discreteness of Γ in G, we can shrink C so that, for γ in Γ, if $\gamma C \cap C \neq \phi$ then $\gamma = 1$, so that

$$\int_C E_s = \int_{\Gamma\backslash\Gamma\cdot C} E_s$$

Unwind:

$$\int_{\Gamma\backslash\Gamma\cdot C} E_s = \int_{\Gamma_\infty\backslash\Gamma\cdot C} \eta(n_x a_y)^\sigma \frac{dx\,dy}{y^\ell} \, le \int_{\Gamma_\infty\backslash Y} \eta(n_x a_y)^\sigma \frac{dx\,dy}{y^\ell}$$
$$= \int_{N\cap\Gamma\backslash N} 1 \cdot \int_0^\mu \eta^\sigma \frac{dy}{y^\ell} \ll \int_0^\mu y^{(\ell-1)\sigma} \frac{dy}{y^\ell} = \int_0^\mu y^{(\ell-1)(\sigma-1)} \frac{dy}{y}$$

This is convergent for $\sigma - 1 > 0$. This argument also proves the uniform convergence on compacts.

To see the moderate growth property, without yet attempting to prove that E_s is smooth, differentiating the summands with $c \neq 0$

$$\frac{\partial}{\partial y}\left|\frac{1}{(|cx+d|^2+|cy|^2)^{s(\ell-1)}}\right| = \frac{\partial}{\partial y}\frac{1}{(|cx+d|^2+|cy|^2)^{\mathrm{Re}\,(s)\cdot(\ell-1)}} < 0$$

shows that they all strictly decrease as $\eta(g)$ increases. Precisely, $|E_s(na_y)| < |E_s(na_{y'})|$ for $0 < y < y'$, for every $n \in N$. Since E_s is *continuous*, it has a bound B on a compact set $\{g \in \mathfrak{S}_{t,C} : \eta(g) \le T\}$. Thus, $|y^{-s}E_s| \le B$ on $\mathfrak{S}_{t,C}$. ///

Of course, we want convergence to a *smooth* function:

[1.9.2] Claim: The series for E_s converges in the C^∞ topology for $\mathrm{Re}\,(s) > 1$, and produces a C^∞ moderate-growth function on $\Gamma\backslash X = \Gamma\backslash G/K$. *(Proof in [11.5].)*

As in [13.5], the *idea* of the C^∞ topology is that it is given by the collection of seminorms given by sups on compacts of all derivatives. One issue is which derivatives to use and how to estimate them. In the present setting, one might be tempted to use derivatives with respect to coordinates x, y, but there is a significant disadvantage: $\partial/\partial x$ and $\partial/\partial y$ do not commute with the action of Γ on X, so that $\partial E_s/\partial x$ is unlikely to be left Γ-invariant. For that matter, the effect of differentiating with respect to y (after removing the common factor $y^{s(\ell-1)}$)

$$\frac{\partial}{\partial y}\frac{1}{(|cx+d|^2+|cy|^2)^{s(\ell-1)}} = -s(\ell-1)\cdot\frac{2y\cdot|c|^2}{(|cx+d|^2+|cy|^2)^{s(\ell-1)+1}}$$

on convergence of the series is difficult to appraise. The less-elementary approach in [11.5] uses left-G-invariant derivatives on G, which preserve left Γ-invariance and stabilize a somewhat larger class of Eisenstein series.

[1.9.3] Corollary: In $\mathrm{Re}\,(s) > 1$, E_s inherits the eigenvalue property from η^s:

$$\Delta E_s = (\ell-1)^2 \cdot s(s-1)\cdot E_s$$

Proof: Granting the convergence in the C^∞ topology, in $\mathrm{Re}(s) > 1$, and using the fact that Δ commutes with translations by Γ, letting $\Delta\eta^s = \lambda_s \cdot \eta^s$,

$$\Delta \sum_\gamma \eta(\gamma g)^s = \sum_\gamma \Delta(\eta(\gamma g)^s) = \sum_\gamma (\Delta\eta^s)(\gamma g) = \lambda_s \cdot \sum_\gamma \eta^s(\gamma g)$$

as claimed. ///

However, as we see subsequently, E_s is *never* in $L^2(\Gamma\backslash G)$.

Granting adequate convergence of E_s, and granting that the differential operator Δ can move inside the integral (see [14.1]) expressing the constant term, the constant term $c_P E_s$ is a Δ-eigenfunction with the same eigenvalue:

$$\begin{aligned}\Delta\big(c_P E_s\big)(g) &= \Delta \int_{N\cap\Gamma\backslash N} E_s(ng)\, dn = \int_{N\cap\Gamma\backslash N} \Delta E_s(ng)\, dn \\ &= \int_{N\cap\Gamma\backslash N} (\ell-1)^2 \cdot s(s-1) \cdot E_s(ng)\, dn \\ &= (\ell-1)^2 \cdot s(s-1) \cdot \int_{N\cap\Gamma\backslash N} E_s(ng)\, dn \\ &= (\ell-1)^2 \cdot s(s-1) \cdot c_P E_s(g)\end{aligned}$$

[1.9.4] Claim: The constant term of the Eisenstein series E_s is of the form

$$c_P E_s = \eta^s + c_s \eta^{1-s}$$

Proof: From the very beginning of this section, at least for $s \neq \frac{1}{2}$, η^s and η^{1-s} are a basis for the space of Δ-eigenfunctions with eigenvalue $(\ell-1)^2 \cdot s(s-1)$ on $N\backslash G/K$. Thus, the constant term $c_P E_s$ is a linear combination of η^s and η^{1-s}. The term η^s comes from the representative $1 \in \Gamma_\infty\backslash\Gamma$. Every other representative $\gamma = \begin{pmatrix} a & b \\ c & d \end{pmatrix}$ has $c \neq 0$. Implicitly recapitulating the computation in the Bruhat decomposition from [1.1],

$$\begin{aligned}&\int_{N\cap\Gamma\backslash N} \sum_{c,d} \left(\frac{y}{|cx+d|^2 + |cy|^2}\right)^{s(\ell-1)} dn_x \\ &= {\sum_{0\neq c}}' \frac{1}{|c|^{s(\ell-1)}} \int_{N\cap\Gamma\backslash N} \sum_d \left(\frac{y}{|x+\frac{d}{c}|^2 + |y|^2}\right)^{s(\ell-1)} dn_x\end{aligned}$$

where the sum over c is over all possible lower-left entries of $\gamma \in \Gamma$, modulo $M \cap \Gamma$, and the inner sum over d is over possible lower-right entries given lower-left entry c. With $\Xi = \{\xi : n_\xi \in \Gamma \cap N\}$, this is

$$\sum_{0\neq c} \frac{1}{|c|^{s(\ell-1)}} \sum_{d \bmod c} \int_{N\cap\Gamma\backslash N} \sum_{\xi\in\Xi} \left(\frac{y}{|x+\xi+\frac{d}{c}|^2 + |y|^2}\right)^{s(\ell-1)} dn_x$$

Note that if $\gamma = \begin{pmatrix} * & * \\ c & d \end{pmatrix} \in \Gamma$ and $\xi \in \Xi$, then

$$\Gamma \ni \gamma \cdot n_\xi = \begin{pmatrix} * & * \\ c & d \end{pmatrix} \cdot \begin{pmatrix} 1 & \xi \\ 0 & 1 \end{pmatrix} = \begin{pmatrix} * & * \\ c & d + c\xi \end{pmatrix}$$

The integral unwinds, giving

$$\sum_{0 \neq c} \frac{1}{|c|^{s(\ell-1)}} \sum_{d \bmod c} \int_N \left(\frac{y}{|x + \frac{d}{c}|^2 + |y|^2} \right)^{s(\ell-1)} dn_x$$

For each suitable d mod c, replace x by $x - \frac{d}{c}$, and let $\nu(c)$ be the number of such d, so the whole becomes

$$\sum_{0 \neq c} \frac{\nu(c)}{|c|^{s(\ell-1)}} \int_{\mathbb{R}^{\ell-1}} \left(\frac{y}{|x|^2 + |y|^2} \right)^{s(\ell-1)} dx$$
$$= y^{(1-s)(\ell-1)} \sum_{0 \neq c} \frac{\nu(c)}{|c|^{s(\ell-1)}} \int_{\mathbb{R}^{\ell-1}} \left(\frac{1}{|x|^2 + 1} \right)^{s(\ell-1)} dx$$

upon replacing x by yx. This demonstrates the asserted shape of the constant term. ///

In fact, as usual, c_s has an Euler product and is identifiable as a ratio of L-functions, as we consider further in the sequel.

[1.9.5] Corollary: The Eisenstein series is not in $L^2(\Gamma \backslash X)$.

Proof: By reduction theory [1.5], it suffices to show that it is not square-integrable on a quotient $(N \cap \Gamma)\backslash\{n_x a_y \in X : y \geq t_o\}$ for t_o large enough. Functions on such a set are left $N \cap \Gamma$-invariant functions on $N \times A^+$, so have *Fourier expansions* on the *product of circles* $(N \cap \Gamma)\backslash N$, with Fourier coefficients depending on $a \in A^+$. Specifically, let Ψ be the collection of $N \cap \Gamma$-invariant continuous group homomorphisms $\psi : N \to \mathbb{C}^\times$. A function f in $L^2((N \cap \Gamma)\backslash N)$ has a Fourier expansion converging (at least) in L^2: the ψth Fourier coefficient is

$$\widehat{f}(\psi) = \int_{(N \cap \Gamma)\backslash N} \overline{\psi}(n) \cdot f(n)\, dn$$

giving $(N \cap \Gamma)\backslash N$ total measure 1, and

$$f(n) = \sum_\psi \widehat{f}(\psi) \cdot \psi(n) \qquad (n \in N, \text{ convergent in an } L^2 \text{ sense})$$

The Plancherel theorem for $L^2((N\cap\Gamma)\backslash N)$ is

$$\int_{(N\cap\Gamma)\backslash N} |f(n)|^2\, dn = \sum_{\psi} |\widehat{f}(\psi)|^2$$

For a function f on $(N\cap\Gamma)\backslash N \times A^+$, the Fourier coefficients are functions of $a \in A^+$, and

$$f(na) = \sum_{\psi} \widehat{f}(\psi)(a)\cdot\psi(n)$$

Plancherel for $L^2((N\cap\Gamma)\backslash N)$ now gives

$$\int_{(N\cap\Gamma)\backslash N} |f(na)|^2\, dn = \sum_{\psi} |\widehat{f}(\psi)(a)|^2$$

In particular, Bessel's inequality applied to the trivial character $\psi = 1$ gives

$$\int_{(N\cap\Gamma)\backslash N} |f(na)|^2\, dn \geq |\widehat{f}(1)(a)|^2$$

For $f(na) = E_s(na)$, that Fourier component for the trivial character in is exactly the constant term $c_P E_s = \eta^s + c_s\eta^{1-s}$, so

$$\int_{(N\cap\Gamma)\backslash N}\int_{t_o}^{\infty} |E_s(na)|^2\, dn\, da \geq \int_{t_o}^{\infty} |\eta^s + c_s\eta^{1-s}|^2\, \frac{dy}{y^\ell}$$

With $\sigma = \mathrm{Re}\,(s) > 1$,

$$\int_{t_o}^{T} |\eta^s + c_s\eta^{1-s}|^2\, \frac{dy}{y^\ell} \gg_s \int_{t_o}^{T} \eta^{2\sigma}\, \frac{dy}{y^\ell} = \int_{t_o}^{T} y^{2\sigma(\ell-1)}\, \frac{dy}{y^\ell}$$
$$= \int_{t_o}^{T} y^{(2\sigma-1)(\ell-1)}\, \frac{dy}{y} = T^{(2\sigma-1)(\ell-1)} - t_o^{(2\sigma-1)(\ell-1)}$$

This blows up as $T \to \infty$ for $2\sigma - 1 > 0$. ///

1.10 Meromorphic Continuation of Eisenstein Series

Although special tricks [2.B] applicable to $\Gamma = SL_2(\mathbb{Z})$ and $\Gamma = SL_2(\mathbb{Z}[i])$ have been known for almost 100 years, those tricks almost immediately fail in any larger context. For example, they do not apply to $\Gamma = Sp^*_{1,1}(\mathfrak{o})$ or $\Gamma = SL_2(\mathbb{Z})$. [Selberg 1956] and [Roelcke 1956b] first approached more general cases.

In [11.4] we will give a proof applying uniformly to our four example cases:

[1.10.1] Theorem: E_s has a *meromorphic continuation* in $s \in \mathbb{C}$, as a smooth function of moderate growth on $\Gamma\backslash G$. As a function of s, $E_s(g)$ it is of at most polynomial growth vertically, uniformly in bounded strips, uniformly for g in compacts. *(Proof in [11.4].)*

Although we give further details in a somewhat different logical order in [11.4], some consequences of the meromorphic continuation can be discussed directly:

[1.10.2] Corollary: The eigenfunction property $\Delta E_s = \lambda_s \cdot E_s$ with $\lambda_s = (\ell - 1)^2 \cdot s(s-1)$ persists under meromorphic continuation.

Proof: Both ΔE_s and $\lambda_s \cdot E_s$ are holomorphic function-valued functions of s, taking values in the topological vector space of smooth functions. They agree in the region of convergence $\mathrm{Re}\,(s) > 1$, so the vector-valued form [15.2] of the Identity Principle from complex analysis gives the result. ///

[1.10.3] Corollary: The meromorphic continuation of E_s implies the meromorphic continuation of the constant term $c_P E_s = \eta^s + c_s \eta^{1-s}$, in particular, of the function c_s.

Proof: Since E_s meromorphically continues at least as a smooth function, the integral over the compact set $(N \cap \Gamma)\backslash N$ expressing a pointwise value $c_P E_s(g)$ of the constant term certainly converges absolutely. In fact, the function-valued function $n \to (g \to E_s(ng))$ is a continuous smooth-function-valued function and has a smooth-function-valued Gelfand-Pettis integral $g \to c_P E_s(g)$ [14.1].

In particular, the constant term $c_P E_s$ of the continuation of E_s must still be of the form $A_s \eta^s + B_s \eta^{1-s}$ for some functions A_s, B_s, since (at least for $s \neq \frac{1}{2}$) η^s and η^{1-s} are the two linearly independent solutions of $\Delta f = \lambda_s \cdot f$ for functions f on $N\backslash G/K \approx A^+$. In the region of convergence $\mathrm{Re}\,(s) > 1$, the linear independence of η^s and η^{1-s} gives $A_s = 1$ and $B_s = c_s$. The vector-valued form of the Identity Principle from complex analysis implies that $A_s = 1$ throughout, and that $B_s = c_s$ throughout. In particular, this gives the meromorphic continuation of c_s. ///

The *theory of the constant term* in [8.1] asserts that a Δ-eigenfunction *of moderate growth* is asymptotic to its constant term. For example,

[1.10.4] Claim: For every s away from poles of $s \to E_s$, in a fixed Siegel set $\mathfrak{S}_{t,C}$,

$$E_s(na_yk) - (\eta^s + c_s \eta^{1-s}) \ll y^{-B}$$

for every $B > 0$, with the implied constant depending on t, s and B. That is, $E_s - c_P E_s$ is *rapidly decreasing* in a Siegel set. More generally, for s_o a pole of

$s \to E_s$ of order ν,

$$(s - s_o)^\nu E_s(na_y k)\big|_{s=s_o} - (s - s_o)^\nu(\eta^s + c_s\eta^{1-s})\big|_{s=s_o} \ll y^{-B}$$

Proof: Since E_s is a Δ-eigenfunction of moderate growth, the theory of the constant term [8.1] exactly ensures that E_s is asymptotic to its constant term, in the sense of the first assertion. Near a pole s_o of order ν, writing a vector-valued Laurent expansion

$$E_s = \frac{C_\nu}{(s - s_o)^\nu} + \frac{C_{\nu-1}}{(s - s_o)^{\nu-1}} + \cdots$$

as in [15.2], where the coefficients C_j are moderate-growth automorphic forms. Application of Δ termwise is justified, for example, by invocation of the vector-valued form of Cauchy's formulas [15.2]: with $\lambda_s = (\ell - 1)^2 \cdot s(s - 1)$,

$$\lambda_s \cdot \left(\frac{C_\nu}{(s - s_o)^\nu} + \frac{C_{\nu-1}}{(s - s_o)^{\nu-1}} + \cdots\right) = \lambda_s \cdot E_s = \Delta E_s$$
$$= \frac{\Delta C_\nu}{(s - s_o)^\nu} + \frac{\Delta C_{\nu-1}}{(s - s_o)^{\nu-1}} + \cdots$$

Multiplying through by $(s - s_o)^\nu$ and evaluating at $s = s_o$, $\lambda_{s_o} \cdot C_\nu = \Delta C_\nu$ as claimed. Apply the theory of the constant term [8.1]. ///

Granting the meromorphic continuation and the asymptotic estimation of the Eisenstein series by its constant term, the functional equation of E_s is determined by its constant term:

[1.10.5] Corollary: E_s has the *functional equation* $E_{1-s} = c_{1-s}E_s$, and $c_s \cdot c_{1-s} = 1$. In particular, $|c_s| = 1$ on $\text{Re}(s) = \frac{1}{2}$.

Proof: Take $\text{Re}(s) = \sigma > \frac{1}{2}$ and $s \notin \mathbb{R}$. Then $f = E_{1-s} - c_{1-s}E_s$ has constant term

$$c_P f = (\eta^{1-s} + c_{1-s}\eta^s) - c_{1-s}(\eta^s + c_s\eta^{1-s}) = (1 - c_{1-s}c_s) \cdot \eta^{1-s}$$

For $\text{Re}(s) > \frac{1}{2}$, η^{1-s} is square-integrable on $\mathfrak{S}_{t,C}$:

$$\int_{\mathfrak{S}_{t,C}} |\eta^{1-s}|^2 \frac{dx\,dy}{y^\ell} = (N\text{-measure } C) \cdot \int_t^\infty |y^{(1-s)(\ell-1)}|^2 \frac{dy}{y^\ell}$$
$$= (N\text{-measure } C) \cdot \int_t^\infty y^{(1-2\sigma)(\ell-1)} \frac{dy}{y}$$

Since $1 - 2\sigma < 0$, the integral is absolutely convergent. By the theory of the constant term [8.1], on a standard Siegel set

$$f = c_P f + (\text{rapidly decreasing}) \ll_s \eta^{1-\sigma} + (\text{rapidly decreasing})$$

Thus, on $\mathfrak{S}_{t,C}$,

$$\begin{aligned}|f|^2 &\ll |\eta^{1-\sigma} + (\text{rapidly decreasing})|^2\\ &= \eta^{2(1-\sigma)} + 2\cdot\eta^{1-\sigma}\cdot(\text{rapidly decreasing}) + (\text{rapidly decreasing})^2\\ &= \eta^{2(1-\sigma)} + (\text{rapidly decreasing})\end{aligned}$$

Thus, $f = E_{1-s} - c_{1-s}E_s \in L^2(\Gamma\backslash X)$. It is a Δ-eigenfunction with eigenvalue $\lambda_s = (\ell-1)^2\cdot s(s-1)$, which is *not real* for $\mathrm{Re}(s) > \frac{1}{2}$ and $s \notin \mathbb{R}$. But

$$\begin{aligned}\lambda_s\cdot\langle f,f\rangle &= \langle\lambda_s f,f\rangle = \langle\Delta f,f\rangle = \overline{\langle f,\Delta f\rangle} = \overline{\langle f,\lambda_s f\rangle}\\ &= \overline{\lambda_s}\cdot\overline{\langle f,f\rangle} = \overline{\lambda_s}\cdot\langle f,f\rangle\end{aligned}$$

Note that we did *not* use symmetry properties of Δ, but only that $\langle f,F\rangle = \overline{\langle F,f\rangle}$. Thus, necessarily $E_{1-s} - c_{1-s}E_s = 0$ for such s. For all $g \in G$, by the Identity Principle applied to the $\mathbb{C}$-valued meromorphic functions $s \longrightarrow (E_{1-s}(g) - c_{1-s}E_s(g))$, the same identity applies for all s away from poles.

The constant term $(1 - c_s c_{1-s})\cdot\eta^{1-s}$ of $E_{1-s} - c_{1-s}E_s = 0$ is identically 0, thus necessarily $c_s c_{1-s} = 1$. Further, $s \to \overline{c_{\overline{s}}}$ is holomorphic and equal to c_s for $\mathrm{Re}(s) \gg 1$, so the Identity Principle gives equality everywhere. Then $\overline{c_{\frac{1}{2}+it}} = c_{1-(\frac{1}{2}+it)} = c_{\frac{1}{2}-it}$, and $|c_{\frac{1}{2}+it}|^2 = c_{\frac{1}{2}+it}c_{\frac{1}{2}-it} = 1$. ///

[1.10.6] Claim: For $\mathrm{Re}(s) \neq \frac{1}{2}$ and $s \notin \mathbb{R}$, so that $\lambda_{s_o} \notin \mathbb{R}$, the poles of $s \to E_s$ are exactly the poles of c_s, and of the same order.

Proof: For s_o such a pole, of order $\nu \geq 1$, corollary [1.10.2] showed that the leading Laurent coefficient is a Δ-eigenfunction with eigenvalue λ_{s_o} and is of moderate growth and so is asymptotic to its constant term. The Laurent expansion of the constant term is the constant term of the Laurent expansion, from the vector-valued version of Cauchy's formula [15.2] and using the good behavior of continuous, compactly supported vector-valued integrals [14.1]. Thus, if c_s failed to have a pole at s_o, then the leading Laurent coefficient of E_s at s_o would have vanishing constant term and so (by the theory of the constant term) would be in $L^2(\Gamma\backslash X)$. Then $\lambda_{s_o} \in \mathbb{R}$, which is impossible for s_o as in the hypotheses. ///

1.11 Truncation and Maaß-Selberg Relations

The genuine Eisenstein series are not in $L^2(\Gamma\backslash X)$, but from the theory of the constant term [8.1] the only obstruction is the constant term, which is subtly altered by *truncation*, sufficiently removing this obstacle. The Maaß-Selberg relations are computation of the L^2 inner products of the resulting truncated Eisenstein series. As corollaries, we show that E_s has only finitely many poles in $\mathrm{Re}(s) \geq \frac{1}{2}$, that these are simple, lie on $(\frac{1}{2}, 1]$, and the *residues* are in $L^2(\Gamma\backslash X)$.

Granting the spectral decomposition of cuspforms [7.1], and from the theory of the constant term [8.1] that the Δ-eigenfunction cuspforms are *of rapid decay*, we prove that these residues of Eisenstein series are *orthogonal* to cuspforms.

The *truncation operators* $\wedge^T$ for large positive real T act on an automorphic form f by killing off f's constant term for large y. Thus, for a right K-invariant function, with a normalized version of *height* given by $\eta(na_yk) = y^{\ell-1}$, one might imagine

$$(\textit{naive } T\text{-truncation of } f)(g) = \begin{cases} f(g) & \text{for } \eta(g) \leq T \\ f(g) - c_P f(g) & \text{for } \eta(g) > T \end{cases}$$

This is not quite right. On a standard Siegel set $\mathfrak{S}_{t,C}$ this description is accurate, but it fails to correctly describe the truncated function on the whole domain X or whole group G, in the sense that the truncation is not properly described as an *automorphic form*, that is, as a left Γ-invariant function. We want truncation to produce automorphic forms. For sufficiently large (depending on the reduction theory) T, we achieve the same effect by first defining the *tail* $c_P^T f$ of the constant term $c_P f$ of f to be

$$c_P^T f(g) = \begin{cases} 0 & (\text{for } \eta(g) \leq T) \\ c_P f(y) & (\text{for } \eta(g) > T) \end{cases}$$

For legibility, we may replace a subscript by an argument in parentheses in the notation for pseudo-Eisenstein series: write

$$\Psi(\varphi) = \Psi_\varphi$$

Although $c_P^T f$ need not be smooth nor compactly supported, by design (that is, for T sufficiently large) its support is sufficiently high so that we have control over the analytical issues:

[1.11.1] Claim: For T sufficiently large, the pseudo-Eisenstein series $\Psi(c_P^T f)$ is a locally finite sum—hence, uniformly convergent on compacts.

Proof: The tail $c_P^T f$ is left N-invariant. The reduction theory of [1.5] shows that, given t_o, for large enough t, a set $\{na_yk : y > t_o\}$ does not meet $\gamma \cdot \{na_yk : y > t\}$ unless $\gamma \in \Gamma_\infty$. Thus, for large enough T, $\{na_yk : y > T\}$ does not meet $\gamma \cdot \{na_yk : y > T\}$ unless $\gamma \in \Gamma_\infty$. Thus, $\gamma_1 \cdot \{na_yk : y > T\}$ does not meet $\gamma_2 \cdot \{na_yk : y > T\}$ unless $\gamma_1\Gamma_\infty = \gamma_2\Gamma_\infty$. ///

Similarly,

[1.11.2] Claim: On a standard Siegel set $\mathfrak{S}_{t,C}$, $\Psi(c_P^T f) = c_P^T f$ for all T sufficiently large depending on t.

Proof: By reduction theory, a set $\{na_yk : y > t_o\}$ does not meet $\gamma \cdot \{na_yk : y > T\}$ unless $\gamma \in \Gamma_\infty$, for large enough T depending on t_o. Thus, for large enough T, $\{na_yk : y > T\}$ does not meet $\mathfrak{S}_{t_o,C}$ unless $\gamma \in \Gamma_\infty$. That is, the only nonzero summand in $\Psi(c_P^T f)$ is the term $c_P^T f$ itself. ///

Thus, we find that the proper definition of the *truncation operator* $\wedge^T$ is

$$\wedge^T f = f - \Psi(c_P^T f)$$

As desired, a critical effect of the truncation procedure is:

[1.11.3] Claim: For s away from poles, the truncated Eisenstein series $\wedge^T E_s$ is of rapid decay in all Siegel sets.

Proof: From the theory of the constant term [8.1], $E_s - c_P E_s$ is of rapid decay in a standard Siegel set. By the previous claim, $(E_s - c_P^T E_s)(g) = (E_s - c_P E_s)(g)$ for $\eta(g) \geq T$, so it is is also of rapid decay. ///

[1.11.4] Theorem: *(Maaß-Selberg relation)* Up to a uniform constant depending on normalization of measure,

$$\frac{1}{\ell - 1}\int_{\Gamma\backslash X} \wedge^T E_s \cdot \wedge^T E_r = \frac{T^{s+r-1}}{s+r-1} + c_s \frac{T^{(1-s)+r-1}}{(1-s)+r-1}$$
$$+ c_r \frac{T^{s+(1-r)-1}}{s+(1-r)-1} + c_s c_r \frac{T^{(1-s)+(1-r)-1}}{(1-s)+(1-r)-1}$$

Proof: First,

$$\int_{\Gamma\backslash X} \wedge^T E_s \cdot \wedge^T E_r = \int_{\Gamma\backslash X} \wedge^T E_s \cdot E_r$$

because the tail of the constant term of E_r is orthogonal to the truncated version $\wedge^T E_s$ of E_s. Then

$$\int_{\Gamma\backslash X} \wedge^T E_s \cdot \wedge^T E_r = \int_{\Gamma\backslash X} \left(\Psi(\eta^s) - \Psi\begin{pmatrix} 0 & (\text{for } \eta < T) \\ \eta^s + c_s\eta^{1-s} & (\text{for } \eta \geq T)\end{pmatrix}\right) \cdot E_r$$
$$= \int_{\Gamma\backslash X} \Psi\begin{pmatrix} \eta^s & (\text{for } \eta < T) \\ -c_s\eta^{1-s} & (\text{for } \eta \geq T)\end{pmatrix} \cdot E_r$$

Unwinding the awkward pseudo-Eisenstein series, noting that Γ_∞ differs from $N \cap \Gamma$ only by the finite group $M \cap K$ which commutes with A^+, and the integrand is right K-invariant,

$$\int_{\Gamma_\infty\backslash X} \begin{Bmatrix} \eta^s & (\text{for } \eta < T) \\ -c_s\eta^{1-s} & (\text{for } \eta \geq T) \end{Bmatrix} \cdot E_r$$

$$= \int_{N\backslash X}\int_{N\cap\Gamma\backslash N} \begin{Bmatrix} \eta^s & (\text{for } \eta < T) \\ -c_s\eta^{1-s} & (\text{for } \eta \geq T) \end{Bmatrix} \cdot E_r$$

$$= \int_{N\backslash X} \begin{Bmatrix} \eta^s & (\text{for } \eta < T) \\ -c_s\eta^{1-s} & (\text{for } \eta \geq T) \end{Bmatrix} \cdot \left(\int_{N\cap\Gamma\backslash N} E_r(ng)\, dn \right) dg$$

The constant term of the Eisenstein series is just $\eta^r + c_r\eta^{1-r}$, so this is

$$\int_{N\backslash X} \begin{Bmatrix} \eta^s & (\text{for } \eta < T) \\ -c_s\eta^{1-s} & (\text{for } \eta \geq T) \end{Bmatrix} \cdot (\eta^r + c_r\eta^{1-r})$$

$$= \int_0^\infty \begin{Bmatrix} \eta^s(\eta^r + c_r\eta^{1-r}) & (\text{for } \eta < T) \\ -c_s\eta^{1-s}(\eta^r + c_r\eta^{1-r}) & (\text{for } \eta \geq T) \end{Bmatrix}$$

$$= \int_0^T \eta^s \cdot (\eta^r + c_r\eta^{1-r}) \frac{dy}{y^\ell} - \int_T^\infty c_s\eta^{1-s}\, (\eta^r + c_r\eta^{1-r}) \frac{dy}{y^\ell}$$

Note that the measure dy/y^ℓ is descended from the right G-invariant measure on $N\backslash G$. Assume that $\mathrm{Re}\,(r)$ is bounded above and below, so that $\mathrm{Re}\,(1-r)$ is also bounded, and take $\mathrm{Re}\,(s)$ sufficiently large so that all the integrals converge. The foregoing becomes

$$\int_0^T \eta^{s+r-1} \frac{dy}{y} + c_r \int_0^T \eta^{s+(1-r)-1} \frac{dy}{y}$$
$$- c_s \int_T^\infty \eta^{(1-s)+r-1} \frac{dy}{y} - c_s c_r \int_T^\infty \eta^{(1-s)+(1-r)-1} \frac{dy}{y}$$

Since $d\eta/\eta = (\ell - 1)\cdot dy/y$, this gives the expression of the theorem. Note that $\ell - 1 = 1$ in the most familiar case of $\Gamma = SL_2(\mathbb{Z})$. By analytic continuation (in s and in r) it is valid everywhere it makes sense. ///

The following corollaries can be proven directly in special cases by use of explicit details about Fourier expansion of the Eisenstein series. However, the arguments here generalize.

[1.11.5] Corollary: There are only finitely many poles of E_s in the region $\mathrm{Re}\,(s) \geq \frac{1}{2}$, all on the segment $(\frac{1}{2}, 1]$, and these are poles of c_s. Any such pole is *simple*, with residue in $L^2(\Gamma\backslash X)$. Specifically, with measure normalized as in the previous proof,

$$\int_{\Gamma\backslash X} |\mathrm{Res}_{\sigma_o} E_s|^2 = (\ell - 1) \cdot \mathrm{Res}_{\sigma_o} c_s < +\infty$$

Such a residue is also *smooth*, and *moderate growth*, and has eigenvalue $\lambda_{s_o} = (\ell - 1)^2 \cdot s_o(s_o - 1)$.

Proof: The Eisenstein series is indeed treated as a meromorphic function-valued function, as in [15.2], so its Laurent coefficients or power series coefficients are functions in the same topological vector space, by the vector-valued form of Cauchy's formulas [15.2]. From the identity principle, since $\overline{E_{\overline{s}}} = E_s$ for $\mathrm{Re}(s) > 1$, we have $\overline{E_s} = E_{\overline{s}}$ for all s away from poles, and similarly for truncated Eisenstein series. Thus, taking $r = \overline{s} = \sigma - it$ in the theorem,

$$\frac{1}{\ell - 1} \int_{\Gamma\backslash X} |\wedge^T E_s|^2 = \frac{T^{2\sigma-1}}{2\sigma - 1} + c_s \frac{T^{-2it}}{-2it} + c_{\overline{s}} \frac{T^{2it}}{2it} + c_s c_{\overline{s}} \frac{T^{1-2\sigma}}{1 - 2\sigma}$$

Suppose E_s has a pole $s_o = \sigma_o + it_o$ of order ν with $t_o \neq 0$ and $\sigma_o > \frac{1}{2}$.

From corollary [1.10.4] to the *theory of the constant term* [8.1], with nonreal eigenvalue, this is equivalent to the assertion that c_s has a pole at s_o of order ν. Also, $c_{\overline{s}} = \overline{c_s}$, so c_s has a pole at $\overline{s_o}$ as well, of the same order, with leading Laurent term the complex conjugate of that at s_o. Thus, the function $\wedge^T E_s$ also has a pole exactly at poles of c_s, of the same order, for nonreal λ_s.

Take $s = \sigma_o + it$ in the foregoing. In the real variable t, the left-hand side of the Maaß-Selberg relation is asymptotic to a *positive* constant multiple of $(t - t_o)^{-2\nu}$ as $t \to t_o$, since the pole is of order ν and inner products are positive. The first term on the right-hand side is bounded as $t \to t_o$, and the second and third terms are asymptotic to nonzero constant multiples of $(t - t_o)^{-\nu}$. Thus, the first three terms on the right can be ignored as $t \to t_o$. The fourth term on the right-hand side is asymptotic to a positive constant multiple of $(t - t_o)^{-2\nu}$ from $c_s c_{\overline{s}}$, multiplied by $T^{1-2\sigma_o}/(1 - 2\sigma_o)$. The denominator is *negative*, so that, altogether, the fourth term on the right-hand side is asymptotic to a *negative* constant multiple of $(t - t_o)^{-2\nu}$. The positivity of the left-hand side of the Maaß-Selberg, and negativity of the right-hand side (as $t \to t_o$), contradict the hypothesized pole. Thus, E_s and c_s have no poles off the real axis in the region $\mathrm{Re}(s) > 1/2$.

Next, let $s_o = \sigma_o$ be a pole of E_s of order $\nu \geq 1$ on $(\frac{1}{2}, 1]$. Take $r = s = \sigma_o + it$, obtaining

$$\frac{1}{\ell - 1} \int_{\Gamma\backslash X} |t^\nu \cdot \wedge^T E_s|^2$$
$$= t^{2\nu} \cdot \left(\frac{T^{2\sigma_o - 1}}{2\sigma_o - 1} + c_s \frac{T^{-2it}}{-2it} + c_{\overline{s}} \frac{T^{2it}}{2it} + c_s c_{\overline{s}} \frac{T^{1-2\sigma_o}}{1 - 2\sigma_o} \right)$$

As $t \to t_o = 0$, the right-hand side goes to 0 unless c_s *also* has a pole of order ν at s_o. The fourth term is *negative*, and if $\nu > 1$ is the only term that survives on the right-hand side as $t \to 0$, contradicting the nonnegativity of the left-hand side. Thus, $\nu = 1$, in which case the second and third terms' blow-up is of the same order as the left-hand side and the fourth term on the right-hand side. This proves that any pole on $(\frac{1}{2}, 1]$ is *simple*.

Letting $t \to 0$,

$$\frac{1}{\ell - 1} \int_{\Gamma\backslash X} |\text{Res}_{\sigma_o} E_s^T|^2 = \frac{\text{Res}_{\sigma_o} c_s}{2} + \frac{\text{Res}_{\sigma_o} c_{\overline{s}}}{2} + \text{Res}_{\sigma_o} c_s \cdot \overline{\text{Res}_{\sigma_o} c_s} \, \frac{T^{1-2\sigma_o}}{1 - 2\sigma_o}$$

Since $1 - 2\sigma_o < 0$, the limit of the last term is 0 as $T \to +\infty$, given the square-integrability of the residue. General considerations about meromorphic vector-valued functions [15.2] and Gelfand-Pettis integrals [14.1] ensure that taking residues commutes with taking the limit as $T \to \infty$. The two remaining terms are equal because the pole is on the real line.

Now suppose s_o is a pole of E_s of order $\nu \geq 1$ on the line $\text{Re}(s) = \frac{1}{2}$ and off $\mathbb{R}$. The leading Laurent coefficient C_ν of E_s at s_o is a Δ-eigenfunction with eigenvalue λ_{s_o} and is of moderate growth, again by the vector-valued form of Cauchy's formulas. Thus, by the theory of the constant term [8.1], C_ν is asymptotic to its constant term $c_P C_\nu$, which, again, is the the leading Laurent coefficient of the constant term $\eta^s + c_s \eta^{1-s}$ of E_s. The property $|c_s| = 1$ on $\text{Re}(s) = \frac{1}{2}$ shows that c_s has no pole there. Then $c_P C_\nu = 0$, so $\wedge^T C_\nu = C_\nu$, and C_ν is in $L^2(\Gamma\backslash X)$. By Maaß-Selberg with $r = \overline{s} = \sigma - it$,

$$\frac{1}{\ell - 1} \int_{\Gamma\backslash X} |C_\nu|^2 = \frac{1}{\ell - 1} \int \left| \lim_{s \to s_o} (s - s_o)^{2\nu} \wedge^T E_s \right|^2$$
$$= \lim_{s \to s_o} (s - s_o)^{2\nu} \cdot \left(\frac{T^{2\sigma - 1}}{2\sigma - 1} + c_s \frac{T^{-2it}}{-2it} + c_{\overline{s}} \frac{T^{2it}}{2it} + c_s c_{\overline{s}} \frac{T^{1-2\sigma}}{1 - 2\sigma} \right)$$

Since $\nu \geq 1$, approaching s_o from *off* the line, the limit of $(s - s_o)^{2\nu}/(2\sigma - 1)$ is 0. Since $|c_s| \to 1$ as $s \to s_o \in \frac{1}{2} + i\mathbb{R}$, the whole limit is 0. Thus, E_s has no pole on $\frac{1}{2} + i\mathbb{R}$.

Finally, we see that the residues are not only in $L^2(\Gamma\backslash X)$ but are also smooth (and moderate growth pointwise) Δ-eigenfunctions with the indicated eigenvalues. By the vector-valued form of Cauchy's formula for residues,

$$\text{Res}_{s=s_o} E_s = \frac{1}{2\pi i} \int_\gamma \frac{E_w}{w-s} dw$$

where γ is a small circle around s_o, traversed counterclockwise. Since $w \to E_w/(s-w)$ is a continuous moderate-growth-function-valued function, Gelfand-Pettis [14.1] ensures that the integral is in the same space. In particular, the residue is *smooth*. Because the pole is *simple*, the function $f_s = (s - s_o)E_s$ has a removable singularity at s_o, and its value there is the residue. In the topology on moderate-growth functions (as in [13.10]), Δ is a continuous map. From the theory of vector-valued holomorphic functions [15.2] and Gelfand-Pettis integrals [14.1], evaluation commutes with continuous linear maps, so

$$\Delta(\text{Res}_{s_o} E_s) = \Delta(f_s|_{s=s_o}) = (\Delta f_s)|_{s=s_o} = \lambda_{s_o} \cdot f_{s_o}$$

demonstrating that residues are (smooth and) Δ-eigenfunctions. ///

For $\Gamma = SL_2(\mathbb{Z})$ and $SL_2(\mathbb{Z}[i])$, there are special arguments that show that the only relevant residues of Eisenstein series are at $s = 1$. The eigenvalue $\lambda_s = s(s-1)$ of Δ at $s = 1$ is 0.

[1.11.6] Claim: Any smooth $f \in L^2(\Gamma\backslash G/K)$ such that $\Delta f = 0$ is *constant*.

Proof: Let ∇ be the tangent-space-valued gradient on $\Gamma\backslash G/K$, as developed in more detail in [10.7],

$$\int_{\Gamma\backslash G} \Delta f \cdot F = -\int_{\Gamma\backslash G} \langle \nabla f, \nabla F\rangle$$

where, for the moment, $\langle -, -\rangle$ is a inner product on the tangent space. For $\Delta f = 0$, this gives

$$0 = \int_{\Gamma\backslash G} 0 \cdot f = \int_{\Gamma\backslash G} \Delta f \cdot f = -\int_{\Gamma\backslash G} \langle \nabla f, \nabla f\rangle$$

Thus, ∇f is identically 0, so f is constant. ///

The other two current examples, $\Gamma = Sp^*_{1,1}(\mathfrak{o})$ and $SL_2(\mathfrak{o})$, do not admit those special arguments to decisively locate poles, although they still do have poles at $s = 1$, with constant residues, by the same argument. To treat residues in $\text{Re}(s) > \frac{1}{2}$ generally:

[1.11.7] Corollary: Residues of Eisenstein series at distinct poles s_1, s_2 in $(\frac{1}{2}, 1]$ are mutually orthogonal.

Proof: Let f_j be the residue at s_j, with eigenvalue λ_j. The eigenvalues are real, since $s_1, s_2 \in \mathbb{R}$. It is reasonable to *think* that Δ has the symmetry $\langle \Delta f_1, f_2 \rangle = \langle f_1, \Delta f_2 \rangle$ so that the usual argument

$$\lambda_1 \cdot \langle f_1, f_2 \rangle = \langle \Delta f_1, f_2 \rangle = \langle f_1, \Delta f_2 \rangle = \lambda_2 \cdot \langle f_1, f_2 \rangle$$

would give $(\lambda_1 - \lambda_2) \cdot \langle f_1, f_2 \rangle = 0$, and then $\langle f_1, f_2 \rangle = 0$. However, the defensible starting-point [6.6] for this symmetry property of Δ is that it holds for functions in $C_c^\infty(\Gamma\backslash G/K)$, in effect avoiding any boundary terms in integration by parts. To preserve symmetry in an *extension* requires care. In fact, the method of [9.10] shows that Δ is *essentially self-adjoint* in the sense of having a unique self-adjoint extension to an unbounded operator densely defined in $L^2(\Gamma\backslash G/K)$. The domain of that extension does include the residues f_j, but demonstration of the latter fact is more a *consequence* than *starting-point*.

Instead, the Maaß-Selberg relation (with $s_1 \neq s_2$ both real, eliminating some complex conjugations) gives

$$\frac{1}{\ell - 1} \int_{\Gamma\backslash X} \wedge^T E_{s_1} \cdot \wedge^T E_{s_2} = \frac{T^{s_1+s_2-1}}{s_1 + s_2 - 1} + c_{s_1} \frac{T^{(1-s_2)+s_2-1}}{(1-s_1) + s_2 - 1}$$
$$+ c_{s_2} \frac{T^{s_1+(1-w_2)-1}}{s_1 + (1-s_2) - 1}$$
$$+ c_{s_1} c_{s_2} \frac{T^{(1-s_1)+(1-s_2)-1}}{(1-s_1) + (1-s_2) - 1}$$

With simple poles of E_s at s_1 and s_2, multiplying through by $(s - s_1)(s' - s_2)$ and taking the limit as $s \to s_1$ and $s' \to s_2$ gives

$$\frac{1}{\ell - 1} \int_{\Gamma\backslash X} \wedge^T \mathrm{Res}_{s_1} E_{s_1} \cdot \wedge^T \mathrm{Res}_{s_2} E_{s_2}$$
$$= 0 + 0 + 0 + \mathrm{Res}_{s=s_1} c_s \cdot \mathrm{Res}_{s=s_2} c_s \cdot \frac{T^{(1-s_1)+(1-s_2)-1}}{(1-s_1) + (1-s_2) - 1}$$

Since $(1 - s_1) + (1 - s_2) - 1 < (1 - \frac{1}{2}) + (1 - \frac{1}{2}) - 1 < 0$, letting $T \to +\infty$ gives the orthogonality. ///

[1.11.8] Corollary: The residues of E_s for $s \in (\frac{1}{2}, 1]$ are orthogonal to cuspforms.

Proof: This uses the spectral decomposition of cuspforms [1.7]: there is an orthonormal basis for L^2 cuspforms consisting of Δ-eigenfunctions, and each eigenspace is finite-dimensional. The theory of the constant term [8.1] shows that any such eigenfunction is asymptotic to its constant term. Since constant terms of cuspforms are 0, cuspform-eigenfunctions are of rapid decay in Siegel sets.

Thus, for a cuspform-eigenfunction f, granting [1.9.1] that Eisenstein series E_s are of *moderate growth* on Siegel sets, the literal integrals $\langle f, E_s\rangle = \int_{\Gamma\backslash X} f \cdot \overline{E_s}$ are absolutely convergent for all s away from poles. These are *not* L^2 inner products because E_s is never in L^2, but we use the same notation for brevity. In the region of convergence $\mathrm{Re}(s) > 1$, any integral $\int_{\Gamma\backslash X} f \cdot E_s$ *unwinds* to compute an integral against the constant term of f, and the latter is 0:

$$\begin{aligned}\int_{\Gamma\backslash X} f \cdot E_s &= \int_{\Gamma_\infty\backslash X} f \cdot \eta^s \\ &= \int_{N\Gamma_\infty\backslash X} \left(\int_{(N\cap\Gamma)\backslash N} f(ng)\, \eta(ng)^s \, dn \right) dg \\ &= \int_{N\Gamma_\infty\backslash X} \eta^s(g) \left(\int_{(N\cap\Gamma)\backslash N} f(ng)\, dn \right) dg \\ &= \int_{N\Gamma_\infty\backslash X} \eta^s(g) \cdot c_P f(g)\, dg = \int_{N\Gamma_\infty\backslash X} \eta^s(g) \cdot 0 \, dg = 0\end{aligned}$$

Because $s \to E_s$ is a meromorphic function-valued function taking values in (at least continuous) functions of moderate growth, the function $s \to \langle f, E_s\rangle$ is meromorphic on $\mathbb{C}$. By the Identity Principle, since this function is 0 on $\mathrm{Re}(s) > 1$, it is identically 0. The vector-valued form of Cauchy's formula expresses the residue at $s = s_o$ as an integral:

$$\mathrm{Res}_{s=s_o} E_s = \frac{1}{2\pi i} \int_\gamma \frac{E_w}{w - s}\, dw$$

where γ is a small circle around s_o, traversed counterclockwise. Then

$$\langle f, \mathrm{Res}_{s_o} E_s\rangle = \left\langle f, \frac{1}{2\pi i} \int_\gamma \frac{E_w}{w - s}\, dw \right\rangle$$

The functional $u \to \langle f, u\rangle$ is a continuous linear functional on functions of moderate growth, and $w \to E_w/(s - w)$ is a continuous, compactly supported moderate-growth-function-valued function, so by Gelfand-Pettis [14.1] the inner product passes inside the integral:

$$\langle f, \mathrm{Res}_{s_o} E_s\rangle = \frac{1}{2\pi i} \int_\gamma \langle f, E_w\rangle \frac{1}{w - s}\, dw = \frac{1}{2\pi i} \int_\gamma 0 \cdot \frac{1}{w - s}\, dw = 0$$

Again, $\langle f, E_s\rangle$ is *not* an L^2 pairing because E_s is not in L^2. Nevertheless, because of the rapid decay of f, the implied integral is absolutely convergent. This proves that the residues of E_s for $s \in (\frac{1}{2}, 1]$, all of which *are* in L^2, are L^2-orthogonal to cuspforms. ///

1.12 Decomposition of Pseudo-Eisenstein Series

We saw in [1.8] that the pseudo-Eisenstein series Ψ_φ with $\varphi \in C_c^\infty(N\backslash G/K)$ generate the orthogonal complement to cuspforms in $L^2(\Gamma\backslash G/K)$: since the orthogonal complement of these pseudo-Eisenstein series is the space of cuspforms, the orthogonal complement to cuspforms is the L^2-closure of the set of these pseudo-Eisenstein series.

To express such pseudo-Eisenstein series as superpositions of Δ-eigenfunctions in the four examples at hand, once we know the meromorphic continuation and functional equation of the genuine Eisenstein series E_s, the essential harmonic analysis is *Fourier transform* on the real line, in coordinates in which it is known as a *Mellin transform*. That is, the noncuspidal part of harmonic analysis on $\Gamma\backslash X$ in each of these four examples reduces to harmonic analysis on $\mathbb{R}$.

For $\varphi \in C_c^\infty(\Gamma\backslash G/K) = C_c^\infty(N\backslash G)^K$, the pseudo-Eisenstein series Ψ_φ is in $C_c^\infty(\Gamma\backslash G)^K$, so its integral against E_s converges absolutely since E_s is continuous, even after meromorphic continuation. Thus, by abuse of notation, we may write

$$\langle \Psi_\varphi, E_s\rangle = \int_{\Gamma\backslash X} \Psi_\varphi \cdot \overline{E_s}$$

even though this $\langle , \rangle$ cannot be the L^2 pairing, since $E_s \notin L^2(\Gamma\backslash X)$. The following is a *preliminary* version of a spectral decomposition of the L^2 closure of the space containing pseudo-Eisenstein series, insofar as it only treats Ψ_φ with test-function φ, only computes point-wise values, so does not consider the integral of genuine Eisenstein series as a function-valued integral, and omits a Plancherel assertion.

[1.12.1] Theorem: *(Numerical form)* Let s_o run over poles of E_s in $\mathrm{Re}(s) > \frac{1}{2}$. For $\varphi \in C_c^\infty(N\backslash G/K)$, the pseudo-Eisenstein series Ψ_φ is expressible in terms of genuine Eisenstein series, by an integral converging absolutely and uniformly on compacts in $\Gamma\backslash G/K$:

$$\Psi_\varphi(g) = \frac{(\ell-1)}{4\pi i}\int_{\frac{1}{2}-i\infty}^{\frac{1}{2}+i\infty} \langle \Psi_\varphi, E_s\rangle \cdot E_s(g)\, ds + (\ell-1)\sum_{s_o} \langle \Psi_\varphi, \mathrm{Res}_{s_o} E_s\rangle \cdot \mathrm{Res}_{s_o} E_s(g)$$

where we abuse notation by writing $\langle \Psi_\varphi, E_s\rangle = \int_{\Gamma\backslash G} \Psi_\varphi \cdot \overline{E}_s$ even though E_s is not in L^2.

Proof: One form of Fourier inversion for Schwartz functions[15] f on the real line is

$$f(x) = \frac{1}{2\pi} \int_{-\infty}^{\infty} \left(\int_{-\infty}^{\infty} f(u)\, e^{-i\xi u}\, du \right) e^{i\xi x}\, d\xi$$

Both outer and inner integrals converge very well, uniformly pointwise. The inner integral is a Schwartz function in ξ. Fourier transforms on $\mathbb{R}$ put into multiplicative coordinates are *Mellin* transforms: for $\varphi \in C_c^\infty(0, +\infty)$, take $f(x) = \varphi(e^x)$. Let $y = e^x$ and $r = e^u$, and rewrite Fourier inversion as

$$\varphi(y) = \frac{1}{2\pi} \int_{-\infty}^{\infty} \left(\int_0^{\infty} \varphi(r)\, r^{-i\xi}\, \frac{dr}{r} \right) y^{i\xi}\, d\xi$$

The Fourier transform in these coordinates is a *Mellin* or *Laplace* transform. For *compactly supported* φ, as we use throughout this discussion, the integral definition extends to all complex s in place of $i\xi$, and $d\xi$ replaced by $-i\,ds$. The variant Fourier inversion identity gives *Mellin inversion*

$$\varphi(y) = \frac{1}{2\pi i} \int_{-i\infty}^{i\infty} \left(\int_0^{\infty} \varphi(r)\, r^{-s}\, \frac{dr}{r} \right) y^s\, ds$$

By an easy part of the *Paley-Wiener* theorem [13.16], for $f \in C_c^\infty(\mathbb{R})$ the Fourier transform is an *entire* function in s, of rapid decay on horizontal lines, uniformly so on strips of finite width, so the Mellin transform of φ has rapid decay *vertically*. This allows movement of the contour: for compactly supported φ, Mellin inversion is, for *any* $\sigma \in \mathbb{R}$,

$$\varphi(y) = \frac{1}{2\pi i} \int_{\sigma-i\infty}^{\sigma+i\infty} \left(\int_0^{\infty} \varphi(r)\, r^{-s}\, \frac{dr}{r} \right) y^s\, ds$$

In the present context, adjust the coordinates so that the Mellin transform is an integral against $\eta(a_y)^s = y^{(\ell-1)s}$, and inversion likewise: replace s by $s(\ell - 1)$ (and readjust the contour):

$$\begin{aligned} \varphi(y) &= \frac{(\ell-1)}{2\pi i} \int_{\sigma-i\infty}^{\sigma+i\infty} \left(\int_0^{\infty} r^{-(\ell-1)s}\, \varphi(r)\, \frac{dr}{r} \right) y^{s(\ell-1)}\, ds \\ &= \frac{(\ell-1)}{2\pi i} \int_{\sigma-i\infty}^{\sigma+i\infty} \left(\int_0^{\infty} \eta(a_r)^{-s}\, \varphi(r)\, \frac{dr}{r} \right) \eta(a_y)^s\, ds \end{aligned}$$

[15] As usual, *Schwartz functions* $\mathscr{S}(\mathbb{R})$ on $\mathbb{R}$ or any copy of it are smooth functions f such that f and all its derivatives are rapidly decreasing, in the sense that $(1 + x^2)^N \cdot |f^{(k)}(x)|$ is *bounded* on $x \in \mathbb{R}$ for every k and N. These sups are a family of seminorms that give $\mathscr{S}(\mathbb{R})$ a Fréchet-space structure. See Chapter 12.

Identifying $N\backslash G/K \approx A^+ \approx (0, +\infty)$, this is

$$\varphi(a_y) = \frac{(\ell - 1)}{2\pi i} \int_{\sigma - i\infty}^{\sigma + i\infty} \left(\int_0^\infty \eta(a_r)^{-s}\, \varphi(a_r)\, \frac{dr}{r} \right) \eta(a_y)^s\, ds$$

Thus, for all $g \in G$,

$$\varphi(g) = \frac{(\ell - 1)}{2\pi i} \int_{\sigma - i\infty}^{\sigma + i\infty} \left(\int_0^\infty \eta(a_r)^{-s}\, \varphi(a_r)\, \frac{dr}{r} \right) \eta(g)^s\, ds$$

The Mellin transform useful here is

$$\mathcal{M}\varphi(s) = \int_0^\infty \eta(a_r)^{-s}\, \varphi(a_r)\, \frac{dr}{r}$$

and the pseudo-Eisenstein series is

$$\begin{aligned}
\Psi_\varphi(g) &= \sum_{\gamma \in \Gamma_\infty \backslash \Gamma} \varphi(\gamma g) = \sum_{\gamma \in \Gamma_\infty \backslash \Gamma} \varphi(a_{\gamma g}) \\
&= \frac{(\ell - 1)}{2\pi i} \sum_{\gamma \in \Gamma_\infty \Gamma} \int_{\sigma - i\infty}^{\sigma + i\infty} \mathcal{M}\varphi(s) \cdot \eta(\gamma g)^s\, ds
\end{aligned}$$

Taking $\sigma = 0$ would be natural, but with $\sigma = 0$, the double integral (sum and integral) is not absolutely convergent, and *the two integrals cannot be interchanged.* For $\sigma > 1$, the Eisenstein series is absolutely convergent, so the rapid vertical decrease of $\mathcal{M}\varphi$ makes the double integral absolutely convergent, and by Fubini the two integrals can be interchanged: with $\sigma > 1$,

$$\Psi_\varphi(g) = \frac{(\ell - 1)}{2\pi i} \int_{\sigma - i\infty}^{\sigma + i\infty} \mathcal{M}\varphi(s) \left(\sum_{\gamma \in \Gamma_\infty \backslash \Gamma} \eta(\gamma g)^s \right) ds$$

The inner sum is the *Eisenstein series* $E_s(g)$, so

$$\Psi_\varphi(g) = \frac{(\ell - 1)}{2\pi i} \int_{\sigma - i\infty}^{\sigma + i\infty} \mathcal{M}\varphi(s) \cdot E_s(g)\, ds \qquad (\text{for } \sigma > 1)$$

Although this does express Ψ_φ as a superposition of Δ-eigenfunctions, it is unsatisfactory because it should refer to $\mathcal{M}c_P\Psi_\varphi$, not to $\mathcal{M}\varphi$, to give a direct decomposition formula for functions in the span of the pseudo-Eisenstein series.

We want to move the line of integration to the left, to $\sigma = 1/2$, stabilized by the functional equation of E_s. From the corollary [1.11.5] to the Maaß-Selberg relations, there are only finitely many poles of E_s in $\mathrm{Re}(s) \geq \frac{1}{2}$, removing one possible obstacle to the contour move. From the theorem [1.10.1] on meromorphic continuation, even the meromorphically continued $E_s(g_o)$ is of polynomial growth vertically in s, uniformly in bounded strips in s, uniformly for g_o in compacts. Thus, we may move the contour, picking up finitely many residues:

$$\Psi_\varphi = \frac{(\ell-1)}{2\pi i}\int_{\frac{1}{2}-i\infty}^{\frac{1}{2}+i\infty} \mathcal{M}\varphi(s) E_s\, ds + (\ell-1)\sum_{s_o} \mathcal{M}\varphi(s_o)\cdot \mathrm{Res}_{s_o} E_s$$

since the poles of E_s are simple and $\mathcal{M}\varphi$ is entire. The $1/2\pi i$ from inversion cancels the $2\pi i$ in the residue formula. By the adjunction/unwinding property of Ψ_φ, on $\mathrm{Re}(s) = \frac{1}{2}$,

$$\begin{aligned}\langle \Psi_\varphi, E_s\rangle &= \int_{\Gamma\backslash X} \Psi_\varphi \cdot E_{1-s} = \int_{\Gamma_\infty\backslash X} \varphi \cdot c_P E_{1-s} \\ &= \int_{\Gamma_\infty\backslash X} (\eta^{1-s} + c_{1-s}\eta^s)\cdot\varphi = \int_0^\infty (\eta^{1-s} + c_{1-s}\eta^s)\cdot\varphi(y)\,\frac{dy}{y^\ell} \\ &= \int_0^\infty (\eta^{-s} + c_{1-s}\eta^{-(1-s)})\cdot\varphi(y)\,\frac{dy}{y} = \mathcal{M}\varphi(s) + c_{1-s}\mathcal{M}\varphi(1-s)\end{aligned}$$

The integral part of the expression of Ψ_φ in terms of Eisenstein series can be *folded in half*, integrating from $\frac{1}{2}+i0$ to $\frac{1}{2}+i\infty$ rather than from $\frac{1}{2}-i\infty$ to $\frac{1}{2}+i\infty$:

$$\begin{aligned}\Psi_\varphi - (\text{residual part}) &= \frac{(\ell-1)}{2\pi i}\int_{\frac{1}{2}-i\infty}^{\frac{1}{2}+i\infty} \mathcal{M}\varphi(s)\cdot E_s(g)\, ds \\ &= \frac{(\ell-1)}{2\pi i}\int_{\frac{1}{2}+i0}^{\frac{1}{2}+i\infty} \mathcal{M}\varphi(s) E_s + \mathcal{M}\varphi(1-s) E_{1-s}\, ds \\ &= \frac{(\ell-1)}{2\pi i}\int_{\frac{1}{2}+i0}^{\frac{1}{2}+i\infty} \mathcal{M}\varphi(s) E_s + \mathcal{M}\varphi(1-s)\, c_{1-s} E_s\, ds \\ &= \frac{(\ell-1)}{2\pi i}\int_{\frac{1}{2}+i0}^{\frac{1}{2}+i\infty} \langle \Psi_\varphi, E_s\rangle\, E_s\, ds\end{aligned}$$

by the immediately preceding functional equation and the computation of $\langle \Psi_\varphi, E_s\rangle$. The integral can be written as an integral over the whole line $\mathrm{Re}(s) = \frac{1}{2}$, by the functional equation of E_s and dividing by 2:

$$\Psi_\varphi - (\text{residual part}) = \frac{(\ell-1)}{4\pi i}\int_{\frac{1}{2}-i\infty}^{\frac{1}{2}+i\infty} \langle \Psi_\varphi, E_s\rangle \cdot E_s\, ds$$

It remains to explicate the finitely many residues that appear. The notation is normalized so that in all these examples there is a pole at $s = 1$. The coefficient

$\mathcal{M}\varphi(1)$ is

$$\mathcal{M}\varphi(1) = \int_0^{+\infty} \varphi(a_y)\,\eta^{-1}\,\frac{dy}{y} = \int_0^{+\infty} \varphi(a_y)\,\frac{dy}{y^\ell}$$
$$= \int_{\Gamma_\infty\backslash X} \varphi(n_x a_y)\,\frac{dx\,dy}{y^\ell}$$

giving $(N \cap \Gamma)\backslash N$ total measure 1. *Winding up,*

$$\mathcal{M}\varphi(1) = \int_{\Gamma\backslash\mathfrak{H}} \sum_{\gamma\in\Gamma_\infty\backslash\Gamma} \varphi(\gamma n_x a_y)\,\frac{dx\,dy}{y^\ell} = \int_{\Gamma\backslash\mathfrak{H}} \Psi_\varphi(n_x a_y)\,\frac{dx\,dy}{y^\ell}$$
$$= \int_{\Gamma\backslash\mathfrak{H}} \Psi_\varphi(n_x a_y)\cdot 1\,\frac{dx\,dy}{y^\ell} = \langle \Psi_\varphi, 1\rangle$$

That is, $\mathcal{M}\varphi(1)$ is the inner product of Ψ_φ with the constant function 1. For $\Gamma = SL_2(\mathbb{Z})$ and $\Gamma = SL_2(\mathbb{Z}[i])$, special arguments [2.B] easily show that the only pole of E_s in the half-plane $\mathrm{Re}(s) \geq 1/2$ is at $s_o = 1$, is simple, and the residue is a *constant* function. However, these special arguments do not easily extend to $Sp^*_{1,1}(\mathfrak{o})$ or $SL_2(\mathfrak{o})$, and, in any case, these are meant to be examples toward a larger context. As the pseudo-Eisenstein series do, E_s fits into an *adjunction*

$$\int_{\Gamma\backslash X} E_s \cdot f = \int_{\Gamma_\infty\backslash X} \eta^s \cdot c_P f \qquad (\text{for } f \text{ on } \Gamma\backslash X)$$

whenever the implied integrals converge absolutely. Via the analytic continuation of E_s, the adjunction asserts that integrals against Eisenstein series are Mellin transforms of constant terms:

$$\int_{\Gamma\backslash X} E_s \cdot f = \int_0^\infty c_P f(a_y)\,\eta^s\,\frac{dy}{y^\ell}$$
$$= \int_0^\infty c_P f(a_y)\,\eta^{-(1-s)}\,\frac{dy}{y} = \mathcal{M}c_P f(1-s)$$

Again, at a pole s_o of E_s in $\mathrm{Re}(s) > \frac{1}{2}$, c_s also has a pole of the same order. Since $c_s \cdot c_{1-s} = 1$, necessarily c_{1-s} has a zero at s_o. Thus, from

$$\mathcal{M}c_P\Psi_\varphi(s) = \mathcal{M}\varphi(s) + c_{1-s}\mathcal{M}\varphi(1-s)$$

at a pole s_o of E_s we have

$$\mathcal{M}c_P\Psi_\varphi(s_o) = \mathcal{M}\varphi(s_o) + c_{1-s_o}\mathcal{M}\varphi(1-s_o)$$
$$= \mathcal{M}\varphi(s_o) + 0\cdot\mathcal{M}\varphi(1-s_o) = \mathcal{M}\varphi(s_o)$$

That is, the value $\mathcal{M}c_P\Psi_\varphi$ at s_o is just the value of $\mathcal{M}\varphi$, so the coefficients appearing in the decomposition of Ψ_φ *are intrinsic*. Thus, the preceding decomposition has an intrinsic form as in the statement of the theorem. ///

To have an L^2 assertion and Plancherel require somewhat more care in the argument, as in the following section.

1.13 Plancherel for Pseudo-Eisenstein Series

A refined form of the previous theorem, proving convergence of the integral as a $C^\infty(\Gamma\backslash G/K)$-valued integral, from a corresponding result for behavior of Fourier inversion integrals, gives an immediate proof of a Plancherel theorem for pseudo-Eisenstein series.

[1.13.1] Theorem: *(Function-valued form)* Let s_o run over poles of E_s in $\mathrm{Re}(s) > \frac{1}{2}$. For $\varphi \in C_c^\infty(N\backslash X) = C_c^\infty(N\backslash G)^K$, the pseudo-Eisenstein series Ψ_φ is expressible in terms of genuine Eisenstein series, by an integral converging as a Gelfand-Pettis $C^\infty(\Gamma\backslash G/K)$-valued integral:

$$\Psi_\varphi = \frac{(\ell-1)}{4\pi i}\int_{\frac{1}{2}-i\infty}^{\frac{1}{2}+i\infty} \langle \Psi_\varphi, E_s\rangle \cdot E_s\, ds + (\ell-1)\sum_{s_o} \langle \Psi_\varphi, \mathrm{Res}_{s_o} E_s\rangle \cdot \mathrm{Res}_{s_o} E_s$$

writing $\langle \Psi_\varphi, E_s\rangle = \int_{\Gamma\backslash G} \Psi_\varphi \cdot \overline{E}_s$ even though E_s is not in L^2.

Proof: Let $\psi_\xi(x) = e^{i\xi x}$. The integral expressing Fourier inversion for Schwartz functions f on the real line

$$\begin{aligned} f(x) &= \frac{1}{2\pi}\int_{-\infty}^{\infty}\left(\int_{-\infty}^{\infty} f(u)\,\overline{\psi}_\xi(u)\,du\right)\psi_\xi(x)\,d\xi \\ &= \frac{1}{2\pi}\int_{-\infty}^{\infty} \psi_\xi(x)\cdot \widehat{f}(\xi)\,d\xi \end{aligned}$$

does *not* express f as a superposition of Schwartz functions, but as a superposition of exponentials $x \to e^{2\pi i\xi x}$. These exponentials are not Schwartz and are not L^2. But the Fourier inversion integral *does* converge as a Gelfand-Pettis integral with values in the Fréchet space $C^\infty(\mathbb{R})$, by [14.3]. Changing coordinates to give Mellin inversion for functions on $(0,+\infty) \approx N\backslash G/K$ gives convergence as Gelfand-Pettis integral with values in $C^\infty(0,+\infty) \approx C^\infty(N\backslash G/K) \subset C^\infty(G)$, with its Fréchet-space structure as in [13.5].

By the same steps as in the proof of the numerical form of the theorem,

$$\Psi_\varphi = \frac{(\ell-1)}{4\pi i}\int_{\frac{1}{2}-i\infty}^{\frac{1}{2}+i\infty} \mathcal{M}\varphi(s) E_s\, ds + (\ell-1)\sum_{s_o} \mathcal{M}\varphi(s_o)\cdot \mathrm{Res}_{s_o} E_s$$

as a $C^\infty(G)$-valued Gelfand-Pettis integral. As in [13.6] and the analogue for G as in [6.2], [6.4], left and right translation by G are continuous maps on $C^\infty(G)$, so the linear operators of left translation by Γ and right translation by K

commute with the integral, so the integral converges as a Gelfand-Pettis integral with values in $C^\infty(\Gamma\backslash G/K)$. Similarly, the rearrangement to the statement of the theorem preserves this convergence. ///

[1.13.2] Corollary: For $\varphi, \psi \in C_c^\infty(N\backslash G/K) \approx C_c^\infty(A^+) \approx C_c^\infty(0, +\infty)$,

$$\langle \Psi_\varphi, \Psi_\psi \rangle = \frac{(\ell-1)}{4\pi i} \int_{\frac{1}{2}-i\infty}^{\frac{1}{2}+i\infty} \langle \Psi_\varphi, E_s\rangle \overline{\langle \Psi_\psi, E_s\rangle}\, ds + (\ell-1)\sum_{s_o} \langle \Psi_\varphi, \mathrm{Res}_{s_o} E_s\rangle \cdot \overline{\langle \Psi_\psi, \mathrm{Res}_{s_o} E_s\rangle}$$

Proof: For $f \in C_c^\infty(\Gamma\backslash G/K)$, the map $F \to \int_{\Gamma\backslash G} F \cdot \overline{f}$ is a continuous linear functional on $F \in C^o(\Gamma\backslash G/K)$, so the Gelfand-Pettis property legitimizes the obvious interchange:

$$\langle \Psi_\varphi, f\rangle = \left\langle \frac{(\ell-1)}{4\pi i} \int_{\frac{1}{2}-i\infty}^{\frac{1}{2}+i\infty} \langle \Psi_\varphi, E_s\rangle\, E_s\, ds + (\ell-1)\sum_{s_o} \langle \Psi_\varphi, \mathrm{Res}_{s_o} E_s\rangle \cdot \mathrm{Res}_{s_o} E_s,\ f\right\rangle$$

$$= \frac{(\ell-1)}{4\pi i} \int_{\frac{1}{2}-i\infty}^{\frac{1}{2}+i\infty} \langle \Psi_\varphi, E_s\rangle\, \overline{\langle f, E_s\rangle}\, ds + \sum_{s_o} \langle \Psi_\varphi, \mathrm{Res}_{s_o} E_s\rangle \cdot \overline{\langle f, \mathrm{Res}_{s_o} E_s\rangle}$$

where $\langle E_s, f\rangle$ converges because $f \in C_c^o(\Gamma\backslash G/K)$. Taking $f = \Psi_\psi$ for $\psi \in C_c^\infty(N\backslash G/K)$ gives the asserted equality. ///

This corollary *looks like* an assertion of a Plancherel theorem, that is, inducing (extending by continuity) an isometry from the L^2 closure of the span of pseudo-Eisenstein series Ψ_φ with test function data φ to spaces of functions on $\frac{1}{2} + i\mathbb{R}$ and a finite-dimensional space generated by residues. What remains to show is *surjectivity* to a clearly specified space, and *orthogonality* of the residues to the integrals on $\frac{1}{2} + i\mathbb{R}$, neither of which is surprising.

[1.13.3] Claim: The residues of E_s in $\mathrm{Re}(s) \geq \frac{1}{2}$ are in the closure of the space of pseudo-Eisenstein series.

Proof: The residues $\mathrm{Res}_{s_j} E_s$ are in L^2 by [1.11.5], mutually orthogonal by [1.11.7], and orthogonal to cuspforms by [1.11.8]. By the adjunction property [1.8.2] they are in the closure of the span of the pseudo-Eisenstein series. ///

Thus, for test function φ and expansion

$$\Psi_\varphi = \frac{(\ell-1)}{4\pi i} \int_{\frac{1}{2}-i\infty}^{\frac{1}{2}+i\infty} \langle \Psi_\varphi, E_s\rangle \cdot E_s\, ds + (\ell-1)\sum_{s_o} \langle \Psi_\varphi, \mathrm{Res}_{s_o} E_s\rangle \cdot \mathrm{Res}_{s_o} E_s$$

the integral is itself in the closure of the span of the pseudo-Eisenstein series. The functions F on $\frac{1}{2} + i\mathbb{R}$ possibly arising as $F(s) = \langle \Psi_\varphi, E_s\rangle$ are constrained by the functional equation $E_{1-s} = c_{1-s}E_s$: on $\mathrm{Re}(s) = \frac{1}{2}$,

$$\langle \Psi_\varphi, E_{1-s} \rangle = \langle \Psi_\varphi, c_{1-s}E_s \rangle = \overline{c_{1-s}} \cdot \langle \Psi_\varphi, E_s \rangle = c_s \cdot \langle \Psi_\varphi, E_s \rangle$$

Let

$$V = \{F \in L^2(\tfrac{1}{2} + i\mathbb{R}) : F(1-s) = c_s F(s)\}$$

[1.13.4] Claim: The images $\langle \Psi_\varphi, E_s \rangle \oplus (\ldots, \langle \Psi_\varphi, \mathrm{Res}_{s_j} E_s \rangle, \ldots)$ are dense in $V \oplus \mathbb{C}^n$.

Proof: The residues are in the closure of pseudo-Eisenstein series, so the integral parts of the spectral decompositions are in the closure, as well, by subtraction. The remaining question is identification of the L^2 closure of the functions $s \to \langle \Psi_\varphi, E_s \rangle$. Test functions φ are dense in the Schwartz space, and the map $\varphi \to \mathcal{M}\varphi$, essentially Fourier transform, is an isomorphism to the Schwartz space on $\frac{1}{2} + i\mathbb{R}$, so the images $\mathcal{M}\varphi$ of test functions are dense in that Schwartz space, which is dense in L^2. Noting that $|c_s| = 1$ on $\mathrm{Re}(s) = \frac{1}{2}$, the averaging map $F(s) \longrightarrow F(s) + c_{1-s}F(1-s)$ is a *surjection* of $L^2(\frac{1}{2} + i\mathbb{R})$ to V, so the images $\langle \Psi_\varphi, E_s \rangle = \mathcal{M}c_P\Psi_\varphi$ are dense there, so are dense in V. ///

A typical polarization argument finishes the proof of Plancherel. Recall

[1.13.5] Lemma: Let V be a Hilbert space with subspaces V_1 and V_2. If $|v_1 + v_2|^2 = |v_1|^2 + v_2|^2$ for every $v_1 \in V_1$ and $v_2 \in V_2$, then V_1 and V_2 are orthogonal.

Proof: We aim to show that $\langle v_1, v_2 \rangle = 0$. Adjusting v_2 by a complex number of absolute value 1, we may suppose that this inner product is *real*. Then

$$4\langle v_1, v_2 \rangle = |v_1 + v_2|^2 - |v_1 - v_2|^2 = |v_1|^2 + |v_2|^2 - \left(|v_1|^2 + |v_2|^2\right) = 0$$

as claimed. ///

Thus, we can distinguish the *residual* part of Ψ_φ by

$$\Psi_\varphi^R = (\ell - 1) \sum_{s_j} \langle \Psi_\varphi, \mathrm{Res}_{s_j} E_s \rangle \cdot \mathrm{Res}_{s_j} E_s$$

and the *continuous* part

$$\Psi_\varphi^C = \Psi_\varphi - \Psi_\varphi^R = \frac{(\ell - 1)}{4\pi i} \int_{\frac{1}{2}-i\infty}^{\frac{1}{2}+i\infty} \langle \Psi_\varphi, E_s \rangle \cdot E_s \, ds$$

Both parts are in the *closure* of the images of foregoing pseudo-Eisenstein series. Extending by continuity the relation [1.13.2],

$$|\Psi_\varphi|_{L^2}^2 = |\Psi_\varphi^C|_{L^2}^2 + |\Psi_\varphi^R|_{L^2}^2$$

and these two parts are *orthogonal*. We have the corresponding Plancherel theorem:

[1.13.6] Corollary: The map $\Psi_\varphi \to \langle \Psi_\varphi, E_s\rangle \oplus (\ldots, \langle \Psi_\varphi, \mathrm{Res}_{s_j} E_s\rangle, \ldots)$ with test functions φ is an L^2 isometry to its image in $V \oplus \mathbb{C}^n$, and that image is dense in $V \oplus \mathbb{C}^n$. Extending by continuity gives an isometry of the L^2 closure of the space of Ψ_φ's to $V \oplus \mathbb{C}^n$. ///

[1.13.7] Remark: Except on smaller subspaces, such as the span of the pseudo-Eisenstein series with test-function data, the preceding integrals are not *literal*, but are the extension-by-continuity of those integrals, as with Fourier transform and Fourier inversion on $L^2(\mathbb{R})$.

1.14 Automorphic Spectral Expansion and Plancherel Theorem

Combining the decomposition of cuspforms and the decomposition of their orthogonal complement: letting s_o run over the poles of E_s in $\mathrm{Re}(s) > \frac{1}{2}$, and letting F run over an orthonormal basis for the space of cuspforms on $\Gamma\backslash G/K$,

[1.14.1] Corollary: Functions $f \in L^2(\Gamma\backslash G/K)$ have expansions

$$f = \sum_{\text{cfm } F} \langle f, F\rangle \cdot F + \frac{(\ell-1)}{4\pi i}\int_{\frac{1}{2}-i\infty}^{\frac{1}{2}+i\infty} \langle f, E_s\rangle \cdot E_s\, ds$$
$$+ (\ell-1)\sum_{s_o} \langle f, \mathrm{Res}_{s_o} E_s\rangle \cdot \mathrm{Res}_{s_o} E_s$$

and Plancherel

$$|f|^2_{L^2(\Gamma\backslash X)} = \sum_{\text{cfm } F} |\langle f, F\rangle|^2 + \frac{(\ell-1)}{4\pi}\int_{-\infty}^{\infty} |\langle f, E_s\rangle|^2\, dt$$
$$+ (\ell-1)\sum_{s_o} |\langle f, \mathrm{Res}_{s_o} E_s\rangle|^2$$

where integrals involving Eisenstein series are *isometric extensions*, as in the previous section. ///

Again, although the discrete part of the expansion converges in a straightforward L^2 sense, the continuous/Eisenstein part only makes sense as an isometric extension of literal integrals. Nevertheless, the Plancherel theorem is a literal equality.

The factor of $(\ell - 1)$ is purely artifactual and could be normalized away.

1.15 Exotic Eigenfunctions, Discreteness of Pseudo-Cuspforms

An important variant approach both to the discrete decomposition of the space of cuspforms already described and to the meromorphic continuation of Eisenstein series, as in [11.5], is the decomposition of spaces of *pseudo-cuspforms*

$$L^2_b(\Gamma\backslash G/K) = \{f \in L^2(\Gamma\backslash G/K) : c_P f(a_y) = 0 \text{ for } y > b\}$$

for fixed $b \geq 1$, with respect to a self-adjoint operator[16] $\widetilde{\Delta}_b$ closely related to Δ, but subtly different. For any $b > 0$, the corresponding space of pseudo-cuspforms contains the space of genuine cuspforms $L^2_o(\Gamma\backslash G/K)$. This operator $\widetilde{\Delta}_b$ is a *pseudo-Laplacian*. The basic, surprising result is

[1.15.1] Theorem: $L^2_b(\Gamma\backslash G/K)$ is a direct sum of eigenspaces for the pseudo-Laplacian $\widetilde{\Delta}_b$, each of finite dimension. In particular, $\widetilde{\Delta}_b$ has *compact resolvent. (Proof in [10.3].)*

Without further information, this does *not* instantly prove that the subspace consisting of genuine cuspforms decomposes discretely for $\widetilde{\Delta}_b$, because the description [9.2] of the domain $\widetilde{\Delta}_b$ puts technical requirements on possible eigenfunctions.

In any case, for $b \gg 1$, the space $L^2_b(\Gamma\backslash G/K)$ contains many functions not in the space of genuine cuspforms, for example, pseudo-Eisenstein series Ψ_φ with data φ supported in the interval $[1, b]$. As in [1.12], these are expressible as integrals of genuine Eisenstein series. However, by the theorem, apparently these pseudo-Eisenstein series are (infinite) sums of L^2-eigenfunctions for $\widetilde{\Delta}_b$ orthogonal to cuspforms. Further, truncations $\wedge^b E_{s_o}$ of genuine Eisenstein series are square-integrable, by [1.11.3] or [1.11.4], for any $s_o \in \mathbb{C}$ away from the poles of $s \to E_s$. By [1.12], these truncations are expressible as integrals of genuine Eisenstein series but, by the theorem here, are also infinite sums of L^2-eigenfunctions for $\widetilde{\Delta}_b$. Thus, evidently, there are many *exotic* eigenfunctions for $\widetilde{\Delta}_b$, pseudo-cuspforms in a strong sense. Indeed,

[1.15.2] Corollary: The eigenfunctions for $\widetilde{\Delta}_b$ in $L^2_b(\Gamma\backslash G/K)$ with eigenvalues $\lambda = s(s-1) < -1/4$ are exactly the truncated Eisenstein series $\wedge^b E_s$ with $c_P E(a_b) = 0$. *(Proof in [10.4].)*

16 This $\widetilde{\Delta}_b$ is the Friedrichs self-adjoint extension [9.2] of the restriction of the unbounded operator Δ to the *test functions* $C^\infty_c(\Gamma\backslash G/K) \cap L^2_b(\Gamma\backslash G/K)$ in the space $L^2_b(\Gamma\backslash G/K)$ of pseudo-cuspforms.

These particular truncated Eisenstein series are indeed *not* smooth. The slightly nonintuitive nature of the operator $\widetilde{\Delta}_b$ explains the situation, in [10.4]. For example, in addition to meeting the Gelfand condition of constant-term vanishing about height b:

[1.15.3] Corollary: An L^2-eigenfunction u for $\widetilde{\Delta}_b$ with eigenvalue λ satisfies $(\widetilde{\Delta}_b - \lambda)u = 0$ *locally* at heights above b. ///

2
The Quotient $Z^+GL_2(k)\backslash GL_2(\mathbb{A})$

This chapter treats a slightly less elementary example, automorphic forms on $GL_2(k)\ GL_2(\mathbb{A})$ for a number field k. The *shape* of the group elements is still two-by-two matrices, but the contents are not the purely archimedean $\mathbb{R}$, $\mathbb{C}$, $\mathbb{H}$ of the previous chapter, now involving p-adic and adelic scalars. Among several advantages, this viewpoint consistently maintains the *unicuspidality* of the quotient. In contrast, a classical approach to congruence subgroups of $SL_2(\mathbb{Z})$ apparently produces an ever-growing number of cusps, and for Hilbert-Blumenthal groups $GL_2(\mathfrak{o})$ for rings of integers $\mathfrak{o}$ of totally real[1] number fields

[1] A finite extension k of $\mathbb{Q}$ is *totally real* when all archimedean completions are isomorphic to $\mathbb{R}$, rather than to $\mathbb{C}$.

k, even at level one, the number of cusps is a *class number*. Miraculously, in the adelic formulation, *there is only one cusp*, regardless of class number or congruence conditions. That is, a single adelic Siegel set covers the quotient, as below in [2.2].

In fact, little subtle information about p-adic numbers or adeles or ideles is used. For most purposes, it is merely the *shape* of matrices, or their structural role, that matters, so things can be cast in a form that scarcely refers to details of the distinctions. The significant result is the compactness of $\mathbb{J}^1/k^\times$, in Appendix [2.A]. Earlier in the chapter, we prove p-adic and archimedean Iwasawa and Cartan decompositions from the most basic features of completions of number fields, with the incidental goal of practicing the relevant physical intuition and noting the truly relevant aspects.

Another point of this example is to accommodate more complicated data in Eisenstein series. With or without congruence conditions, number fields beyond $\mathbb{Q}$ have nontrivial *grossencharacters* (Hecke characters), and apart from complex quadratic extensions, there are always *unramified* grossencharacters. For nontrivial ideal class groups, there are nontrivial ideal class characters. Using $GL_2(\mathbb{A})$ unifies these seemingly disparate features. Thus, the decomposition [2.11–2.12] of pseudo-Eisenstein series involves not only the continuous parameter s, but at least one discrete parameter χ running through grossencharacters with various constraints on ramification. Further, congruence conditions specify further data in Eisenstein series. The functional equation(s) of Eisenstein series will no longer relate one Eisenstein series to itself under $s \to 1 - s$ but must tell how the further data transform. Suggested by physical analogues, the description of the transformation of the further data is often called a *scattering matrix* or *scattering operator*.

Most of the analytical archimedean issues in later chapters are already well illustrated by Chapter 1. The present chapter illustrates interaction of those archimedean issues with p-adic.

2.1 Groups $K_v = GL_2(\mathfrak{o}_v) \subset G_v = GL_2(k_v)$

Throughout this chapter, k is a number field.[2] Let $\mathfrak{o}$ be its ring of algebraic integers. Denote the various archimedean and p-adic (non-archimedean) completions by k_v, where $v < \infty$ means non-archimedean, and $v|\infty$ means archimedean. For non-archimedean v, let $\mathfrak{o}_v$ be the local integers. Normalize all the norms $|\cdot|_v$ so that the product formula $\prod_v |t|_v = 1$ holds for $t \in k^\times$,

[2] The potential conflict with k being an element of a compact subgroup K is avoidable only by other notational infelicities.

preferably by taking the norm in k_v to be the composition of the Galois norm to the corresponding completion $\mathbb{Q}_p$ of $\mathbb{Q}$ and then the standard p-adic norm on $\mathbb{Q}_p$, by $|t|_v = |N_{k_v/\mathbb{Q}_p} t|_p$, and similarly for archimedean places. When useful, ϖ_v will be a local parameter at a non-archimedean place v, that is, a prime element in $\mathfrak{o}_v$. Let $\mathbb{A}, \mathbb{J}$ be the adeles and ideles of k.

Let $G_v = GL_2(k_v)$. Let Z_v be the center of G_v. Temporarily let r be the number of nonisomorphic archimedean completions of k, thus *not* counting a complex completion and its conjugate as 2, but just 1. That is, $[k : \mathbb{Q}] = r_1 + 2r_2$ where r_1 is the number of real completions, and r_2 the number of complex, and $r = r_1 + r_2$. Let Z^+ be the positive real scalar matrices diagonally imbedded across *all* archimedean v, by the map

$$\delta : t \longrightarrow (\ldots, t^{1/r}, \ldots) \qquad (\text{for } t > 0)$$

This map δ gives a *section* of the idele norm map $|t| = \prod_v |t_v|_v$, in that $|\delta(t)| = t$. As usual, let

$$P_v = \{\begin{pmatrix} * & * \\ 0 & * \end{pmatrix} \in G_v\} \qquad N_v = \{\begin{pmatrix} 1 & * \\ 0 & 1 \end{pmatrix} \in G_v\}$$
$$M_v = \{\begin{pmatrix} * & 0 \\ 0 & * \end{pmatrix} \in G_v\}$$

We have already noted the compact subgroups $K_v \approx SO_2(\mathbb{R}) \subset SL_2(\mathbb{R})$ and $K_v \approx SU(2) \subset SL_2(\mathbb{C})$ for archimedean completions $k_v \approx \mathbb{R}$ and $k_v \approx \mathbb{C}$, and the corresponding Iwasawa decompositions [1.3].

[2.1.1] Claim: For $v < \infty$, the subgroup

$$\begin{aligned} K_v &= GL_2(\mathfrak{o}_v) \\ &= \{\text{two-by-two matrices, entries in } \mathfrak{o}_v, \text{ determinant in } \mathfrak{o}_v^\times\} \end{aligned}$$

is a *compact, open* subgroup of $G_v = GL_2(k_v)$. We have *Iwasawa decomposition*

$$G_v = P_v \cdot K_v = N_v \cdot M_v \cdot K_v$$

and *Cartan decomposition*

$$G_v = K_v \cdot M_v \cdot K_v$$

Proof: The local fields k_v are finite-dimensional vectorspaces over respective $\mathbb{Q}_p$ and $\mathbb{R}$, so are locally compact. For non-archimedean v, the local integers are both closed and open:

$$\mathfrak{o}_v = \{x \in k_v : |x|_v \leq 1\} = \{x \in k_v : |x|_v < |\varpi^{-1}|_v\}$$

Similarly for the local units:

$$\mathfrak{o}_v^\times = \{x \in k_v : |x|_v = 1\} = \{x \in k_v : |\varpi_v|_v < |x|_v < |\varpi^{-1}|_v\}$$

From this, the conditions defining the subgroups K_v are both open and closed. Since K_v is a closed subset of the compact set

$$\{\begin{pmatrix} a & b \\ c & d \end{pmatrix} : a, b, c, d \in \mathfrak{o}_v\} \approx \mathfrak{o}_v \times \mathfrak{o}_v \times \mathfrak{o}_v \times \mathfrak{o}_v$$

it is compact. Given $\begin{pmatrix} a & b \\ c & d \end{pmatrix} \in G_v$, either $|c|_v \geq |d|_v$ or $|c|_v \leq |d|_v$, so either $d/c \in \mathfrak{o}_v$ or $c/d \in \mathfrak{o}_v$, respectively. Thus, either

$$\begin{pmatrix} a & b \\ c & d \end{pmatrix} \cdot \begin{pmatrix} -d/c & 1 \\ 1 & 0 \end{pmatrix} = \begin{pmatrix} * & * \\ 0 & * \end{pmatrix} \qquad (\text{when } \begin{pmatrix} -d/c & 1 \\ 1 & 0 \end{pmatrix} \in K_\nu)$$

or

$$\begin{pmatrix} a & b \\ c & d \end{pmatrix} \cdot \begin{pmatrix} 1 & 0 \\ -c/d & 1 \end{pmatrix} = \begin{pmatrix} * & * \\ 0 & * \end{pmatrix} \qquad (\text{when } \begin{pmatrix} 1 & 0 \\ -c/d & 1 \end{pmatrix} \in K_\nu)$$

giving the Iwasawa decomposition.

The Cartan decomposition is a corollary of the structure theorem for finitely generated modules over a principal ideal domain such as $\mathfrak{o}_v$, as follows. Given $g \in G_v$, multiply by a scalar matrix (in M_v) so that all entries of the modified g are in $\mathfrak{o}_v$. (Of course, this does not at all ensure that the determinant is in $\mathfrak{o}_v^\times$.) The columns of such g generate a rank-two $\mathfrak{o}_v$-submodule V of $\mathfrak{o}_v^2$. By the structure theorem, there is an $\mathfrak{o}_v$-basis f_1, f_2 for $\mathfrak{o}_v^2$ and d_1, d_2 in $\mathfrak{o}_v$ such that $V = \mathfrak{o}_v d_1 f_1 + \mathfrak{o}_v d_2 f_2$. Since $\{d_j f_j\}$ is another $\mathfrak{o}_v$-basis for V, there is $h \in K_v$ expressing the two columns of g as $\mathfrak{o}_v$-linear combinations of $d_1 f_1, d_2 f_2$ (and *vice-versa*). That is, in terms of matrix multiplication, writing $d_1 f_1, d_2 f_2$ as column vectors,

$$\begin{pmatrix} a & b \\ c & d \end{pmatrix} = (d_1 f_1 \quad d_2 f_2) \cdot h$$

At the same time, there is $h' \in K_v$ such that $h'e_1 = f_1$ and $h'e_2 = f_2$, where $\{e_j\}$ is the standard $\mathfrak{o}_v$-basis for $\mathfrak{o}_v^2$. Thus,

$$\begin{pmatrix} a & b \\ c & d \end{pmatrix} = h' \cdot \begin{pmatrix} d_1 & 0 \\ 0 & d_2 \end{pmatrix} \cdot h$$

giving the p-adic Cartan decomposition. ///

Unlike the archimedean situation, the compact K_v has substantial intersections with both N_v and M_v, so with P_v. Indeed, since k_v is an ascending union

$k_v = \bigcup_{\ell \geq 0} \varpi^{-\ell} \cdot \mathfrak{o}_v$, the subgroup N_v is an ascending union of compact, open subgroups:

$$N_v = \bigcup_{\ell \geq 0} \begin{pmatrix} 1 & \varpi_v^{-\ell} \mathfrak{o}_v \\ 0 & 1 \end{pmatrix}$$

Again unlike the archimedean situation, K_v has a neighborhood basis at 1 consisting of compact, open subgroups, namely, the (local) *principal congruence subgroups*

$$K_{v,\ell} = \{ \begin{pmatrix} a & b \\ c & d \end{pmatrix} \in GL_2(\mathfrak{o}_v) : a = 1 \bmod \varpi^\ell, b = 0 \bmod \varpi^\ell,$$
$$c = 0 \bmod \varpi^\ell, \text{ and } d = 1 \bmod \varpi^\ell \}$$

The corresponding *adele group* is $G_\mathbb{A} = GL_2(\mathbb{A})$, meaning two-by-two matrices with entries in $\mathbb{A}$, with determinant in the ideles $\mathbb{J}$. This group is also an ascending union (colimit) of *products*

$$G_S = \prod_{v \in S} G_v \times \prod_{v \notin S} K_v$$

where S is a finite set of places v, including archimedean places, ordering finite sets S (of places v) by containment. Similarly, $P_\mathbb{A}$, $M_\mathbb{A}$, $N_\mathbb{A}$, and $Z_\mathbb{A}$ are the adelic forms of those groups, that is, obtained by allowing entries in $\mathbb{A}$, or, equivalently, as colimits of products of local groups. Let $K_\mathbb{A} = \prod_v K_v \subset G_\mathbb{A}$. Let $\delta : (0, \infty) \to \mathbb{J}$ the usual diagonal imbedding of $(0, \infty)$ to the archimedean component of the ideles by $\delta(t) = (\ldots, t^{1/d_v}, \ldots)$ where d_v is the local degree, so that $\delta : (0, \infty) \to \mathbb{J}$ gives a one-sided inverse to the *idele norm* $|\alpha| = \prod_{v \leq \infty} |\alpha|_v$. Then

$$Z_\mathbb{A}/Z^+ Z_k \approx \mathbb{J}/\delta(0, \infty) \cdot k^\times \approx \mathbb{J}^1/k^\times = \text{compact}$$

where Z_k is invertible scalar matrices with entries in k. The compactness is nontrivial [2.A], but standard.

2.2 Discrete Subgroup $GL_2(k) \subset GL_2(\mathbb{A})$, Reduction Theory

Let $G_k = GL_2(k)$, P_k, M_k, N_k be the groups with entries in k. Here we demonstrate that a single (adelic) Siegel set *surjects* to the quotient $G_k \backslash G_\mathbb{A}$, and that (adelic) Siegel sets behave well. First,

[2.2.1] Claim: G_k is a *discrete* subgroup of $G_\mathbb{A}$.

Proof: To show that a subgroup of a topological group is *discrete*, it suffices to show that there is a neighborhood of the identity containing no element of the subgroup except 1, since multiplication $U \to gU$ is homeomorphism of neighborhoods U of 1 to neighborhoods gU of g.

We do this in two steps. First, the subgroup $H = \prod_{v|\infty} G_v \times \prod_{v<\infty} K_v$ is an open neighborhood of $1 \in G_{\mathbb{A}}$, so it suffices to show that the group $G_k \cap H$ is discrete. The condition on H is that entries are locally integral at all finite places, and the determinants are local units. An element of k that is a local integer everywhere is an integer, and an element of $k^\times$ that is a local unit everywhere is a unit in $\mathfrak{o}^\times$. Thus, $G_k \cap H = GL_2(\mathfrak{o})$, and it suffices to show that the projection of $GL_2(\mathfrak{o})$ to $G_\infty = \prod_{v|\infty} G_v$ is discrete in the latter. The topology on G_∞ is the subspace topology from the real vectorspace of 2-by-2 matrices with entries in $k_\infty = \prod_{v|\infty} k_v$, which itself has the product topology. From classical algebraic number theory, $\mathfrak{o}$ is discrete in k_∞, giving the discreteness. ///

On some occasions, one uses

$$G^1 = \{g \in G_{\mathbb{A}} : |\det g| = 1\}$$

noting that $G_{\mathbb{A}} = Z^+ \times G^1$. The *product formula* $\prod_{v\le\infty} |t|_v = 1$ for $t \in k^\times$ shows that $G_k \subset G^1$. In particular, G_k is still *discrete* in $Z^+\backslash G_{\mathbb{A}} \approx G^1$.

Now define local and global *height functions* h_v and h. For v-adic completion $k_v \approx \mathbb{R}$, let h_v be the usual real Hilbert-space norm on $k_v^2 \approx \mathbb{R}^2$. To accommodate the *product formula*, for $k_v \approx \mathbb{C}$, let h_v be the *square* of the usual complex Hilbert-space norm on $k_v^2 \approx \mathbb{C}^2$. For k_v non-archimedean, let $h_v(x_1, x_2) = \max\{|x_1|_v, |x_2|_v\}$. Put $h(x) = \prod_{v\le\infty} h_v(x)$. There is good behavior under scalar multiplication, via the product formula: for $t \in k^\times$,

$$\begin{aligned} h(t \cdot x) &= \prod_{v\le\infty} h_v(t \cdot x) = \prod_{v\le\infty} |t|_v \cdot h_v(x) \\ &= \prod_{v\le\infty} |t|_v \cdot \prod_{v\le\infty} h_v(x) = 1 \cdot \prod_{v\le\infty} h_v(x) \end{aligned}$$

Sufficient conditions are given below for finiteness of the product. By design, the isometry groups of the height functions h_v are the compact subgroups K_v already specified.

For each prime v the group K_v is *transitive* on the collection of vectors in k_v^2 with given norm: at archimedean places, this is because all vectors of a given length can be rotated to each other, while at non-archimedean places the suitable analogue of *length* is *greatest common divisor*.

Let G_k, P_k, N_k, M_k, Z_k be the corresponding groups of matrices with entries in k, and use subscript $\mathbb{A}$ to denote the adelic points.

Now we identify a class of vectors with *finite height*. First, given $x \in k^2 - \{0\}$, for all but finitely many v the components of the vector x are all v-integral and generate the local integers $\mathfrak{o}_v$. In particular, for all but finitely many v the vth local height $h_v(x)$ of $x \in k^r$ is 1, and the infinite product for $h(x)$ is a *finite* product. Write vectors as *row* vectors, and let $G_{\mathbb{A}} = GL_2(\mathbb{A})$ act on the *right* by matrix multiplication. A nonzero vector $x \in \mathbb{A}^2$ is *primitive* when $x \in (k^2 - \{0\}) \cdot G_{\mathbb{A}}$.

[2.2.2] Theorem: For fixed $g \in G_{\mathbb{A}}$ and for fixed $c > 0$,

$$\operatorname{card}(k^\times \backslash \{x \in k^2 - \{0\} : h(x \cdot g) < c\}) < \infty$$

For compact $C \subset G_{\mathbb{A}}$, there are positive implied constants such that, for all $g \in C$, for all primitive x,

$$h(x) \ll_C h(x \cdot g) \ll_C h(x)$$

Proof: Fix $g \in G_{\mathbb{A}}$. Since $K = K_{\mathbb{A}} = \prod_v K_v$ preserves heights, via Iwasawa decompositions locally everywhere, we may suppose that g is in the group $P_{\mathbb{A}}$ of upper triangular matrices in $G_{\mathbb{A}}$. Choose representatives $x = (x_1, x_2)$ for nonzero vectors in $k^\times \backslash k^2$ either of the form $x = (1, x_2)$ or $x = (0, 1)$. There is just one vector of the latter shape, so we consider $x = (1, x_2)$:

$$x \cdot g = (1 \quad x_2)\begin{pmatrix} a & b \\ 0 & d \end{pmatrix} = (a \quad b + x_2 d) \qquad (\text{for } g = \begin{pmatrix} a & b \\ 0 & d \end{pmatrix})$$

At each place v, including archimedean ones, $\max(|a|_v, |b + x_2 d|_v) \leq h_v(xg)$, so

$$|b + x_2 d|_v \prod_{w \neq v} |a|_w \leq \prod_{\text{all } w} h_w(xg) = h(xg)$$

Since g is fixed, a is fixed, and at almost all places $|a|_w = 1$. Thus, for $h(xg) < c$,

$$|b + x_2 d|_v < c \cdot \Big(\prod_{w \neq v} |a|_w \Big)^{-1} \ll_{g,c} 1$$

For the product formula to hold we are using the normalization of norms that $|\varpi_v| = q_v^{-1}$, where ϖ_v is a local parameter at v and q_v is the residue field cardinality at v. There are only finitely many places v with residue field cardinality less than a given constant, so, in fact, $|b + x_2 d|_v \leq 1$ for v outside a finite set depending on g and c. Therefore, $b + x_2 d$ lies in a compact subset Ω of $\mathbb{A}$ depending on g, c. Since b, d are fixed, and since k is discrete (and closed, by [1.5.3]) in $\mathbb{A}$, the collection of images $\{b + x_2 d : x_2 \in k\}$ is discrete in $\mathbb{A}$. The intersection of a closed, discrete set and a compact set is *finite*, so collection of $x_2 \in k$ so that $b + x_2 d$ lies in Ω is finite, proving the first assertion.

For the last assertion, use the *Cartan decompositions* $G_v = K_v \cdot M_v \cdot K_v$ from [2.1]. The map $\theta_1 \times m \times \theta_2 \longrightarrow \theta_1 m \theta_2$, with $\theta_1, \theta_2 \in K_v$ and $m \in M_v$, is *not* an injection, so we cannot immediately infer that for a given compact $C \subset G_v$ the set

$$\{m \in M_v : \text{for some } c \in C, c \in K_v m K_v\}$$

is compact. Since K_v is compact, $C' = K_v \cdot C \cdot K_v$ is compact, and now $\theta_1 m \theta_2 \in C'$ with $\theta_i \in K_v$ implies $m \in C' \cap M_v$, which *is* compact. Thus, any compact subset of $G_{\mathbb{A}}$ is contained in a set $\{\theta_1 m \theta_2 : \theta_1, \theta_2 \in K, m \in C_M\}$ with compact $C_M \subset M_{\mathbb{A}}$. Since K preserves heights and since the set of primitive vectors is stable under K, the set of values $\{h(xg)/h(x) : x \text{ primitive}, g \in C\}$ is contained in a set

$$\left\{\frac{h(x\delta)}{h(x)} : x \text{ primitive}, m \in C_M\right\}$$

for some compact $C_M \subset M$. Letting the diagonal entries of m be m_i,

$$0 < \inf_{m \in C_M} \inf_i |m_i| \leq \frac{h(xm)}{h(x)} \leq \sup_{m \in C_M} \sup_i |m_i| < +\infty$$

This gives the desired bound. ///

To compare to the purely archimedean height functions η used in the four earlier examples, for g upper-triangular,

$$h_v\left((0 \quad 1) \cdot \begin{pmatrix} a & b \\ 0 & d \end{pmatrix}\right) = h_v(0 \quad d) = |d|_v^{-1}$$

For example, with $k_v \approx \mathbb{R}$, for $g = n_x a_y = \begin{pmatrix} 1 & x \\ 0 & 1 \end{pmatrix} \begin{pmatrix} \sqrt{y} & 0 \\ 0 & 1/\sqrt{y} \end{pmatrix} \in SL_2(\mathbb{R})$,

$$h_v((0 \quad 1) \cdot g) = h_v\begin{pmatrix} 0 & \frac{1}{\sqrt{y}} \end{pmatrix} = \frac{1}{\sqrt{y}}$$

However, we want local height functions which are right K_v-invariant and

$$\eta_v \begin{pmatrix} a & b \\ 0 & d \end{pmatrix} = \left|\frac{a}{d}\right|_v$$

so put

$$\eta_v(g) = |\det g|_v \cdot h_v((0 \quad 1) \cdot g)^{-2} \qquad ((\text{for } g \in G_v))$$

and $\eta(g) = \prod_v \eta_v(g)$. This matches earlier use for $SL_2(\mathbb{R})$. Left multiplication by $\gamma \in G_k$ does not change $|\det g|$ (with idele norm) because of the product formula:

$$|\det(\gamma g)| = |\det \gamma \cdot \det g| = |\det \gamma| \cdot |\det g| = 1 \cdot |\det g|$$

[2.2.3] Lemma: η is left P_k-invariant, and $Z_{\mathbb{A}}$-invariant.

Proof: Via the product formula, with $p = \begin{pmatrix} a & b \\ 0 & d \end{pmatrix} \in P_k$,

$$\begin{aligned} \eta(p \cdot g) &= |\det pg| \cdot h((0 \quad 1) \cdot pg)^{-2} = |\det pg| \cdot h((0 \quad d) \cdot g)^{-2} \\ &= |\det p| \cdot |\det g| \cdot |d| \cdot h((0 \quad 1) \cdot g)^{-2} = |\det g| \cdot h((0 \quad 1) \cdot g)^{-2} \end{aligned}$$

For the center-invariance, with $z = \begin{pmatrix} t & 0 \\ 0 & t \end{pmatrix} \in Z_{\mathbb{A}}$,

$$\begin{aligned} \eta(z \cdot g) &= |\det zg| \cdot h((0 \quad 1) \cdot zg)^{-2} = |\det zg| \cdot h((0 \quad t) \cdot g)^{-2} \\ &= |t^2| \cdot |\det g| \cdot |t|^{-2} \cdot h((0 \quad 1) \cdot g)^{-2} = |\det g| \cdot h((0 \quad 1) \cdot g)^{-2} \end{aligned}$$

as claimed. ///

[2.2.4] Corollary: For fixed $g \in G_{\mathbb{A}}$, there are finitely many $\gamma \in P_k \backslash G_k$ such that $\eta(\gamma \cdot g) > \eta(g)$.

Proof: There is a natural bijection

$$k^\times \backslash (k^2 - \{0\}) \longleftrightarrow P_k \backslash G_k \qquad \text{by} \qquad k^\times \cdot (c \quad d) \longleftrightarrow P_k \cdot \begin{pmatrix} * & * \\ c & d \end{pmatrix}$$

for any invertible matrix with bottom row $(c \quad d)$. Indeed, G_k is *transitive* on nonzero vectors, and P_k is the stabilizer, acting on the right, of the line (minus a point) $(0 \quad *) = k^\times \cdot \{(0 \quad 1)\}$. The theorem shows that there are finitely many $x \in k^\times \backslash (k^2 - \{0\})$ such that $h(xg) < c$, that is, such that $h(xg)^{-1} > c^{-1}$. Since $|\det g|$ is G_k-invariant, the bijection just demonstrated gives the assertion of the corollary. ///

[2.2.5] Corollary: $\sup_{\gamma \in G_k} \eta(\gamma \cdot g) < \infty$, and the sup is attained, and

$$G_k \cdot \{g \in G_{\mathbb{A}} : \eta(g) \geq \eta(\gamma \cdot g) \text{ for all } \gamma \in G_k\} = G_{\mathbb{A}}$$

Proof: The previous corollary shows that the sup is finite and that the sup is attained. Thus, the indicated set is a (possibly redundant) collection of representatives for all orbits, by choosing group elements attaining the sup in their G_k-orbit. ///

Critical in legitimizing treatment of truncated Eisenstein series:

[2.2.6] Corollary: Given $t_o > 0$, there is $t_1 \gg 1$ such that, for $\eta(g) \geq t_1$, if $\eta(\gamma \cdot g) \geq t_o$ then $\gamma \in P_k$.

Proof: It suffices to take $g = nm$ since η is right $K_{\mathbb{A}}$-invariant, invoking Iwasawa. Since η is $Z_{\mathbb{A}}$-invariant, it suffices to consider $m = \begin{pmatrix} m_1 & 0 \\ 0 & 1 \end{pmatrix}$. Adjusting on the left by M_k, by the compactness lemma [2.A], we can take m_1 of the special form $m_1 = m_o \cdot \delta(t)$ for $t > 0$ and m_o in a sufficiently large compact subset of $\mathbb{J}$, where $\delta : (0, +\infty) \to \mathbb{J}$ imbeds the ray $(0, \infty)$ at archimedean places. Take compact $C \subset N_{\mathbb{A}}$ such that $N_k \cdot C = N_{\mathbb{A}}$. For $v \in k^2 - \{0\}$,

$$h(v \cdot nm) = h(v \cdot m \cdot m^{-1}nm)$$

For m of the special sort indicated, given $t_1 > 0$, there is compact $C' \subset N_{\mathbb{A}}$ such that if $\eta(m) = |m_1/m_2| \geq t_1$, then $m^{-1}Cm \subset C'$. Let $(c\ d) \in k^2 - \{0\}$. From the second assertion of the theorem, there are constants depending only on C' such that, for all (primitive) $x = v \cdot m$, for all $n \in C'$,

$$h(v \cdot m) = h(x) \ll_{C'} h(x \cdot n) \ll_{C'} h(x) = h(v \cdot m)$$

Thus, it suffices to treat simply $g = m$. In that case, with $v = (c\ d)$ with $c \neq 0$,

$$h(v \cdot m) = h((c \quad d) \cdot m) = h(cm_1 \quad d) \geq |cm_1| = |c| \cdot |m_1| = |m_1|$$

by the product rule, since $c \in k^\times$. Thus, with $\gamma = \begin{pmatrix} a & b \\ c & d \end{pmatrix} \in G_k$, but not in P_k,

$$\eta(\gamma \cdot m) = \frac{|\det \gamma m|}{h((c \quad d) \cdot m)^2} = \frac{|m_1|}{h((c \quad d) \cdot m)^2} \leq \frac{|m_1|}{|m_1|^2} = \frac{1}{|m_1|}$$

With whatever constants are implied in the simplifications in the first part of the proof, a sufficiently high lower bound $\eta(m) = |m_1| \geq t_1$ ensures that $\eta(\gamma \cdot m)$ is below t_o. ///

An element $g \in G_{\mathbb{A}}$ such that $\eta(g) \geq \eta(\gamma \cdot g)$ for all $\gamma \in G_k$ is *reduced*. Given the foregoing preparation, as an application of Dirichlet's pigeon-hole principle, after Minkowski, we can prove

[2.2.7] Theorem: There is a constant $t_o > 0$ depending on k such that $\eta(g) \geq t_o$ for *reduced* $g \in G_{\mathbb{A}}$.

Proof: Since heights are right K-invariant, take $g = nm$ with $n = n_x \in N_{\mathbb{A}}$ and $m \in M_{\mathbb{A}}$. Adjusting by the center, take

$$m = \begin{pmatrix} y & 0 \\ 0 & 1 \end{pmatrix} \qquad n = n_x = \begin{pmatrix} 1 & x \\ 0 & 1 \end{pmatrix}$$

with $y \in \mathbb{J}, x \in \mathbb{A}$. Let $\mathbb{J}^1$ be the ideles of idele-norm 1, and let $\delta : (0, +\infty) \to \mathbb{J}$ by

$$\delta(y_\infty) = (y_\infty^{\frac{1}{n}}, \ldots, y_\infty^{\frac{1}{n}}, 1, 1, 1, \ldots)$$

where, temporarily, $n = [k : \mathbb{Q}]$, with nontrivial values at the archimedean components. Then $\mathbb{J} = \mathbb{J}^1 \times \delta(0, +\infty)$. Let $U = \prod_{v|\infty} k_v^\times \times \prod_{v<\infty} \mathfrak{o}_v^\times$. The quotient $k^\times \backslash \mathbb{J}^1$ is *compact*, by [2.A], so $k^\times \backslash \mathbb{J}^1 / U$ is *finite*.

Thus, adjusting on the left by $\{\begin{pmatrix} m_1 & 0 \\ 0 & 1 \end{pmatrix} : m_1 \in k^\times\}$ and on the right by $\{\begin{pmatrix} m_1 & 0 \\ 0 & 1 \end{pmatrix} : m_1 \in U\}$, we can suppose that $y = \delta(y_\infty) \cdot \theta$ with $y_\infty \in (0, +\infty)$ and θ in a *finite* list Θ of *finite* ideles, essentially representatives for the ideal class group. We can take $\theta \in \Theta$ everywhere locally integral at finite places. Write

$$m = m_\infty \cdot m_{\text{fin}} \qquad \text{with} \qquad m_\infty = \begin{pmatrix} \delta(y_\infty) & 0 \\ 0 & 1 \end{pmatrix}$$

$$m_{\text{fin}} = \begin{pmatrix} \theta & 0 \\ 0 & 1 \end{pmatrix}$$

For fixed $\theta \in \Theta$, with

$$V_\theta = \theta \left(\prod_{v<\infty} \mathfrak{o}_v \right) \theta^{-1} \qquad U_\theta = m_{\text{fin}} \left(N_{\mathbb{A}} \cap \prod_{v<\infty} K_v \right) m_{\text{fin}}^{-1}$$

we have

$$U_\theta = \{\begin{pmatrix} 1 & u \\ 0 & 1 \end{pmatrix} : u \in V_\theta\}$$

Let $\mathbb{A}_\infty = k \otimes_{\mathbb{Q}} \mathbb{R}$ be the archimedean component of the adeles. For each fixed $\theta \in \Theta$, acting on n_x on the left by N_k is equivalent to adjusting $x \in \mathbb{A}$ by k. By additive approximation, we can adjust x by k to be in $\mathbb{A}_\infty + V_\theta$. Right multiplication of $n_x m$ by $N_{\mathbb{A}} \cap \prod_{v<\infty} K_v$ is equivalent to adjusting x by V_θ. Thus, without loss of generality, $x \in \mathbb{A}_\infty$. None of these adjustments changed the height $\eta(nm)$.

The inequality

$$h((0 \quad 1) \cdot nm) \leq h((1 \quad -\alpha) \cdot nm)$$

holds for every $\alpha \in k$, by the *reduced* property of $g = nm$. Letting $h_{\text{fin}} = \prod_{v<\infty} h_v$ and $h_\infty = \prod_{v|\infty} h_v$, since $(0 \quad 1)$ is fixed by $\begin{pmatrix} * & * \\ 0 & 1 \end{pmatrix}$,

$$1 = h((0 \quad 1) \cdot nm) \leq h_\infty(\delta(y_\infty),\ x - \alpha_\infty) \cdot h_{\text{fin}}(\theta,\ -\alpha_{\text{fin}})$$

where α_∞ and α_{fin} are the projections to the archimedean and finite components of the adeles. This is

$$\frac{1}{h_{\text{fin}}(\theta,\ -\alpha_{\text{fin}})} \leq \prod_{v|\infty} \left(|y_\infty|_v^{2/d_v} + |x_v - \alpha_v|_v^{2/d_v} \right)^{d_v/2}$$

where d_v is the local degree at the vth archimedean place. We want to use Dirichlet's principle to choose $\alpha \in k$ so that $|x_v - \alpha_v|_v$ is much smaller than $h_{\text{fin}}(\theta,\ -\alpha_{\text{fin}})$, thereby to give a lower bound on y_∞.

Choose a $\mathbb{Z}$-basis $\omega_1, \omega_2, \ldots$ for $\mathfrak{o}$, and put

$$F = \{\sum_j r_j \cdot \omega_j : \text{each } 0 \leq r_j < 1\} \subset \mathbb{A}_\infty$$

Thus, given $x \in \mathbb{A}_\infty$, there is $\beta \in \mathfrak{o}$ such that $x - \beta \in F$. For fixed large $1 \leq \ell \in \mathbb{Z}$, for each integer $1 \leq a \leq \ell^{[k:\mathbb{Q}]} + 1$, let $b = b_a \in \mathfrak{o}$ such that $ax - b \in F$. Since F is a disjoint union of $\ell^{[k:\mathbb{Q}]}$ translates of $\ell^{-1}F$, by the pigeon-hole principle, there are a, b and a', b' such that $(ax - b) - (a'x - b') \in \ell^{-1}F$. Thus,

$$x - \frac{b - b'}{a - a'} \in \frac{1}{\ell(a - a')} \cdot F$$

Put $p = b - b' \in \mathfrak{o}$, $q = a - a' \in \mathbb{Z}$, and $\alpha = p/q$. Without loss of generality, $q > 0$. With $\mu = \sup_{x \in F,\, v|\infty} |x_v|_v^{2/d_v}$, we have

$$\frac{1}{h_{\text{fin}}(\theta,\ -\frac{p}{q})} \leq \prod_{v|\infty} \left(y_\infty^2 + \frac{\mu}{(\ell \cdot q)^2} \right)^{d_v/2} = \left(y_\infty^2 + \frac{\mu}{(\ell \cdot q)^2} \right)^{[k:\mathbb{Q}]/2}$$

Now

$$\begin{aligned} h_{\text{fin}}(\theta,\ -\frac{p}{q}) &\leq \prod_{v<\infty} \max\{|\theta_v|_v,\ |\frac{p}{q}|_v\} \\ &\leq \prod_{v<\infty} \max\left\{|\theta_v|_v,\ 1\right\} \cdot \prod_{v<\infty} \max\{1,\ |\frac{1}{q}|_v\} = q^n \end{aligned}$$

since $\theta \in \Theta$ is everywhere locally integral. Then

$$\frac{1}{q^n} \leq \left(y_\infty^2 + \frac{\mu}{(\ell \cdot q)^2} \right)^{n/2}$$

or

$$\frac{1}{q^2} \leq y_\infty^2 + \frac{\mu}{(\ell \cdot q)^2}$$

Since $1 \leq q \leq \ell^n$, this implies

$$\frac{1}{\ell^{2n}} \cdot \left(1 - \frac{\mu}{\ell^2}\right) \leq \frac{1}{q^2} \cdot \left(1 - \frac{\mu}{\ell^2}\right) \leq y_\infty^2$$

Taking $\ell^2 \geq 2\mu$ gives a uniform *positive* lower bound $y_\infty \geq t_1 = \frac{1}{2\ell^2} > 0$. For each of the finitely many $\theta \in \Theta$,

$$\eta(m) = \eta(m_\infty \cdot m_{\text{fin}}) = \eta(\delta(y_\infty)) \cdot \eta(\theta) = y_\infty^n \cdot \eta(\theta) \geq t_1^n \cdot \min_{\theta \in \Theta} \eta(\theta)$$

That is, every *reduced* $g \in G_{\mathbb{A}}$ has $\eta(g)$ bounded from below by that (positive) quantity. ///

For compact $C \subset N_{\mathbb{A}}$ and $t > 0$, the corresponding *Siegel set* is

$$\mathfrak{S}_{C,t} = \{nmk : n \in C, k \in K_{\mathbb{A}}, |m_1/m_2| \geq t\}$$

where $m = \begin{pmatrix} m_1 & 0 \\ 0 & m_2 \end{pmatrix}$.

[2.2.8] Corollary: Let C be any compact subset of $N_{\mathbb{A}}$ sufficiently large so that $N_k \cdot C = N_{\mathbb{A}}$. With t_o as in the theorem, $\mathfrak{S}_{C,t_o}$ surjects to the quotient $G_k \backslash G_{\mathbb{A}}$. That is, $G_k \cdot \mathfrak{S}_{C,t_o} = G_{\mathbb{A}}$.

Proof: The theorem asserts that $S = \{g : \eta(g) \geq t_o\}$ surjects to $G_k \backslash G_{\mathbb{A}}$. The set S is left P_k-invariant and left $N_{\mathbb{A}}$-invariant. Thus, we can certainly adjust on the left by N_k so that with $g \in nmK$ in Iwasawa coordinates $n \in C$. ///

2.3 Invariant Measures

We seldom need explicit formulaic evaluation of integrals on groups $G_v = GL_2(k_v)$ or their subgroups. Rather, *qualitative* features of the invariant integrals, such as *uniqueness* and *unwinding* properties, play the main roles.

Locally, from [14.4], up to scalar multiples there is a unique right G_v-invariant measure on G_v, left P_v-invariant measure on P_v, and (left and right) K_v-invariant measure on K_v, for each place v. Even though $P_v \cap K_v$ is nontrivial, given any *two* of the scalar multiple choices, the third is determined, so that

$$\int_{G_v} f = \int_{P_v} \int_{K_v} f(ph)\, dp\, dh$$

The idea of the proof from [5.2] and [14.4] is that the group $H = P_v \times K_v$ acts transitively on G_v by $(p \times k)(g) = p^{-1}gk$, with isotropy group $P_v \times K_v$ at

$1 \in G_v$. Since the modular function of $P_v \times K_v$ is inevitably trivial on the compact $P_v \cap K_v$, there is a unique H-invariant measure on $G_v \approx (P_v \cap K_v)\backslash(P_v \times K_v)$. Since the (left and right) G_v-invariant measure is such, these must be the same, by uniqueness. For example, for f right K_v-invariant,

$$\int_{G_v} f = \int_{P_v} f(p)\, dp$$

left P_v-invariant measure dp on P_v. Even more simply, $P_v = N_v M_v \approx N_v \times M_v$ has a left (or right) invariant measure given by the product of the invariant measures on N_v and M_v. Archimedean examples were already considered in [1.6], and p-adic examples below.

Similarly, *globally*, there is a unique right $G_{\mathbb{A}}$-invariant measure on $G_{\mathbb{A}}$, $Z_{\mathbb{A}}\backslash G_{\mathbb{A}}$, and $Z^+\backslash G_{\mathbb{A}}$. Given these, there are *unique* right $G_{\mathbb{A}}$-invariant measures on $G_k\backslash G_{\mathbb{A}}$, $Z_{\mathbb{A}} G_k\backslash G_{\mathbb{A}}$, and $Z^+ G_k\backslash G_{\mathbb{A}}$ such that the corresponding *unwindings* are correct: for example, for every $f \in C_c^o(Z^+ G_{\mathbb{A}})$,

$$\int_{Z^+G_{\mathbb{A}}} f = \int_{Z^+G_k\backslash G_{\mathbb{A}}} \left(\sum_{\gamma \in G_k} f \circ \gamma \right)$$

Comparisons between global integrals and products of local integrals are as expected: for $f(g) = \prod_v f_v(g_v)$ in $C_c^o(G_{\mathbb{A}})$ expressible as a product of functions $f_v \in C_c^o(G_v)$, up to a scalar depending on all the normalizations,

$$\int_{G_{\mathbb{A}}} f = \prod_v \int_{G_v} f_v$$

despite the fact that the adele group $G_{\mathbb{A}}$ is not the product of the local groups G_v, but only the colimit of the products $G_S = \prod_{v \in S} G_v \times \prod_{v \notin S} K_v$. Indeed, in practice f_v will be K_v-invariant for all but finitely many v, so

$$\int_{G_{\mathbb{A}}} f = \lim_S \int_{G_S} f = \lim_S \left(\prod_{v \in S} \int_{G_v} f_v \cdot \prod_{v \notin S} \int_{K_v} f_v \right)$$
$$= \lim_S \left(\prod_{v \in S} \int_{G_v} f_v \cdot 1 \right) = \prod_{v \in S} \int_{G_v} f_v$$

Nevertheless, on some occasions, explicit computations are useful or necessary. Measures and integrals on $\mathbb{R}$ and $\mathbb{C}$ are familiar. On $\mathbb{Q}_p$ and its finite extensions, somewhat less so. However, the totally disconnected nature of $\mathbb{Q}_p$ and finite extensions makes measure and integration simpler, at least for nice functions. We treat $\mathbb{Q}_p$, and every non-archimedean k_v is a finite cartesian product of such. We do not need to prove *uniqueness*, since this follows for general reasons [14.4].

To *give* a (regular, Borel) measure it suffices to tell the measure of every *open*. Since $\mathbb{Q}_p$ is a group whose group operation and inversion are continuous, for an *invariant* measure it suffices to tell the measure of a local basis at 0 because every translate (coset!) of a given basis element must have the same measure. Such a basis is $p^\ell\mathbb{Z}_p$. Since these are sub*groups*, we can easily compare them: for $1 \le \ell < \ell'$ the subgroup $p^{\ell'}\mathbb{Z}_p$ is of index $p^{\ell'-\ell}$ in $p^\ell\mathbb{Z}_p$. The ratio of measures *must be* the index. Thus, normalizing everything by taking the measure of $\mathbb{Z}_p$ to be 1, the measure of $p^\ell\mathbb{Z}_p$ is its index in $\mathbb{Z}_p$, namely, $p^{-\ell}$. *Larger* opens are unions of translates of sets $p^\ell\mathbb{Z}_p$. This gives the standard additive Haar measure on $\mathbb{Q}_p$ for $p < \infty$.

On finite extensions k_v of $\mathbb{Q}_p$, the same process produces an additive Haar measure giving $\mathfrak{o}_v$ total measure 1. For $k_v/\mathbb{Q}_p$ *unramified*, this is almost always a good normalization. However, for $k_v/\mathbb{Q}_p$ *ramified*, other choices may have advantages, for example, with respect to local Fourier transforms.

A *multiplicative* Haar measure $d^\times x$ on $\mathbb{Q}_p^\times$ can be arranged from the additive d^+x, much as for $\mathbb{R}^\times$ or $\mathbb{C}^\times$, namely, $d^\times x = d^+x/|x|_v$. However, this gives the local units $\mathbb{Z}_p^\times$ measure $\frac{p-1}{p}$, not 1. Since $\mathbb{Z}_p^\times$ is the *unique maximal compact subgroup* of $\mathbb{Q}_p^\times$, we might prefer to give the local units measure 1. A similar device applies to k_v for $v < \infty$. In practice, the superscripts are not used because context explains and determines which measure is meant.

Since $N_v \approx k_v$, the invariant measure on N_v is just the additive Haar measure from k_v. Since $M_v \approx k_v^\times \times k_v^\times$, a product of multiplicative Haar measures is the invariant measure.

Much as in the archimedean cases considered earlier, a left-invariant measure on $P_v = N_vM_v$ is $d(nm) = dn\,dm/|\alpha(m)|_v$, where $\alpha\begin{pmatrix} m_1 & 0 \\ 0 & m_2 \end{pmatrix} = m_1/m_2$. That is,

$$d(\begin{pmatrix} 1 & x \\ 0 & 1 \end{pmatrix}\begin{pmatrix} m_1 & 0 \\ 0 & m_2 \end{pmatrix}) = \frac{d^+x d^\times m_1 d^\times m_2}{|m_1/m_2|_v} = \frac{d^+x d^+m_1 d^+m_2}{|m_1|_v^2}$$

[2.3.1] Claim: Quotients $Z^+M_k\backslash\mathfrak{S}_{C,t}$ of Siegel sets have finite volume.

Proof: The notation has compact $C \subset N_\mathbb{A}$ and $t > 0$. Letting $K_\mathbb{A} = \prod_{v\le\infty} K_v$, up to normalization,

$$\begin{aligned}\int_{Z^+M_k\backslash\mathfrak{S}_{C,t}} 1\,dg &= \int_C 1\,dn \cdot \int_{Z^+M_k\backslash M_\mathbb{A}} 1\frac{dm}{|\alpha(m)|} \cdot \int_K 1\,dk \\ &\asymp \int_{Z^+M_k\backslash M_\mathbb{A}} 1\frac{dm}{|\alpha(m)|}\end{aligned}$$

Further, $M_k\backslash M_{\mathbb{A}} \approx (k^\times\backslash\mathbb{J}) \times (k^\times\backslash\mathbb{J})$, and the integrand is $M_{\mathbb{A}} \cap K_{\mathbb{A}}$-invariant. By [2.A], the group $k^\times\backslash\mathbb{J}$ has *compact* subgroup $k^\times\backslash\mathbb{J}^1$, on which $|\alpha(m)|$ is trivial, and $k^\times\backslash\mathbb{J} \approx \delta(0,\infty) \times k^\times\backslash\mathbb{J}^1$. For brevity, write $\mathbb{R}^+ = \delta(0,\infty)$. In effect, Z^+ is the diagonal copy of $\mathbb{R}^+$ in $\mathbb{J} \times \mathbb{J}$. Thus,

$$Z^+M_k\backslash M_{\mathbb{A}} \approx Z^+\backslash\big((\mathbb{R}^+ \times k^\times\backslash\mathbb{J}^1) \times (\mathbb{R}^+ \times k^\times\backslash\mathbb{J}^1)\big)$$

so, the further quotient by the kernel of $m \to |\alpha(m)|$ has representatives $a_y = \begin{pmatrix} \delta(y) & 0 \\ 0 & 1 \end{pmatrix}$ for $y > 0$. We have $|\alpha(a_y)| = y^{[k:\mathbb{Q}]}$ for $y > 0$. Thus, up to normalization, the integral is

$$\int_{y \geq t} 1 \frac{dy/y}{y^{[k:\mathbb{Q}]}} < \infty$$

The quotient $M_k\backslash\mathfrak{S}_{C,t}$ without that further quotient by Z^+ will *not* have finite volume, because $\mathbb{J}/k^\times$ is noncompact. ///

Thus,

[2.3.2] Corollary: The quotient $Z^+G_k\backslash G_{\mathbb{A}}$ has finite volume. ///

2.4 Hecke Operators, Integral Operators

The simplest non-archimedean analogues of the differential operators on G_v for archimedean v are *integral operators* of the form

$$\varphi \cdot f = \int_{G_v} \varphi(g) g \cdot f \, dg \qquad (\text{for } \varphi \in C_c^o(G_v) \text{ and } f \in V)$$

for any continuous action $G_v \times V \to V$ on a quasi-complete, locally convex topological vectorspace V. The integrand is a continuous, compactly supported V-valued function and so has a Gelfand-Pettis integral [14.1]. Thus, for $f \in V = L^2(Z^+G_k\backslash G_{\mathbb{A}})$, with G_v acting by right translation, pointwise we have

$$(\varphi \cdot f)(x) = \int_{G_v} \varphi(g)(g \cdot f)(x) \, dg = \int_{G_v} \varphi(g) f(xg) \, dg$$

for $\varphi \in C_c^o(G_v)$, at least almost everywhere. Better, for general reasons [6.1], the right-translation action $G_v \times L^2(Z^+G_k\backslash G_{\mathbb{A}}) \to L^2(Z^+G_k\backslash G_{\mathbb{A}})$ is continuous, so the integral converges as an $L^2(Z^+G_k\backslash G_{\mathbb{A}})$-valued integral, and concern about pointwise values is unnecessary. The *composition* of two such operators is readily described as the operator attached to the *convolution*:

for $\varphi, \psi \in C_c^o(G_{\mathbb{A}})$,

$$\varphi \cdot (\psi \cdot f) = \int_{G_{\mathbb{A}}} \varphi(g) g \cdot \left(\int_{G_{\mathbb{A}}} \psi(h) h \cdot f dh \right) dg$$
$$= \int_{G_{\mathbb{A}}} \int_{G_{\mathbb{A}}} \varphi(g) \psi(h)(gh \cdot f) \, dh \, dg$$

because the operation of φ moves inside the Gelfand-Pettis integral. Replacing h by $g^{-1}h$ gives

$$\int_{G_{\mathbb{A}}} \int_{G_{\mathbb{A}}} \varphi(g) \psi(g^{-1}h) h \cdot f \, dh \, dg$$
$$= \int_{G_{\mathbb{A}}} \left(\int_{G_{\mathbb{A}}} \varphi(g) \psi(g^{-1}h) \, dg \right) h \cdot f \, dh$$

by changing the order of integration. The inner integral is one expression for the convolution $\varphi * \psi$.

[2.4.1] Lemma: The *adjoint* to the action of $\varphi \in C_c^o(G_{\mathbb{A}})$ on $L^2(Z^+G_k \backslash G_{\mathbb{A}})$ is given by the action of $\check{\varphi} \in C_c^o(G_{\mathbb{A}})$, where $\check{\varphi}(g) = \overline{\varphi(g^{-1})}$.

Proof: This is a direct computation: for $f, F \in L^2(Z^+G_k \backslash G_{\mathbb{A}})$, by properties of Gelfand-Pettis integrals,

$$\langle \varphi \cdot f, F \rangle = \left\langle \int_{G_{\mathbb{A}}} \varphi(g) g \cdot f \, dg, F \right\rangle = \int_{G_{\mathbb{A}}} \varphi(g) \langle g \cdot f, F \rangle \, dg$$
$$= \int_{G_{\mathbb{A}}} \varphi(g) \langle f, g^{-1} \cdot F \rangle \, dg$$

because the right translation action of $G_{\mathbb{A}}$ is *unitary*:

$$\langle g \cdot f, \, F \rangle = \int_{Z^+G_k \backslash G_{\mathbb{A}}} f(xg) \overline{F(x)} \, dx$$
$$= \int_{Z^+G_k \backslash G_{\mathbb{A}}} f(x) \overline{F(xg^{-1})} \, dx = \langle f, \, g^{-1} \cdot F \rangle$$

by changing variables. This gives

$$\langle \varphi \cdot f, F \rangle = \left\langle f, \int_{G_{\mathbb{A}}} \overline{\varphi(g)} g^{-1} \cdot F dg \right\rangle$$
$$= \left\langle f, \int_{G_{\mathbb{A}}} \overline{\varphi(g^{-1})} g \cdot F dg \right\rangle = \langle f, \, \check{\varphi} \cdot F \rangle$$

by replacing g by g^{-1}. ///

In the four earlier purely archimedean examples, we only considered automorphic forms *invariant* under right translation by the standard compact subgroups. It is reasonable to consider comparable requirements here, for simplicity possibly requiring right K_v-invariance for all places v. It is also reasonable to relax this condition to requiring right K_v-invariance *almost everywhere*, that is, at all but finitely many places.

A somewhat relaxed version of $K_{\mathbb{A}}$-invariance, to cope with the finitely many places where right K_v-invariance is not required, is *K-finiteness* of a function f on $G_{\mathbb{A}}$ or $Z^+G_k\backslash G_{\mathbb{A}}$ or other quotients of $G_{\mathbb{A}}$, namely, the requirement that the vectorspace of functions spanned by $\{x \to f(xh) : h \in K_{\mathbb{A}}\}$ is *finite-dimensional.* At the extreme of $K_{\mathbb{A}}$-invariant f, this space is *one*-dimensional.[3]

[2.4.2] Lemma: For v non-archimedean, K_v-finiteness is equivalent to *invariance* under some finite-index subgroup $K' \subset K_v$.

Proof: Let f be in a topological vectorspace V on which G_v acts continuously. For f invariant under K', the collection of translates of f under K_v is *finite*, given (with possible redundancy) by $g \cdot f$ for representatives g for K_v/K'. On the other hand, when the collection of all right translates of f by K_v is a finite-dimensional (complex) vectorspace $F \subset V$, the map $K_v \to \mathrm{Aut}_{\mathbb{C}}(F)$ is a continuous group homomorphism ρ to some $GL_n(\mathbb{C})$. Given a neighborhood U of $1 \in GL_n(\mathbb{C})$, there is a small-enough neighborhood U' of $1 \in K_v$ such that $\rho(U')$ is inside U. In fact, we can take U' to be a *subgroup*, for example, $\{g = 1_2 \bmod \varpi_v^n\}$ for varying n. Then $\rho(U')$ is a *subgroup* of $GL_n(\mathbb{C})$ inside U. Granting for a moment the *no small subgroups* property of real or complex Lie groups, that a sufficiently small neighborhood of 1 contains no subgroups except $\{1\}$, it must be that $\rho(U') = \{1\}$. Since K_v is compact and U' is open, the cover of K_v by cosets of U' has a finite subcover, so U' is of finite index in K_v. The proof is complete upon proof of the no small subgroups property, following. ///

[2.4.3] Claim: $GL_n(\mathbb{C})$ has the *no small subgroup* property, that a sufficiently small neighborhood of 1 contains no subgroup larger than $\{1\}$.

Proof: For an n-by-n complex matrix x, let $\|x\|$ be the *operator norm*

$$\|x\| = \sup_{v \in \mathbb{C}^n : |v| \le 1} |x \cdot v|$$

[3] In the simplest example, Fourier series on the circle $\mathbb{T}$, *smoothness* is equivalent to rapid decay of Fourier coefficients, while $\mathbb{T}$-finiteness is equivalent to having only finitely many nonzero Fourier coefficients.

where $|(v_1, \ldots, v_n)| = \sqrt{|v_1|^2 + \ldots + |v_n|^2}$. With $r > 0$ small enough so that $\sum_{\ell \geq 2} r^\ell/\ell! < 1 + r$, the matrix exponential $x \to e^x$ is a *bijection* from $E_r = \{x : \|x\| < r\}$ to a neighborhood of $1 \in GL_n(\mathbb{C})$. We claim that $U = \{e^x : x \in E_{r/2}\}$ contains no subgroup other than $\{1\}$. Given $0 \neq x \in E_{r/2}$, there is $1 \leq \ell \in \mathbb{Z}$ such that $\ell \cdot x \in E_{r/2}$ but $(\ell+1) \cdot x \notin E_{r/2}$. Then $\ell \cdot x \in E_{r/2}$, but $(\ell + 1) \cdot x \notin E_{r/2}$. Still, $(\ell+1) \cdot x \in E_r$, so by the injectivity of the exponential on E_r, $e^x \notin U$. ///

Unsurprisingly, it turns out that K-finite functions on $Z^+G_k \backslash G_\mathbb{A}$ are better behaved than arbitrary functions. Of course, most $f \in L^2(Z^+G_k\backslash G_\mathbb{A})$ are *not* K-finite.

For non-archimedean v, the *spherical Hecke operators* for G_v are the integral operators given by left-and-right K_v-invariant $\varphi \in C_c^o(G_v)$, also denoted $C_c^o(K_v\backslash G_v/K_v)$. Since K_v is *open*, such functions are *locally constant*: given $x \in G_v$, $\varphi(xh) = \varphi(x)$ for all $h \in K_v$, but xK_v is a neighborhood of x. Then the compact support implies that such φ takes only finitely many distinct values. Thus, the associated integral operator is really a *finite sum*. Nevertheless, expression as integral operators seems to explain the behavior well.

[2.4.4] Claim: The action of spherical Hecke operators attached to $\varphi \in G_v$ stabilizes K_v-invariant vectors f in any continuous group action $G_v \times V \to V$ for quasi-complete, locally convex V.

Proof: Granting properties of Gelfand-Pettis integrals, this is a direct computation: for $f \in V$ and $h \in K_v$,

$$h \cdot (\varphi \cdot f) = h \cdot \int_{G_v} \varphi(g) g \cdot f \, dg = \int_{G_v} h \cdot (\varphi(g) g \cdot f) \, dg$$
$$= \int_{G_v} \varphi(g) hg \cdot f \, dg = \int_{G_v} \varphi(h^{-1}g) g \cdot f \, dg$$

by replacing g by $h^{-1}g$. Since φ is left K_v-invariant, this is just $\varphi \cdot f$ again. ///

[2.4.5] Claim: For v archimedean or non-archimedean, the *spherical Hecke algebra* $C_c^o(K_v\backslash G_v/K_v)$ with convolution is *commutative*.

Proof: Gelfand's trick is to find an involutive anti-automorphism σ of G_v, that is, $g \to g^\sigma$ such that $(gh)^\sigma = h^\sigma g^\sigma$ and $(g^\sigma)^\sigma = g$, stabilizing double cosets for K_v, that is, using the Cartan decomposition [2.1], such that $(K_v m K_v)^\sigma = K_v m K_v$ for all $m \in M_v$. Here, transpose $g^\sigma = g^\top$ is such an anti-automorphism, since we have a Cartan decomposition $G_v = K_v M_v K_v$, K_v is stabilized by transpose, and the diagonal subgroup M_v of G_v is fixed pointwise by transpose. Then for

φ in the spherical Hecke algebra, with $g = k_1 m k_2$ in Cartan decomposition,

$$\varphi(g^\sigma) = \varphi(k_2^\sigma m^\sigma k_1^\sigma) = \varphi(m^\sigma) = \varphi(m) = \varphi(k_1 m k_2) = \varphi(g)$$

Then the commutativity is a direct computation:

$$\begin{aligned}(\varphi * \psi)(x) &= (\varphi * \psi)(x^\sigma)\\ &= \int_{G_v} \varphi(g)\,\psi(g^{-1}x^\sigma)\,dg = \int_{G_v} \varphi(g^\sigma)\,\psi((g^{-1}x^\sigma)^\sigma)\,dg\\ &= \int_{G_v} \varphi(g^\sigma)\,\psi(x(g^\sigma)^{-1})\,dg = \int_{G_v} \varphi(g)\,\psi(xg^{-1})\,dg\end{aligned}$$

by replacing g by g^σ. Then replace g by $g^{-1}x$, and then by g^{-1}, to obtain

$$\begin{aligned}(\varphi * \psi)(x) &= \int_{G_v} \varphi(gx)\,\psi(g^{-1})\,dg\\ &= \int_{G_v} \varphi(g^{-1}x)\,\psi(g)\,dg = (\psi * \varphi)(x)\end{aligned}$$

as claimed. ///

It is easy to see that the spherical Hecke algebra is stable under adjoints. Thus, it is plausible to ask for *simultaneous eigenvectors* for the spherical Hecke algebra. That is, for $f \in L^2(Z^+G_k\backslash G_\mathbb{A})$, we might additionally try to require that f be a spherical Hecke eigenfunction at almost all non-archimedean v, and be an eigenfunction for invariant Laplacians or Casimir at archimedean places. However, in infinite-dimensional Hilbert spaces, there is no general promise of existence of such simultaneous eigenvectors.

2.5 Decomposition by Central Characters

We have seen that $Z^+G_k\backslash G_\mathbb{A}$ has finite invariant volume, while $G_k\backslash G_\mathbb{A}$ does not. The further quotient $Z_\mathbb{A}G_k\backslash G_\mathbb{A}$ certainly has finite invariant volume.

Functions on $Z_\mathbb{A}G_k\backslash G_\mathbb{A}$ are automorphic forms (or automorphic functions) with *trivial central character*, since they are invariant under the center $Z_\mathbb{A}$ of $G_\mathbb{A}$. Such automorphic forms give a reasonable class to consider, but we can treat a larger class with little further effort. Namely, the *compact* abelian group $Z_\mathbb{A}/Z^+Z_k \approx \mathbb{J}^1/k^\times$, being a quotient of the center $Z_\mathbb{A}$ of $G_\mathbb{A}$, acts on functions on $Z_\mathbb{A}G_k\backslash G_\mathbb{A}$ in a fashion that commutes with right translation by $G_\mathbb{A}$. In particular, the action of $Z_\mathbb{A}/Z^+Z_k$ commutes with the integral operators on G_v for $v < \infty$, and with the Casimir or Laplacians on G_v at archimedean places.

Thus, for each *central character* ω of $Z_\mathbb{A}/Z^+Z_k$, we can consider the space $L^2(Z^+G_k\backslash G_\mathbb{A}, \omega)$ of all left Z^+G_k-invariant f on $G_\mathbb{A}$ such that $|f| \in L^2(Z_\mathbb{A}G_k\backslash G_\mathbb{A})$ and $f(zg) = \omega(a) \cdot f(g)$ for all $z \in Z_\mathbb{A}$.

[2.5.1] Claim: $L^2(Z^+G_k\backslash G_{\mathbb{A}})$ *decomposes by central characters*:

$$L^2(Z^+G_k\backslash G_{\mathbb{A}}) = \text{completion of} \bigoplus_\omega L^2(Z^+G_k\backslash G_{\mathbb{A}}, \omega)$$

Proof: The argument applies to any compact abelian group A acting on a Hilbert space V by *unitary* operators, meaning $\langle a \cdot v, a \cdot w\rangle = \langle v, w\rangle$ for all $a \in A$ and $v, w \in V$. For a character ω of A, let V_ω be the ω-eigenspace:

$$V_\omega = \{v \in V : a \cdot v = \omega(a) \cdot v, \text{ for all } a \in A\}$$

For $\omega \neq \omega'$, V_ω and $V_{\omega'}$ are orthogonal: with $a \in A$ such that $\omega(a) \neq \omega'(a)$ and $v \in V_\omega$, $v' \in V_{\omega'}$,

$$\langle v, v'\rangle = \frac{1}{\omega(a)}\langle a \cdot v, v'\rangle = \frac{1}{\omega(a)}\langle v, a^{-1}v'\rangle = \frac{1}{\omega(a)}\langle v, \omega'(a^{-1})v'\rangle$$
$$= \frac{\overline{\omega'(a^{-1})}}{\omega(a)}\langle v, \omega'(a^{-1})v'\rangle = \frac{\omega'(a)}{\omega(a)}\langle v, v'\rangle$$

giving orthogonality.

Give A an invariant measure with total measure 1. First, $\int_A \omega(a)^{-1}\, a \cdot v da$ exists as a Gelfand-Pettis V-valued integral, so maps $V \to V$ continuously, and in fact maps to V_ω: using the commutativity of the integral with continuous maps, for $b \in A$,

$$b \cdot \int_A \omega(a)^{-1}\, a \cdot v\, da = \int_A \omega(a)^{-1}\, ba \cdot v\, da = \int_A \omega(b^{-1}a)^{-1}\, a \cdot v\, da$$
$$= \omega(b) \cdot \int_A \omega(a)^{-1}\, a \cdot v\, da$$

Take $v \neq 0$ in V. The scalar-valued function $a \to \langle a \cdot v, v\rangle$ is continuous on A, and, since $\langle 1 \cdot v, v\rangle = |v|^2 \neq 0$, is not identically 0. By [6.11], $L^2(A)$ is the completion of the direct sum of the one-dimensional spaces of functions $\mathbb{C} \cdot \omega$ as ω ranges over characters. Thus, in $L^2(A)$,

$$0 \neq \langle av, v\rangle = \sum_\omega \int_A \omega(b)^{-1}\langle bv, v\rangle\, db \cdot \omega(a)$$
$$= \sum_\omega \left\langle \int_A \omega(b)^{-1} bv\, da, v\right\rangle \cdot \omega(a)$$

Thus, not all the coefficients on the right-hand side can be 0, so the projection of nonzero $v \in V$ to *some* V_ω must be nonzero. Thus, the completion of the sum of the V_ω is all of V. ///

2.6 Discrete Decomposition of Cuspforms

Automorphic forms or *automorphic functions* are functions of various sorts on $G_k\backslash G_\mathbb{A}$, with $G_k = GL_2(k)$, $G_\mathbb{A} = GL_2(\mathbb{A})$. Here, because $G_k\backslash G_\mathbb{A}$ has infinite volume, it is reasonable to look at the further quotient $Z^+G_k\backslash G_\mathbb{A}$, for example. Naturally $L^2(Z^+G_k\backslash G_\mathbb{A})$ is the space of square-integrable automorphic forms. The *constant term* of an automorphic form f is

$$c_P f(g) = \int_{N_k\backslash N_\mathbb{A}} f(ng)\, dn$$

[2.6.1] Claim: Constant terms are functions on $Z^+N_\mathbb{A}M_k\backslash G_\mathbb{A}$.

Proof: By changing variables, we can see that $g \to c_P f(g)$ is a left $N_\mathbb{A}$-invariant function on $G_\mathbb{A}$:

$$\begin{aligned} c_P f(n'x) &= \int_{N_k\backslash N_\mathbb{A}} f(n\cdot n'x)\, dn = \int_{N_k\backslash N_\mathbb{A}} f((nn')\cdot x)\, dn \\ &= \int_{N_k\backslash N_\mathbb{A}} f(n\cdot x)\, dn \qquad (\text{for } n' \in N_\mathbb{A}) \end{aligned}$$

Similarly, for $m \in M_k$,

$$\begin{aligned} c_P f(mx) &= \int_{N_k\backslash N_\mathbb{A}} f(n\cdot mx)\, dn \\ &= \int_{N_k\backslash N_\mathbb{A}} f(m\cdot m^{-1}nm\cdot x)\, dn = \int_{N_k\backslash N_\mathbb{A}} f(m^{-1}nm\cdot x)\, dn \end{aligned}$$

since f itself is left M_k-invariant. Then replacing n by mnm^{-1} gives the expression for $c_P f(g)$, noting that conjugation by $m \in M_k$ stabilizes N_k, and by the product formula the change of measure on $N_\mathbb{A}$ is trivial. Invariance under Z^+ is even easier. ///

A *cuspform* is a function f on $Z^+G_k\backslash G_\mathbb{A}$ meeting Gelfand's condition[4] $c_P f = 0$. When f is merely measurable, so does not have well-defined

[4] In fact, the Gelfand condition for f on $G_k\backslash G_\mathbb{A}$ to be a cuspform is that $\int_{N_k^Q N_\mathbb{A}^Q} f(ng)\, dn = 0$ as a function of $g \in G_\mathbb{A}$ for the unipotent radical N^Q of *every* parabolic Q. For $GL_2(k)$, proper parabolic subgroups can be characterized as stabilizers of lines in k^2, and their unipotent radicals as pointwise-fixers of lines. Since $GL_2(k)$ is transitive on lines, all proper parabolics (and their unipotent radicals) are *conjugate*. Thus, vanishing of *one* constant term (as a function on $G_\mathbb{A}$) implies vanishing of *every* constant term, by a change of variables in the integral: for $h \in GL_2(k)$,

$$\begin{aligned} \int_{hN_kh^{-1}\backslash hN_\mathbb{A}h^{-1}} f(ng)\, dn &= \int_{N_k\backslash N_\mathbb{A}} f(hnh^{-1}g)\, dn \\ &= \int_{N_k\backslash N_\mathbb{A}} f(n\cdot h^{-1}g)\, dn = 0 \end{aligned}$$

using the left $GL_2(k)$-invariance. Thus, vanishing of the constant term along N implies vanishing along every conjugate of N.

pointwise values everywhere, this condition is best interpreted *distributionally*, as is clarified in the next section, using pseudo-Eisenstein series. The space of square-integrable cuspforms is

$$L^2_o(Z^+G_k\backslash G_{\mathbb{A}}) = \{f \in L^2(Z^+G_k\backslash G_{\mathbb{A}}) : c_P f = 0\}$$

The fundamental theorem proven in [7.1-7.7] is the *discrete decomposition of spaces of cuspforms.* A simple version addresses the space $L^2_o(Z^+G_k\backslash G_{\mathbb{A}}/K_{\mathbb{A}}, \omega)$ of right-$K_{\mathbb{A}}$-invariant square-integrable cuspforms with central character ω, where $K_{\mathbb{A}} = \prod_{v\leq\infty} K_v$. This space is $\{0\}$ unless ω is *unramified*, that is, is trivial on $Z_{\mathbb{A}} \cap K_{\mathbb{A}}$, since $K_{\mathbb{A}}$-invariance implies $Z_{\mathbb{A}} \cap K_{\mathbb{A}}$-invariance, and we also require $Z_{\mathbb{A}}$, ω-equivariance.

Since the spherical Hecke algebras act by *right* translation, and the Gelfand condition is an integral on the *left*, spaces of cuspforms are *stable* under all these integral operators. It is less clear a priori how they behave with respect to the invariant Laplacians [4.2].

[2.6.2] Theorem: $L^2_o(Z^+G_k\backslash G_{\mathbb{A}}/K_{\mathbb{A}}, \omega)$ has an orthonormal basis of simultaneous eigenfunctions for invariant Laplacians Δ_v at archimedean places, and for spherical Hecke algebras $C^o_c(K_v\backslash G_v/K_v)$ at non-archimedean places. Each simultaneous eigenspace occurs with *finite multiplicity*, that is, is finite-dimensional. *(Proof in [7.1–7.7].)*

In contrast, the full spaces $L^2(Z^+G_k\backslash G_{\mathbb{A}}/K_{\mathbb{A}}, \omega)$ do *not* have bases of simultaneous L^2-eigenfunctions: as in [2.11–2.12], the orthogonal complement of cuspforms in $L^2(Z^+G_k\backslash G_{\mathbb{A}}/K_{\mathbb{A}}, \omega)$ mostly consists of *integrals* of *non-L^2* eigenfunctions for the Laplacians and Hecke operators, the *Eisenstein series* E_s introduced in [2.8].

For spaces of automorphic forms more complicated than being right K_v-invariant for *every* place v, there is generally no decomposition in terms of simultaneous eigenspaces for *commuting* operators. The decomposition argument in [7.7] directly uses the *non-commutative* algebras $C^\infty_c(G_v)$ of test functions on the groups G_v: for v archimedean, $C^\infty_c(G_v)$ consists of the usual smooth, compactly supported functions, and for v non-archimedean, $C^\infty_c(G_v)$ is compactly supported locally constant functions. Both cases are called *smooth.* Letting right translation be $R_g f(x) = f(xg)$ for $x, g \in G_{\mathbb{A}}$, the action of $\varphi \in C^\infty_c(G_v)$ on functions f on $G_k\backslash G_{\mathbb{A}}$ is

$$\varphi \cdot f = \int_{G_v} \varphi(g)\, R_g f\, dg$$

This makes sense not just as a pointwise-value integral, but as a Gelfand-Pettis integral when f lies in any quasi-complete, locally convex topological vectorspace V on which G_v acts so that $G_v \times V \to V$ is continuous. Such V is a *representation* of G_v. The multiplication in $C_c^\infty(G_v)$ compatible with such actions is *convolution*: associativity $\varphi \cdot (\psi \cdot f) = (\varphi * \psi) \cdot f$.

Here, we are mostly interested in actions $G_v \times X \to X$ on *Hilbert-spaces* X. Such a representation is (topologically) *irreducible* when X has no closed, G_v-stable subspace. The convolution algebras $C_c^\infty(G_v)$ are *not commutative*, so, unlike the commutative case, few irreducible representations are one-dimensional. In fact, *typical* irreducible representations of $C_c^\infty(G_v)$ turn out to be *infinite-dimensional*. Fortunately, there is no mandate to attempt to *classify* these irreducibles. Indeed, the spectral theory of compact self-adjoint operators still proves [7.7] discrete decomposition with finite multiplicities, for example, as follows.

For every place v, let K'_v be a compact subgroup of G_v, and for all but a finite set S of places require that $K'_v = K_v$, the standard compact subgroup. For simplicity, we still assume $K'_v = K_v$ at archimedean places. Put $K' = \prod_v K'_v$. Let ω be a central character trivial on $Z_\mathbb{A} \cap K'$, so that the space $L^2_o(Z^+G_k\backslash G_\mathbb{A}/K', \omega)$ of right K'-invariant cuspforms with central character ω is not $\{0\}$ for trivial reasons. For $v \in S$, we have a *subalgebra* $C_v^\infty(K'_v\backslash G_v/K'_v)$ of the convolution algebra of test functions at v, stabilizing $L^2_o(Z^+G_k\backslash G_\mathbb{A}/K', \omega)$.

[2.6.3] Theorem: $L^2_o(Z^+G_k\backslash G_\mathbb{A}/K', \omega)$ is the completion of the orthogonal direct sum of subspaces, each consisting of simultaneous eigenfunctions for invariant Laplacians Δ_v at archimedean places, of simultaneous eigenfunctions for spherical Hecke algebras $C_c^o(K_v\backslash G_v/K_v)$ at non-archimedean places $v \notin S$, and irreducible $C_v^\infty(K'_v\backslash G_v/K'_v)$-representations at $v \in S$. Each occurs with finite multiplicity. *(Proof in [7.1–7.7].)*

The technical features of decomposition with respect to non-commutative rings of operators certainly bear amplification, postponed to [7.2.18] and [7.7]. The notion of *multiplicity* is made precise in [9.D.14]. In anticipation,

[2.6.4] Theorem: $L^2_o(Z^+G_k\backslash G_\mathbb{A}/K_\mathbb{A}, \omega)$ is the completion of the orthogonal direct sum of irreducibles V for the simultaneous action of all algebras $C_c^\infty(G_v)$. Each irreducible occurs with finite multiplicity. *(Proof in [7.7].)*

[2.6.5] Corollary: $L^2_o(Z^+G_k\backslash G_\mathbb{A}, \omega)$ is the completion of the orthogonal direct sum of subspaces, each consisting of simultaneous eigenfunctions for invariant Laplacians Δ_v at archimedean places, of simultaneous eigenfunctions for spherical Hecke algebras $C_c^o(K_v\backslash G_v/K_v)$ at non-archimedean places $v \notin S$, and irreducible $C_v^\infty(K'_v\backslash G_v/K'_v)$-representations at $v \in S$. Each occurs with finite multiplicity. ///

Again, the various sorts of orthogonal complements to spaces of cuspforms are mostly *not* direct sums of irreducibles but are *integrals* of Eisenstein series, as we see subsequently.

2.7 Pseudo-Eisenstein Series

Returning to the larger spaces $L^2(Z^+G_k\backslash G_{\mathbb{A}}/K_{\mathbb{A}})$ or $L^2(Z^+G_k\backslash G_{\mathbb{A}}/K_{\mathbb{A}}, \omega)$ or $L^2(Z^+G_k\backslash G_{\mathbb{A}}/K', \omega)$, we want to express the orthogonal complement of cuspforms in terms of simultaneous eigenfunctions for invariant Laplacians at archimedean places and for spherical Hecke algebras at finite places when possible. To consider larger, noncommutative algebras of operators, the more complicated notion of *irreducible representation* must replace the notion of *simultaneous eigenvector*. Therefore, we emphasize the commutating operators. As it happens, the *pseudo-Eisenstein series* here and the genuine Eisenstein series in the next section avoid some of the subtleties that cuspforms may require.

To exhibit explicit L^2 functions demonstrably spanning the orthogonal complement to cuspforms, we will recast the Gelfand condition that the constant term vanish as a requirement of *vanishing as a distribution* on $Z^+N_{\mathbb{A}}M_k\backslash G_{\mathbb{A}}$ and give an equivalent distributional vanishing condition on $Z^+G_k\backslash G_{\mathbb{A}}$.

Vanishing as a distribution is that

$$\int_{Z^+N_{\mathbb{A}}M_k\backslash G_{\mathbb{A}}} \varphi \cdot c_P f = 0 \qquad (\text{for all } \varphi \in C_c^\infty(Z^+N_{\mathbb{A}}M_k\backslash G_{\mathbb{A}}))$$

where $C_c^\infty(Z^+N_{\mathbb{A}}M_k\backslash G_{\mathbb{A}})$ consists of compactly supported functions on that quotient which are *smooth* in the archimedean coordinates and locally constant in the nonarchimedean coordinates. *Smoothness* of such φ can be described more precisely in a fashion that makes clearer the noninteraction of this property with taking a quotient on the left. Namely, smoothness for archimedean places should mean indefinite differentiability *on the right* with respect to the differential operators coming from the Lie algebra, as in [4.1], and, given the compact support, (uniform) smoothness for non-archimedean places should mean that there exists a compact, open subgroup K' of $\prod_{v<\infty} K_v$ under which φ is right invariant.

As mentioned briefly in the previous section, the nature of $c_P f$ for f merely L^2 is potentially obscure. For example, it is not likely that $c_P f \in L^2(Z^+N_{\mathbb{A}}M_k\backslash G_{\mathbb{A}})$. Instead, for general reasons [6.1], $C_c^o(Z^+G_k\backslash G_{\mathbb{A}})$ is dense in $L^2(Z^+G_k\backslash G_{\mathbb{A}})$ in the L^2 topology, and for general reasons [6.1] the *left* action of $N_k\backslash N_{\mathbb{A}}$ on the Fréchet space $C^o(Z^+N_k\backslash G_{\mathbb{A}})$ is a continuous map $(N_k\backslash N_{\mathbb{A}}) \times C^o(Z^+N_k\backslash G_{\mathbb{A}}) \to C^o(Z^+N_{\mathbb{A}}\backslash G_{\mathbb{A}})$, so $c_P f$ exists as a $C^o(Z^+N_{\mathbb{A}}\backslash G_{\mathbb{A}})$-valued Gelfand-Pettis integral [14.1]. Then one sees directly that $c_P f$ is left

M_k-invariant. For such f, the integral of $c_P f$ against $\varphi \in C_c^\infty(Z^+N_\mathbb{A}M_k\backslash G_\mathbb{A})$ is the integral of a compactly supported, continuous function. There is no immediate necessity of elaborating a general notion of *distribution* on p-adic groups or adele groups, since cuspforms are ordinary functions, essentially having pointwise values.

For $\varphi \in C_c^\infty(Z^+N_\mathbb{A}M_k\backslash G_\mathbb{A})$, the corresponding *pseudo-Eisenstein series* is

$$\Psi_\varphi(g) = \sum_{\gamma \in P_k\backslash G_k} \varphi(\gamma \cdot g)$$

Convergence is good:

[2.7.1] Claim: The series for a pseudo-Eisenstein series Ψ_φ is *locally finite*, meaning that for g in a fixed compact in $G_\mathbb{A}$, there are only finitely many *nonzero* summands in $\Psi_\varphi(g) = \sum_\gamma \varphi(\gamma g)$. Thus, $\Psi_\varphi \in C_c^\infty(Z^+G_k\backslash G_\mathbb{A})$.

Proof: Grant for a moment that there is compact $C \subset G_\mathbb{A}$ such that the image of C in the quotient contains the (compact) support of φ. Fix compact $C_o \subset G$ in which g is constrained to lie. A summand $\varphi(\gamma g)$ is nonzero only if $\gamma g \in Z^+N_\mathbb{A}M_k \cdot C$, which holds only if $\gamma \in Z^+N_\mathbb{A}M_k \cdot C \cdot g^{-1}$, so $\gamma \in G_k \cap (Z^+N_\mathbb{A}M_k \cdot C \cdot C_o^{-1})$.

In the quotient $Z^+N_\mathbb{A}M_k\backslash G_\mathbb{A}$, the image of G_k is $P_k\backslash G_k$ is closed and discrete [1.5.3], while the continuous image of the compact set $C \cdot C_o^{-1}$ is *compact*. Thus, left modulo $Z^+N_\mathbb{A}M_k$, that intersection is the intersection of a *closed* discrete set and a compact set, so *finite*, as in [1.5.3]. Therefore, the series is *locally finite* and defines a smooth function on $Z^+G_k\backslash G_\mathbb{A}$. Summing over left translates certainly retains right $K_\mathbb{A}$-invariance.

Similarly, Ψ_φ has compact support in $Z^+G_k\backslash G_\mathbb{A}$: for a summand $\varphi(\gamma g)$ to be nonzero, it must be that $g \in G_k \cdot C$. The image $G_k\backslash(G_k \cdot C)$ is compact, being the continuous image of the compact set C.

To prove the existence of C, let $q : G \to Z^+N_\mathbb{A}M_k\backslash G_A$ be the quotient map. Let U be a neighborhood of $1 \in G_\mathbb{A}$ having compact closure $\overline{U}$. For each $g \in G_\mathbb{A}$, gU is a neighborhood of g. The images $q(gU)$ are *open*, by the characterization of the quotient topology. The support $\mathrm{spt}(\varphi)$ is covered by the opens $q(gU)$, and admits a finite subcover $q(g_1U), \ldots, q(g_nU)$. The set $C = g_1\overline{U} \cup \cdots \cup g_n\overline{U}$ is compact, and its image covers the support of φ. ///

[2.7.2] Claim: Square-integrable cuspforms $L_o^2(Z^+G_k\backslash G_\mathbb{A})$ are the orthogonal complement in $L^2(Z^+G_k\backslash G_\mathbb{A})$ to the subspace spanned by the pseudo-Eisenstein series Ψ_φ with $\varphi \in C_c^\infty(Z^+N_\mathbb{A}M_k\backslash G_\mathbb{A})$. In particular, the *pseudo-Eisenstein series* Ψ_φ fit into an *adjunction*

$$\int_{Z^+N_{\mathbb{A}}M_k\backslash G_{\mathbb{A}}} \varphi \cdot c_P f = \int_{Z^+G_k\backslash G_{\mathbb{A}}} \Psi_\varphi \cdot f$$

for $f \in L^2(Z^+G_k\backslash G_{\mathbb{A}})$.

Proof: As noted earlier, for general reasons [6.1] $C_c^o(Z^+G_k\backslash G_{\mathbb{A}})$ is dense in $L^2(Z^+G_k\backslash G_{\mathbb{A}})$, and we consider $f \in C_c^o(Z^+G_k\backslash G_{\mathbb{A}})$. This allows *unwinding* as in [5.2]:

$$\int_{Z^+N_{\mathbb{A}}M_k\backslash G_{\mathbb{A}}} \varphi \cdot c_P f = \int_{Z^+N_{\mathbb{A}}M_k\backslash G_{\mathbb{A}}} \varphi(g)\Big(\int_{N_k\backslash N_{\mathbb{A}}} f(ng)\,dn\Big)\,dg$$
$$= \int_{Z^+N_kM_k\backslash G_{\mathbb{A}}} \varphi(g)\,f(g)\,dg$$

Winding up, using the left G_k-invariance of f and $N_kM_k = P_k$,

$$\int_{Z^+P_k\backslash G_{\mathbb{A}}} f(g)\,\varphi(g)\,dg = \int_{Z^+G_k\backslash G_{\mathbb{A}}} \sum_{\gamma\in P_k\backslash G_k} f(\gamma\cdot g)\,\varphi(\gamma\cdot g)\,dg$$
$$= \int_{Z^+G_k\backslash G_{\mathbb{A}}} f(g)\Big(\sum_{\gamma\in P_k\backslash G_k} \varphi(\gamma g)\Big)\,dg$$

The inner sum in the last integral is the pseudo-Eisenstein series attached to φ. By Cauchy-Schwarz-Bunyakowsky,

$$\left|\int_{Z^+P_k\backslash G_{\mathbb{A}}} f\,\varphi\right| = \left|\int_{Z^+G_k\backslash G_{\mathbb{A}}} f\Psi_\varphi\right| \le |f|_{L^2}\cdot|\Psi_\varphi|_{L^2}$$

which proves that the functional $f \to \int_{Z^+P_k\backslash G_{\mathbb{A}}} f\,\varphi$ on $C_c^o(Z^+G_k\backslash G_{\mathbb{A}})$ is continuous in the L^2 topology, so extends by continuity to a continuous linear functional on $L^2(Z^+G_k\backslash G_{\mathbb{A}})$. Indeed, this inequality asserts continuity of $f \to c_P f$ as a linear map from $L^2(Z^+G_k\backslash G_{\mathbb{A}})$ to distributions on $Z^+N_{\mathbb{A}}M_k\backslash G_{\mathbb{A}}$ with the weak dual topology as in [13.14]. ///

Similarly, with $C_c^\infty(Z^+N_{\mathbb{A}}M_k\backslash G_{\mathbb{A}}, \omega)$ the collection of functions $\varphi \in C_c^\infty(Z^+N_{\mathbb{A}}M_k\backslash G_{\mathbb{A}})$ such that $\varphi(zg) = \omega(z)\cdot\varphi(g)$ for all $z \in Z_{\mathbb{A}}$, $g \in G$, we have the comparable assertion, now keeping track of complex conjugations:

[2.7.3] Claim: Square-integrable cuspforms $L_o^2(Z^+G_k\backslash G_{\mathbb{A}}, \omega)$ with central character ω are the orthogonal complement in $L^2(Z^+G_k\backslash G_{\mathbb{A}}, \omega)$ to the subspace spanned by the pseudo-Eisenstein series Ψ_φ with $\varphi \in C_c^\infty(Z^+N_{\mathbb{A}}M_k\backslash G_{\mathbb{A}}, \omega)$.

The *pseudo-Eisenstein series* Ψ_φ fit into an *adjunction*

$$\int_{Z^+N_{\mathbb{A}}M_k\backslash G_{\mathbb{A}}} \overline{\varphi} \cdot c_P f = \int_{Z^+G_k\backslash G_{\mathbb{A}}} \overline{\Psi_\varphi} \cdot f$$

for $f \in L^2(Z^+G_k\backslash G_{\mathbb{A}}, \omega)$. *(Formation of pseudo-Eisenstein series respects central characters.)* ///

It is useful to understand simpler subfamilies of pseudo-Eisenstein series, toward their spectral decomposition in terms of *genuine* Eisenstein series in [2.11, 2.12, 2.13].

With

$$M^1 = \{\begin{pmatrix} m_1 & 0 \\ 0 & m_2 \end{pmatrix} : m_1, m_2 \in \mathbb{J}, |m_1| = 1 = |m_2|\}$$

the group $M_k\backslash M^1$ is *compact*, because $\mathbb{J}^1/k^\times$ is compact [2.A]. Certainly $C_c^\infty(Z^+N_{\mathbb{A}}M_k\backslash G_{\mathbb{A}})$ is inside $L^2(Z^+N_{\mathbb{A}}M_k\backslash G_{\mathbb{A}})$, so such functions φ admit decompositions in $L^2(Z^+N_{\mathbb{A}}M_k\backslash G_{\mathbb{A}})$ by characters χ of the compact abelian group $M_k\backslash M^1$ acting on the left, as in [6.11]. The integral expressing the χth component

$$\varphi^\chi(g) = \int_{M_k\backslash M^1} \chi(m)^{-1}\varphi(mg)dm$$

is a Gelfand-Pettis integral converging in $C_c^\infty(Z^+N_{\mathbb{A}}M_k\backslash G_{\mathbb{A}})$ for any quasi-complete [14.7] locally convex [13.11] topology on this space. That is, the Fourier components φ^χ of a compactly supported smooth function along $M_k\backslash M^1$ are again compactly supported smooth, and their sum converges to the original in $L^2(Z^+N_{\mathbb{A}}M_k\backslash G_{\mathbb{A}})$, at least. The support of φ^χ is worst $(M_k\backslash M^1) \times \operatorname{spt}\varphi$.

[2.7.4] Lemma: A function $f \in L^2(Z^+G_k\backslash G_{\mathbb{A}})$ has constant term $c_P f$ integrating to 0 against φ in $C_c^\infty(Z^+N_{\mathbb{A}}M_k\backslash G_{\mathbb{A}})$ if and only if $c_P f$ integrates to 0 against every $M_k\backslash M^1$-component φ^χ of φ.

Proof: The technicality is that there is no claim that constant terms of functions in $L^2(Z^+G_k\backslash G_{\mathbb{A}})$ are in $L^2(Z^+N_{\mathbb{A}}M_k\backslash G_{\mathbb{A}})$. Fortunately, this is not an obstacle: as earlier, it suffices to consider $f \in C_c^o(Z^+G_k\backslash G_{\mathbb{A}})$, so $c_P f \in C^o(Z^+N_{\mathbb{A}}M_k\backslash G_{\mathbb{A}})$. With u the characteristic function of $(M_k\backslash M^1) \times \operatorname{spt}\varphi$, the truncation $u \cdot c_P f$ is in $L^2(Z^+N_{\mathbb{A}}M_k\backslash G_{\mathbb{A}})$, and truncation does not alter the integrals against φ^χ or φ. Letting $\langle , \rangle$ be the inner product in $L^2(Z^+N_{\mathbb{A}}M_k\backslash G_{\mathbb{A}})$, since $\varphi = \sum_\chi \varphi^\chi$ in $L^2(Z^+N_{\mathbb{A}}M_k\backslash G_{\mathbb{A}})$,

$$\langle c_P f, \varphi\rangle = \langle u \cdot c_P f, \varphi\rangle = \sum_\chi \langle u \cdot c_P f, \varphi^\chi\rangle = \sum_\chi \langle c_P f, \varphi^\chi\rangle$$

giving the assertion. ///

For central character ω and character χ extending ω to $M_k\backslash M^1$, define a space of functions on $G_\mathbb{A}$ by[5]

$$J_\chi = \{\varphi \in C_c^\infty(Z^+N_\mathbb{A}M_k\backslash G_\mathbb{A}) : \varphi(mg) = \chi(m)\cdot\varphi(g) \text{ for all } m \in M^1, g \in G_\mathbb{A}\}$$

[2.7.5] Remark: In [2.13.5], we will show that pseudo-Eisenstein series made from J_χ and $J_{\chi'}$ with distinct characters $\chi' \neq \chi$ and $\chi' \neq \chi^w$ are mutually *orthogonal.*

[2.7.6] Corollary: Square-integrable cuspforms $L^2_o(Z^+G_k\backslash G_\mathbb{A}, \omega)$ with central character ω are the orthogonal complement in $L^2(Z^+G_k\backslash G_\mathbb{A}, \omega)$ to the subspace spanned by the pseudo-Eisenstein series Ψ_φ with $\varphi \in J_\chi$, as χ ranges over characters of M^1 extending ω.

Proof: The lemma shows that it suffices to form pseudo-Eisenstein series from the $M_k\backslash M^1$-components φ^χ, and each φ^χ is in J_χ. ///

[2.7.7] Claim: For any compact subgroup $K' \subset K_\mathbb{A}$, right K'-invariant square-integrable cuspforms $L^2_o(Z^+G_k\backslash G_\mathbb{A}/K')$ are the orthogonal complement in $L^2(Z^+G_k\backslash G_\mathbb{A}/K')$ to the subspace spanned by the pseudo-Eisenstein series Ψ_φ with $\varphi \in C_c^\infty(N_\mathbb{A}M_k\backslash G_\mathbb{A}/K')$.

Proof: The point is that for f right K'_v-invariant, $c_P f$ remains K'_v-invariant, so we need only test against test functions φ with the same right K'_v invariance as f, at all places v because integration against more general φ has the same effect as integrating against right K'_v-invariant ones: giving K'_v total measure one for convenience,

$$\begin{aligned}
\int_{Z^+N_\mathbb{A}M_k\backslash G_\mathbb{A}} \varphi\cdot c_P f &= \int_{Z^+N_\mathbb{A}M_k\backslash G_\mathbb{A}} \varphi(g)\left(\int_{K'_v} c_P f(gh)\,dh\right) dg \\
&= \int_{K'_v}\int_{Z^+N_\mathbb{A}M_k\backslash G_\mathbb{A}} c_P f(gh)\varphi(g)\,dg\,dh \\
&= \int_{K'_v}\int_{Z^+N_\mathbb{A}M_k\backslash G_\mathbb{A}} c_P f(g)\varphi(gh^{-1})\,dg\,dh \\
&= \int_{Z^+N_\mathbb{A}M_k\backslash G_\mathbb{A}} c_P f(g)\left(\int_{K'_v} \varphi(gh^{-1})\,dh\right) dg
\end{aligned}$$

as claimed. ///

[5] This space J_χ is an instance of an *induced representation*, but we use no properties of such. Rather, the natural appearance of this function space explains attention to induced representations.

Right K_v-invariance requires that $\chi|_{M_v}$ be right $(M_v \cap K_v)$-invariant, so χ is *unramified* at v, as is ω. That is, the set of right K_v-invariant elements of J_χ is just $\{0\}$ unless χ is unramified at v.

[2.7.8] Claim: Fix a central character ω, and character χ of $M_k\backslash M^1$ extending ω. Fix a place v. The space of right K_v-invariant pseudo-Eisenstein series Ψ_φ with $\varphi \in J_\chi$ is *stable* under the invariant Laplacians for archimedean v, or under spherical Hecke operators for non-archimedean places v: $\Delta_v \Psi_\varphi = E_{\Delta_v\varphi}$ for archimedean v and $\eta \cdot \Psi_\varphi = E_{\eta\cdot\varphi}$ for $\eta \in C_c^\infty(K_v\backslash G_v/K_v)$ for non-archimedean v.

Proof: Since the Laplacians Δ_v commute with the group action, the effect of Δ_v on a pseudo-Eisenstein series is reflected entirely in its effect on the data: the sum is locally finite, so interchange of the operator and the sum is easy, giving

$$\begin{aligned}\Delta_v \Psi_\varphi &= \Delta_v \sum_{\gamma\in\Gamma_\infty\backslash\Gamma} \varphi\circ\gamma = \sum_{\gamma\in\Gamma_\infty\backslash\Gamma} \Delta_v(\varphi\circ\gamma) \\ &= \sum_{\gamma\in\Gamma_\infty\backslash\Gamma} (\Delta_v\varphi)\circ\gamma = E_{\Delta_v\varphi}\end{aligned}$$

Similarly, the action of the spherical Hecke algebra is on the *right*, while the winding-up to form a pseudo-Eisenstein series is on the *left*:

$$\eta\cdot\Psi_\varphi = \eta\cdot \sum_{\gamma\in\Gamma_\infty\backslash\Gamma} \varphi\circ\gamma = \sum_{\gamma\in\Gamma_\infty\backslash\Gamma} \eta\cdot(\varphi\circ\gamma) = \sum_{\gamma\in\Gamma_\infty\backslash\Gamma} (\eta\cdot\varphi)\circ\gamma = E_{\eta\cdot\varphi}$$

as claimed. ///

As a simple special situation, consider cuspforms f right invariant under the standard compact subgroup K_v for all v. Thus, we can invoke the Iwasawa decomposition $G_v = P_v K_v$ everywhere locally, and the constant term $c_P f$ is a function on

$$\begin{aligned}Z^+N_{\mathbb{A}}M_k\backslash G_{\mathbb{A}}/K_{\mathbb{A}} &\approx Z^+N_{\mathbb{A}}M_k\backslash N_{\mathbb{A}}M_{\mathbb{A}}K_{\mathbb{A}}/K_{\mathbb{A}} \\ &\approx Z^+M_k\backslash M_{\mathbb{A}}/(M_{\mathbb{A}}\cap K_{\mathbb{A}})\end{aligned}$$

The quotient $Z^+M_k\backslash M_{\mathbb{A}}$ is the quotient of $k^\times\backslash\mathbb{J} \times k^\times\backslash\mathbb{J}$ by a diagonal copy of the *ray* $\mathbb{R}^+ = \delta(0, +\infty)$, as above, thus, with representatives of the form

$$\begin{pmatrix} \mathbb{R}^+ \times \mathbb{J}^1/k^\times & 0 \\ 0 & \mathbb{J}^1/k^\times \end{pmatrix}$$

Thus, for fixed central character ω and character χ on $M_k\backslash M^1$ extending ω, a test function φ on $Z^+N_{\mathbb{A}}M_k\backslash G_{\mathbb{A}}/K_{\mathbb{A}} \approx Z^+M_k\backslash M_{\mathbb{A}}$ that is in J_χ is entirely specified by a test function φ_∞ on the ray $\delta(0,\infty)$:

$$\varphi(a_y)\cdot m = \varphi_\infty(y)\cdot\chi(m)$$

for $a_y = \begin{pmatrix} \delta(y) & 0 \\ 0 & 1\end{pmatrix}$, and $m \in M^1$, $y > 0$.

2.8 Eisenstein Series

We can attempt to make a pseudo-Eisenstein series Ψ_φ which is an *eigenfunction* for an invariant Laplacian Δ_v (or Casimir operator) at archimedean v, or for Hecke operators at non-archimedean v, by using a right K_v-invariant φ which is such an eigenfunction. However, we already saw in [1.9] that left N_v-invariant right K_v-invariant eigenfunctions on G_v with trivial central character are

$$Z_vN_v\begin{pmatrix} y & 0 \\ 0 & 1\end{pmatrix}K_v \longrightarrow y^s \qquad (\text{for } y>0, \text{ for suitable } s\in\mathbb{C})$$

with eigenvalues $s(s-1)$ (up to normalization). That is, these are *characters* on M_v and are *not* compactly supported modulo Z_v. At non-archimedean places, a parallel computation, but now of the effect of spherical Hecke operators, gives a parallel result, illustrating the constraints on eigenfunctions for spherical Hecke algebras:

[2.8.1] Claim: Let f be a function on $N_v\backslash G_v/K_v$, with (unramified) central character ω_v, which is an eigenfunction for the spherical Hecke algebra $C_c^\infty(K_v\backslash G_v/K_v)$. Then there is a character χ_v on M_v extending ω_v such that f is a linear combination of two Hecke eigenfunctions of the special form $f_1(nmk) = \chi_v(m)$ and $f_2(nmk) = \chi_v(m)^{-1}\cdot\omega(m)|m_1/m_2|_v$ for $n \in N_v$, $m \in M_v$, $k \in K_v$ and character χ_v on M_v extending ω_v on Z_v.

Proof: By the Iwasawa decomposition, the right K_v-invariance and left N_v-invariance of f, and central character, determine f completely by its values on elements $g^\ell = \begin{pmatrix} \varpi^\ell & 0 \\ 0 & 1\end{pmatrix}$. We need just a single Hecke operator, the one attached to the characteristic function η of the set $C = K_v\begin{pmatrix} \varpi & 0 \\ 0 & 1\end{pmatrix}K_v$. Let $\eta\cdot f = \lambda f$ for $\lambda\in\mathbb{C}$. Then

$$\lambda\cdot f(g) = (\eta\cdot f)(g) = \int_{G_v}\eta(h)\,f(gh)dh = \int_C f(gh)dh$$

By the p-adic Cartan decomposition [2.1], C is exactly the collection of two-by-two matrices with entries in $\mathfrak{o}_v$ and determinant in $\varpi\mathfrak{o}_v^\times$ with local parameter $\varpi = \varpi_v$. By p-adic Iwasawa decomposition [2.1], C is the disjoint union of right K_v-cosets

$$C = \begin{pmatrix} 1 & 0 \\ 0 & \varpi \end{pmatrix} K_v \cup \bigcup_{b \bmod \varpi} \begin{pmatrix} \varpi & b \\ 0 & 1 \end{pmatrix} K_v$$

Giving K_v measure 1 and letting q_v be the residue field cardinality,

$$\begin{aligned} \lambda \cdot f(g^\ell) &= \int_C f(g^\ell h)dh = f(\begin{pmatrix} \varpi^\ell & 0 \\ 0 & 1 \end{pmatrix}\begin{pmatrix} 1 & 0 \\ 0 & \varpi \end{pmatrix}) \\ &\quad + \sum_b f(\begin{pmatrix} \varpi^\ell & 0 \\ 0 & 1 \end{pmatrix}\begin{pmatrix} \varpi & b \\ 0 & 1 \end{pmatrix}) \\ &= f\begin{pmatrix} \varpi^\ell & 0 \\ 0 & \varpi \end{pmatrix} + \sum_b f\begin{pmatrix} \varpi^{\ell+1} & b \\ 0 & 1 \end{pmatrix} \\ &= \omega(\varpi) f\begin{pmatrix} \varpi^{\ell-1} & 0 \\ 0 & 1 \end{pmatrix} + q_v \cdot f\begin{pmatrix} \varpi^{\ell+1} & 0 \\ 0 & 1 \end{pmatrix} \\ &= \omega(\varpi)\cdot f(g^{\ell-1}) + q_v \cdot f(g^{\ell+1}) \end{aligned}$$

This gives the recursion

$$\begin{pmatrix} f(g^{\ell+1}) \\ f(g^\ell) \end{pmatrix} = \begin{pmatrix} \lambda/q_v & -\omega(\varpi)/q_v \\ 1 & 0 \end{pmatrix} \begin{pmatrix} f(g^\ell) \\ f(g^{\ell-1}) \end{pmatrix}$$

The eigenvalues of that two-by-two matrix are

$$\{\alpha, \beta\} = \left\{ \frac{\lambda \pm \sqrt{\lambda^2 - 4q_v}}{2q_v} \right\}$$

with eigenvectors $\begin{pmatrix} \alpha \\ 1 \end{pmatrix}$ and $\begin{pmatrix} \beta \\ 1 \end{pmatrix}$. Thus, there are two such eigenfunctions, both with value 1 at 1:

$$f_1(g^\ell) = \alpha^\ell \qquad f_2(g^\ell) = \beta^\ell$$

extended to $M_\mathbb{A}$ to have central character ω. That is, on $M_\mathbb{A}$, these two functions are *characters*. Since $\alpha \cdot \beta = \omega(\varpi)/q$, the two characters are related as asserted. ///

This last claim shows the impossibility of making Hecke eigenfunction pseudo-Eisenstein series with φ in $C_c^\infty(Z^+N_\mathbb{A}M_k\backslash G_\mathbb{A})$. However, it does illustrate a systematic device to make Hecke eigenfunctions:

[2.8.2] Claim: For non-archimedean v, any function f on G_v of the form $f(nmk) = \chi(m)$ for unramified character χ on M_v is an eigenfunction for the spherical Hecke algebra.

Proof: Let I_χ be the space of smooth functions f on G_v with the property $f(nmk) = \chi(m) \cdot f(k)$ for all $n \in N_v$, $m \in M_v$, and $k \in K_v$.[6] Here the smoothness means that, for each f, there is a compact open subgroup $K' \subset K_v$ such that f is right K'-invariant. Thus, by p-adic Iwasawa decomposition, I_χ is a colimit of finite-dimensional spaces (compare [13.8]). The action of G_v on I_χ by right translation $(g \cdot f)(h) = f(hg)$ is *continuous*, so η in the spherical Hecke algebra $C_c^\infty(K_v\backslash G/K_v)$ acts by the integrated version of the action:

$$(\eta \cdot f)(h) = \int_{G_v} \eta(g)\, f(hg)\, dg$$

By changing variables in the integral, the action of such η preserves right K_v-invariance. By p-adic Iwasawa decomposition, the subspace of I_χ of right K_v-invariant functions is *one-dimensional*, spanned by $f(nmk) = \chi(m)$ itself. Since this one-dimensional space is stabilized by the spherical Hecke algebra, this f is inevitably an eigenfunction for the Hecke algebra. ///

We wish to decompose pseudo-Eisenstein series Ψ_φ into Δ_v-eigenfunctions and spherical Hecke algebra eigenfunctions to the extent possible. We have already seen that we can take φ in the spaces J_χ of [2.7], for χ a character on $M_k\backslash M^1$. The previous two claims suggest taking this further: every character on $Z^+M_k\backslash M_{\mathbb{A}}$ can be written in the form

$$\nu^s\chi : a_y \cdot m \longrightarrow |y|^s \cdot \chi(m)$$

for $a_y = \begin{pmatrix} \delta(y) & 0 \\ 0 & 1 \end{pmatrix}$, and $m \in M^1$, $y > 0$, for suitable complex s and character χ on $M_k\backslash M^1$. Let

$$I_{s,\chi} = \{f \in C^\infty(Z^+N_{\mathbb{A}}M_k\backslash G_{\mathbb{A}}) : f(nmg) = (\nu^s\chi)(m) \cdot f(g) \quad \text{for all } n \in N_{\mathbb{A}}, m \in M_{\mathbb{A}}\}$$

A *genuine* Eisenstein series E_f for $f \in I_{s,\chi}$ is

$$E_f(g) = \sum_{\gamma \in P_k\backslash G_k} f(\gamma \cdot g) \qquad (\text{for } f \in I_{s,\chi})$$

[6] This space I_χ is an example of an *unramified principal series* representation of G_v, meaning that it is *induced* from $\tilde{\chi}(nm) = \chi(m)$ on P_v, with χ unramified. The previous two claims touch on the importance of principal series representations in the application of representation theory of p-adic groups to automorphic forms. A strong form of the generalization of these claims to a wide class of p-adic groups is in [Borel 1976], [Matsumoto 1977], [Casselman 1980].

One immediate issue is *convergence*: unlike pseudo-Eisenstein series Ψ_φ where φ has controlled support, the sum for E_f is *not* locally finite. Ignoring convergence for a moment, E_f is *genuine* in the sense that it is a spherical Hecke algebra eigenfunction at all but (at worst) finitely many non-archimedean places, since *smoothness* at finite places requires that f is right K'-invariant for some compact *open* subgroup $K' = \prod_{v<\infty} K'_v$ of $\prod_{v<\infty} K_v$, and the product topology requires that $K'_v = K_v$ for all but finitely many $v < \infty$.

The extreme, simplest case is that $f \in I_{s,\chi}$ is right K_v-invariant at all places v, that is, is *spherical* everywhere locally. From all the Iwasawa decompositions for groups G_v, up to a scalar there is a unique such f, namely, $f(nmk) = f(m) = (\nu^s\chi)(m)$. The *everywhere spherical* Eisenstein series attached to an unramified grossencharacter χ is

$$E_{s,\chi}(g) = \sum_{\gamma \in P_k\backslash G_k} f(\gamma \cdot g) \qquad (\text{for } f(nmk) = f(m) = (\nu^s\chi)(m))$$

[2.8.3] Claim: Assuming the series expression for the everywhere-spherical Eisenstein series $E_{s,\chi}$ is convergent, it is an eigenfunction for the invariant Laplacians at archimedean places and for the spherical Hecke algebras at non-archimedean places.

Proof: Assuming convergence, the invariance of Laplacians and spherical Hecke operators under *left* translation implies that we need merely check that the function $f(nmk) = f(m) = \widetilde{\chi}(m)$ itself is an eigenfunction. In [1.9] we saw the archimedean assertion, and the preceding claim proves the non-archimedean assertion. ///

[2.8.4] Claim: Assuming the series expression for the everywhere-spherical Eisenstein series $E_{s,\chi}$ is convergent, its *constant term* is

$$c_P E_{s,\chi}(znmk) = (\nu^s\chi)(m) + c_{s,\chi} \cdot (\nu^{1-s}\chi^w)(m)$$

for $z \in Z^+, n \in N_\mathbb{A}, m \in M_\mathbb{A}, k \in K_\mathbb{A}$, where $\chi^w(m) = \chi(wmw^{-1})$ with long Weyl element $w = \begin{pmatrix} 0 & 1 \\ 1 & 0 \end{pmatrix}$, and

$$c_{s,\chi} = \frac{\Lambda(2s-1, \chi_1/\chi_2)}{\Lambda(2s, \chi_1/\chi_2)} \qquad \left(\text{with } \chi\begin{pmatrix} m_1 & 0 \\ 0 & m_2 \end{pmatrix} = \chi_1(m_1)\chi_2(m_2)\right)$$

where $\Lambda(s, \chi_1/\chi_2)$ is the Hecke grossencharacter L-function *completed* by multiplying by the appropriate Gamma factors.

Proof: Via the Bruhat decomposition $G_k = P_k \sqcup P_k w N_k$,

$$P_k\backslash G_k = P_k\backslash P_k \sqcup P_k\backslash P_k w N_k \approx \{1\} \sqcup wN_k$$

The small Bruhat cell P_k produces the first summand in the constant term:

$$\int_{N_k\backslash N_{\mathbb{A}}} \sum_{\gamma\in P_k\backslash P_k} f(\gamma ng)\,dn = \int_{N_k\backslash N_{\mathbb{A}}} f(ng)\,dn = f(g)\cdot\int_{N_k\backslash N_{\mathbb{A}}} 1\,dn$$

The large Bruhat cell $P_k wN_k$ gives

$$\int_{N_k\backslash N_{\mathbb{A}}} \sum_{\gamma\in wN_k} f(\gamma ng)\,dn = \int_{N_k\backslash N_{\mathbb{A}}} \sum_{\gamma\in N_k} f(w\gamma ng)\,dn$$

$$= \int_{N_{\mathbb{A}}} f(wng)\,dn$$

by unwinding, as in [5.2]. Since $c_P f$ will be left $N_{\mathbb{A}}$-invariant and right $K_{\mathbb{A}}$-invariant, it suffices to evaluate this integral on $g = m \in M_{\mathbb{A}}$. Then

$$\int_{N_{\mathbb{A}}} f(wnm)\,dn = \int_{N_{\mathbb{A}}} f(wm\cdot m^{-1}nm)\,dn$$

$$= \int_{N_{\mathbb{A}}} f(wm\cdot n)\nu(m)\,dn$$

by replacing n by mnm^{-1}, with $\nu(m)$ resulting from change-of-measure. This is

$$\int_{N_{\mathbb{A}}} f(wmw^{-1}\cdot wn)\nu(m)\,dn$$

$$= \nu(m)\nu^s(wmw^{-1})\chi(wmw^{-1})\cdot\int_{N_{\mathbb{A}}} f(wn)\,dn$$

$$= \nu(m)^{1-s}\chi(wmw^{-1})\cdot\int_{N_{\mathbb{A}}} f(wn)\,dn$$

Since the right $K_{\mathbb{A}}$-invariance is preserved by integrating on the left, this is (unique up to constant) the spherical function in I_{1-s,χ^w}. That normalization constant is significant, being a ratio of L-functions, as follows.

Let $f_v^o(nmk) = (\nu_v^s\chi_v)(m)$ be the normalized spherical vector on G_v, where $\nu_v^s\chi_v$ is the vth local factor of $\nu^s\chi$,

$$\begin{pmatrix} m_1 & 0 & m_2 \end{pmatrix} \longrightarrow |m_1/m_2|^s\cdot\chi_1(m_1)\chi_2(m_2) \qquad (\text{on } M_v)$$

The integral giving the normalizing constant factors over primes:

$$\int_{N_{\mathbb{A}}} f(wn)\,dn = \prod_{v\le\infty}\int_{N_v} f_v^o(wn)\,dn$$

To evaluate the vth factor, we must determine the local Iwasawa decomposition of wn for $n \in N_v$. At $k_v \approx \mathbb{R}$, as in [1.3]

$$w\begin{pmatrix} 1 & x \\ 0 & 1 \end{pmatrix} = \begin{pmatrix} 0 & 1 \\ 1 & x \end{pmatrix} = \begin{pmatrix} \frac{-1}{\sqrt{1+x^2}} & * \\ 0 & \sqrt{1+x^2} \end{pmatrix} \cdot \begin{pmatrix} \frac{x}{\sqrt{1+x^2}} & \frac{-1}{\sqrt{1+x^2}} \\ \frac{1}{\sqrt{1+x^2}} & \frac{x}{\sqrt{1+x^2}} \end{pmatrix}$$

with that last matrix in $SO_2(\mathbb{R})$. Unramified unitary characters on $k_v^\times \approx \mathbb{R}^\times$ are of the form $\alpha_v(y) = |y|^{it_v}$ for some purely imaginary it_v. With $|y|^{it_v} = \chi_1(y)/\chi_2(y)$, the corresponding local integral is evaluated via the standard trick $\int_0^\infty e^{-ty}\, t^s\, dt/t = y^{-s}\,\Gamma(s)$: first, with $it_v = 0$,

$$\begin{aligned}\int_{\mathbb{R}} f_v^o(wn)\, dn &= \int_{\mathbb{R}} \frac{1}{(1+x^2)^s}\, dx \\ &= \frac{1}{\pi^{-s}\Gamma(s)} \int_0^\infty \int_{\mathbb{R}} e^{-\pi t(1+x^2)} t^s \frac{dt}{t} \\ &= \frac{1}{\pi^{-s}\Gamma(s)} \int_0^\infty \int_{\mathbb{R}} e^{-\pi t - \pi x^2} t^{s-\frac{1}{2}} \frac{dt}{t} = \frac{\pi^{-(s-\frac{1}{2})}\Gamma(s-\frac{1}{2})}{\pi^{-s}\Gamma(s)}\end{aligned}$$

Replacing s by $s - it_v$, the general unramified case is

$$\int_{\mathbb{R}} f_v^o(wn)\, dn = \int_{\mathbb{R}} \frac{1}{(1+x^2)^{s-it_v}}\, dx = \frac{\pi^{-(s-it_v-\frac{1}{2})}\Gamma(s-it_v-\frac{1}{2})}{\pi^{-(s-it_v)}\Gamma(s-it_v)}$$

Similarly, at $k_v \approx \mathbb{C}$, with trivial χ_1, χ_2,

$$w\begin{pmatrix} 1 & x \\ 0 & 1 \end{pmatrix} = \begin{pmatrix} 0 & 1 \\ 1 & x \end{pmatrix} = \begin{pmatrix} \frac{-1}{\sqrt{1+|x|^2}} & * \\ 0 & \sqrt{1+|x|^2} \end{pmatrix} \cdot \begin{pmatrix} \frac{x}{\sqrt{1+|x|^2}} & \frac{-1}{\sqrt{1+|x|^2}} \\ \frac{1}{\sqrt{1+|x|^2}} & \frac{\overline{x}}{\sqrt{1+|x|^2}} \end{pmatrix}$$

With the normalization of local norms $|t|_{\mathbb{C}} = |N_{\mathbb{R}}^{\mathbb{C}} t|$ for the product formula, up to measure constants the local integral is

$$\begin{aligned}\int_{\mathbb{C}} f_v^o(wn)\, dn &= \int_{\mathbb{C}} \frac{1}{(1+|x|^2)^{2s}}\, dx \\ &= \frac{1}{\pi^{-2s}\Gamma(2s)} \int_0^\infty \int_{\mathbb{R}} e^{-\pi t(1+|x|^2)} t^{2s} \frac{dt}{t} \\ &= \frac{1}{\pi^{-2s}\Gamma(2s)} \int_0^\infty \int_{\mathbb{R}} e^{-\pi t - \pi |x|^2)} t^{2s-1} \frac{dt}{t} = \frac{\pi^{-(2s-1)}\Gamma(2s-1)}{\pi^{-2s}\Gamma(2s)}\end{aligned}$$

In the general unramified case, with $\chi_1(t)/\chi_2(t) = |t|_{\mathbb{C}}^{it_v} = |t|^{2it_v}$, again there is a shift $s \to s - it_v$.

At non-archimedean places, the Iwasawa decomposition has a different nature:

$$w\begin{pmatrix}1 & x\\ 0 & 1\end{pmatrix} = \begin{pmatrix}0 & 1\\ 1 & x\end{pmatrix} = \begin{cases}\begin{pmatrix}\frac{-1}{x} & 1\\ 0 & x\end{pmatrix}\begin{pmatrix}1 & 0\\ \frac{1}{x} & 1\end{pmatrix} & \text{for } |x|_v \geq 1\\[2ex] \begin{pmatrix}1 & 0\\ 0 & 1\end{pmatrix}\begin{pmatrix}0 & 1\\ 1 & x\end{pmatrix} & \text{for } |x|_v \leq 1\end{cases}$$

There is a non-archimedean analogue of the Gamma trick: with ch the characteristic function of $\mathfrak{o}_v$, with multiplicative Haar measure giving $\mathfrak{o}_v^\times$ total measure 1,

[2.8.5] Lemma:

$$\int_{k_v^\times} \mathrm{ch}(ty)\,\mathrm{ch}(t)\,|t|_v^s\,dt = \begin{cases}\dfrac{1}{1-q_v^{-s}} & (\text{for } y \in \mathfrak{o}_v)\\[2ex] \dfrac{|y|^{-s}}{1-q_v^{-s}} & (\text{for } y \notin \mathfrak{o}_v)\end{cases}$$

Proof: For $y \in \mathfrak{o}_v$, the integral becomes an Iwasawa-Tate local zeta integral

$$\int_{k_v^\times} \mathrm{ch}(t)\,|t|^s\,dt = \sum_{\ell\geq 0}\int_{\varpi^\ell \mathfrak{o}_v^\times} q_v^{-\ell s} = \sum_{\ell\geq 0} 1\cdot q_v^{-\ell s} = \frac{1}{1-q_v^{-s}}$$

For $y \notin \mathfrak{o}_v$, replace t by t/y in the integral, producing the $|y|_v^{-s}$ factor and then the integral just evaluated. ///

Returning to the evaluation of the non-archimedean local factor in the constant term, let

$$\gamma(s,y) = \int_{k_v^\times} \mathrm{ch}(ty)\,\mathrm{ch}(t)\,|t|_v^s\,d^\times t$$

emphasizing that it is multiplicative Haar measure. With trivial χ_1, χ_2, the lemma gives

$$\begin{aligned}\int_{N_v} f_v^o(wn)\,dn &= \left(1-q_v^{-2s}\right)\int_{k_v} \gamma(2s,x)\,dx\\ &= \left(1-q_v^{-2s}\right)\int_{k_v^\times}\int_{k_v} \mathrm{ch}(tx)\,\mathrm{ch}(t)\,|t|_v^{2s}\,dx\,d^\times t\\ &= \left(1-q_v^{-2s}\right)\int_{k_v^\times}\int_{k_v} \mathrm{ch}(x)\,\mathrm{ch}(t)\,|t|_v^{2s-1}\,dx\,d^\times t\end{aligned}$$

by replacing x by x/t. The integral in x is just meas $(\mathfrak{o}_v)$, which at k_v unramified over the corresponding $\mathbb{Q}_p$ is reasonably taken to be 1. Thus,

$$\int_{N_v} f_v^o(wn)\,dn = \left(1 - q_v^{-2s}\right) \int_{k_v^\times} \mathrm{ch}(t)\,|t|_v^{2s-1} d^\times t = \frac{1 - q_v^{-2s}}{1 - q_v^{-(2s-1)}}$$

$$= \frac{1/(1 - q_v^{-(2s-1)})}{1/(1 - q_v^{-2s})}$$

The adjustment for a general unramified character again shifts s to $s - it_v$. The products over all places give the indicated ratios of completed L-functions, apart from ratios of powers of the *conductor*, which correspond to the additive measure normalization at ramified places. A ratio of a value and a shift only leaves a constant, which is not immediately important here. ///

[2.8.6] Claim: For $\mathrm{Re}(s) > 1$, the series expression for E_f with (continuous) $f \in I_{s,\chi}$ converges absolutely and uniformly on compacts, to a *continuous* function on $Z^+G_k\backslash G_{\mathbb{A}}$.

Proof: The function f is dominated by the spherical vector, since $|f(znmk)| = |(\nu^s\chi)(m)| \cdot |f(k)|$ and the continuous function f is *bounded* on the compact $K_{\mathbb{A}}$. Also, χ has absolute value 1, so we may as well take χ trivial. It suffices to treat $s = \sigma \in \mathbb{R}$. Use the *height* functions h_v on k_v^2 and h on $\mathbb{A}^2$, and $\eta(g) = |\det g|/h(v_o g)^2$. In particular, $\eta(znmk) = |m_1/m_2|$ for $m = \begin{pmatrix} m_1 & 0 \\ 0 & m_2 \end{pmatrix}$. Also, $\nu(m)^{-1}\,dn\,dm$ is *left* Haar measure dp on $P_{\mathbb{A}}$. Thus, it suffices to prove convergence of

$$\sum_{\gamma \in P_k\backslash G_k} \eta(\gamma g)^\sigma = \sum_{\gamma \in P_k\backslash G_k} |\det \gamma g|^\sigma \cdot h(v_o\gamma g)^{-2\sigma}$$

$$= |\det g|^\sigma \sum_{\gamma \in P_k\backslash G_k} h(v_o\gamma g)^{-2\sigma}$$

By reduction theory [2.2], for compact $C \subset G_{\mathbb{A}}$, there are constants $0 < c \leq c' < +\infty$ such that

$$c \cdot h(v) \leq h(vg) \leq c' \cdot h(v)$$

for all $g \in C$, for all primitive $v \in \mathbb{A}^2$, so

$$h(v_o\gamma g) \leq c' \cdot h(v_o\gamma) \leq \frac{c'}{c} \cdot c \cdot h(v_o\gamma g')$$

for all $g, g' \in C$, for all $\gamma \in G_k$. Thus, convergence of the series is equivalent to convergence of an *averaged* integral $\int_C E_\sigma$. By discreteness of G_k in $G_{\mathbb{A}}$, we can shrink C so that, for γ in G_k, if $\gamma C \cap C \neq \phi$ then $\gamma = 1$. Then

$$\int_C E_\sigma = \int_C \sum_{\gamma \in P_k \backslash G_k} |\det \gamma g|^\sigma h(v_o \gamma g)^{-2\sigma} \, dg$$

$$= |\det g|^\sigma \int_{P_k \backslash G_k \cdot C} h(v_o g)^{-2\sigma} \, dg$$

Let μ be the infimum of $h(v)$ over nonzero primitive v in $\mathbb{A}^2$. From reduction theory [2.2], this infimum is attained, so $\mu > 0$, and $c \cdot \mu \leq h(v_o \gamma g)$ for all $g \in C$ and $\gamma \in G_k$, and $G_k \cdot C$ is contained in a set

$$Y = \{g \in G_\mathbb{A} : h(v_o g) \geq c \cdot \mu \quad \text{and} \quad c_1 \leq |\det g| \leq c_2\}$$

with $0 < c_1$ and $c_2 < +\infty$. The set Y is right $K_\mathbb{A}$-stable, since h is $K_\mathbb{A}$-invariant. Using Iwasawa decompositions, with left Haar measure dp on $P_\mathbb{A}$,

$$\int_{P_k \backslash G_k \cdot C} |\det g|^\sigma h(v_o g)^{-2\sigma} \, dg \leq \int_{P_k \backslash Y} |\det g|^\sigma h(v_o g)^{-2\sigma} \, dg$$

$$= \int_{P_k \backslash (P_\mathbb{A} \cap Y)} |\det p|^\sigma h(v_o p)^{-2\sigma} \, dp$$

The set Y is left $N_\mathbb{A}$-stable, and the induced measure on the compact quotient $N_k \backslash N_\mathbb{A}$ is *finite*, so up to a constant the integral is

$$\int_{M_k \backslash (M_\mathbb{A} \cap Y)} |\det m|^\sigma h(v_o m)^{-2\sigma} \nu(m)^{-1} \, dm$$

$$= \int_{M_k \backslash (M_\mathbb{A} \cap Y)} \nu(m)^{\sigma - 1} \, dm$$

From

$$M_\mathbb{A} \cap Y \subset \{\begin{pmatrix} m_1 & 0 \\ 0 & m_2 \end{pmatrix} : |m_1/m_2| \geq c\mu, c_1 \leq |m_1 m_2| \leq c_2\}$$

and compactness of $\mathbb{J}^1/k^\times$,

$$M_k \backslash (M_\mathbb{A} \cap Y) = \text{compact} \times \{\begin{pmatrix} \delta(y_1) & 0 \\ 0 & \delta(y_2) \end{pmatrix}\}$$

with $y_2 \geq c\mu$ and $c_1 \leq y_1 y_2 \leq c_2$. Thus,

$$\int_{M_k \backslash (M_\mathbb{A} \cap Y)} \nu(m)^{\sigma - 1} dm$$

$$= \int_{M_k \backslash (M^1 \cap Y)} dm \cdot \int_{c\mu}^{\infty} \int_{c_1/y_2}^{c_2/y_2} \left(\frac{y_1}{y_2}\right)^{\sigma - 1} \frac{dy_1}{y_1} \frac{dy_2}{y_2}$$

Replacing y_1 by y_1/y_2, the latter elementary integral becomes

$$\int_{c\mu}^{\infty}\int_{c_1}^{c_2}\left(\frac{y_1}{y_2^2}\right)^{\sigma-1}\frac{dy_1}{y_1}\frac{dy_2}{y_2} = (\text{constant})\cdot\int_{c\mu}^{\infty}\left(y_2^2\right)^{1-\sigma}\frac{dy_2}{y_2}$$

which converges for $\sigma > 1$. This also proves the uniform convergence on compacts. ///

We also want *moderate growth* on Siegel sets: for $n \in N_{\mathbb{A}}$, $k \in K_{\mathbb{A}}$, $z \in Z^+$, and $m = a_y \cdot m'$ with $m' \in M^1$,

$$|E_f(znmk)| \ll_{t,C} y^{\mathrm{Re}\,(s)}$$

on $\mathfrak{S}_{t,C}$, implied constant depending on t, C. And we want convergence to a *smooth* function:

[2.8.7] Claim: The series for E_s converges in the C^∞ topology for $\mathrm{Re}\,(s) > 1$, and produces a C^∞ moderate-growth function on $Z^+G_k\backslash G_{\mathbb{A}}$. *(Proofs in [2.B], [11.5].)*

As in [13.5], the *idea* of the archimedean aspect of the C^∞ topology is that it is given by the collection of seminorms given by sups on compacts of all derivatives, for example left-G-invariant derivatives on $G_\infty = \prod_{v|infty} G_v$ from the Lie algebra, preserving left G_k-invariance, and stabilizing a useful class of Eisenstein series. The non-archimedean smoothness is simpler, being *right invariance* under some compact open subgroup of $K' \subset \prod_{v<\infty} K_v$, which leads to taking an ascending union (colimit) over such K'.

The everywhere spherical computation of *constant terms* applies to computation of local components of $c_P f$ at *good primes* for general $f \in I_{s,\chi}$, that is, places v where f is right K_v-invariant and χ is unramified. However, at the other, *bad*, primes for $f \in I_{s,\chi}$, where f is right K_v'-invariant only for $K_v' \subset K_v$ of high index, the local integrals

$$f \longrightarrow \left(g \longrightarrow \int_{N_v} f(wng)\,dn\right)$$

are naturally more complicated. Still, these maps visibly commute with the *right* translation action of G_v and have predictable left-equivariance under M_v for the same reason as in the simpler computation:

[2.8.8] Claim: The constant term of the Eisenstein series E_f for $f \in I_{s,\chi}$ is

$$c_P E_f = f + C_{s,\chi}(f) \qquad (\text{with } C_{s,\chi}f(g) = \textstyle\int_{N_{\mathbb{A}}} f(wng)\,dn)$$

The map $C_{s,\chi}$ is a $G_{\mathbb{A}}$ map in the sense that $g\cdot(C_{s,\chi}f) = C_{s,\chi}(g\cdot f)$ where $g\cdot f$ is right translation.

Proof: Integration on the left certainly commutes with right translation. As in the earlier, simpler, case, the small Bruhat cell gives

$$\int_{N_k \backslash N_{\mathbb{A}}} f(ng)\,dn = f(g) \cdot \int_{N_k \backslash N_{\mathbb{A}}} 1\,dn$$

and the volume of $N_k \backslash N_{\mathbb{A}}$ is reasonably normalized to 1. The big Bruhat cell integral unwinds:

$$\int_{N_k \backslash N_{\mathbb{A}}} \sum_{\gamma \in N_k} f(w\gamma ng)\,dn = \int_{N_{\mathbb{A}}} f(wng)\,dn$$

For $m \in M_{\mathbb{A}}$,

$$\begin{aligned}\int_{N_{\mathbb{A}}} f(wnmg)\,dn &= \int_{N_{\mathbb{A}}} f(wm \cdot m^{-1}nmg)\,dn \\ &= \int_{N_{\mathbb{A}}} \nu(m) f(wmw^{-1} \cdot wng)\,dn \\ &= \nu(m)^{1-s} \chi(wmw^{-1}) \cdot \int_{N_{\mathbb{A}}} f(wng)\,dn\end{aligned}$$

showing that this part of the constant term is in I_{1-s,χ^w}. ///

The *scattering matrix/operator* is the map[7]

$$f \longrightarrow \left(g \longrightarrow \int_{N_{\mathbb{A}}} f(wng)\,dn\right) \qquad \text{(from } I_{s,\chi} \text{ to } I_{1-s,\chi^w}\text{)}$$

Since the unwound integral over $N_{\mathbb{A}}$ is a (limit of) product(s) of integrals over N_v, it is a (tensor) product of *local* maps $C_{s,\chi,v}$ among corresponding local spaces

$$\begin{aligned}I_{s,\chi,v} = \{f \in C^\infty(G_v) : f(nmg) = (\nu^s \chi)(m) \cdot f(g), \\ \text{for all } n \in N_v,\ m \in M_v,\ g \in G_v\}\end{aligned}$$

in the natural way. For example, for *monomial* $f \in I_{s,\chi}$, that is, of the form $f(g) = \prod_{v \le \infty} f_v(g_v)$, with f_v the *spherical* vector for v outside a finite set S of places,

$$\int_{N_{\mathbb{A}}} f(wng)\,dn = \int_{N_{\mathbb{A}}} \prod_v f_v(wng)\,dn = \prod_v \int_{N_v} f_v(wng)\,dn$$

[7] Since this map respects the right translation action of G_v and/or of $G_{\mathbb{A}}$ on functions, it is an instance of an *intertwining operator* among representations of G_v and/or $G_{\mathbb{A}}$.

The earlier, simple constant term computation shows that for places $v \notin S$, the local operator sends the spherical vector in $I_{s,\chi,v}$ to the spherical vector in $I_{1-s,\chi^w,v}$, multiplied by the vth Euler factor of $\Lambda(2s-1,\chi^w)/\Lambda(2s,\chi)$.

[2.8.9] Remark: The space of functions $I_{s,\chi}$ has *central character*

$$\begin{pmatrix} z & 0 \\ 0 & z \end{pmatrix} \longrightarrow \nu^s \begin{pmatrix} z & 0 \\ 0 & z \end{pmatrix} \cdot \chi_1(z)\,\chi_2(z) = \chi_1(z)\,\chi_2(z)$$

The condition $\chi^w = \chi$ is that $\chi_1 = \chi_2$. Thus, for $f \in I_{s,\chi}$, the normalized function $g \to \chi_1(\det g)^{-1} f(g)$ has *trivial* central character, and, further, is in $I_{s,1}$.

2.9 Meromorphic Continuation of Eisenstein Series

This is an issue of showing that a family of Eisenstein series E_{f_s} with $f_s \in I_{s,\chi}$ has a meromorphic continuation in s beyond the range of convergence $\mathrm{Re}(s) > 1$. So certainly E_{f_s} must be *holomorphic* in $\mathrm{Re}(s) > 1$, and the dependence of f_s on s must be constrained for this to be plausible. The simplest example is to take f_s to be the everywhere-spherical vector in $I_{s,\chi}$, with χ unramified everywhere. A special argument applicable to GL_2 is in [2.B], and instantiation of a more general approach is in [11.5]. The basic theorem is

[2.9.1] Theorem: The everywhere-spherical Eisenstein series $E_{s,\chi}$ has a *meromorphic continuation* in $s \in \mathbb{C}$, as a smooth function of moderate growth on $Z^+G_k\backslash G_{\mathbb{A}}$. As a function of s, $E_{s,\chi}(g)$ is of at most polynomial growth vertically, uniformly in bounded strips, uniformly for g in compacts. *(Proofs in [2.B] and [11.5].)*

[2.9.2] Corollary: At archimedean v, let $t_v \in \mathbb{R}$ be associated to the character χ, as in the proof of [2.8.4], so that $E_{s,\chi}$ is an eigenfunction for the vth invariant Laplacian Δ_v, with eigenvalue $\lambda_{s,\chi} = (s - it_v)(s - it_v - 1)$. This eigenfunction property persists under meromorphic continuation.

Proof: Both $\Delta_v E_{s,\chi}$ and $\lambda_{s,\chi} \cdot E_{s,\chi}$ are holomorphic function-valued functions of s, taking values in the topological vector space of smooth moderate-growth functions. They agree in the region of convergence $\mathrm{Re}(s) > 1$, then apply the vector-valued form [15.2] of the Identity Principle from complex analysis. ///

[2.9.3] Corollary: The meromorphic continuation of $E_{s,\chi}$ implies the meromorphic continuation of the constant term $c_P E_{s,\chi}(m) = (\nu^s\chi)(m) + c_{s,\chi} \cdot (\nu^{1-s}\chi^w)(m)$, and, in particular, of the function

$$c_{s,\chi} = \frac{\Lambda(2s-1,\chi_1/\chi_2)}{\Lambda(2s,\chi_1/\chi_2)}$$

with $\chi\begin{pmatrix} m_1 & 0 \\ 0 & m_2 \end{pmatrix} = \chi_1(m_1)\chi_2(m_2)$.

Proof: Since $E_{s,\chi}$ meromorphically continues at least as a smooth function, the integral over the compact set $N_k\backslash N_{\mathbb{A}}$ expressing a pointwise value $c_P E_{s,\chi}(g)$ of the constant term certainly converges absolutely. In fact, the integral converges as a continuous-function-valued function $n \to (g \to E_{s,\chi}(nm))$, so has a continuous-function-valued Gelfand-Pettis integral $m \to c_P E_{s,\chi}(m)$. In brief, the constant term has a meromorphic continuation. Then the vector-valued form of the Identity Principle from complex analysis implies that the form of the constant term persists outside the region of convergence $\mathrm{Re}(s) > 1$. In particular, this gives the meromorphic continuation of $c_{s,\chi}$. ///

The *theory of the constant term* in [8.2] yields

[2.9.4] Claim: For every s away from poles of $s \to E_{s,\chi}$, in a fixed Siegel set $\mathfrak{S}_{t,C}$,

$$E_{s,\chi}(nmk) - \big((\nu^s\chi)(m) + c_{s,\chi}(\nu^{1-s}\chi^{w})(m)\big) \ll_B \nu(m)^{-B}$$

That is, $E_{s,\chi} - c_P E_{s,\chi}$ is *rapidly decreasing* in standard Siegel sets.

Proof: Since $E_{s,\chi}$ is an eigenfunction for invariant Laplacians and for spherical Hecke algebras, and is of moderate growth, the theory of the constant term [8.2] exactly ensures that $E_{s,\chi}$ is asymptotic to its constant term, in the sense of the assertion. ///

[2.9.5] Corollary: The poles of $E_{s,\chi}$ are exactly the poles of the constant term $c_{s,\chi}$. ///

Granting the meromorphic continuation and the asymptotic estimation of the Eisenstein series by its constant term, the *functional equation* is determined by its constant term:

[2.9.6] Corollary: $E_{s,\chi}$ has the *functional equation* $E_{1-s,\chi} = c_{1-s,\chi}E_{s,\chi^{w}}$, and $c_{s,\chi} \cdot c_{1-s,\chi^{w}} = 1$.

[2.9.7] Remark: Note that the functional equation does not generally relate $E_{1-s,\chi}$ to $E_{s,\chi}$, but to $E_{s,\chi^{w}}$.

Proof: Take $\mathrm{Re}(s) > \frac{1}{2}$ and s off the real line. The function $f = E_{1-s,\chi} - c_{1-s,\chi}E_{s,\chi^{w}}$ has constant term

$$\begin{aligned} c_P f(m) &= \big((\nu^{1-s}\chi)(m) + c_{1-s,\chi} \cdot (\nu^s\chi^{w})(m)\big) \\ &\quad - c_{1-s,\chi} \cdot \big((\nu^s\chi)(m) + c_{s,\chi^{w}} \cdot (\nu^{1-s}\chi)(m)\big) \\ &= \nu^{1-s}(m) \cdot \big(1 - c_{1-s,\chi}c_{s,\chi^{w}}\big) \cdot \chi(m) \end{aligned}$$

For $\sigma = \mathrm{Re}(s) > \frac{1}{2}$, ν^{1-s} is square-integrable on $\mathfrak{S}_{t,C}$: via an Iwasawa decomposition, noting that $\nu(m)^{-1}\, dn\, dm$ is left Haar measure on $P_{\mathbb{A}}$,

$$\int_{\mathfrak{S}_{t,C}} |\nu^{1-s}|^2 = \int_{\mathfrak{S}_{t,C}} \nu^{2-2\sigma}(m)\nu(m)^{-1}\, dn\, dm\, dk$$

$$\ll \int_{m\in Z^+M_k\backslash M_{\mathbb{A}}\,:\,\nu(m)\geq t} \nu(m)^{1-2\sigma}\, dm \ll \int_t^\infty y^{1-2\sigma}\frac{dy}{y}$$

By the theory of the constant term [8.2], on a standard Siegel set

$$\begin{aligned} f &= c_P f + (\text{rapidly decreasing}) \\ &\ll {}_s\eta^{1-\sigma} + (\text{rapidly decreasing}) \end{aligned}$$

Thus, on $\mathfrak{S}_{t,C}$,

$$\begin{aligned} |f|^2 &\ll |\nu^{1-\sigma} + (\text{rapidly decreasing})|^2 \\ &= \nu^{2(1-\sigma)} + 2\cdot\eta^{1-\sigma}\cdot(\text{rapidly decreasing}) + (\text{rapidly decreasing})^2 \\ &= \nu^{2(1-\sigma)} + (\text{rapidly decreasing}) \end{aligned}$$

Thus, $f \in L^2(Z^+G_k\backslash G_{\mathbb{A}})$. For archimedean v, f is a Δ_v-eigenfunction, with eigenvalue of the form $\lambda = (s - it_v)(s - it_v - 1)$ for it_v purely imaginary, depending on χ. This eigenvalue is *not real* for $\mathrm{Re}(s) > \frac{1}{2}$ and $s \notin \mathbb{R}$. But

$$\begin{aligned} \lambda\cdot\langle f, f\rangle &= \langle\lambda f, f\rangle = \langle\Delta_v f, f\rangle = \overline{\langle f, \Delta_v f\rangle} = \overline{\langle f, \lambda f\rangle} \\ &= \overline{\lambda}\cdot\overline{\langle f, f\rangle} = \overline{\lambda}\cdot\langle f, f\rangle \end{aligned}$$

We did *not* use symmetry properties of Δ_v, but only that $\langle f, F\rangle = \overline{\langle F, f\rangle}$. Necessarily $E_{1-s,\chi} - c_{1-s,\chi}E_{s,\chi^w} = 0$ for such s. For all $g \in G_{\mathbb{A}}$, by the Identity Principle applied to the $\mathbb{C}$-valued meromorphic functions $s \longrightarrow (E_{1-s,\chi}(g) - c_{1-s,\chi}E_{s,\chi^w}(g))$, the same identity applies for all s away from poles. Since the constant term is identically 0, necessarily $c_{1-s,\chi}c_{s,\chi^w} = 1$. ///

The more general scenario needs some restrictions to stay near enough to the simple case to apply the same causal mechanisms. In particular, generalizing right $K_{\mathbb{A}}$-*invariance*, the function $f \in I_{s,\chi}$ must be right $K_{\mathbb{A}}$-*finite*, in the sense that the collection of right translates of f by $K_{\mathbb{A}}$ spans a finite-dimensional space of functions. For non-archimedean places, this is *equivalent* to being fixed by a finite-index subgroup in $\prod_{v<\infty} K_v$, but for archimedean places, there is no such equivalence. Also, unsurprisingly, the dependence of f on the complex parameter s must also be controlled: take $f(nmk) = (\nu^s\chi)(m)f_o(k)$ with the function f_o on $K_{\mathbb{A}}$ *independent of* s, and *right* $K_{\mathbb{A}}$*-finite*, and write $E(s, \chi, f_o) = E_f$. Of course, to avoid the potential ambiguity due to the nontriviality of $M_{\mathbb{A}} \cap K_{\mathbb{A}}$, it must be that χ is trivial on $M^1 \cap K_{\mathbb{A}}$, and $f_o(mk) = \chi(m) f_o(k)$ for $m \in M^1$ and $k \in K_{\mathbb{A}}$, or else $f = 0$. The scattering operator $C_{s,\chi}$ not only flips $s \to 1 - s$ and $\chi \to \chi^w$, but also acts on the function f_o by (possibly a meromorphic continuation of)

$$(C_{s,\chi} f_o)(k) = \int_{N_{\mathbb{A}}} f(wnk)dk \qquad (\text{for } k \in K_{\mathbb{A}})$$

The constant term becomes

$$c_P E_f = c_P E(s, \chi, f_o) = \nu^s \chi \otimes f_o + \nu^{1-s} \chi^w \otimes C_{s,\chi}(f_o)$$

[2.9.8] Theorem: $E(s, \chi, f_o)$ has a *meromorphic continuation* in $s \in \mathbb{C}$, as a smooth function of moderate growth on $\Gamma \backslash G$. As a function of s, $E(s, \chi, f_o)(g)$ is of at most polynomial growth vertically, uniformly in bounded strips, uniformly for g in compacts. *(Proofs in [2.B] and [11.5].)*

The general analogue of the argument in the special case proves meromorphic continuation of scattering matrix/operators, with the qualification that they be restricted to $K_{\mathbb{A}}$*-finite* functions $I^{\text{fin}}_{s,\chi}$ in $I_{s\chi}$, commensurate with the conditions for meromorphic continuation of Eisenstein series.

[2.9.9] Corollary: The scattering matrix/operator $C_{s,\chi}$ restricted to a map $C^{\text{fin}}_{s,\chi} : I^{\text{fin}}_{s,\chi} \to I^{\text{fin}}_{1-s,\chi^w}$, has a meromorphic continuation.

Proof: The appropriate sense of meromorphic continuation is that $C_{s,\chi} f$ has a meromorphic continuation as a $I^{\text{fin}}_{1-s,\chi^w}$-valued function for every $f \in I^{\text{fin}}_{s,\chi}$. The meromorphic continuation of E_f gives the meromorphic continuation of $c_P E_f = f + C_{s,\chi} f$, and the special form of f ensures that $s \to f$ is *entire*, so $C_{s,\chi} f$ has a meromorphic continuation. ///

[2.9.10] Corollary: For f_o as in the theorem, the *functional equation* $E(1 - s, \chi, f_o) = E(s, \chi^w, C_{1-s,\chi^w} f_o)$ holds, and $C_{1-s,\chi^w} \circ C_{s,\chi} = 1$. The operator $C^{\text{fin}}_{s,\chi}$ has poles exactly where E_f has a pole for some $f \in I^{\text{fin}}_{s,\chi}$.

Proof: Arranging to cancel the ν^s part of the constant terms,

$$\begin{aligned} &c_P \big(E(1 - s, \chi, f_o) - E(s, \chi^w, C_{1-s,\chi^w} f_o)\big) \\ &= \big(\nu^{1-s} \chi \otimes f_o + \nu^s \chi^w \otimes C_{1-s,\chi} f_o\big) \\ &\quad - \big(\nu^s \chi^w \otimes C_{1-s,\chi} f_o + \nu^{1-s} \chi \otimes C_{1-s,\chi^w} C_{1-s,\chi} f_o\big) \\ &= \nu^{1-s} \chi \otimes \big(f_o - C_{1-s,\chi^w} C_{1-s,\chi} f_o\big) \end{aligned}$$

The theory of the constant term [8.2] implies that Eisenstein series $E(s, \chi, f_o)$ are asymptotic to their constant terms. In $\text{Re}(s) > \frac{1}{2}$, the function ν^{1-s} is in L^2 on Siegel sets, so $E(1-s, \chi, f_o) - E(s, \chi^w, C_{1-s,\chi^w} f_o)$ is in L^2. However, the eigenvalues of the invariant Casimir operators Ω_v at archimedean places are not real in $\text{Re}(s) > \frac{1}{2}$ off the real line, so this difference must be 0. This holds for all f_o. ///

2.10 Truncation and Maaß-Selberg Relations

The genuine Eisenstein series are not in $L^2(Z^+G_k\backslash G_{\mathbb{A}})$, but from the theory of the constant term [8.2], the only obstruction is the constant term, which is sufficiently altered by *truncation*. The Maaß-Selberg relations are computation of the L^2 inner products of the resulting truncated Eisenstein series.

The *truncation operators* $\wedge^T$ for large positive real T act on an automorphic form f by killing off f's constant term on $g = nmk$ for large $\nu(m)$. Thus, for a right $K_{\mathbb{A}}$-invariant function, one might imagine that

$$(\textit{naive } T\text{-truncation of } f)(nmk) = \begin{cases} f(g) & \text{for } \nu(m) \le T \\ f(g) - c_P f(g) & \text{for } \nu(m) > T \end{cases}$$

This is flawed. On a standard Siegel set $\mathfrak{S}_{t,C}$, this description is good, but it fails to describe the truncated function on the whole group $G_{\mathbb{A}}$, in the sense that this failed truncation is not an *automorphic form*, that is, as a left Z^+G_k-invariant function. Truncation should produce automorphic forms. For sufficiently large T the same effect is achieved by first defining the *tail* $c_P^T f$ of the constant term $c_P f$ of f:

$$c_P^T f(nmk) = \begin{cases} 0 & (\text{for } \nu(m) \le T) \\ c_P f(nmk) & (\text{for } \nu(m) > T) \end{cases}$$

Although $c_P^T f$ need not be smooth, nor compactly supported, by design, for T large, its support is sufficiently high to control analytical issues: writing $\Psi(\varphi) = \Psi_\varphi$ for legibility,

[2.10.1] Claim: For T sufficiently large, the pseudo-Eisenstein series $\Psi(c_P^T f)$ is a locally finite sum, hence, uniformly convergent on compacts.

Proof: The tail $c_P^T f$ is left $N_{\mathbb{A}}$-invariant. The reduction theory of [2.2] shows that a set $\{nmk : \nu(m) \ge t_o\}$ does not meet $\gamma \cdot \{nmk : \nu(m)y \ge t\}$ for $\gamma \in G_k$ unless $\gamma \in P_k$, for large-enough t depending on t_o. Thus, for large-enough T, the set $S = \{nmk : \nu(m) \ge T\}$ does not meet its translate $\gamma \cdot S$ unless $\gamma \in P_k$. Thus, $\gamma_1 \cdot S$ does not meet $\gamma_2 \cdot S$ unless $\gamma_1 P_k = \gamma_2 P_k$. ///

Similarly,

[2.10.2] Claim: On a standard Siegel set $\mathfrak{S}_{t,C}$, $\Psi(c_P^T f) = c_P^T f$ for all T sufficiently large depending on t.

Proof: By reduction theory [2.2], for large-enough T depending on t_o, a set $\{nmk : \nu(m) > t_o\}$ does not meet $\gamma \cdot \{nmk : \nu(m) > T\}$ unless $\gamma \in P_k$. Thus, for large-enough T, $\{namk : \nu(m) > T\}$ does not meet $\mathfrak{S}_{t_o,C}$ unless $\gamma \in P_k$, and the sole nonzero summand is $c_P^T f$. ///

A proper definition of the *truncation operator* $\wedge^T$ is

$$\wedge^T f = f - \Psi\big(c_P^T f\big)$$

The critical effect of the truncation procedure is to have

[2.10.3] Corollary: For $K_{\mathbb{A}}$-finite $f \in I_{s,\chi}$, for s away from poles, the truncated Eisenstein series $\wedge^T E_f$ is of rapid decay in all Siegel sets.

Proof: By the previous claim and by the theory of the constant term [8.2], $E_f - c_P E_f$ is of rapid decay in standard Siegel sets. (Meromorphic continuation uses $K_{\mathbb{A}}$-finiteness.) ///

Surprisingly, inner products of truncated Eisenstein series have a useful explication. Let

$$X^- = \{g \in Z^+ N_{\mathbb{A}} M_k \backslash G_{\mathbb{A}} : \eta(g) < T\}$$
$$X^+ = \{g \in Z^+ N_{\mathbb{A}} M_k \backslash G_{\mathbb{A}} : \eta(g) \geq T\}$$

[2.10.4] Theorem: *(Maaß-Selberg relation)* Given χ, χ' characters of $M_k \backslash M_{\mathbb{A}}$ and $f \in I_{s,\chi}$ and $f' \in I_{r,\chi'}$

$$\int_{Z^+ G_k \backslash G_{\mathbb{A}}} \wedge^T E_f \cdot \overline{\wedge^T E_{f'}}$$

$$= \int_{X^-} f \cdot \overline{f'} + \int_{X^-} f \cdot \overline{C_{r,\chi'}(f')} - \int_{X^+} C_{s,\chi}(f) \cdot \overline{f'} - \int_{X^+} C_{s,\chi}(f) \cdot \overline{C_{r,\chi'}(f')}$$

$$= \frac{T^{s+\overline{r}-1}}{s+\overline{r}-1} \int_{M_k \backslash M^1} \chi\overline{\chi'} \int_{K_{\mathbb{A}}} f \cdot \overline{f'} + \frac{T^{(1-s)+\overline{r}-1}}{(1-s)+\overline{r}-1} \int_{M_k \backslash M^1} \chi^w \overline{\chi'} \int_{K_{\mathbb{A}}} C_{s,\chi}(f) \cdot \overline{f'}$$

$$+ \frac{T^{s+(1-\overline{r})-1}}{s+(1-\overline{r})-1} \int_{M_k \backslash M^1} \chi \overline{\chi'^w} \int_{K_{\mathbb{A}}} f \cdot \overline{C_{r,\chi'}(f')}$$

$$+ \frac{T^{(1-s)+(1-\overline{r})-1}}{(1-s)+(1-\overline{r})-1} \int_{M_k \backslash M^1} \chi^w \overline{\chi'^w} \int_{K_{\mathbb{A}}} C_{s,\chi}(f) \cdot \overline{C_{r,\chi'}(f')}$$

[2.10.5] Remark: The integrals over $M_k\backslash M^1$ are 0 unless the integrand is the trivial character on M^1.

Proof: Because the tail of the constant term of $E_{f'}$ is orthogonal to the truncation $\wedge^T E_f$ of E_f,

$$\int_{Z^+G_k\backslash G_\mathbb{A}} \wedge^T E_f \cdot \overline{\wedge^T E_{f'}} = \int_{Z^+G_k\backslash G_\mathbb{A}} \wedge^T E_f \cdot \overline{E_{f'}}$$

This is

$$\int_{Z^+G_k\backslash G_\mathbb{A}} \left(E_f - \Psi\begin{pmatrix} 0 & (\text{for } \eta < T) \\ f + C_{s,\chi}(f) & (\text{for } \eta \geq T)\end{pmatrix}\right) \cdot \overline{E_{f'}}$$

$$= \int_{Z^+G_k\backslash G_\mathbb{A}} \Psi\begin{pmatrix} f & (\text{for } \eta < T) \\ -C_{s,\chi}(f) & (\text{for } \eta \geq T)\end{pmatrix} \cdot \overline{E_{f'}}$$

Unwinding the awkward pseudo-Eisenstein series gives

$$\int_{Z^+N_\mathbb{A}M_k\backslash G_\mathbb{A}} \int_{N_k\backslash N_\mathbb{A}} \begin{cases} f & (\text{for } \eta < T) \\ -C_{s,\chi}(f) & (\text{for } \eta \geq T)\end{cases} \cdot \overline{E_{f'}}$$

$$= \int_{Z^+N_\mathbb{A}M_k\backslash G_A} \begin{cases} f & (\text{for } \eta < T) \\ -C_{s,\chi}(f) & (\text{for } \eta \geq T)\end{cases} \cdot \left(\int_{N_k\backslash N_\mathbb{A}} \overline{E_{f'}}(ng)\, dn\right) dg$$

$$= \int_{Z^+N_\mathbb{A}M_k\backslash G_A} \begin{cases} f & (\text{for } \eta < T) \\ -C_{s,\chi}(f) & (\text{for } \eta \geq T)\end{cases} \cdot \overline{(f' + C_{r,\chi'}(f'))}$$

$$= \int_{X^-} f \cdot \overline{f'} + \int_{X^-} f \cdot \overline{C_{r,\chi'}(f')} - \int_{X^+} C_{s,\chi}(f) \cdot \overline{f'} - \int_{X^+} C_{s,\chi}(f) \cdot \overline{C_{r,\chi'}(f')}$$

The sets $X^\pm$ are stable under the left action of $M_k\backslash M^1$. Since f is left χ-equivariant for $M_k\backslash M^1$ and $\overline{f'}$ is left $\overline{\chi'}$-equivariant, via the Iwasawa decomposition, noting that $\nu(m)^{-1}\, dn\, dm$ is left Haar measure on $P_\mathbb{A}$,

$$\int_{X^-} f \cdot \overline{f'} = \int_{Z^+M_k\backslash M_\mathbb{A}:\nu<T} \int_{K_\mathbb{A}} f(mk) \cdot \overline{f'(mk)} \quad \nu^{-1}(m)\, dm\, dk$$

$$= \int_{Z^+M_k\backslash M_\mathbb{A}:\nu<T} (\nu^s\chi)(m) \cdot \overline{(\nu^r\chi)(m)}\nu^{-1}(m)\, dm \cdot \int_{K_\mathbb{A}} f(k) \cdot \overline{f'(k)}\, dk$$

The left-most integral is left $M_k\backslash M^1$-equivariant by $\chi\overline{\chi'}$. When this is a nontrivial character, the integral is 0, by the usual *cancellation trick*: with

$m_o \in M^1$ such that $\chi(m_o) \neq \chi'(m_o)$, by replacing m by $m_o m$ in the integral,

$$\int_{Z^+M_k\backslash M_{\mathbb{A}}:\nu<T} (\nu^s\chi)(m) \cdot \overline{\chi'(m)}dm$$

$$= \int_{Z^+M_k\backslash M_{\mathbb{A}}:\nu<T} (\nu^s\chi)(m_o m) \cdot \overline{\chi'(m_o m)}dm$$

$$= \chi\overline{\chi'}(m_o) \int_{Z^+M_k\backslash M_{\mathbb{A}}:\nu<T} (\nu^s\chi)(m) \cdot \overline{\chi'(m)}dm$$

For $\chi' = \chi$, the integral over $M_k\backslash M^1$ gives a volume. What remains is the integral over the image of the fragment $(0, T)$ of the ray $(0, \infty)$, giving

$$\int_0^T y^{s+\overline{r}-1}\frac{dy}{y} = \frac{T^{s+\overline{r}-1}}{s+\overline{r}-1}$$

The other three summands are similarly evaluated. ///

[2.10.6] Corollary: Unless $\chi' = \chi$ or $\chi' = \chi^w$, for $f \in I_{s,\chi}$ and $f' \in I_{r,\chi'}$, the corresponding truncated Eisenstein series $\wedge^T E_f$ and $\wedge^T E_{f'}$ are *orthogonal.*

Proof: For $\chi' \neq \chi$ and $\chi' \neq \chi^w$, all four integrals over $M_k\backslash M^1$ vanish. ///

The situation $\chi^w = \chi$ can be adjusted, by multiplying by $\chi_1(\det g)^{-1}$, to have *trivial central character.* Thus, the following corollary refers essentially to the case of trivial central character:

[2.10.7] Corollary: For characters $\chi' = \chi = \chi^w$ of $M_k\backslash M_{\mathbb{A}}$, and simplest Eisenstein series $E_{s,\chi}$, $E_{r,\chi}$ attached to the everywhere-spherical elements of $I_{s,\chi}$ and $I_{r,\chi}$,

$$\int_{Z^+G_k\backslash G_{\mathbb{A}}} \wedge^T E_{s,\chi}\overline{\wedge^T E_{r,\chi}} = \frac{T^{s+\overline{r}-1}}{s+\overline{r}-1} + \frac{c_{s,\chi}T^{(1-s)+\overline{r}-1}}{(1-s)+\overline{r}-1}$$
$$+ \frac{c_{\overline{r},\overline{\chi}}T^{s+(1-\overline{r})-1}}{s+(1-\overline{r})-1} + \frac{c_{s,\chi}c_{\overline{r},\overline{\chi}}T^{(1-s)+(1-\overline{r})-1}}{(1-s)+(1-\overline{r})-1} \quad ///$$

The following result has a special, more direct argument, but the proof mechanism used here is more broadly applicable.

[2.10.8] Corollary: For $f \in I_{s,\chi}$ of the form $f(nmk) = (\nu^s\chi)(m) \cdot f(k)$ for $n \in N_{\mathbb{A}}$, $m \in M_{\mathbb{A}}$, and $k \in K_{\mathbb{A}}$, with $f|_{K_{\mathbb{A}}}$ independent of s, neither the Eisenstein series E_f nor the scattering operator $C_{s,\chi}$ has any poles in $\mathrm{Re}(s) \geq \frac{1}{2}$ off the interval $(\frac{1}{2}, 1]$. The poles on $(\frac{1}{2}, 1]$, if any, are *simple*. When $\chi \neq \chi^w$, there are *no* poles on $(\frac{1}{2}, 1]$. Any residues in $\text{Re}(s) > \frac{1}{2}$ are square-integrable.

Proof: Suppose E_f has a pole $s_o = \sigma_o + it_o$ of order $\ell > 0$ with $t_o \neq 0$ and $\sigma_o > \frac{1}{2}$. Certainly the order of pole of the constant term can be no greater than that of E_f, so the second summand $C_{s,\chi}(f)$ has a pole of order at most ℓ at $s = s_o$. The first summand, f itself, as a function of s is entire, by the assumptions about the dependence of f on s. Take $r = s = \sigma_o + it$ in the theorem, giving an equality of the form

$$\int_{Z^+G_k\backslash G_{\mathbb{A}}} |\wedge^T E_f|^2 = \frac{T^{2\sigma_o-1}}{2\sigma_o - 1}A_1 + \frac{T^{-2it}}{-2it}A_2 + \frac{T^{2it}}{2it}A_3 + \frac{T^{1-2\sigma_o}}{1-2\sigma_o}A_4$$

The left-hand side of the Maaß-Selberg relation blows up like $(t - t_o)^{-2\ell}$ as $t \to t_o$ on $\mathbb{R}$. The second and third terms blow up at most like $C_{s,\chi}(f)$ does, which is at worst $(t - t_o)^{-\ell}$. The fourth term blows up at worst like $|C_{s,\chi}(f)|^2$, which is at worst $(t - t_o)^{-2\ell}$. Thus, as $t \to t_o$, the left-hand side and the fourth term on the right dominate. However, the left-hand side is *positive*, while the fourth term is *negative*, since $1 - 2\sigma < 0$. That is, there can be no such pole.

Next, let $s_o = \sigma_o$ be a pole of E_f of order $\ell \geq 1$ on $(\frac{1}{2}, 1]$. Looking at the same expression, again, A_1 does not blow up as $t \to t_o = 0$, unlike the previous case the second and third terms blow up at most like $t^{-(\ell+1)}$ since $t_o = 0$, and the fourth again at most like $t^{-2\ell}$. Again, the fourth term is *negative*, and if $\ell > 1$ dominates the right-hand side as $t \to 0$, contradicting the positivity of the left-hand side. Thus, $\ell = 1$, in which case the second and third terms' blow-up may be the same order as the left-hand side, and as the fourth term on the right-hand side. This proves that any pole on $(\frac{1}{2}, 1]$ is *simple*. Further, when $\chi \neq \chi^w$, the second and third terms are identically 0, so there can be no pole on $(\frac{1}{2}, 1]$ in that case.

To prove square-integrability of a residue at $\sigma_o \in (\frac{1}{2}, 1]$, treat the Eisenstein series as a meromorphic function-valued function, as in [15.2]. Its Laurent coefficients coefficients are functions in the same topological vector space, by the vector-valued form of Cauchy's formulas [15.2]. From the Maaß-Selberg expression again, at $r = s = \sigma_o + it$, multiplying through by t^2 and letting $t \to 0$, the first term on the right-hand side disappears, the powers of T in the second and third terms become T^0, giving

$$\int_{Z^+G_k\backslash G_{\mathbb{A}}} |\text{Res}_{\sigma_o} E_f^T|^2 = \frac{\text{Res}_{\sigma_o} A_2}{2} + \frac{\text{Res}_{\sigma_o} A_3}{2} + \frac{T^{1-2\sigma_o}}{1-2\sigma_o} \lim_{t\to t_o} t^2 A_4$$

Since $1 - 2\sigma_o < 0$, the limit of the last term is 0 as $T \to +\infty$, given the square-integrability of the residue. Properties of meromorphic vector-valued functions [15.2] and Gelfand-Pettis integrals [14.1] ensure that taking residues commutes with taking the limit as $T \to \infty$. The two remaining terms are equal, since the pole is on the real line.

Suppose $s_o = \frac{1}{2} + it_o$ is a pole of E_f of order $\ell \geq 1$ with $t \neq 0$. Take $r = s = \sigma + it_o$ with $\sigma > \frac{1}{2}$ in the theorem, giving an equality of the form

$$\int_{Z^+G_k\backslash G_{\mathbb{A}}} |\wedge^T E_f|^2 = \frac{T^{2\sigma-1}}{2\sigma-1}A_1 + \frac{T^{-2it_o}}{-2it_o}A_2 + \frac{T^{2it_o}}{2it_o}A_3 + \frac{T^{1-2\sigma}}{1-2\sigma}A_4$$

The left-hand side is *positive*, and blows up like $(\sigma - \frac{1}{2})^{-2\ell}$ as $\sigma \to \frac{1}{2}^+$, while the first three terms on the right blow up with orders at most 1, and the fourth term is *negative*, impossible.

For a possible pole at $s_o = \frac{1}{2}$, take $r = s = \frac{1}{2} + \frac{\varepsilon}{2}(1+i)$ with $\varepsilon > 0$, giving an equality of the form

$$\int_{Z^+G_k\backslash G_{\mathbb{A}}} |\wedge^T E_f|^2 = \frac{T^{\varepsilon}}{\varepsilon}A_1 + \frac{T^{-i\varepsilon}}{-i\varepsilon}A_2 + \frac{T^{i\varepsilon}}{i\varepsilon}A_3 + \frac{T^{-\varepsilon}}{-\varepsilon}A_4$$

with the second and third terms *absent* unless $\chi^w = \overline{\chi}$. Thus, for $\chi^w \neq \overline{\chi}$, the left-hand side is positive and blows up like $\varepsilon^{-2\ell}$ as $\varepsilon \to 0^+$, while the first term on the right blows up like ε^{-1}, and the fourth term is negative, so this is impossible. For $\chi^w = \overline{\chi}$, for $\mathrm{Re}(s) = \frac{1}{2}$, the functional equation $1 = C^{\mathrm{fin}}_{s,\chi} \circ C^{\mathrm{fin}}_{1-s,\chi^w}$ becomes

$$1 = C^{\mathrm{fin}}_{s,\chi} \circ C^{\mathrm{fin}}_{1-s,\chi^w} = C^{\mathrm{fin}}_{s,\chi} \circ C^{\mathrm{fin}}_{\overline{s},\overline{\chi}}$$

Thus, $(C^{\mathrm{fin}}_{\frac{1}{2},\chi})^2 = 1$, so $C^{\mathrm{fin}}_{s,\chi}$ has neither pole nor zero at $s = \frac{1}{2}$. Thus, the first three terms on the right blow up like ε^{-1}, while the last is negative, impossible. ///

2.11 Decomposition of Pseudo-Eisenstein Series: Level One

From [2.7], the pseudo-Eisenstein series Ψ_φ with $\varphi \in J_\chi$ and varying χ generate the orthogonal complement to cuspforms in $L^2(Z^+G_k\backslash G_{\mathbb{A}})$. Thus, the orthogonal complement to cuspforms is the L^2-closure of the set of these pseudo-Eisenstein series. For this section, we take *trivial central character* and consider only the simplest case, right $K_{\mathbb{A}}$-*invariant* pseudo-Eisenstein series. These are *everywhere-spherical* case, or *level one*. As earlier,

$$\chi\begin{pmatrix} m_1 & 0 \\ 0 & m_2 \end{pmatrix} = \chi_1(m_1) \cdot \chi_2(m_2) \qquad (\text{for } m_1, m_2 \in \mathbb{J}^1)$$

so $\chi_2 = \chi_1^{-1}$, and $\chi^w = \chi^{-1}$. Since $M_k\backslash M^1$ is *compact*, $\overline{\chi} = \chi^{-1}$. Thus, $\overline{\chi} = \chi^{-1} = \chi^w$. The potential ambiguity in the decomposition $g = nmk$ must be accommodated in χ, or else $f = 0$, so χ is *unramified* everywhere locally. Thus, for genuine Eisenstein series E_f, we take $f \in I_{s,\chi}$ right $K_{\mathbb{A}}$-invariant, so necessarily of the form $f(nmk) = (\nu^s\chi)(m)$ for $n \in N_{\mathbb{A}}$, $m \in M^1$, and $k \in K_{\mathbb{A}}$, up to a constant multiple. In principle, the constant multiple could depend on s, but we want $f|_{K_{\mathbb{A}}}$ to be independent of s, for meromorphic continuation. Thus, take $f|_{K_{\mathbb{A}}} = 1$.

The essential harmonic analysis is *Fourier transform* on the real line, as *Mellin transform* on functions on the *ray* $(0, +\infty)$.

From [2.7], pseudo-Eisenstein series Ψ_φ are in $C_c^\infty(Z^+G_k\backslash G_{\mathbb{A}})$, so their integrals against genuine Eisenstein series E_f converge absolutely, since E_f is continuous, even after meromorphic continuation. Thus, even though this $\langle , \rangle$ cannot be the L^2 pairing, since $E_f \notin L^2(Z^+G_k\backslash G_{\mathbb{A}})$, write

$$\langle \Psi_\varphi, E_s \rangle = \int_{Z^+G_k\backslash G_{\mathbb{A}}} \Psi_\varphi \cdot \overline{E_f}$$

First consider $\chi^w = \chi$.

[2.11.1] Theorem: Fix unramified χ with $\chi^w = \chi$. Let $\varphi \in J_\chi$ be right $K_{\mathbb{A}}$-invariant, with trivial central character. Let s_o run over poles of $E_{s,\chi}$ in $\text{Re}(s) \geq \frac{1}{2}$. The pseudo-Eisenstein series Ψ_φ is expressible in terms of genuine Eisenstein series $E_{s,\chi}$ by an integral converging absolutely and uniformly on compacts in $Z^+G_k\backslash G_{\mathbb{A}}$: *pointwise*, uniformly on compacts,

$$\Psi_\varphi = \frac{1}{4\pi i}\int_{\frac{1}{2}-i\infty}^{\frac{1}{2}+i\infty} \langle \Psi_\varphi, E_{s,\chi}\rangle \cdot E_{s,\chi}\, ds + \sum_{\text{Re}(s_o)\geq\frac{1}{2}} \langle \Psi_\varphi, \text{Res}_{s_o} E_{s,\chi}\rangle \cdot \text{Res}_{s_o} E_{s,\chi}(g)$$

$$= \frac{1}{4\pi i}\int_{\frac{1}{2}-i\infty}^{\frac{1}{2}+i\infty} \langle \Psi_\varphi, E_{s,\chi}\rangle \cdot E_{s,\chi}\, ds + \langle \Psi_\varphi,\ \chi_1 \circ \det\rangle \cdot \frac{\chi_1 \circ \det}{|\chi_1 \circ \det|_{L^2}^2}$$

[2.11.2] Remark: By various devices, for example, the Poisson summation argument of [2.B], the only possible pole in $\text{Re}(s) \geq \frac{1}{2}$ is at $s_o = 1$, with residue a constant multiple of $\chi_1 \circ \det$. However, the general pattern of argument does not depend on our fortuitous knowledge of these further details.

Proof: By an easy part of the *Paley-Wiener* theorem, the Mellin transform of $\varphi_\infty \in C_c^\infty(0, \infty)$ is *entire* and has rapid decay *vertically*, and Mellin inversion is

$$\varphi_\infty(y) = \frac{1}{2\pi i}\int_{\sigma-i\infty}^{\sigma+i\infty} \left(\int_0^\infty \varphi_\infty(r)\, r^{-s}\, \frac{dr}{r}\right) y^s\, ds$$

for *any* real σ. With

$$\varphi(za_y nmk) = \varphi_\infty(y)\chi(m)$$

for $z \in Z^+$, $a_y = \begin{pmatrix} \delta(y) & 0 \\ 0 & 1 \end{pmatrix}$, $y > 0, n \in N_{\mathbb{A}}, m \in M^1, k \in K_{\mathbb{A}}$, define a kind of Mellin transform $J_\chi \to I_{s,\chi}$ by

$$\mathcal{M}\varphi(s)(g) = \int_0^\infty r^{-s}\,\varphi(a_r g)\frac{dr}{r}$$

with $a_r = \begin{pmatrix} \delta(r) & 0 \\ 0 & 1 \end{pmatrix}$ with $r > 0$. We decompose the pseudo-Eisenstein series Ψ_φ along the ray $(0, \infty)$:

$$\Psi_\varphi(g) = \sum_{\gamma \in P_k \backslash G_k} \varphi(\gamma g) = \frac{1}{2\pi i} \sum_{\gamma \in P_k \backslash G_k} \int_{\sigma - i\infty}^{\sigma + i\infty} \mathcal{M}\varphi(s)(\gamma g)\,ds$$

Since χ is specified, and φ is right $K_{\mathbb{A}}$-invariant, in fact

$$\mathcal{M}\varphi(s)(za_y nmk) = \mathcal{M}\varphi(s)(1) \cdot y^s\,\chi(m)$$

Thus, although $\mathcal{M}\varphi(s)$ is a function on $G_{\mathbb{A}}$ for each s, it is simply a scalar multiple of the everywhere-spherical function in $I_{s,\chi}$. Thus, for subsequent computations, suppress the argument $g \in G_{\mathbb{A}}$, and just write

$$\mathcal{M}\varphi(s) = \int_0^\infty r^{-s}\,\varphi(a_r)\frac{dr}{r}$$

with $a_r = \begin{pmatrix} \delta(r) & 0 \\ 0 & 1 \end{pmatrix}$ with $r > 0$, and, commensurately,

$$\Psi_\varphi(g) = \sum_{\gamma \in P_k \backslash G_k} \varphi(\gamma g) = \frac{1}{2\pi i} \sum_{\gamma \in P_k \backslash G_k} \int_{\sigma - i\infty}^{\sigma + i\infty} \mathcal{M}\varphi(s) \cdot y_{\gamma g}^s\,\chi(m_{\gamma g})ds$$

Taking $\sigma = 0$ would be natural, but with $\sigma = 0$ the double integral (sum and integral) is not absolutely convergent, and *the two integrals cannot be interchanged.* For $\sigma > 1$, the Eisenstein series is absolutely convergent, so the rapid vertical decrease of $\mathcal{M}\varphi$ in s makes the double integral absolutely convergent, and by Fubini the two integrals can be interchanged:

$$\Psi_\varphi(g) = \frac{1}{2\pi i} \int_{\sigma - i\infty}^{\sigma + i\infty} \mathcal{M}\varphi(s) \cdot \left(\sum_{\gamma \in P_k \backslash G_k} y_{\gamma g}^s\,\chi(m_{\gamma g}) \right) ds$$

with $\sigma > 1$. The inner sum is the everywhere spherical *Eisenstein series* $E_{s,\chi}$, so, pointwise in $g \in G_{\mathbb{A}}$,

$$\Psi_\varphi = \frac{1}{2\pi i}\int_{\sigma-i\infty}^{\sigma+i\infty} \mathcal{M}\varphi(s)\cdot E_{s,\chi}\,ds \qquad (\text{for } \sigma > 1)$$

Although this does express Ψ_φ as a superposition of eigenfunctions $E_{s,\chi}$ for invariant Laplacians and for spherical Hecke operators, it is unsatisfactory, because it should not refer to $\mathcal{M}\varphi$, but to Ψ_φ, to have an *intrinsic* integral formula. Elimination of this issue is the remainder of the argument.

We move the line of integration in

$$\Psi_\varphi = \frac{1}{2\pi i}\int_{\sigma-i\infty}^{\sigma+i\infty} \mathcal{M}\varphi(s)\cdot E_{s,\chi}\,ds \qquad (\text{for } \sigma > 1)$$

to the left, to $\sigma = 1/2$, which is stabilized by the functional equation of $E_{s,\chi}$. From the corollary [2.10.8] to the Maaß-Selberg relations, there are only finitely many poles of E_s in $\mathrm{Re}(s) \geq \frac{1}{2}$, removing one possible obstacle to the contour move. From the theorem [2.B], [11.5] on meromorphic continuation, we know that even the meromorphically continued $E_{s,\chi}$ is of polynomial growth vertically in s, uniformly in bounded strips in s, uniformly for g in compacts. Thus, we may move the contour, picking up finitely many residues:

$$\Psi_\varphi = \frac{1}{2\pi i}\int_{\frac{1}{2}-i\infty}^{\frac{1}{2}+i\infty} \mathcal{M}\varphi(s)\cdot E_{s,\chi}\,ds + \sum_{s_o} \mathcal{M}\varphi(s_o)\cdot \mathrm{Res}_{s_o} E_{s,\chi}$$

since the poles of $E_{s,\chi}$ in $\mathrm{Re}(s) > \frac{1}{2}$ are simple and $\mathcal{M}\varphi(s)$ is entire in s. The $1/2\pi i$ from inversion cancels the $2\pi i$ in the residue formula. The integral in the expression of Ψ_φ in terms of $E_{s,\chi}$ can be *folded in half*, integrating from $\frac{1}{2} + i0$ to $\frac{1}{2} + i\infty$ rather than from $\frac{1}{2} - i\infty$ to $\frac{1}{2} + i\infty$:

$$\Psi_\varphi - (\text{residual part}) = \frac{1}{2\pi i}\int_{\frac{1}{2}-i\infty}^{\frac{1}{2}+i\infty} \mathcal{M}\varphi(s)\cdot E_{s,\chi}(g)\,ds$$

$$= \frac{1}{2\pi i}\int_{\frac{1}{2}+i0}^{\frac{1}{2}+i\infty} \mathcal{M}\varphi(s)\,E_{s,\chi} + \mathcal{M}\varphi(1-s)\,E_{1-s,\chi}\,ds$$

The functional equation is $E_{1-s,\chi} = c_{1-s,\chi} E_{s,\chi^w}$, and we are assuming $\chi^w = \chi$, so

$$\Psi_\varphi - (\text{residual}) = \frac{1}{2\pi i} \int_{\frac{1}{2}+i0}^{\frac{1}{2}+i\infty} \mathcal{M}\varphi(s) E_{s,\chi} + \mathcal{M}\varphi(1-s) c_{1-s,\chi} E_{s,\chi} \, ds$$

To rewrite this in terms of Ψ_φ, use the adjunction/unwinding property of Ψ_φ:

$$\begin{aligned}
\langle \Psi_\varphi, E_{s,\chi} \rangle &= \int_{Z^+ N_{\mathbb{A}} M_k \backslash G_{\mathbb{A}}} \varphi \cdot \overline{c_P E_{s,\chi}} \\
&= \int_{Z^+ N_{\mathbb{A}} M_k \backslash G_{\mathbb{A}}} \varphi(g) \cdot \overline{(y_g^s \chi(m_g) + c_{s,\chi} y_g^{1-s} \chi^w(m_g))} \, dg \\
&= \int_{Z^+ N_{\mathbb{A}} M_k \backslash P_{\mathbb{A}}} \int_{K_{\mathbb{A}}} \varphi(pk) \cdot \overline{(y_{pk}^s \chi(m_{pk} + c_{s,\chi} y_{pk}^{1-s} \chi^w(m_{pk}))} dp \, dk \\
&= \int_{K_{\mathbb{A}}} 1 dk \int_0^\infty \int_{M_k \backslash M^1} \varphi(a_y) \chi(m) \overline{(y^s \chi(m) + c_{s,\chi} y^{1-s} \chi^w(m))} dm \frac{dy}{y^2} \\
&= \int_0^\infty \int_{M_k \backslash M^1} \varphi(a_y) \chi(m) \cdot \overline{(y^s \chi(m) + c_{s,\chi} y^{1-s} \chi^w(m))} dm \frac{dy}{y^2} \\
&= \int_0^\infty \varphi(a_y) \cdot \overline{(y^s + c_{s,\chi} y^{1-s})} \frac{dy}{y^2}
\end{aligned}$$

by using the Iwasawa decomposition, the right $K_{\mathbb{A}}$-invariance, and $\chi^w = \chi$. On $\mathrm{Re}(s) = \frac{1}{2}$, this is

$$\begin{aligned}
\int_0^\infty \varphi(a_y) \cdot (y^{1-s} + c_{1-s,\overline{\chi}} y^s) \frac{dy}{y^2} &= \int_0^\infty \varphi(a_y) \cdot (y^{-s} + c_{1-s,\overline{\chi}} y^{-(1-s)}) \frac{dy}{y} \\
&= \mathcal{M}\varphi(s) + c_{1-s,\overline{\chi}} \mathcal{M}\varphi(1-s)
\end{aligned}$$

Using $\overline{\chi} = \chi$,

$$\langle \Psi_\varphi, E_{s,\chi} \rangle = \mathcal{M}\varphi(s) + c_{1-s,\chi} \mathcal{M}\varphi(1-s)$$

Thus,

$$\Psi_\varphi - (\text{residual}) = \frac{1}{2\pi i} \int_{\frac{1}{2}+i0}^{\frac{1}{2}+i\infty} \langle \Psi_\varphi, E_{s,\chi} \rangle \cdot E_{s,\chi} \, ds$$

The integral can be restored to be over the whole line $\mathrm{Re}(s) = \frac{1}{2}$, since the integrand is invariant under $s \to 1-s$: by the functional equations of $E_{s,\chi}$

and $c_{s,\chi}$,

$$\begin{aligned}\langle \Psi_\varphi, E_{1-s,\chi}\rangle \cdot E_{1-s,\chi} &= \langle \Psi_\varphi, c_{1-s,\chi}E_{s,\chi}\rangle \cdot c_{1-s,\chi}E_{s,\chi} \\ &= \overline{c_{1-s,\chi}}c_{1-s,\chi}\langle \Psi_\varphi, E_{s,\chi}\rangle \cdot E_{s,\chi} = \langle \Psi_\varphi, E_{s,\chi}\rangle \cdot E_{s,\chi}\end{aligned}$$

Thus, dividing by 2,

$$\Psi_\varphi - (\text{residual part}) = \frac{1}{4\pi i}\int_{\frac{1}{2}-i\infty}^{\frac{1}{2}+i\infty} \langle \Psi_\varphi, E_{s,\chi}\rangle \cdot E_{s,\chi}\, ds$$

It remains to explicate the finitely many residual contributions $\mathcal{M}\varphi(s_o) \cdot \mathrm{Res}_{s_o} E_{s,\chi}$. In fact, by [2.10.8] or [2.B], there are no poles unless $\chi_1/\chi_2 = 1$ on $\mathbb{J}^1$, and then the only pole in that region is at $s_o = 1$, with residue $\chi_1 \circ \det$. However, we want to illustrate more widely applicable methods, as follows.

As do the pseudo-Eisenstein series, $E_{s,\overline{\chi}}$ fits into an *adjunction*

$$\int_{Z^+G_k\backslash G_\mathbb{A}} f \cdot E_{s,\overline{\chi}} = \int_{Z^+N_\mathbb{A}M_k\backslash G_\mathbb{A}} c_P f(g) \cdot y_g^s\, \overline{\chi}(m_g)\, dg$$

for f on $Z^+G_k\backslash G_\mathbb{A}$, whenever the implied integrals converge absolutely. By the identity principle from complex analysis, the same formula holds for the meromorphic continuation of $E_{s,\overline{\chi}}$ for s away from poles. For right $K_\mathbb{A}$-invariant f, via Iwasawa decomposition,

$$\int_{Z^+N_\mathbb{A}M_k\backslash G_\mathbb{A}} c_P f(g) \cdot y_g^s\, \overline{\chi}(m_g)\, dg = \int_0^\infty \int_{M_k\backslash M^1} c_P f(a_y m) \cdot y^s\, \overline{\chi}(m)\, \frac{dy}{y^2}$$

even though $P_\mathbb{A} \cap K_\mathbb{A}$ is not simply $\{1\}$. The integration over the compact group $M_k\backslash M^1$ computes the χ-component $(c_P f)^\chi$ of $c_P f$ with respect to the left action of M^1:

$$\begin{aligned}\int_0^\infty \int_{M_k\backslash M^1} c_P f(a_y m) \cdot y^s\, \overline{\chi}(m)\, \frac{dy}{y^2} &= \int_0^\infty (c_P f)^\chi(a_y) \cdot y^{s-1}\, \frac{dy}{y} \\ &= \mathcal{M}(c_P f)^\chi(1-s)\end{aligned}$$

On $\mathrm{Re}(s) = \frac{1}{2}$, where $\overline{s} = 1 - s$, using $1 - (1-s) = s$,

$$\begin{aligned}\langle f, E_{s,\chi}\rangle &= \int_{Z^+G_k\backslash G_\mathbb{A}} f \cdot \overline{E_{s,\chi}} = \int_{Z^+G_k\backslash G_\mathbb{A}} f \cdot E_{\overline{s},\overline{\chi}} \\ &= \int_{Z^+G_k\backslash G_\mathbb{A}} f \cdot E_{1-s,\overline{\chi}} = \mathcal{M}(c_P f)^\chi(s)\end{aligned}$$

Taking f to be the pseudo-Eisenstein series Ψ_φ,

$$\langle \Psi_\varphi, E_{s,\chi}\rangle = \mathcal{M}(c_P \Psi_\varphi)^\chi(s) \qquad (\text{on } \mathrm{Re}(s) = \tfrac{1}{2})$$

At a pole s_o of $E_{s,\chi}$ in $\text{Re}(s) \geq \frac{1}{2}$, $c_{s,\chi}$ also has a pole of the same order. Since $c_{s,\chi} \cdot c_{1-s,\chi} = 1$ for $\chi^w = \chi$, necessarily $c_{1-s,\chi}$ has a zero at s_o. Thus, from

$$\mathcal{M}c_P\Psi_\varphi(s) = \langle \Psi_\varphi, E_{s,\chi}\rangle = \mathcal{M}\varphi(s) + c_{1-s,\chi}\mathcal{M}\varphi(1-s)$$

at a pole s_o of E_s

$$\begin{aligned}\mathcal{M}c_P\Psi_\varphi(s_o) &= \mathcal{M}\varphi(s_o) + c_{1-s_o,\chi}\mathcal{M}\varphi(1-s_o)\\ &= \mathcal{M}\varphi(s_o) + 0 \cdot \mathcal{M}\varphi(1-s_o) = \mathcal{M}\varphi(s_o)\end{aligned}$$

That is, the value $\mathcal{M}c_P\Psi_\varphi$ at s_o is just the value of $\mathcal{M}\varphi$ there, so the coefficients appearing in the decomposition of Ψ_φ *are intrinsic*. Thus, the preceding decomposition has an intrinsic form as in the statement of the theorem. This completes the argument for the decomposition of right $K_\mathbb{A}$-invariant pseudo-Eisenstein series Ψ_φ with $\varphi \in J_\chi$, with trivial central character, and $\chi^w \neq \chi$. In fact, the residues at poles are constant multiples of $\chi_1(\det g)$, from [2.B].///

Still with trivial central character and right $K_\mathbb{A}$-invariance, consider $\chi^w \neq \chi$:

[2.11.3] Theorem: Fix unramified χ with $\chi^w \neq \chi$. Let $\varphi \in J_\chi$ be right $K_\mathbb{A}$-invariant, with trivial central character. The pseudo-Eisenstein series Ψ_φ is expressible in terms of genuine Eisenstein series $E_{s,\chi}$ by an integral converging absolutely and uniformly on compacts in $Z^+G_k\backslash G_\mathbb{A}$:

$$\Psi_\varphi(g) = \frac{1}{4\pi i}\int_{\frac{1}{2}-i\infty}^{\frac{1}{2}+i\infty} \langle \Psi_\varphi, E_{s,\chi}\rangle \cdot E_{s,\chi}\, ds$$

[2.11.4] Remark: As in the corollary [2.10.8] to the Maaß-Selberg relation, there is no pole at all unless $\chi_1/\chi_2 = 1$ on $\mathbb{J}^1$. This absence of poles is also visible by the Poisson summation argument [2.B].

Proof: As in the situation of the previous theorem, pointwise in $g \in G_\mathbb{A}$,

$$\Psi_\varphi = \frac{1}{2\pi i}\int_{\sigma-i\infty}^{\sigma+i\infty} \mathcal{M}\varphi(s) \cdot E_{s,\chi}\, ds \qquad (\text{for } \sigma > 1)$$

Move the line of integration to the left, to $\sigma = 1/2$, using the lack of poles for these Eisenstein series in $\text{Re}(s)$,

$$\Psi_\varphi = \frac{1}{2\pi i}\int_{\frac{1}{2}-i\infty}^{\frac{1}{2}+i\infty} \mathcal{M}\varphi(s) \cdot E_{s,\chi} ds$$

To rewrite this in terms of Ψ_φ, by the adjunction/unwinding property of Ψ_φ,

$$\begin{aligned}\langle \Psi_\varphi, E_{s,\chi}\rangle &= \int_{Z^+N_{\mathbb{A}}M_k\backslash G_{\mathbb{A}}} \varphi \cdot \overline{c_P E_{s,\chi}} \\ &= \int_{Z^+N_{\mathbb{A}}M_k\backslash G_{\mathbb{A}}} \varphi(g)\cdot \overline{(y_g^s\chi(m_g) + c_{s,\chi}y_g^{1-s}\chi^w(m_g))}\,dg \\ &= \int_0^\infty \int_{M_k\backslash M^1} \varphi(a_y)\chi(m)\cdot \overline{(y^s\chi(m) + c_{s,\chi}y^{1-s}\chi^w(m))}dm\frac{dy}{y^2} \\ &= \int_0^\infty \varphi(a_y)\cdot \overline{y^s}\frac{dy}{y^2}\end{aligned}$$

by using the Iwasawa decomposition and the right $K_{\mathbb{A}}$-invariance, since φ is left χ-equivariant under M^1 and $\chi^w \neq \chi$. On $\mathrm{Re}(s) = \frac{1}{2}$, this is

$$\langle \Psi_\varphi, E_{s,\chi}\rangle = \int_0^\infty \varphi(a_y)\cdot y^{1-s}\frac{dy}{y^2} = \int_0^\infty \varphi(a_y)\cdot y^{-s}\frac{dy}{y} = \mathcal{M}\varphi(s)$$

Since the functional equation $E_{1-s,\chi} = c_{1-s,\chi}E_{s,\chi^w}$ involves E_{s,χ^w} and $\chi^w \neq \chi$, we anticipate needing the complementary computation

$$\begin{aligned}\langle \Psi_\varphi, E_{s,\chi^w}\rangle &= \int_{Z^+N_{\mathbb{A}}M_k\backslash G_{\mathbb{A}}} \varphi \cdot \overline{c_P E_{s,\chi^w}} \\ &= \int_{Z^+N_{\mathbb{A}}M_k\backslash G_{\mathbb{A}}} \varphi(g)\cdot \overline{(y_g^s\chi^w(m_g) + c_{s,\chi^w}y_g^{1-s}\chi(m_g))} \\ &= \int_0^\infty \int_{M_k\backslash M^1} \varphi(a_y)\chi(m)\cdot \overline{(y^s\chi^w(m) + c_{s,\chi^w}y^{1-s}\chi(m))}dm\frac{dy}{y^2} \\ &= \int_0^\infty \varphi(a_y)\cdot \overline{c_{s,\chi^w}y^{1-s}}\frac{dy}{y^2}\end{aligned}$$

using the Iwasawa decomposition and the right $K_{\mathbb{A}}$-invariance, since φ is left χ-equivariant under M^1 and $\chi^w \neq \chi$. On $\mathrm{Re}(s) = \frac{1}{2}$, using $\chi^w = \overline{\chi}$ due to the trivial central character, this is

$$\begin{aligned}\langle \Psi_\varphi, E_{s,\chi^w}\rangle &= \int_0^\infty \varphi(a_y)\cdot c_{1-s,\chi}y^s\frac{dy}{y^2} = \int_0^\infty \varphi(a_y)\cdot c_{1-s,\chi}y^{-(1-s)}\frac{dy}{y} \\ &= c_{1-s,\chi}\mathcal{M}\varphi(1-s)\end{aligned}$$

The integral in the expression of Ψ_φ in terms of $E_{s,\chi}$ can be *folded in half*, integrating from $\frac{1}{2}+i0$ to $\frac{1}{2}+i\infty$ rather than from $\frac{1}{2}-i\infty$ to $\frac{1}{2}+i\infty$:

$$\Psi_\varphi = \frac{1}{2\pi i}\int_{\frac{1}{2}-i\infty}^{\frac{1}{2}+i\infty} \mathcal{M}\varphi(s)\cdot E_{s,\chi}(g)\,ds$$

$$= \frac{1}{2\pi i}\int_{\frac{1}{2}+i0}^{\frac{1}{2}+i\infty} \mathcal{M}\varphi(s)\,E_{s,\chi} + \mathcal{M}\varphi(1-s)\,E_{1-s,\chi}\,ds$$

The functional equation is $E_{1-s,\chi} = c_{1-s,\chi}E_{s,\chi^w}$, and $\chi^w \neq \chi$, so

$$\Psi_\varphi = \frac{1}{2\pi i}\int_{\frac{1}{2}+i0}^{\frac{1}{2}+i\infty} \mathcal{M}\varphi(s)\,E_{s,\chi} + \mathcal{M}\varphi(1-s)\,c_{1-s,\chi}E_{s,\chi^w}\,ds$$

$$= \frac{1}{2\pi i}\int_{\frac{1}{2}+i0}^{\frac{1}{2}+i\infty} \langle\Psi_\varphi, E_{s,\chi}\rangle\cdot E_{s,\chi}\,ds + \frac{1}{2\pi i}\int_{\frac{1}{2}+i0}^{\frac{1}{2}+i\infty} \langle\Psi_\varphi, E_{s,\chi^w}\rangle\cdot E_{s,\chi^w}\,ds$$

using $\langle\Psi_\varphi, E_{s,\chi}\rangle = \mathcal{M}\varphi(s)$ and $\langle\Psi_\varphi, E_{s,\chi^w}\rangle = c_{1-s,\chi}\mathcal{M}\varphi(1-s)$. The integrals can be restored to be over the whole line $\mathrm{Re}(s)=\frac{1}{2}$, since the two integrals are interchanged under $s \to 1-s$:

$$\begin{aligned}
&\langle\Psi_\varphi, E_{1-s,\chi}\rangle\cdot E_{1-s,\chi}\\
&= \langle\Psi_\varphi, c_{1-s,\chi}E_{s,\chi^w}\rangle\cdot c_{1-s,\chi}E_{s,\chi^w}\\
&= \overline{c_{1-s,\chi}}c_{1-s,\chi}\langle\Psi_\varphi, E_{s,\chi^w}\rangle\cdot E_{s,\chi^w}\\
&= c_{s,\chi^w}c_{1-s,\chi}\langle\Psi_\varphi, E_{s,\chi^w}\rangle\cdot E_{s,\chi^w} = 1\cdot\langle\Psi_\varphi, E_{s,\chi^w}\rangle\cdot E_{s,\chi^w}
\end{aligned}$$

and similarly for E_{s,χ^w}. Dividing by 2,

$$\Psi_\varphi = \frac{1}{4\pi i}\int_{\frac{1}{2}-i\infty}^{\frac{1}{2}+i\infty} \langle\Psi_\varphi, E_{s,\chi}\rangle\cdot E_{s,\chi}\,ds$$

$$+ \frac{1}{4\pi i}\int_{\frac{1}{2}-i\infty}^{\frac{1}{2}+i\infty} \langle\Psi_\varphi, E_{s,\chi^w}\rangle\cdot E_{s,\chi^w}\,ds$$

This completes the argument for the decomposition of right $K_{\mathbb{A}}$-invariant pseudo-Eisenstein series Ψ_φ with $\varphi \in J_\chi$, with trivial central character, and $\chi^w \neq \chi$. ///

2.12 Decomposition of Pseudo-Eisenstein Series: Higher Level

A similar argument applies to decomposition of pseudo-Eisenstein series *without* the everywhere-spherical constraint, necessarily tracking the additional information about the restriction of $f \in I_{s,\chi}$ to $(P_{\mathbb{A}} \cap K_{\mathbb{A}})\backslash K_{\mathbb{A}}$. We retain the trivial central character condition, for simplicity. For meromorphic continuation and analytical properties of Eisenstein series, $f|_{K_{\mathbb{A}}}$ should be $K_{\mathbb{A}}$-finite, and $f_{K_{\mathbb{A}}}$ should not depend on s. For simplicity, invariance under $K_\infty = \prod_{v|\infty} K_v$ is still required, so the relaxation of conditions will be at non-archimedean places. Putting $K_{\text{fin}} = \prod_{v<\infty} K_v$, we require K_{fin}-finiteness.

Let $\Theta_v = P_v \cap K_v$, and $\Theta_{\text{fin}} = \prod_{v<\infty} \Theta_v$. With fixed χ, restrictions of $\varphi \in J_\chi$ or $f \in I_{s,\chi}$ to K_{fin} necessarily lie in[8]

$$\Phi = \{u \in C^\infty(K_{\text{fin}}) : u(\theta k) = \chi(\theta) \cdot u(k), \text{for} \theta \in \Theta_{\text{fin}}, k \in K_{\text{fin}}\}$$

where *smooth* in this context means means *locally constant*. There is a natural right K_{fin}-invariant inner product on Φ by

$$\langle u_1, u_2 \rangle = \int_{K_{\mathbb{A}}} u_1 \cdot \overline{u_2}$$

For each irreducible representation ρ of the compact group K_{fin}, let Φ^ρ be the ρ isotypic component in Φ, namely, the sum of all isomorphic copies of ρ inside Φ [9.D.14]. The dimension of the space $\text{Hom}_{K_{\text{fin}}}(\rho, \Phi)$ of K_{fin}-homomorphisms of ρ to Φ is the *multiplicity* of ρ in Φ_{fin}. It is not obvious that the following claim is *true*, nor that it will be *needed* in the proof:

[2.12.1] Claim: Φ is a direct sum of irreducible representations of K_{fin}, and is *multiplicity-free*, in the sense that multiplicity of any irreducible ρ in Φ is at most 1. *(Proof after proof of the theorem.)*

Similarly, let J_χ^ρ be the elements of J_χ which restrict to Φ_{fin}^ρ on K_{fin}, and are right K_∞-invariant. Let $I_{s,\chi}^\rho$ be the collection of elements in $I_{s,\chi}$ which restrict to Φ_{fin}^ρ on K_{fin}, and are right K_∞-invariant. Indeed, for each fixed s, $I_{s,\chi}^\rho$ is in bijection with the space Φ_{fin}^ρ by extending $u \in \Phi_{\text{fin}}^\rho$ by

$$u_{s,\chi}(za_y mk) = y^s \chi(m) u(k_{\text{fin}})$$

for $z \in Z^+$, $n \in N_{\mathbb{A}}$, $y > 0$, $m \in M^1$, $k_o \in K_{\text{fin}}$, $k_\infty \in K_\infty$, with corresponding Eisenstein series

$$E(s, \chi, u)(g) = \sum_{\gamma \in P_k \backslash G_k} u_{s,\chi}(\gamma \cdot g)$$

[8] Again, this Φ is an *induced representation*, but we do not immediately need any properties of such.

Let $u_1, \ldots, u_\ell$ be an orthonormal basis for Φ^ρ, and let $u_{j,s,\chi} \in I^\rho_{s,\chi}$ be the corresponding extensions. We use the fact from [2.B] that the only poles of Eisenstein series are at $s = 1$, and the residues are constant multiples of $\chi_1 \circ \det$.

[2.12.2] Theorem: For $\varphi \in J^\rho_\chi$ and $\chi^w = \chi$, the pseudo-Eisenstein series Ψ_φ is expressible in terms of genuine Eisenstein series $E(s, \chi, u_j)$ by an integral converging absolutely and uniformly on compacts in $Z^+G_k\backslash G_{\mathbb{A}}$: *pointwise*, uniformly on compacts,

$$\Psi_\varphi = \sum_j \frac{1}{4\pi i}\int_{\frac{1}{2}-i\infty}^{\frac{1}{2}+i\infty} \langle \Psi_\varphi, E(s, \chi, u_j)\rangle \cdot E(s, \chi, u_j)ds$$

$$+ \langle \Psi_\varphi, \chi_1 \circ \det\rangle \cdot \frac{\chi_1 \circ \det}{|\chi_1 \circ \det|^2_{L^2}}$$

Proof: To track the dependence on $u \in \Phi$, modify the earlier notation slightly: let

$$\mathcal{M}_s\varphi(g) = \int_0^\infty r^{-s}\, \varphi(a_r g)\frac{dr}{r}$$

with $a_r = \begin{pmatrix} \delta(r) & 0 \\ 0 & 1 \end{pmatrix}$ with $r > 0$. Thus, $\mathcal{M}_s$ is a map $J^\rho_\chi \to I^\rho_{s,\chi}$. By Mellin inversion,

$$\varphi(g) = \frac{1}{2\pi i}\int_{\sigma-i\infty}^{\sigma+i\infty} \mathcal{M}_s\varphi(s)(g)\, ds \qquad (\text{for } \varphi \in J^\rho_\chi)$$

Further, $\mathcal{M}_s\varphi = \sum_{j=1}^{\ell} \langle \mathcal{M}_s\varphi|_{K_{\mathbb{A}}}, u_j\rangle \cdot u_{j,s,\chi}$, so

$$\Psi_\varphi(g) = \sum_{\gamma\in P_k\backslash G_k} \varphi(\gamma\cdot g) = \sum_{\gamma\in P_k\backslash G_k} \frac{1}{2\pi i}\int_{\sigma-i\infty}^{\sigma+i\infty} \mathcal{M}_s\varphi(\gamma g)\, ds$$

$$= \sum_{\gamma\in P_k\backslash G_k}\sum_{j=1}^{\ell} \frac{1}{2\pi i}\int_{\sigma-i\infty}^{\sigma+i\infty} \langle \mathcal{M}_s\varphi|_{K_{\mathbb{A}}}, u_j\rangle \cdot u_{j,s,\chi}(\gamma g)\, ds$$

For $\sigma > 1$, the integral and the infinite sum can be interchanged, giving

$$\sum_{j=1}^{\ell} \frac{1}{2\pi i}\int_{\sigma-i\infty}^{\sigma+i\infty} \langle \mathcal{M}\varphi(s)|_{K_{\mathbb{A}}}, u_j\rangle \cdot \Big(\sum_{\gamma\in P_k\backslash G_k} u_{j,s,\chi}(\gamma g)\Big)\, ds$$

$$= \sum_{j=1}^{\ell} \frac{1}{2\pi i}\int_{\sigma-i\infty}^{\sigma+i\infty} \langle \mathcal{M}_s\varphi|_{K_{\mathbb{A}}}, u_j\rangle \cdot E(s, \chi, u_j)\, ds$$

As in the simpler cases, *fold up* the integral:

$$\int_{\sigma+i0}^{\sigma+i\infty} \langle \mathcal{M}\varphi(s)|_{K_\mathbb{A}}, u_j\rangle E(s,\chi,u_j) + \langle \mathcal{M}\varphi(1-s)|_{K_\mathbb{A}}, u_j\rangle E(1-s,\chi,u_j)\, ds$$

The functional equations of such Eisenstein series can be written

$$E(1-s,\chi,u) = E\big(s,\chi^w, C_{1-s,\chi}(u_{1-s,\chi})\big|_{K_\mathbb{A}}\big)$$

so

$$\Psi_\varphi = \sum_j \int_{\sigma+i0}^{\sigma+i\infty} \langle \mathcal{M}_s\varphi|_{K_\mathbb{A}}, u_j\rangle \cdot E(s,\chi,u_j)$$
$$+ \langle \mathcal{M}_{1-s}\varphi|_{K_\mathbb{A}}, u_j\rangle \cdot E\big(s,\chi^w, C_{1-s,\chi}(u_{j,1-s,\chi})\big|_{K_\mathbb{A}}\big)\, ds$$

Here $\chi^w = \chi$, but it is not obvious that $C_{1-s,\chi}(u_{j,1-s,\chi})|_{K_\mathbb{A}}$ is simply related to u_j, unlike the simple case of right $K_\mathbb{A}$-*invariant* functions. The maps

$$u \longrightarrow u_{1-s,\chi} \longrightarrow C_{1-s,\chi}(u_{1-s,\chi}) \longrightarrow C_{1-s,\chi}(u_{1-s,\chi})\big|_{K_\mathbb{A}} \in \Phi$$

for $u \in \Phi^\rho$ all respect the right translation action of $K_\mathbb{A}$, so the image is again in Φ^ρ. Now we use the following claim: Φ^ρ consists of a single copy of ρ, so the composition of these maps is an automorphism of ρ respecting the action of K_{fin}. Irreducible Hilbert-space representations of compact groups are finite-dimensional [9.C.7], so by a suitable form [9.D.12] of Schur's lemma, $u \to C_{1-s,\chi}(u_{1-s,\chi})|_{K_{\text{fin}}}$ is a *scalar* $c^\rho_{1-s,\chi}$ depending on s, χ, and ρ, but *not* on $u \in \Phi^\rho$:

$$\Psi_\varphi = \sum_j \int_{\sigma+i0}^{\sigma+i\infty} \big(\langle \mathcal{M}_s\varphi|_{K_\mathbb{A}}, u_j\rangle + \langle \mathcal{M}_{1-s}\varphi|_{K_\mathbb{A}}, u_j\rangle c^\rho_{1-s,\chi}\big) \cdot E(s,\chi,u_j)\, ds$$

As in the simpler cases, to express this in terms of Ψ_φ itself, not φ, unwind and use the Iwasawa decomposition:

$$\begin{aligned}
\langle \Psi_\varphi, E(s,\chi,u)\rangle &= \int_{Z^+N_\mathbb{A}M_k\backslash G_\mathbb{A}} \varphi \cdot \overline{c_P E(s,\chi,u)} \\
&= \int_{Z^+N_\mathbb{A}M_k\backslash G_\mathbb{A}} \varphi \cdot \overline{(u_{s,\chi} + C_{s,\chi}u_{s,\chi})} \\
&= \int_{Z^+M_k\backslash M_\mathbb{A}} \int_{K_\mathbb{A}} \varphi(pk) \cdot \overline{(u_{s,\chi}(pk) + C_{s,\chi}u_{s,\chi}(pk)} dp\, dk \\
&= \int_0^\infty \int_{M_k\backslash M^1} \int_{K_\mathbb{A}} \chi(m)\varphi(a_y k) \cdot \overline{(y^s\chi(m)u_{s,\chi}(k)} \\
&\quad \overline{+ y^{1-s}\chi^w(m)C_{s,\chi}(u_{s,\chi})(k))} \frac{dy}{y^2}\, dm\, dk
\end{aligned}$$

On $\mathrm{Re}(s) = \frac{1}{2}$, and with $\chi^w = \chi = \overline{\chi}$, we have $\overline{C_{s,\chi}} = C_{1-s,\overline{\chi}} = C_{1-s,\chi}$, and $\overline{u_{s,\chi}} = \overline{u}_{1-s,\overline{\chi}} = \overline{u}_{1-s,\chi}$. Also, of course, $u_{s,\chi}$ is just u itself on $K_{\mathbb{A}}$. Again use the fact that $u \to C_{1-s,\chi}(u_{1-s,\chi})\big|_{K_{\mathrm{fin}}}$ is a *scalar* $c^{\rho}_{1-s,\chi}$, so this is

$$\int_0^\infty \int_{K_{\mathbb{A}}} \varphi(a_y k) \cdot y^{-s}\overline{u}(k) + \varphi(a_y k) \cdot y^{-(1-s)} C_{1-s,\chi}(\overline{u}_{1-s,\overline{\chi}})(k) \frac{dy}{y}\, dk$$

$$= \int_{K_{\mathbb{A}}} \mathcal{M}_s\varphi(k) \cdot \overline{u}(k)\, dk + \int_{K_{\mathbb{A}}} \mathcal{M}_{1-s}\varphi \cdot C_{1-s,\chi}(\overline{u}_{1-s,\overline{\chi}})(k)\, dk$$

$$= \int_{K_{\mathbb{A}}} \mathcal{M}_s\varphi(k) \cdot \overline{u}(k)\, dk + \int_{K_{\mathbb{A}}} \mathcal{M}_{1-s}\varphi(k) \cdot c^{\rho}_{1-s,\chi} \cdot \overline{u}(k)\, dk$$

$$= \langle \mathcal{M}_s\varphi|_{K_{\mathbb{A}}} + \mathcal{M}_{1-s}\varphi|_{K_{\mathbb{A}}} \cdot c^{\rho}_{1-s,\chi}, u \rangle$$

Thus, the coefficients in the expression for Ψ_φ are these inner products, apart from the residue picked up by moving the contour from $\sigma > 1$ to $\sigma = \frac{1}{2}$. Regardless of choice of the basis u_j, the residues are all constant multiples of $\chi_1 \circ \det$, so their sum must be as indicated. ///

[2.12.3] Theorem: For $\varphi \in J^{\rho}_{\chi}$ and $\chi^w \neq \chi$, the pseudo-Eisenstein series Ψ_φ is expressible in terms of genuine Eisenstein series $E(s, \chi, u_j)$ by an integral converging absolutely and uniformly on compacts in $Z^+G_k\backslash G_{\mathbb{A}}$: *pointwise*, uniformly on compacts,

$$\Psi_\varphi = \sum_j \frac{1}{4\pi i} \int_{\frac{1}{2}-i\infty}^{\frac{1}{2}+i\infty} \langle \Psi_\varphi, E(s, \chi, u_j)\rangle \cdot E(s, \chi, u_j)\, ds$$
$$+ \langle \Psi_\varphi, E(s, \chi^w, u_j)\rangle \cdot E(s, \chi^w, u_j)\, ds$$

(The proof combines the argument from the level-one analogue with the multiplicity-free claim.) ///

Now we prove the multiplicity-free property $\dim_{\mathbb{C}} \mathrm{Hom}_{K_{\mathrm{fin}}}(\rho, \Phi) \leq 1$: First, *continuity* of ρ requires that it restricts to 1 on all but finitely many K_v. Irreducibles of finite products of compact groups are (external) tensor products of irreducibles of the factors [9.C.8]. Thus, it suffices to prove a *local* fact, that

$$\Phi_v = \{u \in C^\infty(K_v) : u(\theta k) = \chi(\theta) \cdot u(k), \text{for } \theta \in \Theta_v, k \in K_v\}$$

is multiplicity-free. By the Gelfand-Kazhdan criterion [6.11], it suffices to find an involutive anti-automorphism σ of K_v such that every left and right Θ_v-invariant *distribution* u on K_v is invariant under σ, $u^\sigma = u$, where $u^\sigma(\varphi) = u(\varphi^\sigma)$, for all $\varphi \in C^\infty_c(K_v)$, where $(\varphi \circ \sigma)(k) = u(k^\sigma)$. We will use $g^\sigma = w(g^\top)w^{-1}$ with w the Weyl element, which stabilizes Θ_v. We find representatives for $\Theta_v\backslash K_v/\Theta_v$, for v non-archimedean. Suppress the subscript v in what follows.

Given $g = \begin{pmatrix} a & b \\ c & d \end{pmatrix} \in K$, for $c = 0$, we have representative 1. For $c \in \mathfrak{o}^\times$, left multiplication by Θ can make $c = 1$, and also $a = 0$ by subtracting an integer multiple of the lower row from the upper. Then (the modified version of) b is in $\mathfrak{o}^\times$, so can be made 1, and right multiplication by Θ makes $d = 0$, by subtracting an integer multiple of the left column from the right. This gives representatives w. For the intermediate cases $0 < \text{ord}c = \ell < \infty$, both a, d must be units for the determinant to be a unit. Right multiplication by Θ makes $b = 0$, by subtracting an integer of the left column from the right, and then $a = d = 1$ by left or right multiplication by Θ. Then left and right multiplication by $\begin{pmatrix} u & 0 \\ 0 & 1 \end{pmatrix}$ with $u \in \mathfrak{o}^\times$ does not disturb $a = d = 1$, and makes $c = \varpi^\ell$ with chosen local parameter ϖ. Thus, there are representatives

$$r_\infty = \begin{pmatrix} 1 & 0 \\ 0 & 1 \end{pmatrix}, r_o = \begin{pmatrix} 0 & 1 \\ 1 & 0 \end{pmatrix}, r_1 = \begin{pmatrix} 1 & 0 \\ \varpi & 1 \end{pmatrix},$$
$$r_2 = \begin{pmatrix} 1 & 0 \\ \varpi^2 & 1 \end{pmatrix}, r_3 = \begin{pmatrix} 1 & 0 \\ \varpi^3 & 1 \end{pmatrix}, \ldots$$

Each representative is fixed under σ, so the double cosets are stabilized by σ. Every double coset $\Theta r \Theta$ is *closed*, being continuous images of compacts. The double coset $\Theta r_\infty \Theta = \Theta$ is not open, but all other double cosets $\Theta r_\ell \Theta$ are open, being defined by the open condition $|\varpi^{\ell+1}| < |c| < |\varpi^{\ell-1}|$. The characteristic function φ_ℓ of $\Theta r_\ell \Theta$ is a test function for $\ell < \infty$. The integrations

$$u_\infty(\varphi) = \int_\Theta \varphi \qquad u_o(\varphi) = \int_{\Theta r_o \Theta} \varphi \qquad u_1(\varphi) = \int_{\Theta r_1 \Theta} \varphi \qquad \ldots$$

for $\varphi \in C_c^\infty(K)$ are left and right Θ-invariant, and σ-invariant.

The uniqueness of invariant functionals [14.4] shows that u_ℓ is the unique $\Theta \times \Theta$-invariant distribution on K supported on the compact, open set $\Theta r_\ell \Theta$ for $\ell < \infty$, up to constants. Left and right Θ-invariant distributions factor through the two-sided averaging map $\varphi \to \int_{\Theta\times\Theta} \varphi(\theta k \theta')\, d\theta\, d\theta'$. The space $\mathcal{D} = C_c^\infty(K)$ is the colimit over compact open subgroups H of the finite-dimensional spaces $\mathcal{D}^H$ of test functions left and right H-invariant: indeed, by smoothness, $u \in \mathcal{D}$ is left H_1-invariant and right H_2-invariant for some compact-open subgroups H_i, and take $H = H_1 \cap H_2$. For $\varphi \in \mathcal{D}^H$, the $\Theta \times \Theta$-averaged φ is *constant* on $\Theta H \Theta$. The representatives r_ℓ approach $r_\infty = 1_2 \in K$. Thus, $\Theta r_\ell \Theta$ for every $\ell \geq \ell_o$ with $\ell_o = \ell_o(H)$ depending on H. Letting $\text{ch}_{\Theta H \Theta}$ be the characteristic function of $\Theta H \Theta$,

$$\begin{aligned} u(\varphi) &= u(\varphi - \varphi(0) \cdot \text{ch}_{\Theta H \Theta}) + \varphi(0) \cdot u(\text{ch}_{\Theta H \Theta}) \\ &= u(\varphi - \varphi(0) \cdot \text{ch}_{\Theta H \Theta}) + u_\infty(\varphi) \cdot u(\text{ch}_{\Theta H \Theta}) \end{aligned}$$

The test function $\varphi - \varphi(0) \cdot \mathrm{ch}_{\Theta H \Theta})$ is supported on the finitely many double cosets $\Theta r_\ell \Theta$ with $0 \le \ell < \ell_o(H)$. The restriction of u to test functions supported on this finite union of compact, open double cosets is a linear combination of $u_0, u_1, \ldots, u_{\ell_o - 1}$, and the constant $u(\mathrm{ch}_{\Theta H \Theta})$ does not depend on the individual $\varphi \in \mathcal{D}^H$. Thus, the restriction of u to $\mathcal{D}^H$ is σ-invariant. Thus, u is σ-invariant on $\mathrm{colim}_H \mathcal{D}^H$, the ascending union of the spaces $\mathcal{D}^H$. This verifies the hypothesis for application of the Gelfand-Kazhdan criterion, so Φ_v is multiplicity-free. ///

2.13 Plancherel for Pseudo-Eisenstein Series: Level One

The previous decompositions can be refined to prove convergence of the integral as a $C^\infty(Z^+G_k\backslash G_\mathbb{A})$-valued integral, from a corresponding result for behavior of Fourier inversion integrals. This refinement gives a Plancherel theorem for pseudo-Eisenstein series. For simplicity, we treat trivial central character. This entails $\chi_2 = \chi_1^{-1}$, so $\chi^w = \overline{\chi} = \chi^{-1}$. More significantly, we restrict our attention to *level one*, that is, right $K_\mathbb{A}$-invariant, pseudo-Eisenstein series Ψ_φ in this section. One corollary, awkward to obtain otherwise, is the mutual orthogonality of pairs of pseudo-Eisenstein series made from data in J_χ and $J_{\chi'}$ with $\chi' \neq \chi$ and $\chi' \neq \chi^w$ on $M_k\backslash M^1$.

[2.13.1] Claim: With $\chi^w = \chi$, the integral in

$$\Psi_\varphi = \frac{1}{4\pi i}\int_{\frac12 - i\infty}^{\frac12 + i\infty} \langle \Psi_\varphi, E_{s,\chi}\rangle \cdot E_{s,\chi}\, ds + \langle \Psi_\varphi,\, \chi_1 \circ \det\rangle \cdot \frac{\chi_1 \circ \det}{|\chi_1 \circ \det|_{L^2}^2}$$

converges as a *vector-valued* integral, taking values in the Fréchet space $C^o(Z^+G_k\backslash G_\mathbb{A})$ of continuous functions on $Z^+G_k\backslash G_\mathbb{A}$.

Proof: Let $\psi_\xi(x) = e^{i\xi x}$ on $\mathbb{R}$. From [14.3], and as already applied in [1.13], the integral expressing Fourier inversion for Schwartz functions f on the real line

$$\begin{aligned} f(x) &= \frac{1}{2\pi}\int_{-\infty}^{\infty}\left(\int_{-\infty}^{\infty} f(u)\overline{\psi}_\xi(u)du\right)\psi_\xi(x)\, d\xi \\ &= \frac{1}{2\pi}\int_{-\infty}^{\infty} \psi_\xi(x)\cdot \widehat{f}(\xi)\, d\xi \end{aligned}$$

converges as a Gelfand-Pettis integral with values in the Fréchet space $C^o(\mathbb{R})$. Changing coordinates, Mellin inversion gives convergence as Gelfand-Pettis integral with values in smooth functions $C^o(0, +\infty)$. With

$$\varphi(za_y nmk) = \varphi_\infty(y)\chi(m)$$

for $z \in Z^+$, $a_y = \begin{pmatrix} \delta(y) & 0 \\ 0 & 1 \end{pmatrix}$, $y > 0$, $n \in N_{\mathbb{A}}$, $m \in M^1$, $k \in K_{\mathbb{A}}$, define a transform $J_\chi \to I_{s,\chi}$ by

$$\mathcal{M}\varphi(s)(g) = \int_0^\infty r^{-s}\, \varphi(a_r g) \frac{dr}{r}$$

with $a_r = \begin{pmatrix} \delta(r) & 0 \\ 0 & 1 \end{pmatrix}$ with $r > 0$. Because φ is completely determined except as a function on the ray $(0, +\infty)$, the inversion integral

$$\varphi(g) = \frac{1}{2\pi i} \int_{\sigma - i\infty}^{\sigma + i\infty} \mathcal{M}\varphi(s)(g)\, ds$$

converges as a vector-valued integral with values in the Fréchet space $C^o(G_{\mathbb{A}})$. From [6.1], left and right translation by $G_{\mathbb{A}}$ are continuous maps on $C^o(G_{\mathbb{A}})$, so the linear operators of left translation by G_k commute with the integral, and in the region of convergence, the expression of the pseudo-Eisenstein series Ψ_φ

$$\Psi_\varphi(g) = \sum_{\gamma \in P_k \backslash G_k} \varphi(\gamma g) = \frac{1}{2\pi i} \sum_{\gamma \in P_k \backslash G_k} \int_{\sigma - i\infty}^{\sigma + i\infty} \mathcal{M}\varphi(s)(\gamma g)\, ds$$

converges as a vector-valued integral with values in that Fréchet space. By the same steps as in the proof of the numerical form of the theorem,

$$\Psi_\varphi = \frac{1}{4\pi i} \int_{\frac{1}{2} - i\infty}^{\frac{1}{2} + i\infty} \langle \Psi_\varphi, E_{s,\chi} \rangle\, E_{s,\chi}\, ds + \langle \Psi_\varphi,\ \chi_1 \circ \det \rangle \cdot \frac{\chi_1 \circ \det}{|\chi_1 \circ \det|_{L^2}^2}$$

as a $C^o(G_{\mathbb{A}})$-valued Gelfand-Pettis integral. Since the integrand is in $C^o(Z^+G_k\backslash G_{\mathbb{A}})$-valued and the topology on this subspace is the restriction of that from $C^o(G_{\mathbb{A}})$, the integral converges in $C^o(Z^+G_k\backslash G_{\mathbb{A}})$. ///

[2.13.2] Corollary: For φ, ψ right $K_{\mathbb{A}}$-invariant functions in J_χ with $\chi^w = \chi$,

$$\langle \Psi_\varphi, \Psi_\psi \rangle = \frac{1}{4\pi i} \int_{\frac{1}{2} - i\infty}^{\frac{1}{2} + i\infty} \langle \Psi_\varphi, E_{s,\chi} \rangle\, \overline{\langle \Psi_\psi, E_{s,\chi} \rangle}\, ds$$

$$+ \frac{\langle \Psi_\varphi,\ \chi_1 \circ \det \rangle \cdot \overline{\langle \Psi_\psi,\ \chi_1 \circ \det \rangle}}{|\chi_1 \circ \det|_{L^2}^2}$$

Proof: For $f \in C_c^o(Z^+G_k\backslash G_{\mathbb{A}})$, the map $F \to \int_{Z^+G_k\backslash G_{\mathbb{A}}} F \cdot \overline{f}$ is a continuous linear functional on $F \in C^o(Z^+G_k\backslash G_{\mathbb{A}})$, so the Gelfand-Pettis property legitimizes the obvious interchange:

$$\langle \Psi_\varphi, f\rangle = \left\langle \frac{1}{4\pi i}\int_{\frac{1}{2}-i\infty}^{\frac{1}{2}+i\infty} \langle \Psi_\varphi, E_{s,\chi}\rangle\, E_{s,\chi}\, ds + \langle \Psi_\varphi,\ \chi_1 \circ \det\rangle \cdot \frac{\chi_1\circ\det}{|\chi_1 \circ \det|_{L^2}^2}, f\right\rangle$$

$$= \frac{1}{4\pi i}\int_{\frac{1}{2}-i\infty}^{\frac{1}{2}+i\infty} \langle \Psi_\varphi, E_{s,\chi}\rangle \langle E_{s,\chi}, f\rangle\, ds + \frac{\langle \Psi_\varphi, \chi_1\circ\det\rangle \cdot \langle \chi_1 \circ \det, f\rangle}{|\chi_1 \circ \det|_{L^2}^2}$$

where $\langle E_{s,\chi}, f\rangle$ converges because $f \in C_c^o(Z^+G_k\backslash G_{\mathbb{A}})$. Taking $f = E_\psi$ for ψ right $K_{\mathbb{A}}$-invariant in J_χ, this gives the asserted *isometry*. ///

The discussion for trivial central character, $\chi^w \neq \chi$, and right $K_{\mathbb{A}}$-invariance proceeds along similar lines:

[2.13.3] Claim: With $\chi^w \neq \chi$, the integral

$$\Psi_\varphi = \frac{1}{4\pi i}\int_{\frac{1}{2}-i\infty}^{\frac{1}{2}+i\infty} \langle \Psi_\varphi, E_{s,\chi}\rangle \cdot E_{s,\chi} + \langle \Psi_\varphi, E_{s,\chi^w}\rangle \cdot E_{s,\chi^w}\, ds$$

converges as a *vector-valued* integral, taking values in the Fréchet space $C^o(Z^+G_k\backslash G_{\mathbb{A}})$ of continuous functions on $Z^+G_k\backslash G_{\mathbb{A}}$. ///

[2.13.4] Corollary: For φ, ψ right $K_{\mathbb{A}}$-invariant functions in J_χ with $\chi^w \neq \chi$,

$$\langle \Psi_\varphi, \Psi_\psi\rangle = \frac{1}{4\pi i}\int_{\frac{1}{2}-i\infty}^{\frac{1}{2}+i\infty} \langle \Psi_\varphi, E_{s,\chi}\rangle\, \overline{\langle \Psi_\psi, E_{s,\chi}\rangle} + \langle \Psi_\varphi, E_{s,\chi^w}\rangle\, \overline{\langle \Psi_\psi, E_{s,\chi^w}\rangle}\, ds \qquad ///$$

These decomposition formulas facilitate comparison of pseudo-Eisenstein series:

[2.13.5] Corollary: For $\chi' \neq \chi$ and $\chi' \neq \chi^w$, pseudo-Eisenstein series made from J_χ are orthogonal to those made from $J_{\chi'}$.

Proof: For $\varphi \in J_\chi$, for $\chi^w = \chi$ or not, we have a convergent $C^\infty(Z^+G_k\backslash G_{\mathbb{A}})$-valued integral

$$\Psi_\varphi = \frac{1}{2\pi i}\int_{\frac{1}{2}-i\infty}^{\frac{1}{2}+i\infty} \mathcal{M}\varphi(s) \cdot E_{s,\chi}\, ds + \sum_{s_o} \mathcal{M}\varphi(s_o) \cdot \mathrm{Res}_{s_o} E_{s,\chi}$$

where s_o runs over poles of $E_{s,\chi}$ in $\mathrm{Re}(s) \geq \frac{1}{2}$. Inner product with the compactly supported Ψ_ψ is a continuous functional, so this inner product passes inside the integral by [14.1], giving

$$\langle \Psi_\varphi, \Psi_\psi \rangle = \frac{1}{2\pi i} \int_{\frac{1}{2}-i\infty}^{\frac{1}{2}+i\infty} \mathcal{M}\varphi(s) \cdot \langle E_{s,\chi}, \Psi_\psi \rangle \, ds$$
$$+ \sum_{s_o} \mathcal{M}\varphi(s_o) \cdot \langle \mathrm{Res}_{s_o} E_{s,\chi}, \Psi_\psi \rangle$$

Similarly, from [15.2], taking residues commutes with application of the functional. Unwinding,

$$\langle E_{s,\chi}, \Psi_\psi \rangle = \int_{Z^+N_{\mathbb{A}}M_k\backslash G_{\mathbb{A}}} c_P E_{s,\chi} \cdot \overline{\psi}$$
$$= \int_{Z^+N_{\mathbb{A}}M_k\backslash G_{\mathbb{A}}} (\varphi_{s,\chi} + c_{s,\chi}\varphi_{1-s,\chi^w}) \cdot \overline{\psi}$$

Under left multiplication by M^1, the function $\varphi_{s,\chi}$ is equivariant by χ, and φ_{1-s,χ^w} by χ^w, while ψ is left equivariant by χ'. Thus, the integral is 0. ///

As in [1.13], these decomposition formulas *suggest* the form of a Plancherel theorem for the χth fragment of the complement to cuspforms. As in [1.13], for each fixed χ we must identify the closure of the image of

$$\Psi_\varphi \longrightarrow \begin{cases} \langle \Psi_\varphi, E_{s,\chi} \rangle \oplus \langle \Psi_\varphi, \mathrm{Res}_{s=1} E_{s,\chi} \rangle & (\text{when } \chi^w = \chi) \\ \langle \Psi_\varphi, E_{s,\chi} \rangle \oplus \langle \Psi_\varphi, E_{s,\chi^w} \rangle & (\text{when } \chi^w \neq \chi) \end{cases}$$

with $K_{\mathbb{A}}$-invariant $\varphi \in J_\chi$, and certify that residues behave compatibly with the simplest outcome. The functional equation of $E_{s,\chi}$ constrains the functions $s \to \langle \Psi_\varphi, E_{s,\chi} \rangle$ and $s \to \langle \Psi_\varphi, E_{s,\chi^w} \rangle$:

$$\langle \Psi_\varphi, E_{1-s,\chi} \rangle = \langle \Psi_\varphi, c_{1-s,\chi} E_{s,\overline{\chi}} \rangle = \overline{c_{1-s,\chi}} \cdot \langle \Psi_\varphi, E_{s,\overline{\chi}} \rangle$$
$$= c_{s,\chi^w} \cdot \langle \Psi_\varphi, E_{s,\chi^w} \rangle$$

We will show that the L^2 closure of the set of images is as large as possible, given these obvious constraints. In both cases, the map $\varphi \to \mathcal{M}\varphi$ is essentially Fourier transform, so maps test functions to a space of functions dense in the Schwartz functions on $L^2(\frac{1}{2} + i\mathbb{R})$. Then we proceed differently depending on whether $\chi^w = \chi$ or not. The case $\chi^w = \chi$ is much like [1.13]: for $\chi^w = \chi$, let V_χ be the subspace of $L^2(\frac{1}{2} + i\mathbb{R})$ functions meeting $f(1-s) = c_{s,\chi} \cdot f(s)$.

[2.13.6] Claim: With fixed $\chi^w = \chi$, the map

$$\Psi_\varphi \longrightarrow \langle \Psi_\varphi, E_{s,\chi} \rangle \oplus \langle \Psi_\varphi, \mathrm{Res}_{s=1} E_{s,\chi} \rangle$$

has dense image in $V \oplus \mathbb{C}$, and is an L^2-isometry.

Proof: In this case, $\langle \Psi_\varphi, E_{s,\chi}\rangle = \mathcal{M}c_P\Psi_\varphi(s) = \mathcal{M}\eta(s) + c_{1-s,\chi}\mathcal{M}\eta(1-s)$. For F in the Schwartz space on $\frac{1}{2} + i\mathbb{R}$, the averaging $F(s) + c_{1-s,\chi}F(1-s)$ maps to a dense subspace of V. Thus, ignoring for a moment the residual summand, the images $\langle \Psi_\varphi, E_{s,\chi}\rangle$ are dense in V, as desired.

The residue is $\chi_1 \circ \det$, and this should be orthogonal to Ψ_ψ with $\varphi' \in J_{\chi'}$ and $\chi' \neq \chi$. Indeed, unwinding the pseudo-Eisenstein series and using Iwasawa,

$$\begin{aligned}\langle \Psi_\psi, \chi_1 \circ \det\rangle &= \int_{Z^+N_\mathbb{A}M_k\backslash G_\mathbb{A}} \psi(g)\cdot c_P\overline{\chi}_1(\det(g))\,dg\\ &= \int_{Z^+M_k\backslash M_\mathbb{A}} \psi(m)\cdot \overline{\chi}_1(\det(m))\frac{dm}{\delta(m)}\end{aligned}$$

where δ is the modular function of $P_\mathbb{A}$. Let r be the number of isomorphism classes of archimedean completions of k, and let

$$A^+ = \{\begin{pmatrix} t^{1/r} & 0\\ 0 & 1\end{pmatrix} : t > 0\} \qquad (\text{diagonally in } M_\infty = \prod_{v|\infty} M_v)$$

Using $Z^+M_k\backslash M_\mathbb{A} \approx A^+ \times M_k\backslash M^1$, the integrand is equivariant by a nontrivial character of $M_k\backslash M^1$, so is 0. Even more simply, the various functions $\chi_1 \circ \det$ are mutually orthogonal.

Since $\chi_1 \circ \det$ is in the orthogonal complement to cuspforms, it is in the closure of the space generated by pseudo-Eisenstein series. We have just shown that it is orthogonal to all of these except those with data from J_χ, so it must be in the closure of the images from J_χ alone. By subtraction, the integral part of the decomposition is also in the closure of the pseudo-Eisenstein series, so the images are L^2 dense in $V \oplus \mathbb{C}$, as claimed.

Then the spectral-coefficient map extends by continuity, to give an L^2 isometry, the statement of a Plancherel theorem for this fragment of L^2. ///

For the $\chi^w \neq \chi$ case, let

$$\begin{aligned} V &= \{f = f_1 \oplus f_2 \in L^2(\tfrac{1}{2} + i\mathbb{R}) \oplus L^2(\tfrac{1}{2} + i\mathbb{R}) : f_1(1-s) = f_2(s)\}\\ &\subset L^2(\tfrac{1}{2} + i\mathbb{R}) \oplus L^2(\tfrac{1}{2} + i\mathbb{R})\end{aligned}$$

[2.13.7] Claim: With fixed $\chi^w \neq \chi$, the map

$$\Psi_\varphi \longrightarrow \langle \Psi_\varphi, E_{s,\chi}\rangle \oplus \langle \Psi_\varphi, E_{s,\chi^w}\rangle$$

has dense image in V, and is an L^2-isometry.

Proof: In this case, $\langle \Psi_\varphi, E_{s,\chi}\rangle = \mathcal{M}c_P\Psi_\varphi = \mathcal{M}\eta(s)$, and this is in the Schwartz space on $\frac{1}{2} + i\mathbb{R}$, which is dense in L^2. The functional equation relating $E_{s,\chi}$ determines $\langle \Psi_\varphi, E_{1-s,\chi^w}\rangle$. Thus, the images are dense in V, as desired.

Then the spectral-coefficient map extends by continuity, to give an L^2 isometry, the statement of a Plancherel theorem for this fragment of L^2. ///

2.14 Spectral Expansion, Plancherel Theorem: Level One

From [2.7], the collection of right $K_\mathbb{A}$-invariant pseudo-Eisenstein series Ψ_φ with $\varphi \in J_\chi$ and χ running over pairs χ of unramified characters χ_1, χ_2 with $\chi_2 = \chi^{-1}$ (due to trivial central character) is *dense* in the orthogonal complement in $L^2(Z^+G_k\backslash G_\mathbb{A})$ to right $K_\mathbb{A}$-invariant *cuspforms*.

For unramified χ_1 on $\mathbb{J}$, let $\chi\begin{pmatrix} m_1 & 0 \\ 0 & m_2 \end{pmatrix} = \chi_1(m_1/m_2)$. For $\chi_1' \neq \chi_1^{\pm 1}$, the adjunctions [2.7] satisfied by pseudo-Eisenstein series show that $\langle \Psi_\varphi, E_{\varphi'} \rangle = 0$ for $\varphi \in J_\chi$ and $\varphi' \in J_{\chi'}$. Letting F run over an orthonormal basis for the space of cuspforms on $Z^+G_k\backslash G_\mathbb{A}/K_\mathbb{A}$ with trivial central character, we have an automorphic Plancherel theorem at level one:

[2.14.1] Theorem: For f in $L^2(Z^+G_k\backslash G_\mathbb{A}/K_\mathbb{A})$, with trivial central character, with convergence in an L^2,

$$f = \sum_{\text{cfm}F} \langle f, F\rangle \cdot F$$
$$+ \sum_{\chi_1} \left(\frac{1}{4\pi i} \int_{\frac{1}{2}-i\infty}^{\frac{1}{2}+i\infty} \langle f, E_{s,\chi}\rangle \cdot E_{s,\chi}\, ds + \frac{\langle f,\ \chi_1 \circ \det\rangle \cdot \chi_1 \circ \det}{|\chi_1 \circ \det|_{L^2}^2} \right)$$
$$+ \sum_{\chi:\chi^w \neq \chi} \left(\frac{1}{4\pi i} \int_{\frac{1}{2}-i\infty}^{\frac{1}{2}+i\infty} \langle \Psi_\varphi, E_{s,\chi}\rangle \cdot E_{s,\chi} + \langle \Psi_\varphi, E_{s,\chi^w}\rangle \cdot E_{s,\chi^w} ds \right)$$

and

$$|f|_{L^2}^2 = \sum_{\text{cfm}F} |\langle f, F\rangle|^2$$
$$+ \sum_{\chi:\chi^w = \chi} \left(\frac{1}{4\pi} \int_{-\infty}^{\infty} |\langle f, E_{\frac{1}{2}+it,\chi}\rangle|^2\, dt + \frac{|\langle f,\ \chi_1 \circ \det\rangle|^2}{|\chi_1 \circ \det|_{L^2}^2} \right)$$
$$+ \sum_{\chi:\chi^w \neq \chi} \left(\frac{1}{4\pi} \int_{-\infty}^{\infty} |\langle \Psi_\varphi, E_{\frac{1}{2}+it,\chi}\rangle|^2 + |\langle \Psi_\varphi, E_{\frac{1}{2}+it,\chi^w}\rangle|^2\, dt \right)$$

The integrals suggested by the notation are not *literal* integrals but are the extension-by-continuity of the corresponding literal integrals, as with Fourier transform and Fourier inversion on $L^2(\mathbb{R})$. ///

Combining the decomposition of right $K_{\mathbb{A}}$-invariant L^2 cuspforms (with trivial central character) and the decomposition of their orthogonal complement:

[2.14.2] Corollary: Functions $f \in L^2(Z^+G_k\backslash G_{\mathbb{A}}/K_{\mathbb{A}})$ with trivial central character have L^2 expansions

$$f = \sum_{\text{cfm}F} \langle f, F\rangle \cdot F$$

$$+ \sum_{\chi:\chi^w=\chi} \left(\frac{1}{4\pi i} \int_{\frac{1}{2}-i\infty}^{\frac{1}{2}+i\infty} \langle f, E_{s,\chi}\rangle \cdot E_{s,\chi}\, ds + \frac{\langle f,\ \chi_1 \circ \det\rangle \cdot \chi_1 \circ \det}{|\chi_1 \circ \det|^2_{L^2}} \right)$$

$$+ \sum_{\chi:\chi^w\neq\chi} \left(\frac{1}{4\pi i} \int_{\frac{1}{2}-i\infty}^{\frac{1}{2}+i\infty} \langle \Psi_\varphi, E_{s,\chi}\rangle \cdot E_{s,\chi} + \langle \Psi_\varphi, E_{s,\chi^w}\rangle \cdot E_{s,\chi^w} ds \right)$$

converging in an L^2 sense, with corresponding equality of L^2 norms, where integrals involving Eisenstein series are *isometric extensions*, as in the previous section. ///

2.15 Exotic Eigenfunctions, Discreteness of Pseudo-Cuspforms

An important variant approach, both to the discrete decomposition of the space of cuspforms [1.7] and [2.6] and to the meromorphic continuation of Eisenstein series as in [11.5], is the notion of *pseudo-cuspform*. The largest space of pseudo-cuspforms with cutoff height $b \geq 0$ is

$$L^2_b(Z^+G_k\backslash G_{\mathbb{A}}) = \{f \in L^2(Z^+G_k\backslash G_{\mathbb{A}}) : c_P f(m'a_y k) = 0 \text{ for } m' \in M^1,\ b < y \in \mathbb{R},\ k \in K\}$$

The idea is that the constant terms vanish above height b. With $b = 0$, this is the space of square-integrable cuspforms. More precisely, via the adjunction [1.7.3], $L^2_b(Z^+G_k\backslash G_{\mathbb{A}})$ is the orthogonal complement in $L^2(Z^+G_k\backslash G_{\mathbb{A}})$ to all pseudo-Eisenstein series Ψ_φ with data $\varphi \in C_c^\infty(Z^+N_{\mathbb{A}}M_k\backslash G_{\mathbb{A}})$ supported on

$$Z^+N_{\mathbb{A}}M_k\{znma_yk : z \in Z^+, n \in N_{\mathbb{A}}, m \in M^1, b < y \in \mathbb{R}, k \in K_{\mathbb{A}}\}$$

That is, these are pseudo-Eisenstein series Ψ_φ with data φ supported *above height* $y = b$.

However, as throughout this chapter, right $K_{\mathbb{A}}$-finiteness assumptions avoid some relatively uninteresting secondary complications. Thus, for simplicity, we consider only right $K_\infty K'$-fixed functions for K' a fixed open subgroup of K_{fin}.

The *pseudo-Laplacian* $\widetilde{\Delta}_b$ is the *Friedrichs self-adjoint extension* [9.2] of the sum $\Delta = \sum_{v|\infty} \Delta_v$ of the invariant Laplacians on the archimedean quotients G_v/K_v, restricted to the *test functions* in the space $L^2_b(Z^+G_k\backslash G_\mathbb{A}/K_\infty K')$ of pseudo-cuspforms. For any $b > 0$, the corresponding space of square-integrable pseudo-cuspforms contains the space of genuine cuspforms $L^2_o(Z^+G_k\backslash G_\mathbb{A})$. The basic, unexpected result is

[2.15.1] Theorem: $L^2_b(Z^+G_k\backslash G_\mathbb{A})$ is a direct sum of eigenspaces for $\widetilde{\Delta}_b$, each of finite dimension. In particular, $\widetilde{\Delta}_b$ has *compact resolvent. (Proof in [10.3].)*

Without further information, this does *not* immediately prove that the subspace consisting of genuine cuspforms decomposes discretely for $\widetilde{\Delta}_b$, because the description [9.2] of $\widetilde{\Delta}_b$ imposes technical conditions on possible eigenfunctions, and one should check that the *smooth* cuspforms are dense in the space of L^2 cuspforms.

In any case, for $b \gg 1$, the space $L^2_b(Z^+G_k\backslash G_\mathbb{A})$ contains many functions not in the space of genuine cuspforms, for example, pseudo-Eisenstein series Ψ_φ with data φ supported in the interval $[1, b]$. As in [2.11] and [2.12], these are expressible as integrals of genuine Eisenstein series. However, by the theorem, apparently these pseudo-Eisenstein series are (infinite) sums of L^2-eigenfunctions for $\widetilde{\Delta}_b$ orthogonal. Similarly, by [2.10.3] and [2.10.4], truncated Eisenstein series $\wedge^b E_f$ are in $L^2_b(Z^+G_k\backslash G_\mathbb{A})$. Because they are in the span of pseudo-Eisenstein series with compactly supported data, by [2.11] and [2.12] they are integrals of genuine Eisenstein series. Again, however, by the theorem, they are also (infinite) sums of $\widetilde{\Delta}_b$-eigenfunctions. Evidently, there are many *exotic* eigenfunctions for $\widetilde{\Delta}_b$. Indeed,

[2.15.2] Corollary: In $L^2_b(Z^+G_k\backslash G_\mathbb{A}/K_{\text{fin}}K')$, the eigenfunctions for $\widetilde{\Delta}_b$ with eigenvalues $\lambda = s(s-1) < -1/4$ are exactly the truncated Eisenstein series $\wedge^b E_f$ with $c_P E_f(a_b) = 0$, for right K_∞-invariant right K'-right-invariant $f \in I_{s,\chi}$, for $s(s-1) < 0$ and character χ on $M_k\backslash M^1/(M^1 \cap K_\infty K')$. *(Proof in [10.4].)*

These truncated Eisenstein series are *not* smooth. The slightly nonintuitive nature of the operator $\widetilde{\Delta}_b$ explains the situation, in [10.4]. For example, in addition to meeting the Gelfand condition of constant-term vanishing about height b, eigenfunctions of the pseudo-Laplacian $\widetilde{\Delta}_b$ are pseudo-cuspforms in a stronger sense:

[2.15.3] Corollary: An L^2-eigenfunction u for $\widetilde{\Delta}_b$ with eigenvalue λ satisfies $(\widetilde{\Delta}_b - \lambda)u = 0$ *locally* at heights above b. ///

2.A Appendix: Compactness of $\mathbb{J}^1/k^\times$

The following compactness result has both *finiteness of class numbers* and *Dirichlet's units theorem* as corollaries. Indeed, the compactness can be proven as a consequence of these two results. However, the compactness can be proven directly and is what proves useful here.

[2.A.1] Theorem: $\mathbb{J}^1/k^\times$ is compact.

Proof: As in [5.2], Haar measure on $\mathbb{A} = \mathbb{A}_k$ and Haar measure on the (topological group) quotient $\mathbb{A}/k$ are interrelated in the sense that choice of one uniquely determines the other by the relation

$$\int_{\mathbb{A}} f(x)\,dx = \int_{\mathbb{A}/k} \sum_{\gamma\in k} f(\gamma + x)\,dx \qquad (\text{for } f \in C_c^o(\mathbb{A}))$$

Normalize the measure on $\mathbb{A}$ so that, mediated by this relation, $\mathbb{A}/k$ has measure 1. We have a Minkowski-like claim, a measure-theoretic *pigeon-hole principle*, that a compact subset C of $\mathbb{A}$ with measure greater than 1 cannot *inject* to the quotient $\mathbb{A}/k$. Suppose, to the contrary, that C injects to the quotient. With f the characteristic function of C,

$$1 < \int_{\mathbb{A}} f(x)\,dx = \int_{\mathbb{A}/k} \sum_{\gamma\in k} f(\gamma + x)\,dx \le \int_{\mathbb{A}/k} 1\,dx = 1$$

with the last inequality by injectivity. Contradiction. For *idele* α, the change-of-measure on $\mathbb{A}$ is given conveniently by

$$\frac{\text{meas}\,(\alpha E)}{\text{meas}\,(E)} = |\alpha| \qquad (\text{for measurable } E \subset \mathbb{A})$$

Given $\alpha \in \mathbb{J}^1$, we will adjust α by $k^\times$ to lie in a compact subset of $\mathbb{J}^1$. Fix compact $C \subset \mathbb{A}$ with measure > 1. The topology on $\mathbb{J}$ is *strictly finer* than the subspace topology with $\mathbb{J} \subset \mathbb{A}$: the genuine topology is by imbedding $\mathbb{J} \to \mathbb{A} \times \mathbb{A}$ by $\alpha \to (\alpha, \alpha^{-1})$. For $\alpha \in \mathbb{J}^1$, both αC and $\alpha^{-1}C$ have measure > 1, neither injects to the quotient $k\backslash\mathbb{A}$. So there are $x \neq y$ in k so that $x + \alpha C = y + \alpha C$. Subtracting,

$$0 \neq a = x - y \in \alpha(C - C) \cap k$$

That is, $a \cdot \alpha^{-1} \in C - C$. Likewise, there is $0 \neq b \in \alpha^{-1}(C - C) \cap k$, and $b \cdot \alpha \in C - C$. There is an obvious constraint: observing that $ab = (a \cdot \alpha)(b \cdot \alpha^{-1})$, by [1.5.3]

$$(a \cdot \alpha)(b \cdot \alpha^{-1}) \in (C - C)^2 \cap k^\times = \text{compact} \cap (\text{discrete subgroup}) = \text{finite}$$

as in [1.5.3]. Let $\Xi = (C-C)^2 \cap k^\times$ be this finite set. Paraphrasing: given $\alpha \in \mathbb{J}^1$, there are $a \in k^\times$ and $\xi \in \Xi$ ($\xi = ab$ above) such that $(a \cdot \alpha^{-1}, (a \cdot \alpha^{-1})^{-1}) \in (C-C) \times \xi^{-1}(C-C)$. That is, α^{-1} can be adjusted by $a \in k^\times$ to be in the compact $C-C$, and, simultaneously, for one of the finitely-many $\xi \in \Xi$, $(a \cdot \alpha^{-1})^{-1} \in \xi \cdot (C-C)$. In the topology on $\mathbb{J}$, for each $\xi \in \Xi$,

$$((C-C) \times \xi^{-1}(C-C)) \cap \mathbb{J} = \text{compact in } \mathbb{J}$$

The continuous image in $\mathbb{J}/k^\times$ of each of these finitely many compacts is compact. Their union covers the *closed* subset $\mathbb{J}^1/k^\times$, so the latter is compact. ///

2.B Appendix: Meromorphic Continuation

A somewhat special argument gives precise information about the meromorphic continuation of certain Eisenstein series for GL_2, in particular about possible poles. Analogous arguments are possible in a few other situations. As already demonstrated, with χ denoting a pair of characters χ_1, χ_2 on $\mathbb{J}^1$, take

$$f(zna_r m \cdot k) = |r|^s \cdot \chi_1(m_1)\,\chi_2(m_2) \cdot f_o(k)$$

with $m = \begin{pmatrix} m_1 & 0 \\ 0 & m_2 \end{pmatrix} \in M^1$, with f_o independent of s. In particular, this argument for meromorphic continuation uses the following expression for f.

With diagonal map $\delta : (0, +\infty) \to \mathbb{J}$ as earlier, abuse notation by extending χ_1, χ_2 to characters on $\mathbb{J}$ by extending trivially on $\delta(0, +\infty)$:

$$\chi_j(\delta(r) \cdot \theta) = \chi_j(\theta) \qquad (\text{for } r > 0 \text{ and } \theta \in \mathbb{J}^1,\ j = 1, 2)$$

By changing variables in the integral, one finds that any function f expressed as

$$f(g) = \frac{|\det g|^s \chi_1(\det g)}{\zeta(2s, \frac{\chi_1}{\chi_2}, \Phi(0, -))} \int_{\mathbb{J}} |t|^{2s} \frac{\chi_1}{\chi_2}(t) \cdot \Phi(t \cdot e \cdot g)\, dt$$

with $e = (0 \quad 1)$, with Schwartz function Φ on $\mathbb{A}^2$ and global Iwasawa-Tate zeta integral

$$\zeta(2s, \frac{\chi_1}{\chi_2}, \Phi(0, -)) = \int_{\mathbb{J}} |t|^{2s} \frac{\chi_1}{\chi_2}(t)\, \Phi(0, t)\, dt$$

is in $I_{s,\chi}$. The division by $\zeta(2s, \frac{\chi_1}{\chi_2}, \Phi(0, -))$ normalizes $f(1) = 1$. We do not consider the issue of exactly which $f \in I_{s,\chi}$ can be expressed in this form.

Every Schwartz function Φ is a finite sum of monomial functions $\Phi = \bigotimes_v \Phi_v$, so permissible functions f are finite sums of monomial functions

$f = \bigotimes_v f_v$ with the local functions f_v on G_v expressible as

$$f_v(g) = \frac{|\det g|^s \chi_1(\det g)}{\zeta_v(2s, \frac{\chi_1}{\chi_2}, \Phi_v(0, -))} \int_{k_v^\times} |t|_v^{2s} \frac{\chi_1}{\chi_2}(t) \cdot \Phi_v(t \cdot e \cdot g)\, dt$$

with $e = (0 \quad 1)$. The product formula makes f left P_k-invariant. The corresponding Eisenstein series is

$$E_f(g) = E(s, \chi, \Phi)(g) = \sum_{\gamma \in P_k \backslash G_k} f(\gamma \cdot g)$$

Let $w = \begin{pmatrix} 0 & -1 \\ 1 & 0 \end{pmatrix}$.

[2.B.1] Theorem: These Eisenstein series admit meromorphic continuations to $\mathbb{C}$ and have no poles in $\mathrm{Re}(s) \geq \frac{1}{2}$ unless $\chi_1/\chi_2 = 1$, in which case there is a unique pole in $\mathrm{Re}(s) \geq \frac{1}{2}$, at $s = 1$, with residue a constant multiple of $\chi_1(\det g)$. The functional equation is

$$\frac{\zeta(2s, \rho, \Phi(0, -))}{\chi_1 \circ \det} \cdot E(s, \chi, \Phi)$$

$$= \frac{\zeta(2 - 2s, \rho^{-1}, (\widehat{\Phi} \circ w)(0, -))}{\chi_2 \circ \det} \cdot E(1 - s, \chi^w, \widehat{\Phi} \circ w)$$

[2.B.2] Corollary: For trivial central character ω, these Eisenstein series E_f have no poles in $\mathrm{Re}(s) \geq \frac{1}{2}$ unless $\chi_1^2 = 1$, in which case there is a unique pole in $\mathrm{Re}(s) \geq \frac{1}{2}$, at $s = 1$, with residue a constant multiple of $\chi_1(\det g)$.

Proof: (*of Corollary*) With trivial central character, $\chi_2 = \chi_1^{-1}$. ///

Proof: To isolate the Poisson summation effect, for unitary character $\rho = \chi_1/\chi_2$ on $\mathbb{J}/k^\times$, let

$$\mathcal{E}(s, \rho, \Phi)(g) = \frac{\zeta(2s, \rho, \Phi(0, -))}{|\det g|^s \chi_1(\det g)} \cdot E(s, \chi, \Phi)(g)$$

$$= \int_{\mathbb{J}} |t|^{2s} \rho(t) \sum_{0 \neq x \in k^2} \Phi(t \cdot x \cdot g)\, dt$$

With $\mathbb{J}^+ = \{t \in \mathbb{J} : |t| \geq 1\}$ and $\mathbb{J}^- = \{t \in \mathbb{J} : |t| \leq 1\}$, we follow Riemann, Hecke, Iwasawa, and Tate and break the integral into two pieces:

$$\mathcal{E}(s, \rho, \Phi)(g) = \int_{\mathbb{J}^+} |t|^{2s} \rho(t) \sum_{0 \neq x \in k^2} \Phi(t \cdot x \cdot g)\, dt$$

$$+ \int_{\mathbb{J}^-} |t|^{2s} \rho(t) \sum_{0 \neq x \in k^2} \Phi(t \cdot x \cdot g)\, dt$$

By the first lemma below, the integral over $\mathbb{J}^+$ is absolutely convergent for all $s \in \mathbb{C}$, so is entire. Adelic Poisson summation converts the integral over $\mathbb{J}^-$ to an integral over $\mathbb{J}^+$, plus two elementary terms: first,

$$\begin{aligned}\sum_{0\neq x\in k^2} \Phi(txg) &= \sum_{x\in k^2} \Phi(txg) - \Phi(0)\\ &= |t|^{-2}|\det g|^{-1}\sum_{x\in k^2} \widehat{\Phi}(t^{-1}x(g^\top)^{-1}) - \Phi(0)\\ &= |t|^{-2}|\det g|^{-1}\cdot \sum_{0\neq x\in k^2} \widehat{\Phi}(t^{-1}x(g^\top)^{-1}) - \Phi(0)\\ &\quad + |t|^{-2}|\det g|^{-1}\cdot\widehat{\Phi}(0)\end{aligned}$$

Thus, inverting t to replace the integral over $\mathbb{J}^-$ by one over $\mathbb{J}^+$,

$$\begin{aligned}&\int_{\mathbb{J}^-} |t|^{2s}\rho(t)\sum_{0\neq x\in k^2} \Phi(txg)\,dt\\ &= \int_{\mathbb{J}^-} |t|^{2s}\rho(t)\Big(|t|^{-2}|\det g|^{-1}\sum_{0\neq x\in k^2} \widehat{\Phi}(t^{-1}x(g^\top)^{-1}) - \Phi(0)\\ &\quad + |t|^{-2}|\det g|^{-1}\cdot\widehat{\Phi}(0)\Big)\,dt\\ &= |\det g|^{-1}\Big(\int_{\mathbb{J}^+} |t|^{2-2s}\rho(t)\sum_{0\neq x\in k^2}\widehat{\Phi}(tx(g^\top)^{-1})\,dt\\ &\quad + \widehat{\Phi}(0)\int_{\mathbb{J}^-}|t|^{2s}\rho^{-1}(t)\,|t|^{-2}\,dt\Big) - \Phi(0)\int_{\mathbb{J}^-}|t|^{2s}\rho(t)\,dt\end{aligned}$$

Altogether,

$$\begin{aligned}\mathcal{E}(s,\rho,\Phi)(g) &= \int_{\mathbb{J}^+} |t|^{2s}\rho(t)\sum_{0\neq x\in k^2}\Phi(t\cdot x\cdot g)\,dt - \Phi(0)\int_{\mathbb{J}^-}|t|^{2s}\rho(t)\,dt\\ &\quad + |\det g|^{-1}\Big(\int_{\mathbb{J}^+}|t|^{2-2s}\rho^{-1}(t)\sum_{0\neq x\in k^2}\widehat{\Phi}(tx(g^\top)^{-1})\,dt\\ &\quad + \widehat{\Phi}(0)\int_{\mathbb{J}^-}|t|^{2s}\rho(t)\,|t|^{-2}\,dt\Big)\end{aligned}$$

Multiplying through by $|\det g|^{\frac{1}{2}}$ symmetrizes this:

$$\begin{aligned}|\det g|^{\frac{1}{2}} \cdot \mathcal{E}(s, \rho, \Phi)(g) &= |\det g|^{\frac{1}{2}} \cdot \int_{\mathbb{J}^+} |t|^{2s} \rho(t) \sum_{0 \neq x \in k^2} \Phi(t \cdot x \cdot g)\, dt \\ &\quad - |\det g|^{\frac{1}{2}} \cdot \Phi(0) \int_{\mathbb{J}^-} |t|^{2s} \rho(t)\, dt \\ &\quad + |\det(g^\top)^{\frac{1}{2}} \Big(\int_{\mathbb{J}^+} |t|^{2-2s} \rho^{-1}(t) \sum_{0 \neq x \in k^2} \widehat{\Phi}(tx(g^\top)^{-1})\, dt \\ &\quad + |\det(g^\top)^{\frac{1}{2}} \widehat{\Phi}(0) \int_{\mathbb{J}^-} |t|^{2s} \rho(t)\, |t|^{-2}\, dt \Big) \\ &= |\det(g^\top)^{-1}|^{\frac{1}{2}} \cdot \mathcal{E}(1-s, \rho^{-1}, \widehat{\Phi})((g^\top)^{-1})\end{aligned}$$

We examine the two elementary integrals which, if nonzero, give the poles. If $\rho(\tau) \neq 1$ for some $\tau \in \mathbb{J}^1$, then by changing variables,

$$\int_{\mathbb{J}^-} |t|^{2s} \rho(t)\, dt = \int_{\mathbb{J}^-} |\tau t|^{2s} \rho(\tau t)\, dt = \rho(\tau) \int_{\mathbb{J}^-} |t|^{2s} \rho(t)\, dt$$

so the integral must vanish. On the other hand, when $\rho(\tau) = 1$ on $\mathbb{J}^1$, we give the compact group $\mathbb{J}^1/k^\times$ measure 1, and

$$\int_{\mathbb{J}^-} |t|^{2s}\, dt = \int_0^1 y^{2s} \frac{dy}{y} = \frac{1}{2s}$$

and

$$\int_{\mathbb{J}^-} |t|^{2s-2}\, dt = \int_0^1 y^{2s-2} \frac{dy}{y} = \frac{1}{2s-2}$$

Thus, when χ_1/χ_2 is not identically 1 on $\mathbb{J}^1$, there are *no poles*. When $\rho = \chi_1/\chi_2$ is identically 1 on $\mathbb{J}^1$, there are polar terms in $|\det g|^{\frac{1}{2}} \cdot \mathcal{E}(s, \chi, \Phi)(g)$, and they are symmetrical:

$$-\frac{\Phi(0) \cdot |\det g|^{\frac{1}{2}}}{2s} - \frac{\widehat{\Phi}(0) \cdot |\det(g^T)^{-1}|^{\frac{1}{2}}}{2(1-s)}$$

Thus, the preliminary form of the functional equation:

$$|\det g|^{\frac{1}{2}} \cdot \mathcal{E}(s, \chi, \Phi)(g) = |\det(g^\top)^{-1}|^{\frac{1}{2}} \cdot \mathcal{E}(1-s, \chi^w, \widehat{\Phi})((g^\top)^{-1})$$

We would prefer not to have a relationship involving $(g^\top)^{-1}$. Fortunately, $w^{-1}(g^\top)^{-1}w = g/(\det g)$. Thus, replacing x by xw^{-1} in the sum, replacing $\widehat{\Phi}$

by $\widehat{\Phi} \circ w$, and replacing t by $t \cdot \det g$, in the region for $\mathrm{Re}(1-s) > 1$ for convergence,

$$|\det(g^\top)^{-1}|^{\frac{1}{2}} \cdot \mathcal{E}(1-s, \chi^w, \widehat{\Phi})((g^\top)^{-1})$$
$$= |\det g|^{-\frac{1}{2}} \cdot |\det g|^{2-2s} \frac{\chi_2}{\chi_1}(\det g) \cdot \mathcal{E}(1-s, \chi^w, \widehat{\Phi} \circ w)(g)$$

Thus, by the identity principle,

$$|\det g|^{\frac{1}{2}} \cdot \mathcal{E}(s, \chi, \Phi)(g)$$
$$= |\det g|^{-\frac{1}{2}} \cdot |\det g|^{2-2s} \frac{\chi_2}{\chi_1}(\det g) \cdot \mathcal{E}(1-s, \chi^w, \widehat{\Phi} \circ w)(g)$$

and

$$|\det g|^{\frac{1}{2}} \cdot \frac{\zeta(2s, \rho, \Phi(0,-))}{|\det g|^s \chi_1(\det g)} \cdot E(s, \chi, \Phi)(g)$$
$$= |\det g|^{-\frac{1}{2}} \cdot |\det g|^{2-2s} \frac{\chi_2}{\chi_1}(\det g) \cdot \frac{\zeta(2-2s, \rho^{-1}, \widehat{\Phi} \circ w(0,-))}{|\det g|^{1-s} \chi_2(\det g)}$$
$$\times E(1-s, \chi^w, \widehat{\Phi} \circ w)(g)$$

simplifying to

$$\frac{\zeta(2s, \rho, \Phi(0,-))}{\chi_1(\det g)} \cdot E(s, \chi, \Phi)(g)$$
$$= \frac{\zeta(2-2s, \rho^{-1}, \widehat{\Phi} \circ w(0,-))}{\chi_2(\det g)} \cdot E(1-s, \chi^w, \widehat{\Phi} \circ w)(g) \qquad ///$$

[2.B.3] Lemma: Half-zeta integrals over $\mathbb{J}^+$ are absolutely convergent for all $s \in \mathbb{C}$.

Proof: Fix $g \in G_{\mathbb{A}}$, let $\varphi(t) = \Phi(teg)$ and $\varphi_v(t) = \Phi_v(teg)$. By the lemma below,

$$|\varphi(t)| \ll_N \prod_v \sup(|t_v|_v, 1)^{-2N} \qquad (\text{for adele } t = \{t_v\}, \text{ for all } N)$$

For idele t let $\nu(t) = \prod_v \sup(|t_v|_v, |t_v|_v^{-1})$. Almost all factors on the right-hand side are 1, so there is no problem with convergence. Apply

$$(\sup(a, 1))^2 = \sup(a^2, 1) = a \cdot \sup(a, \frac{1}{a}) \qquad (\text{for } a > 0)$$

to every factor:

$$\prod_v \sup(|t_v|_v, 1)^{-2N} = |t|^{-N} \prod_v \sup(|t_v|_v, |t_v^{-1}|_v)^{-N} = |t|^{-N} \nu(t)^{-N}$$

Thus, on $\mathbb{J}^+$,

$$\prod_v \sup(|t_v|_v, 1)^{-2N} = |t|^{-N}\nu(t)^{-N} \le \nu(t)^{-N}$$

when $t \in \mathbb{J}^+$, $N \ge 0$. With $\sigma = \operatorname{Re} s$, for every $N \ge 0$

$$\left|\int_{\mathbb{J}^1} |y|^s\, \varphi(t)\, dt\right| \ll \int_{\mathbb{J}^1} |t|^\sigma\, \nu(t)^{-N}\, dt \ll \int_{\mathbb{J}} |t|^\sigma\, \nu(t)^{-N}\, dt$$

$$= \prod_v \left(\int_{k_v^\times} |t|^\sigma\, \sup(|t|, \frac{1}{|t|})^{-N}\, dt\right)$$

For $N > |\sigma|$, the non-archimedean local integrals are absolutely convergent:

$$\int_{k_v^\times} |t|^\sigma\, \sup(|t|, \frac{1}{|t|})^{-N}\, dt = \sum_{\ell=0}^{\infty} q_v^{-\sigma-N} + \sum_{\ell=1}^{\infty} q_v^{\sigma-N}$$

$$= \frac{1}{1-q^{-\sigma-N}} + \frac{q^{\sigma-N}}{1-q^{\sigma-N}}$$

$$= \frac{1-q^{-2N}}{(1-q^{-\sigma-N})(1-q^{\sigma-N})}$$

The archimedean local integrals are convergent for similar reasons. For $2N > 1$ and $N > |\sigma| + 1$, the Euler product is dominated by the Euler product for the expression $\zeta_k(N+\sigma)\zeta_k(N-\sigma)/\zeta_k(2N)$ in terms of the zeta function $\zeta_k(s)$ of k, which converges absolutely. ///

[2.B.4] Lemma: For all N, a Schwartz function φ on $\mathbb{A}$ satisfies

$$|\varphi(t)| \ll_{\varphi,N} \prod_v \sup(|t_v|_v, 1)^{-2N} \qquad (\text{for } t \in \mathbb{A})$$

Proof: By definition, a Schwartz function φ on $\mathbb{A}$ is a finite sum of *monomials* $\varphi = \bigotimes_v \varphi_v$. Thus, it suffices to consider monomial φ, and to prove the *local* assertion that for $\varphi_v \in \mathscr{S}(k_v)$

$$|\varphi_v(t)| \ll_N \sup(|t_v|_v, 1)^{-2N} \qquad (\text{for } t \in k_v)$$

At archimedean places, the definition of the Schwartz space requires that

$$\sup_{t\in k_v} (1+|t|_v)^N \cdot |\varphi_v(t)| < \infty \qquad (\text{archimedean } k_v, \text{ for all } N)$$

so

$$|\varphi_v(t)| \ll_N (1+|t|_v)^{-2N} \le \sup(|t|_v, 1)^{-2N}$$

Almost everywhere, φ_v is the characteristic function of the local integers. At such places,

$$|\varphi_v(t)| = \left\{ \begin{matrix} 1 & (\text{for } |t|_v \leq 1) \\ 0 & (\text{for } |t|_v > 1) \end{matrix} \right\} \leq \sup(|t|_v, 1)^{-2N} \qquad (\text{for all } N)$$

At the remaining *bad* finite primes, $\varphi_v \in \mathscr{S}(k_v)$ is compactly supported and locally compact. Then, similar to the *good* prime case,

$$|\varphi_v(t)| \ll_{\varphi_v} \left\{ \begin{matrix} 1 & (t \in \operatorname{spt} \varphi_v) \\ 0 & (t \notin \operatorname{spt} \varphi_v) \end{matrix} \right\} \ll_{\varphi_v, N} \sup(|t|_v, 1)^{-2N} \qquad (\text{for all } N)$$

This proves the lemma. ///

2.C Appendix: Hecke-Maaß Periods of Eisenstein Series

These examples, essentially due to Hecke and Maaß, include as special cases both sums of values at Heegner points and integrals over hyperbolic geodesics. The setup of the previous appendix allows a simple computation for GL_2 over a number field k.

Let ℓ be a quadratic field extension of a number field k. Let $G = GL_2(k)$, and $H \subset G$ a copy of $\ell^\times$ inside G, by specifying the isomorphism in

$$\ell^\times \subset \operatorname{Aut}_k(\ell) \approx \operatorname{Aut}_k(k^2) = GL_2(k)$$

Let P be the standard parabolic of upper-triangular elements in G. Factor the idele class group $\mathbb{J}_\ell/\ell^\times = (0, \infty) \times \mathbb{J}^1_\ell/\ell^\times$ where the ray $(0, \infty)$ is imbedded on the diagonal in the archimedean factors. Let χ be a character on $\mathbb{J}_k/k^\times$ trivial on the ray $(0, \infty)$, and define a character on $P_\mathbb{A}$ by

$$\chi_s \begin{pmatrix} a & * \\ 0 & d \end{pmatrix} = \left|\frac{a}{d}\right|^s \cdot \chi(a) \cdot \chi^{-1}(d)$$

Let $\varphi_{s,\chi}$ be a left χ_s-equivariant function on $G_\mathbb{A}$: $\varphi_{s,\chi} h(pg) = \chi_s(p) \cdot \varphi_{s,\chi}(g)$ for all $p \in P_\mathbb{A}$ and $g \in G_\mathbb{A}$. At places v where χ is unramified, we may as well take $\varphi_{s,\chi}$ to be right K_v-invariant, where K_v is the standard maximal compact in G_v. This function has trivial central character. Ignoring the ambiguity at bad primes, put

$$E_{s,\chi}(g) = \sum_{\gamma \in P_k \backslash G_k} \varphi_{s,\chi}(\gamma \cdot g)$$

Let Z be the center of G, and ω the quadratic character of ℓ/k. Let S be a (finite) set of places including those ramified in ℓ/k or ramified for χ.

[2.C.1] Theorem:

$$\int_{Z_{\mathbb{A}} H_k \backslash H_{\mathbb{A}}} E_{s,\chi} = \frac{L^S(s,\chi) \cdot L^S(s,\chi\cdot\omega)}{L^S(2s,\chi^2)} \times \text{(bad prime factors)}$$

where $L^S(s,\alpha)$ denotes the L-function attached to a Hecke character α over k dropping the local factors attached to places $v \in S$.

[2.C.2] Remark: In fact, as in the proof, the numerator arises as an L-function over the quadratic extension ℓ, namely $L_\ell^S(s,\chi \circ N)$, where N is the norm $\ell \to k$. Then quadratic reciprocity gives the indicated factorization into L-functions over the base field k.

Proof: The subgroup P_k is the isotropy group of a k-line $k \cdot e$ for a fixed nonzero $e \in k^2 \approx \ell$. The group G_k is transitive on these k-lines, so $P_k\backslash G_k \approx \{k-\text{lines}\}$. The critical-but-trivial point is that the action of $\ell^\times$ on ℓ is transitive on nonzero elements. Thus, $P_k \cdot H_k = GL_2(k)$. That is, the period integral unwinds

$$\int_{Z_{\mathbb{A}} H_k \backslash H_{\mathbb{A}}} E_{s,\chi} = \int_{Z_{\mathbb{A}}(P_k \cap H_k)\backslash H_{\mathbb{A}}} \varphi_{s,\chi} = \int_{Z_{\mathbb{A}}\backslash H_{\mathbb{A}}} \varphi_{s,\chi}$$

since $H \cap P = Z$. With $\varphi_{s,\chi}$ chosen to factor over primes $\varphi_{s,\chi} = \bigotimes_v \varphi_{s,\chi,v}$, the unwound period integral likewise factors over primes

$$\int_{Z_{\mathbb{A}} H_k \backslash H_{\mathbb{A}}} E_{s,\chi} = \int_{Z_{\mathbb{A}}\backslash H_{\mathbb{A}}} \varphi_{s,\chi} = \prod_v \int_{Z_v \backslash H_v} \varphi_{s,\chi,v}$$

A graceful way to evaluate the local integrals is to use an integral representation of the local vectors $\varphi_{s,\chi,v}$ akin to well-known archimedean devices involving the Gamma function. That is, express φ_v in terms of Iwasawa-Tate local zeta integrals

$$L_{k,v}(2s,\chi^2) = \int_{k_v^\times} |t|_v^{2s} \chi^2(t)\, \Phi_v(t\,e)\, dt$$

as

$$\varphi_{s,\chi,v}(g) = \frac{1}{L_{k,v}(2s,\chi_v^2)} \cdot |\det g|_v^s\, \chi_v(\det g) \cdot \int_{k_v^\times} |t|_v^{2s} \chi_v^2(t) \cdot \Phi_v(t \cdot e \cdot g)\, dt$$

for suitable Schwartz functions Φ_v on $k_v^2 \approx \ell_v = \ell \otimes_k k_v$. The leading local zeta integral factor gives the normalization $\varphi_v(1) = 1$ at $g = 1$. Then

$$\int_{Z_v\backslash H_v} \varphi_{s,\chi,v}$$

$$= \frac{1}{L_{k,v}(2s,\chi_v^2)} \cdot \int_{k_v^\times \backslash \ell_v^\times} |\det h|_v^s\, \chi_v^2(\det h) \cdot \int_{k_v^\times} |t|_v^{2s} \chi_v^2(t) \cdot \Phi_v(t \cdot e \cdot h)\, dt\, dh$$

Let N_v be the k_v-extension $\ell \otimes_k k_v \to k_v$ of the norm map $\ell \to k$. Since

$$|\det h|_{k_v} = |Nh|_{k_v} = |h|_{\ell_v}$$

and since $\chi_v(\det h) = \chi_v(Nh)$, the local factor of the period becomes

$$\frac{1}{L_{k,v}(2s)} \cdot \int_{k_v^\times \backslash \ell_v^\times} |h|_{\ell_v}^s \chi_v(Nh) \cdot \int_{k_v^\times} |t|_{\ell_v}^s \chi_v^2(t) \cdot \Phi_v(t \cdot e \cdot h)\, dt\, dh$$

$$= \frac{1}{L_{k,v}(2s, \chi_v^2)} \cdot \int_{k_v^\times \backslash \ell_v^\times} \int_{k_v^\times} |t \cdot h|_{\ell_v}^s \chi_v(N(th)) \cdot \Phi_v(t \cdot e \cdot h)\, dt\, dh$$

$$= \frac{1}{L_{k,v}(2s)} \cdot \int_{\ell_v^\times} |h|_{\ell_v}^s \chi_v(Nh) \cdot \Phi_v(e \cdot h)\, dt\, dh$$

$$= \frac{1}{L_{k,v}(2s, \chi_v^2)} \cdot L_{\ell,v}(s, \chi_v \circ N)$$

where the local L-function $L_{\ell,v}$ is the *product* of the finitely many local factors $L_{\ell,w}$ for places w of ℓ lying over v.

Let S be the collection of ramified primes for χ and primes ramified in ℓ/k. Let ω be the quadratic character attached to ℓ/k, with local characters ω_v: at finite primes v *splitting* in ℓ/k, the character is trivial. Let q_v be the residue field cardinality at a finite place $v \notin S$, and ϖ_v a local parameter.

At a finite place v of k *splitting* in ℓ/k, we immediately have

$$L_{\ell,v}(s, \chi_v \circ N) = \frac{1}{1 - \chi_v(\varpi_v) q_v^{-s}} \cdot \frac{1}{1 - \chi_v(\varpi_v) q_v^{-s}} = L_{k,v}(s, \chi_v)^2$$

Since local L-functions of unramified local characters are determined by their values on local parameters, at a finite place v of k *inert* in ℓ/k, in which case $N : \ell \otimes_k k_v \to k_v$ is the local field norm and ϖ_v remains prime in $\ell \otimes_k k_v$, we similarly have

$$L_{\ell,v}(s, \chi_v \circ N) = \frac{1}{1 - \chi_v(N\varpi_v)(q_v^2)^{-s}} = \frac{1}{1 - \chi_v(\varpi_v^2)(q_v^2)^{-s}}$$

$$= \frac{1}{1 - \chi_v(\varpi_v) q_v^{-s}} \cdot \frac{1}{1 + \chi_v(\varpi_v) q_v^{-s}}$$

$$= \frac{1}{1 - \chi_v(\varpi_v) q_v^{-s}} \cdot \frac{1}{1 - \chi_v \omega_v(\varpi_v) q_v^{-s}}$$

$$= L_{k,v}(s, \chi_v) \cdot L_{k,v}(s, \chi_v \cdot \omega_v)$$

Thus, the good-prime part of the global L-function is

$$L_\ell^S(s, \chi \circ N) = \prod_{v \notin S} L_{\ell,v}(s, \chi_v \circ N) = \prod_{v \notin S} L_{k,v}(s, \chi_v) \cdot L_{k,v}(s, \chi_v \cdot \omega_v)$$

Thus,

$$\int_{Z_{\mathbb{A}}H_k\backslash H_{\mathbb{A}}} E_{s,\chi} = \frac{L_k^S(s,\chi)\cdot L_k^S(s,\chi\cdot\omega)\cdot}{L_k^S(2s,\chi^2)} \times \text{(bad prime factors)}$$

as claimed. ///

[2.C.3] Remark: In this normalization, the unitary line is $\mathrm{Re}(s) = \frac{1}{2}$, and

$$\int_{Z_{\mathbb{A}}H_k\backslash H_{\mathbb{A}}} E_{\frac{1}{2}+it,\chi}$$

$$= \frac{L^S(\frac{1}{2}+it,\chi)\cdot L^S(\frac{1}{2}+it,\chi\cdot\omega)}{L^S(1+2it,\chi^2)} \times \text{(bad prime factors)}$$

[2.C.4] Remark: The basic remaining issue about the finitely many bad-prime local integrals is to be sure that we can choose local data $\varphi_{s,\chi,v}$ so that the local integrals do not vanish identically. This can be accomplished, for example, by taking the bad-prime local functions $\varphi_{s,\chi,v}$ to be 0 off a very small neighborhood of the local points P_v of the parabolic P.

3

$SL_3(\mathbb{Z})$, $SL_4(\mathbb{Z})$, $SL_5(\mathbb{Z})$, ...

We keep most of the conventions and context of the previous chapter, except now G is the group GL_r of r-by-r matrices. The novelties originate in the greater variety of *parabolic subgroups* in GL_r, the latter explicated in the first section. This variety increases the subtlety of the Gelfand condition defining the space of *cuspforms*, with pursuant proliferation of *types* of pseudo-Eisenstein series and Eisenstein series on GL_r corresponding to the various parabolic subgroups. One new phenomenon is the necessary formation of pseudo-Eisenstein series and Eisenstein series using *cuspforms* on smaller groups $GL_{r'}$.

To narrow somewhat the scope of complications, later in the chapter we mostly treat *level one* automorphic forms, that is, right $K_{\mathbb{A}}$-invariant ones, for $K_{\mathbb{A}}$ a maximal compact subgroup of $GL_r(\mathbb{A})$. This specializes to automorphic forms for $GL_r(\mathbb{Z})$ when the ground field is $\mathbb{Q}$.

3.1 Parabolic Subgroups of GL_r

It is perhaps impossible to anticipate the significance of these subgroups.[1] Nevertheless, they subsequently prove their importance. Let $G = GL_r(k)$ with an *arbitrary* field k, acting on k^r by matrix multiplication. A *flag* F in k^n is a nested sequence of one or more nonzero k-subspaces (with proper containments) $V_1 \subset \cdots \subset V_\ell \subset k^r$. The corresponding *parabolic subgroup* $P = P^F$ is the stabilizer of the flag F. The whole group G stabilizes the *improper* flag k^r and so is a parabolic subgroup of itself. The *proper* parabolics are stabilizers of flags $V_1 \subset \cdots \subset V_\ell \subset k^r$ with $\ell \geq 1$.

The *maximal* proper parabolic subgroups are stabilizers $P^{V \subset k^r}$ of flags consisting of single *proper* subspaces $V \subset k^r$. Every proper parabolic subgroup P^F for flag $F = (V_1 \subset \cdots \subset V_\ell \subset k^r)$ is the *intersection* of the maximal proper parabolics $P^{V_i \subset k^r}$. A *minimal* parabolic, stabilizing a *maximal* flag, is a *Borel* subgroup.

With $e_1, e_2, \ldots, e_r$ the standard basis for k^r, identify $k^d = ke_1 + \cdots + ke_d$. By transitivity of G on ordered bases of k^r, every orbit in the action of G on flags has a unique representative among the *standard flags*, namely, for some ordered partition $d_1 + d_2 + \cdots + d_\ell = r$ with $0 < d_j \in \mathbb{Z}$, the corresponding standard flag is

$$F^{d_1,\ldots,d_\ell} = \left(k^{d_1} \subset k^{d_1+d_2} \subset k^{d_1+d_2+d_3} \subset \cdots \subset k^{d_1+\cdots+d_\ell}\right)$$

The stabilizer of $F^{d_1,\ldots,d_\ell}$ is the *standard* (proper) parabolic subgroup $P^{d_1,\ldots,d_\ell}$ of G and is the *intersection* of the maximal proper parabolics containing it, namely

$$P^{d_1,\ldots,d_\ell} = \bigcap_{1 \leq i \leq \ell-1} P^{(d_1+\cdots+d_i),\,(d_{i+1}+\cdots+d_\ell)}$$

Two standard parabolics $P^{d_1,\ldots,d_\ell}$ and $P^{d'_1,\ldots,d'_{\ell'}}$ are *associate* when $\ell = \ell'$ and the lists $d_1, \ldots, d_\ell$ and $d'_1, \ldots, d'_\ell$ merely differ by being permutations of each other. A parabolic $P^{d_1,\ldots,d_\ell}$ is *self-associate* when $d_i = d_j$ for some $i \neq j$. These notions are important for discussion of constant terms of Eisenstein series, meromorphic continuations, and functional equations. The standard maximal

[1] Also, the terminology itself has a long and complicated history and admits a much larger context, inessential to the present illustrative discussion.

proper parabolics are block-upper-triangular, in the sense

$$P^{r',r-r'} = \left\{ \begin{pmatrix} a & b \\ 0 & d \end{pmatrix} : a \in GL_{r'},\ b = r' \times (r-r'),\ d \in GL_{r-r'} \right\}$$

That is, the diagonal blocks are $r' \times r'$ and $(r-r') \times (r-r')$, and the off-diagonal blocks are sized to fit. Next-to-maximal proper parabolics have the shape

$$P^{r_1,r_2,r_3} = \left\{ \begin{pmatrix} m_1 & * & * \\ 0 & m_2 & * \\ 0 & 0 & m_3 \end{pmatrix} : m_1 \in GL_{r_1}, m_2 \in GL_{r_2}, m_3 \in GL_{r_3} \right\}$$

for $r_1 + r_2 + r_3 = r$, with off-diagonal blocks to fit. The general standard proper parabolic $P^{d_1,\ldots,d_\ell}$ consists of block-upper-triangular matrices with diagonal blocks of sizes $d_1 \times d_1, d_2 \times d_2, \ldots, d_\ell \times d_\ell$. The standard Borel subgroup is the subgroup of upper triangular matrices.

The *unipotent radical* N^P of a parabolic P stabilizing a flag $F = (V_1 \subset \cdots \subset V_\ell \subset k^r)$ is the subgroup that fixes the quotients $V_\ell / V_{\ell-1}$ *pointwise*. This characterization shows that N^P is a *normal* subgroup of P. For the standard maximal parabolic $P = P^{r',r-r'}$, the unipotent radical is

$$N = N^P = N^{r',r-r'} = \left\{ \begin{pmatrix} 1_{r'} & b \\ 0 & 1_{r-r'} \end{pmatrix} : b = r' \times (r-r') \right\}$$

Containment of parabolics *reverses* the containment of unipotent radicals: $P \subset Q$ implies $N^P \supset N^Q$. For example, for next-to-maximal standard proper parabolic $P = P^{r_1,r_2,r_3}$, the unipotent radical is

$$N = N^P = N^{r_1,r_2,r_3} = \left\{ \begin{pmatrix} 1_{r_1} & * & * \\ 0 & 1_{r_2} & * \\ 0 & 0 & 1_{r_3} \end{pmatrix} \right\}$$

The *standard Levi component* (or standard *Levi-Malcev component*) $M = M^P = M^{d_1,\ldots,d_\ell}$ of the standard parabolic $P = P^{d_1,\ldots,d_\ell}$ is the subgroup of $P = P^{d_1,\ldots,d_\ell}$ with all the blocks *above* the diagonal 0, namely, with $m_j \in GL_{d_j}$,

$$M = M^P = M^{d_1,\ldots,d_\ell} = \left\{ \begin{pmatrix} m_1 & 0 & 0 & \ldots & 0 \\ 0 & m_2 & 0 & \ldots & 0 \\ \vdots & 0 & \ddots & & \vdots \\ & & & \ddots & 0 \\ 0 & \ldots & & 0 & m_\ell \end{pmatrix} \right\}$$

Unlike the unipotent radical, the standard Levi component is *not* normal in the standard parabolic. Nevertheless, we have the *Levi-Malcev decomposition* $P = N^P \cdot M^P$ for matrices with entries in any field. For the standard parabolics and standard Levi components, this is simply an expression of the behavior of matrix multiplication in block decompositions. For example,

$$\begin{pmatrix} a & b \\ 0 & d \end{pmatrix} = \begin{pmatrix} 1_{r'} & bd^{-1} \\ 0 & 1_{r-r'} \end{pmatrix} \begin{pmatrix} a & 0 \\ 0 & d \end{pmatrix} \qquad \text{(in blocks)}$$

The standard *maximal split torus* in G is the subgroup of diagonal matrices, which is also the Levi component $M^{\min}$ of the standard minimal-parabolic $P^{\min} = P^{1,1,\ldots,1}$. The standard *Weyl group* W can be identified with permutation matrices in G, namely, matrices with exactly one nonzero entry in each row and column, and that entry is 1. The Weyl group normalizes $M^{\min}$.[2] The simplest *Bruhat decomposition* is

[3.1.1] Claim: With $P^{\min}$ the standard minimal parabolic and $N^{\min}$ its unipotent radical, we have a *disjoint* union

$$GL_r(k) = \bigsqcup_{w \in W} P^{\min}\, wP^{\min} = \bigsqcup_{w \in W} P^{\min}\, wN^{\min}$$

Proof: The second equality follows from the first, by the Levi decomposition: letting $P = P^{\min}$ and $N = N^P$ and $M = M^P$,

$$PwP = Pw(MN) = P \cdot wMm^{-1} \cdot wN = PM \cdot wN = P \cdot wN$$

For the first equality, given $g \in GL_r(k)$, left multiplication by N can add or subtract multiples of a row of g to or from any *higher* row. Similarly, right multiplication by N adds or subtracts multiples of a *column* of g to or from any column farther to the right. Thus, letting $g_{i_1,1}$ be the lowest nonzero entry in the first column (maximal index i_1), left multiplication by N makes all other entries in the first column 0. Right multiplication by N makes all other entries in the i_1^{th} row 0. Next, let $g_{i_2,2}$ be the lowest (maximal index i_2) nonzero entry in the (new) second column. Without disturbing the effects of the previous step, all higher entries in the second column, and all entries in the i_2^{th} row to the right can be made 0 by left and right action of N. An induction produces a *monomial* matrix, that is, one with a single nonzero entry in each row and column. Then left multiplication by M normalizes all nonzero entries to 1. Thus, $MNgN \in W$.

In fact, the positions of the lowest nonzero entries $g_{i_j,j}$ in each column are completely determined by this procedure, *and* there is no other way to reach a monomial matrix by left and right multiplication by N. This explains the *disjointness* of the decomposition.

[2] A more extensible form of the definition of Weyl group is as the *normalizer* of $M^{\min}$ modulo the *centralizer* of $M^{\min}$, that is, *monomial* matrices (with one nonzero entry in each row and column) modulo diagonal matrices. However, for $G = GL_r$, it is often convenient to fix a set of representatives in G.

Rather than set up notation for the general case, the induction is better illustrated by an example: writing $*$ for unknown entries and $\times$ for nonzero entries, with suitable values in the elements of N, the first stage, using the nonzero 2, 1 entry, is

$$\begin{pmatrix} * & * & * \\ \times & * & * \\ 0 & \times & * \end{pmatrix} \longrightarrow \begin{pmatrix} 1 & * & 0 \\ 0 & 1 & 0 \\ 0 & 0 & 1 \end{pmatrix} \begin{pmatrix} * & * & * \\ \times & * & * \\ 0 & \times & * \end{pmatrix} = \begin{pmatrix} 0 & * & * \\ \times & * & * \\ 0 & \times & * \end{pmatrix}$$

$$\longrightarrow \begin{pmatrix} 0 & * & * \\ \times & * & * \\ 0 & \times & * \end{pmatrix} \begin{pmatrix} 1 & * & * \\ 0 & 1 & 0 \\ 0 & 0 & 1 \end{pmatrix} = \begin{pmatrix} 0 & * & * \\ \times & 0 & 0 \\ 0 & \times & * \end{pmatrix}$$

The second stage, using the nonzero 3, 2 entry, is

$$\begin{pmatrix} 0 & * & * \\ \times & 0 & 0 \\ 0 & \times & * \end{pmatrix} \longrightarrow \begin{pmatrix} 1 & 0 & * \\ 0 & 1 & 0 \\ 0 & 0 & 1 \end{pmatrix} \begin{pmatrix} 0 & * & * \\ \times & 0 & 0 \\ 0 & \times & * \end{pmatrix} = \begin{pmatrix} 0 & 0 & * \\ \times & 0 & 0 \\ 0 & \times & * \end{pmatrix}$$

$$\longrightarrow \begin{pmatrix} 0 & 0 & * \\ \times & 0 & 0 \\ 0 & \times & * \end{pmatrix} \begin{pmatrix} 1 & 0 & 0 \\ 0 & 1 & * \\ 0 & 0 & 1 \end{pmatrix} = \begin{pmatrix} 0 & 0 & * \\ \times & 0 & 0 \\ 0 & \times & 0 \end{pmatrix}$$

The upper-right entry is invertible, since the original is. ///

[3.1.2] Corollary: $G = \bigcup_{w \in W} PwQ$ for any standard parabolics P, Q. ///

[3.1.3] Remark: Letting $W^P = W \cap P$ and $W^Q = W \cap Q$, we have a *disjoint* union $G = \bigsqcup_{w \in W^P \backslash W / W^Q} PwQ$. However, except for minimal or maximal-proper parabolics, this requires a subtler proof. Our subsequent examples will not need this precise form of the general case, although it would become relevant in other examples.

To distinguish matrices with entries in k from entries in k_v, in what follows we will write P_k, M_k, and N_k in place of the foregoing unadorned P, M, N.

3.2 Groups $K_v = GL_r(\mathfrak{o}_v) \subset G_v = GL_r(k_v)$

Now k is again a number field with integers $\mathfrak{o}$, completions k_v, and local rings of integers $\emptyset_v$ at non-archimedean places. Let $G_v = GL_r(k_v)$, and let Z_v be the center of G_v. At non-archimedean places v, let $K_v = GL_r(\mathfrak{o}_v)$. At real v, let K_v be the standard orthogonal group $O_n(\mathbb{R}) = \{g \in GL_n(\mathbb{R}) : g^\top g = 1_r\}$, and at complex v let K_v be the standard unitary group $U_n = \{g \in GL_n(\mathbb{C}) : g^* g = 1_r\}$.

Temporarily, let ℓ be the number of nonisomorphic archimedean completions of k, thus *not* counting a complex completion and its conjugate as 2, but just 1. That is, $[k : \mathbb{Q}] = \ell_1 + 2\ell_2$ where ℓ_1 is the number of real completions, and

ℓ_2 the number of complex, and $\ell = \ell_1 + \ell_2$. Let Z^+ be the positive real scalar matrices diagonally imbedded across *all* archimedean v, by the map

$$t \longrightarrow (\ldots, t^{1/\ell}, \ldots) \qquad (\text{for } t > 0)$$

This map δ gives a *section* of the idele norm map $|t| = \prod_v |t_v|_v$, in that $|\delta(t)| = t$.

The group P_v of v-adic points of a standard parabolic $P = P^F = P^{d_1,\ldots,d_\ell}$ is the stabilizer in G_v with the same shape as the k-rational points P_k in the previous section, but with entries in k_v rather than k. That is, the v-adic version of the flag $F = (k^{d_1} \subset k^{d_1+d_2} \subset \cdots \subset k^r)$ is the natural $F_v = (k_v^{d_1} \subset k_v^{d_1+d_2} \subset \cdots \subset k_v^r)$, and $P_v = P_v^{d_1,\ldots,d_\ell}$ is its stabilizer.[3] Similarly, $N_v = N_v^P$ is the v-adic points of the unipotent radical $N = N^P$ of P, and $M_v = M_v^P$ is the v-adic points of the *standard* Levi component of a *standard* parabolic P. That is, again, the *shapes* of the matrices are the same as in the previous section, but with entries in k_v rather than k.

Iwasawa decompositions are analogous to the previous chapter's, with proofs merely iterations of the arguments there:

[3.2.1] Claim: $G_v = P_v \cdot K_v$ for standard minimal-parabolic P.

Proof: For archimedean v, the right action of K_v *rotates* the rows of given $g \in G_v$. The bottom row can be rotated to be of the form $(0, 0, \ldots, 0, 0, *)$. Without disturbing this effect, the second-to-bottom row can be rotated to be of the form $(0, 0, \ldots, 0, *, *)$. Continuing with higher rows puts the result in $P_v^{\min}$.

For non-archimedean v, right multiplication by K_v can subtract local-integer multiples of the largest entry in the bottom row from all others, to put the bottom row into the form $(0, 0, \ldots, 0, *, 0, \ldots 0)$ with a nonzero entry at just one position. Then a permutation matrix (in K_v) can move the nonzero entry to the far right. Without disturbing this effect, the first $r-1$ entries of the second-to-bottom row can be dealt with similarly, putting it into the form $(0, 0, \ldots, 0, *, *)$. Because the determinant is nonzero, the second-to-right entry in the new second-to-bottom row is nonzero. Continuing to modify higher rows puts the result in P_v. ///

As in [1.2] and [2.1], *Cartan decompositions* follow from the spectral theorem for symmetric or Hermitian operators at archimedean places, and from the structure theorem for finitely generated modules at finite places:

[3.2.2] Claim: $G_v = K_v M_v K_v$ with $M = M^{\min}$ the standard Levi component of the minimal parabolic.

[3] The k_v-vectorspace k_v^r has many subspaces and flags *not* obtained by extending scalars from subspaces and flags in k^r, but these will play no role here.

Proof: For archimedean v, letting $g \to g^*$ be either transpose for $k_v \approx \mathbb{R}$, or conjugate-transpose for $k_v \approx \mathbb{C}$, the matrix gg^* is positive-definite symmetric or Hermitian-symmetric. The spectral theorem for such operators gives an orthogonal or unitary matrix k such that $k(gg^*)k^* = \delta$ is *diagonal* with strictly positive real diagonal entries. Let $\sqrt{\delta}$ be the positive-definite diagonal square root of δ. Then $h = k^*\sqrt{\delta}k$ is a positive-definite Hermitian/symmetric square root of gg^*. Then

$$(h^{-1} \cdot g) \cdot (h^{-1} \cdot g)^* = h^{-1} \cdot gg^* \cdot h^{-1} = h^{-1} \cdot h^2 \cdot h^{-1} = 1_r$$

Inverting $k^*k = 1_r$ gives $k^{-1}(k^*)^{-1} = 1_r$, and then $1_r = kk^*$, so the latter condition also defines K_v. That is, $h^{-1}g \in K_v$, and $g \in k^*\sqrt{\delta}k \cdot K_v \subset K_v\sqrt{\delta}K_v$.

For non-archimedean v, multiply through by scalar matrix $c \cdot 1_r$ so that all entries of cg are in $\mathfrak{o}_v$, although of course the determinant may fail to be a local unit. The rows of $R_1, \ldots, R_r$ of $g \in G_v$ are linearly independent, and generate a free $\mathfrak{o}_v$-module F of rank r inside $\mathfrak{o}_v^r$. Observe that $K_v = GL_r(\mathfrak{o}_v)$ is the stabilizer of $\mathfrak{o}_v^r$. Since $\mathfrak{o}_v$ has a unique nonzero prime ideal, the applicable form of the structure theorem for finitely generated modules over principal ideal domains is even simpler, and produces an $\mathfrak{o}_v$-basis $f_1, \ldots, f_r$ of $\mathfrak{o}_v^r$ and elementary divisors $d_1| \ldots |d_r$ such that $\{d_i f_i\}$ is an $\mathfrak{o}_v$-basis of F. Let $k_1 \in K_v$ be the change-of-basis element such that the jth row of $k_1 \cdot g$ is $d_j f_j$. Let $\{e_i\}$ be the standard basis of $\mathfrak{o}_v^r$, and $k_2 \in K_v$ such that $f_j \cdot g = e_j$ for all j. Then $\delta = k_1 g k_2$ is diagonal with entries $d_1, \ldots, d_r$. We can undo the initial multiplication to get $g \in k_1^{-1}c^{-1}\delta k_2^{-1} \in K_v M_v K_v$. ///

As in GL_2, unlike the archimedean situation, for non-archimedean v the compact K_v has substantial intersections with N_v^P and M_v^P for every standard parabolic P. As for GL_2, unipotent radicals N_v^P of standard parabolics P are ascending unions of compact, open subgroups:

$$N_v^P = \bigcup_{\ell \geq 0} \{n \in N_v^P : n = 1_r \bmod \varpi^\ell\}$$

Again, unlike the archimedean situation, K_v has a basis at 1 consisting of compact, open subgroups, namely, the (local) *principal congruence subgroups*

$$K_{v,\ell} = \{g \in K_v = GL_r(\mathfrak{o}_v) : g = 1_r \bmod \varpi^\ell\}$$

The corresponding *adele group* is $G_{\mathbb{A}} = GL_r(\mathbb{A})$, meaning r-by-r matrices with entries in $\mathbb{A}$, with determinant in the ideles $\mathbb{J}$. This group is also an ascending union (colimit) of *products*

$$G_S = \prod_{v \in S} G_v \times \prod_{v \notin S} K_v$$

for S a finite set of places v, including archimedean places, ordering finite sets S (of places v) by containment. Similarly, $P_{\mathbb{A}}, M_{\mathbb{A}}^P, N_{\mathbb{A}}^P$, and $Z_{\mathbb{A}}$ are the adelic forms of those groups. Let $K_{\mathbb{A}} = \prod_v K_v \subset G_{\mathbb{A}}$. With the usual one-sided inverse to $\delta : (0,\infty) \to \mathbb{J}$ the *idele norm* $|\cdot| : \mathbb{J} \to (0,\infty)$,

$$Z_{\mathbb{A}}/Z^+Z_k \;\approx\; \mathbb{J}/\delta(0,\infty)\cdot k^\times \;\approx\; \mathbb{J}^1/k^\times = \text{compact}$$

where Z_k is the invertible scalar matrices with entries in k, with compactness demonstrated in [2.15].

3.3 Discrete Subgroup $G_k = GL_r(k)$, Reduction Theory

As expected, $G_k = GL_r(k)$, and P_k, M_k^P, N_k^P are the corresponding groups with entries in k. Proof of the *discreteness* of G_k in $G_{\mathbb{A}}$ is essentially identical to that for GL_2 in [2.2], and we will not repeat it. Let

$$G^1 = \{g \in G_{\mathbb{A}} : |\det g| = 1\}$$

and $G_{\mathbb{A}} = Z^+ \times G^1$. The *product formula* $\prod_{v\le\infty} |t|_v = 1$ for $t \in k^\times$ shows that $G_k \subset G^1$. In particular, G_k is still *discrete* in $Z^+\backslash G_{\mathbb{A}} \approx G^1$. As in the simpler cases [1.5] and [2.2], *reduction theory* should show that the quotient $G_k\backslash G_{\mathbb{A}}$ is covered by a suitable notion of *Siegel set*, and that these Siegel sets interact well with each other. We prove that a single Siegel set covers the quotient but omit the discussion of their interaction.

The notion of (standard) *Siegel set* becomes somewhat more complicated. The notion of a single numerical *height* as in [1.5] and [2.2] is replaced by a *family*: the standard positive *simple roots*[4] are characters on $M = M^P = M^{\min}$ with $P = P^{1,\dots,1} = P^{\min}$ the standard minimal parabolic:

$$\alpha_i\begin{pmatrix} m_1 & & \\ & \ddots & \\ & & m_r \end{pmatrix} = \frac{m_i}{m_{i+1}} \qquad (\text{for } 1 \le i < r)$$

These simple roots make sense on M_k or M_v or $M_{\mathbb{A}}$, taking values in $k^\times$, $k_v^\times$, and $\mathbb{J}$. A standard Siegel set *adapted to* or *aligned with* $P = P^{\min}$ is a set of the form

$$\begin{aligned}\mathfrak{S}^P &= \mathfrak{S}^P_{t,C}\\ &= \{g = nmk : n \in C, m \in M_{\mathbb{A}}, k \in K_{\mathbb{A}}, |\alpha_i(m)| \ge t \text{ for } 1 \le i < r\}\end{aligned}$$

with idele norm $|\cdot|$, for $0 < t \in \mathbb{R}$ and compact $C \subset N_{\mathbb{A}}^P$. Let $N^{\min} = N^P$ with $P = P^{\min}$.

[4] As with nomenclature for other objects, the terminology *simple root* is the correct name, but the origins, general definitions, and abstracted properties of simple roots would not help us here.

[3.3.1] Theorem: For given k, there is $t > 0$ and a compact subgroup $C \subset N_{\mathbb{A}}^{\min}$ such that $G_k \cdot \mathfrak{S}_{t,C}^P = G_{\mathbb{A}}$. That is, $G_k \backslash G_{\mathbb{A}}$ is covered by a single, sufficiently large Siegel set.

Proof: We need a notion of *height* on $\mathbb{A}^r$ as in [2.2] for $r = 2$. Let $G_{\mathbb{A}}$ act on the right on $\mathbb{A}^r$ by matrix multiplication. For *real* primes v of k the *local height* function h_v on $x = (x_1, \ldots, x_r) \in k_v^r$ is $h_v(x) = \sqrt{x_1^2 + \cdots + x_n^2}$. For *complex* v, take $h_v(x) = |x_1|_{\mathbb{C}} + \cdots + |x_r|_{\mathbb{C}}$ with $|z|_{\mathbb{C}} = |N_{\mathbb{R}}^{\mathbb{C}} z|_{\mathbb{R}}$ to not disturb the product formula. For *non-archimedean* v, $h_v(x) = \sup_i |x_i|_v$.

A vector $x \in \mathbb{A}^r$ is *primitive* when it is of the form $x_o g$ for $g \in G_{\mathbb{A}}$ and $x_o \in k^r$. For $x = (x_1, \ldots, x_r) \in k^r$, at almost all non-archimedean primes v the x_i's are in $\mathfrak{o}_v$ and have local greatest common divisor 1. Elements of the adele group $g \in G_{\mathbb{A}}$ are in K_v almost everywhere, so this is not changed by multiplication by g. That is, a primitive vector x has the property that at almost all v the components of x are locally integral and have local greatest common divisor 1. For primitive x the *global height* is $h(x) = \prod_v h_v(x_v)$. Since x is primitive, at almost all finite primes the local height is 1, so this product has only finitely many non-1 factors. The proof of the following is mostly identical to the $r = 2$ case [2.2.2]:

[3.3.2] Claim: For fixed $g \in GL_r(\mathbb{A})$ and for fixed $c > 0$,

$$\text{card}\,(k^\times \backslash \{x \in k^r - \{0\} : h(x \cdot g) < c\}) < \infty$$

For compact $C \subset G_{\mathbb{A}}$ there are positive implied constants such that

$$h(x) \ll_C h(x \cdot g) \ll_C h(x) \quad \text{(for all } g \in C\text{, for all primitive } x)$$

Proof: Fix $g \in G_{\mathbb{A}}$. Since $K = K_{\mathbb{A}} = \prod_v K_v$ preserves heights, via Iwasawa decompositions locally everywhere, we may suppose that g is in the group $P_{\mathbb{A}}$ of upper triangular matrices in $G_{\mathbb{A}}$. Let g_{ij} be the $(i, j)^{\text{th}}$ entry of g. Choose representatives $x = (x_1, \ldots, x_r)$ for nonzero vectors in k^r modulo $k^\times$ such that, with μ the first index with $x_\mu \neq 0$, $x_\mu = 1$. That is, x is of the form $x = (0, \ldots, 0, 1, x_{\mu+1}, \ldots, x_r)$.

The more easily written-out case $r = 2$ of the first assertion was treated in [2.2.2]. For $x \in k^r - \{0\}$ such that $h(xg) < c$, let $\mu - 1$ be the least index such that $x_\mu \neq 0$. Adjust by $k^\times$ such that $x_\mu = 1$. For each v, from $h(xg) < c$,

$$|g_{\mu-1,\mu} + x_\mu g_{\mu,\mu}|_v \prod_{w \neq v} |g_{\mu-1,\mu-1}|_w \leq h(gx) < c$$

For almost all v we have $|g_{\mu-1,\mu-1}|_v = 1$, so there is a uniform c' such that

$$|g_{\mu-1,\mu} + x_\mu g_{\mu,\mu}|_v < c' \qquad \text{(for all } v)$$

For almost all v the residue field cardinality q_v is strictly greater than c', so for almost all v

$$|g_{\mu-1,\mu} + x_\mu g_{\mu,\mu}|_v \le 1$$

Therefore, $g_{\mu-1,\mu} + x_\mu g_{\mu,\mu}$ lies in a compact subset C of $\mathbb{A}$. Since k is discrete, the collection of x_μ is finite.

Continuing, there are only finitely many choices for the other entries of x. Inductively, suppose $x_i = 0$ for $i < \mu - 1$, and $x_\mu, \ldots, x_{\nu-1}$ fixed, and show that x_ν has only finitely many possibilities. Looking at the νth component $(xg)_\nu$ of xg,

$$|g_{\mu-1,\nu} + x_\mu g_{\mu,\nu} + \cdots + x_{\nu-1} g_{\nu-1,\nu} + x_\nu g_{\nu,\nu}|_v \prod_{w \ne v} |g_{\mu-1,\mu-1}|_w$$
$$\le h(xg) \le c$$

For almost all places v we have $|g_{\mu-1,\mu-1}|_w = 1$, so there is a uniform c' such that for all v

$$|(xg)_\nu|_v = |g_{\mu-1,\nu} + x_\mu g_{\mu,\nu} + \cdots + x_{\nu-1} g_{\nu-1,\nu} + x_\nu g_{\nu,\nu}|_v < c'$$

For almost all v the residue field cardinality q_v is strictly greater than c', so for almost all v

$$|g_{\mu-1,\nu} + x_\mu g_{\mu,\nu} + \cdots + x_{\nu-1} g_{\nu-1,\nu} + x_\nu g_{\nu,\nu}|_v \le 1$$

Therefore,

$$g_{\mu-1,\nu} + x_\mu g_{\mu,\nu} + \cdots + x_{\nu-1} g_{\nu-1,\nu} + x_\nu g_{\nu,\nu}$$

lies in the intersection of a compact subset C of $\mathbb{A}$ with a closed discrete set, so lies in a finite set. Thus, the number of possibilities for x_ν is finite. Induction gives the first assertion of the claim.

For the second assertion of the claim, let E be a compact subset of $G_\mathbb{A}$, and let $K = \prod_v K_v$. Then $K \cdot E \cdot K$ is compact, being the continuous image of a compact set. So without loss of generality E is left and right K-stable. By Cartan decompositions the compact set E of $G_\mathbb{A}$ is contained in a set KCK where $C \subset M_\mathbb{A}$ is compact. Let $g = \theta_1 m \theta_2$ with $\theta_i \in K$, $m \in C$, and x a primitive vector. By the K-invariance of the height,

$$\frac{h(xg)}{h(x)} = \frac{h(x\theta_1 m\theta_2)}{h(x)} = \frac{h(x\theta_1 m)}{\theta(x)} = \frac{h((x\theta_1)m)}{h((x\theta))}$$

Thus, the set of ratios $h(xg)/h(x)$ for g in a compact set and x ranging over primitive vectors is exactly the set of values $h(xm)/h(x)$ where m ranges over a compact set and x varies over primitives. With diagonal entries m_i of m, by compactness of C,

$$0 < \inf_{m \in C} \inf_i |m_i| \le \frac{h(xm)}{h(x)} \le \sup_{m \in C} \sup_i |m_i| < \infty$$

for all primitive x. This proves the second assertion of the claim.

Analogous to [2.2] for $r = 2$, we could put $\eta(g) = |\det g| \cdot h(e_r \cdot g)^{-r}$, where $\{e_i\}$ is the standard basis for k^r. The parabolic $Q = P^{r-1,1}$ is the stabilizer of the line $k \cdot e_r$. This modification makes η invariant under $Z_{\mathbb{A}}$, as well as left Q_k-invariant and right $K_{\mathbb{A}}$-invariant.

[3.3.3] Corollary: *(of claim)* Given $g \in G_{\mathbb{A}}$, there are finitely many $\gamma \in Q_k\backslash G_k$ such that $\eta(\gamma \cdot g) > \eta(g)$. Thus, the supremum $\sup_\gamma \eta(\gamma \cdot g)$ is attained and is finite.

Proof: There is a natural bijection $Q_k\backslash G_k \longleftrightarrow k^\times\backslash(k^r - \{0\})$ mapping a matrix to its bottom row. The claim shows that there are finitely many $x \in k^\times\backslash(k^r - \{0\})$ such that $h(xg) < c$, that is, such that $h(xg)^{-1} > c^{-1}$. Since $|\det g|$ is G_k-invariant, the bijection gives the assertion. ///

Now we prove the theorem by induction on r. Given $g \in G_{\mathbb{A}}$, by the corollary there is $x \in k^r - \{0\}$ such that $h(xg) > 0$ is minimal among values $h(x'g)$ with $x' \in k^r - \{0\}$. Take $\gamma_o \in G_k$ so that $e_r\gamma_o = x$, so $h(xg) = h(e_r\gamma_o g)$ is minimal, and $\eta(\gamma_o g)$ is *maximal* among all values $\eta(\gamma \cdot \gamma_o g)$ for $\gamma \in G_k$. By Iwasawa, there is $\theta \in K$ such that $q = \gamma_o g\theta \in Q_{\mathbb{A}}$. Then $h(\gamma g\theta) = |q_{rr}|$ where q_{ij} is the ij^{th} entry of q, and $\eta(q)$ is maximal among all values $\eta(\gamma \cdot q)$ for $\gamma \in G_k$. Let $H \subset M^Q$ be the subgroup of $G_{\mathbb{A}}$ fixing e_r and stabilizing the subspace spanned by $e_1, \ldots, e_{r-1}$, so $H \approx GL_{r-1}(\mathbb{A})$. By induction on r, beginning at $r = 2$ in [2.2], by acting on $q = \gamma g\theta$ on the left by H_k and on the right by $H_{\mathbb{A}} \cap K_{\mathbb{A}}$, we can suppose that $q \in P^{\min}_{\mathbb{A}}$ and $|q_{ii}/q_{i+1,i+1}| \ge t$ for $i < r - 1$, without altering $\eta(q)$. The induction step reduces to the case $r = 2$. The extremal property $h(e_r q) \le h(x' \cdot q)$ for all $x' \in k^r - \{0\}$ certainly implies $h(e_r q) \le h(x' \cdot q)$ with x' ranging over the smaller set of vectors of the form $x' = (0 \ldots 0\ x_{r-1}\ x_r)$. Thus, the lower right 2-by-2 block q' of q is *reduced* as an element of $GL_2(\mathbb{A})$. This reduces to the $r = 2$ case treated in [2.2.7], giving $|q_{r-1,r-1}|/|q_{rr}| \ge t$ for sufficiently small t and proving the theorem. ///

3.4 Invariant Differential Operators and Integral Operators

For archimedean G_v, for some purposes, such as meromorphic continuation of Eisenstein series [11.5], [11.10], [11.12], the Casimir operator or Laplacian as in [4.2], [4.4] suffices.[5] Beyond that, the tractability of *integral operators*, as in the rewriting of non-archimedean Hecke operators as such, suggests using integral operators at archimedean places, as well, especially in light of the commutativity result [3.4.3].

As usual, for a continuous action $G_v \times V \to V$ on a quasi-complete, locally convex topological vectorspace V, the corresponding *integral operators* are

$$\varphi \cdot f = \int_{G_v} \varphi(g)\, g \cdot f\, dg \qquad (\text{for } \varphi \in C_c^o(G_v) \text{ and } f \in V)$$

The integrand is a continuous, compactly supported V-valued function, so has a Gelfand-Pettis integral [14.1]. Thus, for $f \in V = L^2(Z^+G_k\backslash G_{\mathbb{A}})$, with G_v acting by right translation, pointwise

$$(\varphi \cdot f)(x) = \int_{G_v} \varphi(g)\,(g \cdot f)(x)\, dg = \int_{G_v} \varphi(g)\, f(xg)\, dg$$

for $\varphi \in C_c^o(G_v)$ and $f \in V$. In fact, for general reasons [6.1] the right-translation action $G_v \times L^2(Z^+G_k\backslash G_{\mathbb{A}}) \to L^2(Z^+G_k\backslash G_{\mathbb{A}})$ is continuous, so the integral converges as an $L^2(Z^+G_k\backslash G_{\mathbb{A}})$-valued integral, obviating concern about pointwise values. The *composition* of two such operators is the operator attached to the *convolution*: for $\varphi, \psi \in C_c^o(G_{\mathbb{A}})$, by the same computation as in [2.4],

$$\varphi \cdot (\psi \cdot f) = \int_{G_v} \varphi(g)\, g \cdot \left(\int_{G_v} \psi(h)\, h \cdot f\, dh \right) dg = \int_{G_v}\int_{G_v} \varphi(g)\, \psi(h)(gh \cdot f)\, dh\, dg$$

because the operation of φ moves inside the Gelfand-Pettis integral. Replacing h by $g^{-1}h$ gives

$$\int_{G_v}\int_{G_v} \varphi(g)\, \psi(g^{-1}h)\, h \cdot f\, dh\, dg = \int_{G_v} \left(\int_{G_v} \varphi(g)\, \psi(g^{-1}h)\, dg \right) h \cdot f\, dh$$

by changing the order of integration.

[3.4.1] Lemma: The *adjoint* to the action of $\varphi \in C_c^o(G_v)$ on $L^2(Z^+G_k\backslash G_{\mathbb{A}})$ is the action of $\check{\varphi} \in C_c^o(G_v)$, where $\check{\varphi}(g) = \overline{\varphi(g^{-1})}$. *(Proof identical to [2.4.1].)* ///

For simplicity of discussion, we restrict attention to functions on $Z^+G_k\backslash G_{\mathbb{A}}$ right K_v-invariant for archimedean v. In that situation, for archimedean v,

[5] For $SL_r(\mathbb{R})$ or $SL_r(\mathbb{C})$, with $r \geq 3$, the center of the universal enveloping algebra $U\mathfrak{g}$ of the corresponding algebra $\mathfrak{g}$ is generated by $r - 1$ commuting operators.

the integral operators given by left-and-right K_v-invariant $\varphi \in C_c^o(G_v)$, also denoted $C_c^o(K_v\backslash G_v/K_v)$, act on right K_v-invariant functions on $Z^+G_k\backslash G_\mathbb{A}$.

[3.4.2] Claim: The action of integral operators attached to $\varphi \in C_c^o(K_v\backslash G_v/\,K_v)$ stabilizes K_v-invariant vectors f in any continuous group action $G_v \times V \to V$ for quasi-complete, locally convex V. *(Proof identical to [2.4.3].)* ///

Invoking Gelfand's trick [2.4.5],

[3.4.3] Claim: The action of integral operators attached to $\varphi \in C_c^o(K_v\backslash G_v/K_v)$ with convolution is *commutative*, for both non-archimedean and archimedean v.

Proof: To apply [2.4.5], we need an involutive anti-automorphism σ of G_v, that is, $g \to g^\sigma$ such that $(gh)^\sigma = h^\sigma g^\sigma$ and $(g^\sigma)^\sigma = g$, stabilizing K_v and acting trivially on representatives for double cosets $K_v\backslash G_v/K_v$. Use the Cartan decomposition [3.2] $G_v = K_v M_v K_v$ and use transpose $g^\sigma = g^\top$. Transpose stabilizes K_v, and acts trivially on M_v. ///

As in other cases, the algebra of integral operators attached to $\varphi \in C_c^o(K_v\backslash G_v/K_v)$ is stable under adjoints. Thus, it is plausible to ask for *simultaneous eigenvectors* for this commutative algebra of integral operators.

3.5 Hecke Operators and Integral Operators

For non-archimedean G_v, for any continuous action $G_v \times V \to V$ on a quasi-complete, locally convex topological vectorspace V, the corresponding *integral operators* are

$$\varphi \cdot f = \int_{G_v} \varphi(g)\, g \cdot f\, dg \qquad (\text{for } \varphi \in C_c^o(G_v) \text{ and } f \in V)$$

The integrand is a continuous, compactly supported V-valued function, so has a Gelfand-Pettis integral [14.1]. Thus, for $f \in V = L^2(Z^+G_k\backslash G_\mathbb{A})$, with G_v acting by right translation,

$$(\varphi \cdot f)(x) = \int_{G_v} \varphi(g)\,(g \cdot f)(x)\, dg = \int_{G_v} \varphi(g)\, f(xg)\, dg$$

for $\varphi \in C_c^o(G_v)$ and $f \in V$, and for general reasons [6.1] the right-translation action $G_v \times L^2(Z^+G_k\backslash G_\mathbb{A}) \to L^2(Z^+G_k\backslash G_\mathbb{A})$ is continuous, so the integral converges as an $L^2(Z^+G_k\backslash G_\mathbb{A})$-valued integral. The *composition* of two such operators is the operator attached to the *convolution*: as in [2.4] and [3.4], for $\varphi, \psi \in C_c^o(G_\mathbb{A})$,

$$\varphi \cdot (\psi \cdot f) = \int_{G_v} \left(\int_{G_v} \varphi(g)\, \psi(g^{-1}h)\, dg \right) h \cdot f\, dh$$

[3.5.1] Lemma: The *adjoint* to the action of $\varphi \in C_c^o(G_v)$ on $L^2(Z^+G_k\backslash G_\mathbb{A})$ is the action of $\check{\varphi} \in C_c^o(G_v)$, where $\check{\varphi}(g) = \overline{\varphi(g^{-1})}$. *(Proof identical to [2.4.1].)* ///

It is reasonable to restrict attention to functions on $Z^+G_k\backslash G_\mathbb{A}$ right K_v-invariant for all v. But it is also reasonable to relax this condition to require right K_v-invariance *almost everywhere*, that is, at all but finitely many places. A variant of $K_\mathbb{A}$-invariance, to cope with the finitely many places where right K_v-invariance is not required, is $K_\mathbb{A}$*-finiteness* of a function f on $G_\mathbb{A}$ or $Z^+G_k\backslash G_\mathbb{A}$ or other quotients of $G_\mathbb{A}$, namely, the requirement that the vectorspace of functions spanned by $\{x \to f(xh) : h \in K_\mathbb{A}\}$ is *finite-dimensional*. At the extreme of $K_\mathbb{A}$-invariant f, this space is *one*-dimensional.

[3.5.2] Lemma: For v non-archimedean, K_v-finiteness is equivalent to *invariance* under some finite-index subgroup $K' \subset K_v$. *(Proof identical to [2.4.3].)* ///

Unsurprisingly, it turns out that $K_\mathbb{A}$-finite functions on $Z^+G_k\backslash G_\mathbb{A}$ are better behaved than arbitrary functions. Of course, most $f \in L^2(Z^+G_k\backslash G_\mathbb{A})$ are *not* $K_\mathbb{A}$-finite.

For non-archimedean v, the *spherical Hecke operators* for G_v are the integral operators given by left-and-right K_v-invariant $\varphi \in C_c^o(G_v)$, also denoted $C_c^o(K_v\backslash G_v/K_v)$. Since K_v is *open*, such functions are *locally constant*: given $x \in G_v$, $\varphi(xh) = \varphi(x)$ for all $h \in K_v$, but xK_v is a neighborhood of x. Then the compact support implies that such φ takes only finitely many distinct values. Thus, the associated integral operator is really a *finite sum*. Nevertheless, expression as integral operators explains the behavior well.

[3.5.3] Claim: The action of spherical Hecke operators attached to $\varphi \in G_v$ stabilizes K_v-invariant vectors f in any continuous group action $G_v \times V \to V$ for quasi-complete, locally convex V. *(Proof identical to [2.4.3].)* ///

[3.5.4] Claim: For non-archimedean v, the *spherical Hecke algebra* $C_c^o(K_v\backslash G_v/K_v)$ with convolution is *commutative*. *(Again, Gelfand's trick [2.4.5].)* ///

It is easy to see that the spherical Hecke algebra is stable under adjoints. Thus, it is plausible to ask for *simultaneous eigenvectors* for the spherical Hecke algebra. That is, for $f \in L^2(Z^+G_k\backslash G_\mathbb{A})$, we might try to require that f be a spherical Hecke eigenfunction at almost all non-archimedean v, in addition to conditions at archimedean places. However, it bears repeating that, in infinite-dimensional Hilbert spaces, there is no general promise of existence of such simultaneous eigenvectors.

3.6 Decomposition by Central Characters

While $Z^+G_k\backslash G_{\mathbb{A}}$ has finite invariant volume, $G_k\backslash G_{\mathbb{A}}$ does not. The further quotient $Z_{\mathbb{A}}G_k\backslash G_{\mathbb{A}}$ certainly has finite invariant volume. Functions on $Z_{\mathbb{A}}G_k\backslash G_{\mathbb{A}}$ are automorphic forms (or automorphic functions) with *trivial central character* because they are invariant under the center $Z_{\mathbb{A}}$ of $G_{\mathbb{A}}$. We can treat a larger class with little further effort. Namely, the *compact* abelian group $Z_{\mathbb{A}}/Z^+Z_k \approx \mathbb{J}^1/k^\times$, being a quotient of the center $Z_{\mathbb{A}}$ of $G_{\mathbb{A}}$, acts on functions on $Z_{\mathbb{A}}G_k\backslash G_{\mathbb{A}}$ and commutes with right translation by $G_{\mathbb{A}}$. In particular, the action of $Z_{\mathbb{A}}/Z^+Z_k$ commutes with the integral operators on G_v for $v < \infty$, and with differential operators coming from the Lie algebra $\mathfrak{g}_v$ of G_v at archimedean places. Thus, for this chapter, an *automorphic form* or *automorphic function* is a function on $Z^+G_k\backslash G_{\mathbb{A}}$. For each character ω of $Z_{\mathbb{A}}/Z^+Z_k$, the space $L^2(Z^+G_k\backslash G_{\mathbb{A}}, \omega)$ of all left Z^+G_k-invariant f on $G_{\mathbb{A}}$ such that $|f| \in L^2(Z_{\mathbb{A}}G_k\backslash G_{\mathbb{A}})$ and $f(zg) = \omega(a) \cdot f(g)$ for all $z \in Z_{\mathbb{A}}$ is the space of L^2 automorphic forms *with central character* ω.

[3.6.1] Claim: *$L^2(Z^+G_k\backslash G_{\mathbb{A}})$ decomposes by central characters:*

$$L^2(Z^+G_k\backslash G_{\mathbb{A}}) = \text{completion of } \bigoplus_{\omega} L^2(Z^+G_k\backslash G_{\mathbb{A}}, \omega)$$

(Proof identical to that in [2.5].) ///

3.7 Discrete Decomposition of Cuspforms

For a standard parabolic subgroup P of G with unipotent radical $N = N^P$, the *constant term* of an automorphic form f along P is

$$c_P f(g) = \int_{N_k\backslash N_{\mathbb{A}}} f(ng)\, dn$$

For general reasons [6.1], the group $N_{\mathbb{A}}$ acts continuously on the Fréchet space $C^o(Z^+N_k\backslash G_{\mathbb{A}})$, and $N_k\backslash N_{\mathbb{A}}$ is compact, so for $f \in C^o(Z^+N_k\backslash G_{\mathbb{A}})$, the constant-term integral exists as a Gelfand-Pettis integral and is a continuous function.

[3.7.1] Claim: Constant terms are functions on $Z^+N_{\mathbb{A}}M_k\backslash G_{\mathbb{A}}$.

Proof: By changing variables, $g \to c_P f(g)$ is a left $N_{\mathbb{A}}$-invariant function on $G_{\mathbb{A}}$:

$$c_P f(n'x) = \int_{N_k\backslash N_{\mathbb{A}}} f(n \cdot n'x)\, dn = \int_{N_k\backslash N_{\mathbb{A}}} f((nn') \cdot x)\, dn$$

$$= \int_{N_k\backslash N_{\mathbb{A}}} f(n \cdot x)\, dn \qquad (\text{for } n' \in N_{\mathbb{A}})$$

Similarly, for $m \in M_k$,

$$c_P f(mx) = \int_{N_k \backslash N_{\mathbb{A}}} f(n \cdot mx)\, dn = \int_{N_k \backslash N_{\mathbb{A}}} f(m \cdot m^{-1} nm \cdot x)\, dn$$

$$= \int_{N_k \backslash N_{\mathbb{A}}} f(m^{-1} nm \cdot x)\, dn$$

since f itself is left M_k-invariant. Replacing n by mnm^{-1} gives the expression for $c_P f(g)$, noting that conjugation by $m \in M_k$ stabilizes N_k, and by the product formula the change of measure on $N_{\mathbb{A}}$ is trivial. Invariance under Z^+ is even easier. ///

A *cuspform* is a function f on $Z^+ G_k \backslash G_{\mathbb{A}}$ meeting Gelfand's condition $c_P f = 0$ for every standard parabolic P. When f is merely *measurable*, so does not have well-defined pointwise values everywhere, this condition is best interpreted *distributionally*, as in [2.7] for GL_2, and addressed in [3.8], in terms of pseudo-Eisenstein series. The space of square-integrable cuspforms is

$$L^2_o(Z^+ G_k \backslash G_{\mathbb{A}}) = \{f \in L^2(Z^+ G_k \backslash G_{\mathbb{A}}) : c_P f = 0, \text{ for all } P\}$$

The fundamental theorem proven in [7.1–7.7] is the *discrete decomposition of spaces of cuspforms.* A simple version addresses the space

$$L^2_o(Z^+ G_k \backslash G_{\mathbb{A}} / K_{\mathbb{A}}, \omega)$$
$$= \{\text{right-}K_{\mathbb{A}}\text{-invariant } L^2 \text{ cuspforms, central character } \omega\}$$

where $K_{\mathbb{A}} = \prod_{v \le \infty} K_v$. This space is $\{0\}$ unless ω is *unramified*, that is, is trivial on $Z_{\mathbb{A}} \cap K_{\mathbb{A}}$, since $K_{\mathbb{A}}$-invariance implies $Z_{\mathbb{A}} \cap K_{\mathbb{A}}$-invariance, and we also require $Z_{\mathbb{A}}$, ω-equivariance.

The spherical Hecke algebras $C^o_c(K_v \backslash G_v / K_v)$ act by *right* translation, and the Gelfand condition is an integral on the *left*, so spaces of cuspforms are *stable* under all these integral operators. The everywhere-spherical form of the decomposition result is

[3.7.2] Theorem: $L^2_o(Z^+ G_k \backslash G_{\mathbb{A}} / K_{\mathbb{A}}, \omega)$ has an orthonormal basis of simultaneous eigenfunctions for $C^o_c(K_v \backslash G_v / K_v)$, the local spherical Hecke algebras. Each simultaneous eigenspace occurs with *finite multiplicity*, that is, is finite-dimensional. *(Proof in [7.1–7.7].)*

In contrast, the full spaces $L^2(Z^+ G_k \backslash G_{\mathbb{A}} / K_{\mathbb{A}}, \omega)$ do *not* have bases of simultaneous L^2-eigenfunctions. Instead, as in [2.11–2.12] and [3.15–3.16], the orthogonal complement of cuspforms in the space $L^2(Z^+ G_k \backslash G_{\mathbb{A}} / K_{\mathbb{A}}, \omega)$ mostly consists of *integrals* of *non* L^2 eigenfunctions for the Laplacians and Hecke operators, the *Eisenstein series* introduced later in [3.9].

For spaces of automorphic forms more complicated than being right K_v-invariant for *every* place v, there is generally no decomposition in terms of simultaneous eigenspaces for *commuting* operators. The decomposition argument in [7.7] directly uses the larger *noncommutative* algebras of test functions on the groups G_v:

$$C_c^\infty(G_v) = \begin{cases} \text{compactly supported smooth, } v \text{ archimedean} \\ \text{compactly supported locally constant, } v \text{ non-archimedean} \end{cases}$$

Both cases are called *smooth*. With right translation $R_g f(x) = f(xg)$ for $x, g \in G_{\mathbb{A}}$, the action of $\varphi \in C_c^\infty(G_v)$ on functions f on $G_k \backslash G_{\mathbb{A}}$ is

$$\varphi \cdot f = \int_{G_v} \varphi(g) R_g f \, dg$$

This makes sense not just as a pointwise-value integral but also as a Gelfand-Pettis integral [14.1] when f lies in any quasi-complete, locally convex topological vectorspace V on which G_v acts so that $G_v \times V \to V$ is continuous. Such V is a *representation* of G_v. The multiplication in $C_c^\infty(G_v)$ compatible with such actions is *convolution*: associativity $\varphi \cdot (\psi \cdot f) = (\varphi * \psi) \cdot f$.

Here, we are mostly interested in actions $G_v \times X \to X$ on *Hilbert-spaces* X. Such a representation is (topologically) *irreducible* when X has no closed, G_v-stable subspace. The convolution algebras $C_c^\infty(G_v)$ are *not commutative*, so, unlike the commutative case, few irreducible representations are one-dimensional. In fact, *typical* irreducible representations of $C_c^\infty(G_v)$ turn out to be *infinite-dimensional*. There is no mandate to attempt to *classify* these irreducibles. Indeed, the spectral theory of compact self-adjoint operators still proves [7.7] discrete decomposition with finite multiplicities, for example, formulated as follows.

For every place v, let K'_v be a compact subgroup of G_v, and for all but a finite set S of places require that $K'_v = K_v$, the standard compact subgroup. For simplicity, we still assume $K'_v = K_v$ at archimedean places. Put $K' = \prod_v K'_v$. Let ω be a central character trivial on $Z_{\mathbb{A}} \cap K'$, so that the space $L^2_o(Z^+ G_k \backslash G_{\mathbb{A}}/K', \omega)$ of right K'-invariant cuspforms with central character ω is not $\{0\}$ for trivial reasons. For $v \in S$, we have a *subalgebra* $C_c^\infty(K'_v \backslash G_v / K'_v)$ of the convolution algebra of test functions at v, stabilizing $L^2_o(Z^+ G_k \backslash G_{\mathbb{A}}/K', \omega)$.

[3.7.3] Theorem: $L^2_o(Z^+ G_k \backslash G_{\mathbb{A}}/K', \omega)$ is the completion of the orthogonal direct sum of subspaces, each consisting of simultaneous eigenfunctions for spherical algebras $C_c^o(K_v\backslash G_v/K_v)$ at $v \notin S$, and irreducible $C_c^\infty(K_v'\backslash G_v/K_v')$-representations at $v \in S$. Each occurs with finite multiplicity. *(Proof in [7.1–7.7].)*

The technical features of decomposition with respect to non-commutative rings of operators certainly bear amplification, postponed to [7.7]. In anticipation,

[3.7.4] Theorem: *(Gelfand and Piatetski-Shapiro)* The Hilbert space $L_o^2(Z^+G_k\backslash G_\mathbb{A}/K_\mathbb{A}, \omega)$ is the completion of the orthogonal direct sum of irreducibles V for the simultaneous action of all algebras $C_c^\infty(G_v)$. Each irreducible occurs with finite multiplicity. *(Proof in [7.7].)*

Again, the various sorts of orthogonal complements to spaces of cuspforms are mostly *not* direct sums of irreducibles but are *integrals* of Eisenstein series, as shown subsequently, with a relatively small number of square-integrable *residues* of Eisenstein series. For GL_2 or GL_3 the square-integrable residues of Eisenstein series are relatively boring, but for GL_4 and larger, there are highly nontrivial square-integrable residues, namely, the *Speh forms*, since for $GL_4(\mathbb{R})$ the relevant unitary representations appear in [Speh 1981/1982]. The general pattern for residual spectrum for GL_n was conjectured in [Jacquet 1982/1983] and proven in [Moeglin-Waldspurger 1989].

3.8 Pseudo-Eisenstein Series

We want to express the orthogonal complement of cuspforms in the larger spaces $L^2(Z^+G_k\backslash G_\mathbb{A}/K_\mathbb{A})$ or $L^2(Z^+G_k\backslash G_\mathbb{A}/K_\mathbb{A}, \omega)$ or $L^2(Z^+G_k\backslash G_\mathbb{A}/K', \omega)$ in terms of simultaneous eigenfunctions for spherical Hecke algebras almost everywhere. Therefore, we emphasize the commutative algebras of integral operators attached to left-and-right K_v-invariant test functions on G_v. To exhibit explicit L^2 functions demonstrably spanning the orthogonal complement to cuspforms, recast the Gelfand condition that all constant terms c_Pf vanish as a requirement that *c_Pf vanishes as a distribution* on $Z^+N_\mathbb{A}M_k\backslash G_\mathbb{A}$, and give an equivalent distributional vanishing condition on $Z^+G_k\backslash G_\mathbb{A}$.

For each standard parabolic P, with $N = N^P$, the condition that *c_Pf vanishes as a distribution* is that

$$\int_{Z^+N_\mathbb{A}M_k\backslash G_\mathbb{A}} \varphi \cdot c_Pf = 0 \qquad (\text{for all } \varphi \in C_c^\infty(N_\mathbb{A}M_k\backslash G_\mathbb{A}))$$

where, again, $C_c^\infty(N_\mathbb{A}M_k\backslash G_\mathbb{A})$ is compactly supported functions on that quotient, *smooth* in the archimedean coordinates and locally constant in the non-archimedean coordinates. Smoothness for archimedean places should mean indefinite differentiability *on the right* with respect to the differential operators coming from the Lie algebra, as in [4.1]. Given the compact support, (uniform) smoothness for non-archimedean places should mean that there exists a compact, open subgroup K' of $\prod_{v<\infty} K_v$ under which φ is right invariant.

Beyond perhaps having pointwise values almost everywhere, the nature of $c_P f$ for f merely L^2 is potentially obscure. For example, it is not likely that $c_P f \in L^2(Z^+ N_{\mathbb{A}} M_k \backslash G_{\mathbb{A}})$. Instead, for general reasons [6.1], $C_c^o(Z^+ G_k \backslash G_{\mathbb{A}})$ is dense in $L^2(Z^+ G_k \backslash G_{\mathbb{A}})$ in the L^2 topology, and for general reasons [6.1] the *left* action of $N_k \backslash N_{\mathbb{A}}$ on $C^o(Z^+ P_k \backslash G_{\mathbb{A}})$ is a continuous map $N_k \times C^o(Z^+ P_k \backslash G_{\mathbb{A}}) \to C^o(Z^+ N_{\mathbb{A}} M_k \backslash G_{\mathbb{A}})$, so $c_P f$ exists as a $C^o(Z^+ N_{\mathbb{A}} M_k \backslash G_{\mathbb{A}})$-valued Gelfand-Pettis integral [14.1]. For such f, the integral of $c_P f$ against $\varphi \in C_c^\infty(Z^+ N_{\mathbb{A}} M_k \backslash G_{\mathbb{A}})$ is the integral of a compactly supported, continuous function.

The simplest type of *pseudo-Eisenstein series* is

$$\Psi_\varphi(g) = \Psi_\varphi^P(g) = \sum_{\gamma \in P_k \backslash G_k} \varphi(\gamma \cdot g)$$

for $\varphi \in C_c^\infty(Z^+ N_{\mathbb{A}} M_k \backslash G_{\mathbb{A}})$. Convergence is good:

[3.8.1] Claim: The series for a pseudo-Eisenstein series Ψ_φ with $\varphi \in C_c^\infty(Z^+ N_{\mathbb{A}} M_k \backslash G_{\mathbb{A}})$ is *locally finite*, meaning that for g in a fixed compact in $G_{\mathbb{A}}$, there are only finitely many *nonzero* summands in $\Psi_\varphi(g) = \sum_\gamma \varphi(\gamma g)$. Further, $\Psi_\varphi \in C_c^\infty(Z^+ G_k \backslash G_{\mathbb{A}})$, so these pseudo-Eisenstein series are in $L^2(Z^+ G_k \backslash G_{\mathbb{Q}})$. *(Proof identical to [2.7.1].)* ///

[3.8.2] Claim: For $f \in L_o^2(Z^+ G_k \backslash G_{\mathbb{A}})$, for a standard parabolic P, pseudo-Eisenstein series $\Psi_\varphi = \Psi_\varphi^P$ with $\varphi \in C_c^\infty(Z^+ N_{\mathbb{A}} M_k \backslash G_{\mathbb{A}})$ fit into an *adjunction*

$$\int_{Z^+ N_{\mathbb{A}} M_k \backslash G_{\mathbb{A}}} \varphi \cdot c_P f = \int_{Z^+ G_k \backslash G_{\mathbb{A}}} \Psi_\varphi \cdot f$$

for $f \in L^2(Z^+ G_k \backslash G_{\mathbb{A}})$. In particular, $c_P f = 0$ if and only if $\int_{Z^+ G_k \backslash G_{\mathbb{A}}} \Psi_\varphi \cdot f = 0$ for all $\varphi \in C_c^\infty(Z^+ N_{\mathbb{A}} M_k \backslash G_{\mathbb{A}})$.

Proof: The mechanism of the proof is that of [2.7.2]. For general reasons [6.1] $C_c^o(Z^+ G_k \backslash G_{\mathbb{A}})$ is dense in $L^2(Z^+ G_k \backslash G_{\mathbb{A}})$, and we consider $f \in C_c^o(Z^+ G_k \backslash G_{\mathbb{A}})$. This allows *unwinding* as in [5.2]:

$$\int_{Z^+ N_{\mathbb{A}} M_k \backslash G_{\mathbb{A}}} \varphi \cdot c_P f = \int_{Z^+ N_{\mathbb{A}} M_k \backslash G_{\mathbb{A}}} \varphi(g) \left(\int_{N_k \backslash N_{\mathbb{A}}} f(ng)\, dn \right) dg$$
$$= \int_{Z^+ N_k M_k \backslash G_{\mathbb{A}}} \varphi(g)\, f(g)\, dg$$

Winding up, using the left G_k-invariance of f and $N_k M_k = P_k$,

$$\int_{Z^+P_k\backslash G_{\mathbb{A}}} f(g)\,\varphi(g)\,dg = \int_{Z^+G_k\backslash G_{\mathbb{A}}} \sum_{\gamma\in P_k\backslash G_k} f(\gamma\cdot g)\,\varphi(\gamma\cdot g)\,dg$$

$$= \int_{Z^+G_k\backslash G_{\mathbb{A}}} f(g)\left(\sum_{\gamma\in P_k\backslash G_k}\varphi(\gamma g)\right)dg$$

The inner sum in the last integral is the pseudo-Eisenstein series attached to φ. By Cauchy-Schwarz-Bunyakowsky,

$$\Big|\int_{Z^+P_k\backslash G_{\mathbb{A}}} f\,\varphi\Big| = \Big|\int_{Z^+G_k\backslash G_{\mathbb{A}}} f\,\Psi_\varphi\Big| \le |f|_{L^2}\cdot|\Psi_\varphi|_{L^2}$$

which proves that the functional $f \to \int_{Z^+P_k\backslash G_{\mathbb{A}}} f\,\varphi$ on $C_c^\infty(Z^+G_k\backslash G_{\mathbb{A}})$ is continuous in the L^2 topology, so extends by continuity to a continuous linear functional on $L^2(Z^+G_k\backslash G_{\mathbb{A}})$. Indeed, this inequality asserts continuity of $f \to c_P f$ as a linear map from $L^2(Z^+G_k\backslash G_{\mathbb{A}})$ to distributions on $Z^+N_{\mathbb{A}}M_k\backslash G_{\mathbb{A}}$ with the weak dual topology as in [13.14]. ///

Similarly, with

$$C_c^\infty(Z^+N_{\mathbb{A}}M_k\backslash G_{\mathbb{A}},\omega)$$
$$= \{\varphi\in C_c^\infty(N_{\mathbb{A}}M_k\backslash G_{\mathbb{A}}) : \varphi(zg) = \omega(z)\cdot\varphi(g),\ \text{for all } z\in Z_{\mathbb{A}},\ g\in G\}$$

analogously, keeping track of complex conjugations:

[3.8.3] Claim: Let $N = N^P$. For $f\in L^2(Z^+G_k\backslash G_{\mathbb{A}},\omega)$, with φ in $C_c^\infty(N_{\mathbb{A}}M_k\backslash G_{\mathbb{A}},\omega)$,

$$\int_{Z^+N_{\mathbb{A}}M_k\backslash G_{\mathbb{A}}} \overline{\varphi}\cdot c_P f = \int_{Z^+G_k\backslash G_{\mathbb{A}}} \overline{\Psi_\varphi}\cdot f$$

Thus, $c_P f = 0$ if and only if $\int_{Z^+G_k\backslash G_{\mathbb{A}}} \overline{\Psi_\varphi}\cdot f = 0$ for all $\varphi\in C_c^\infty(N_{\mathbb{A}}M_k\backslash G_{\mathbb{A}},\omega)$. ///

For $P = P^{\min}$ the *minimal* standard parabolic, especially for right $K_{\mathbb{A}}$-invariant functions, as in [2.7.4] for GL_2, minimal-parabolic pseudo-Eisenstein series with test-function data can be broken up into subfamilies parametrized by (tuples of) Hecke characters, as follows. With $P = P^{\min}$, let

$$M_P^1 = \{\begin{pmatrix} m_1 & & \\ & \ddots & \\ & & m_r\end{pmatrix} : m_1,\ldots,m_r\in\mathbb{J},\ |m_1| = \cdots = |m_r| = 1\}$$

The group $M_k\backslash M^1$ is *compact*, because $\mathbb{J}^1/k^\times$ is compact [2.A]. Certainly $C_c^\infty(Z^+N_{\mathbb{A}}M_k\backslash G_{\mathbb{A}})$ is inside $L^2(Z^+N_{\mathbb{A}}M_k\backslash G_{\mathbb{A}})$, so such functions φ admit decompositions in $L^2(Z^+N_{\mathbb{A}}M_k\backslash G_{\mathbb{A}})$ by characters χ of the compact abelian group $M_k\backslash M^1$ acting on the left, as in [6.11]. The integral expressing the χth component

$$\varphi^\chi(g) = \int_{M_k\backslash M^1} \chi(m)^{-1}\, \varphi(mg)\, dm$$

is a Gelfand-Pettis integral converging in $C_c^\infty(Z^+N_{\mathbb{A}}M_k\backslash G_{\mathbb{A}})$ for any quasi-complete [14.7] locally convex [13.11] topology on this space. That is, the Fourier components φ^χ of a compactly supported smooth function along $M_k\backslash M^1$ are again compactly supported smooth, and their sum converges to the original in $L^2(Z^+N_{\mathbb{A}}M_k\backslash G_{\mathbb{A}})$, at least. The support of φ^χ is worst $(M_k\backslash M^1) \times \operatorname{spt}\varphi$.

[3.8.4] Lemma: A function $f \in L^2(Z^+G_k\backslash G_{\mathbb{A}})$ has constant term c_Pf integrating to 0 against φ in $C_c^\infty(Z^+N_{\mathbb{A}}M_k\backslash G_{\mathbb{A}})$ if and only if c_Pf integrates to 0 against every $M_k\backslash M^1$-component φ^χ of φ.

Proof: The potential pitfall is that there is no claim that constant terms of functions in $L^2(Z^+G_k\backslash G_{\mathbb{A}})$ are in $L^2(Z^+N_{\mathbb{A}}M_k\backslash G_{\mathbb{A}})$. Fortunately, this is not an obstacle: as earlier, it suffices to consider $f \in C_c^o(Z^+G_k\backslash G_{\mathbb{A}})$, so $c_Pf \in C^o(Z^+N_{\mathbb{A}}M_k\backslash G_{\mathbb{A}})$. With u the characteristic function of $(M_k\backslash M^1) \times \operatorname{spt}\varphi$, the truncation $u \cdot c_Pf$ is in $L^2(Z^+N_{\mathbb{A}}M_k\backslash G_{\mathbb{A}})$, and truncation does not alter the integrals against φ^χ or φ. Letting $\langle , \rangle$ be the inner product in $L^2(Z^+N_{\mathbb{A}}M_k\backslash G_{\mathbb{A}})$, since $\varphi = \sum_\chi \varphi^\chi$ in $L^2(Z^+N_{\mathbb{A}}M_k\backslash G_{\mathbb{A}})$,

$$\langle c_Pf, \varphi\rangle = \langle u \cdot c_Pf,\, \varphi\rangle = \sum_\chi \langle u \cdot c_Pf,\, \varphi^\chi\rangle = \sum_\chi \langle c_Pf,\, \varphi^\chi\rangle$$

giving the assertion. ///

[3.8.5] Corollary: With $P = P^{\min}$ and $M = M^P$, to know $c_Pf = 0$ it suffices to know orthogonality to Ψ_φ for φ in

$$\{\varphi \in C_c^\infty(Z^+N_{\mathbb{A}}M_k\backslash G_{\mathbb{A}}) : \varphi(mg) = \chi(m)\varphi(g), \text{ for all } m \in M^1\}$$

with χ ranging over characters of the compact group $M_k\backslash M^1$. ///

3.9 Cuspidal-Data Pseudo-Eisenstein Series

The simplest pseudo-Eisenstein series Ψ_φ^P, with φ in $C_c^\infty(Z^+N_{\mathbb{A}}M_k\backslash G_{\mathbb{Q}})$ having compact support on the relevant quotient, behave well, as in [3.8.1]. For *minimal-parabolic P*, such pseudo-Eisenstein series suffice for the corresponding part of spectral theory, as they have good decompositions in terms of corresponding genuine Eisenstein series, much as in [2.11] and [2.12], and as in [3.15]. However, for $r \geq 3$ and for *nonminimal* P, genuine Eisenstein series with best behavior involve *cuspforms* on the Levi component M^P. Anticipating this, we want pseudo-Eisenstein series Ψ_φ^P that facilitate this part [3.16] of the spectral decomposition. This entails minor analytical complications, since the data φ can no longer have *compact support*.

Let $\delta : (0, +\infty) \to \mathbb{J}$ be the usual imbedding of the ray in the archimedean factors of the ideles, so that $|\delta(t)| = t$ for $t > 0$. The centers of the factors $GL_{d_i}(\mathbb{A})$ of the standard Levi component M^P are copies of $\mathbb{J}$. The standard *archimedean split component* A_P^+ of a parabolic $P = P^{d_1,\ldots,d_\ell}$ is the product of the copies of $\delta(0, \infty)$ in the product of the centers of the factors of $M^P(\mathbb{A})$. Another important subgroup of $M_\mathbb{A}^P$ is

$$M_P^1 = \{\begin{pmatrix} m_1 & & \\ & \ddots & \\ & & m_\ell \end{pmatrix} : |\det m_1| = \cdots = |\det m_\ell| = 1\}$$

with $m_i \in GL_{d_i}(\mathbb{A})$. By design, $M_\mathbb{A}^P = A_P^+ \cdot M_P^1$, and $M_k^P \subset M_P^1$. As already in GL_2, the center of M^P is larger than the center of G, so $Z_\mathbb{A} M_k \backslash M_\mathbb{A}$ is not quite a product of quotients of the form $Z^+ G_k \backslash G_\mathbb{A}$ or $Z_\mathbb{A} G_k \backslash G_\mathbb{A}$. This discrepancy necessitates looking at test functions on the archimedean split components A_P^+ or their quotients $Z^+ \backslash \mathbb{A}_P^+$, in addition to automorphic data on $M_k \backslash M_P^1$.

In brief, the data φ on $Z^+ N_\mathbb{A} M_k \backslash G_\mathbb{A}$ appropriate for spectral decompositions of pseudo-Eisenstein series Ψ_φ^P in terms of genuine Eisenstein series with good behavior, must specify *test function* data on the split component A_P^+, and *cuspforms* on the $M_k \backslash M_P^1$. For the minimal parabolic, the cuspidal data is vacuous, since the Levi component is a product of copies of GL_1, and test function data and specification of *character* on the compact abelian group $M_k \backslash M_P^1 \approx (k^\times \backslash \mathbb{J}^1)^r$ suffices for the spectral decomposition [3.15] of *minimal*-parabolic pseudo-Eisenstein series in terms of genuine Eisenstein series with analytic continuations and functional equations. In contrast, for a *nonminimal* parabolic, some factor of the Levi component is $GL_{r'}$ with $r' > 1$, and the cuspform condition is nonvacuous.

Further, we only consider *everywhere spherical* automorphic forms, that is, right $K_\mathbb{A}$-invariant and left $Z_\mathbb{A}$-invariant functions. This has the convenient simplification, via Iwasawa decomposition, that constant terms $c_P f$ are identifiable with functions on the quotient of the Levi component of P:

$$\begin{aligned} Z_\mathbb{A} N_\mathbb{A} M_k \backslash G_\mathbb{A} / K_\mathbb{A} &= Z_\mathbb{A} N_\mathbb{A} M_k \backslash N_\mathbb{A} M_\mathbb{A} K_\mathbb{A} / K_\mathbb{A} \\ &\approx Z_\mathbb{A} M_k \backslash M_\mathbb{A} / (M_\mathbb{A} \cap / K_\mathbb{A}) \longleftarrow Z_\mathbb{A} M_k \backslash M_\mathbb{A} \end{aligned}$$

This allows easier description of the cuspidal data, as follows. Let f_1, f_2 be cuspforms on $GL_{r_1}(\mathbb{A})$ and $GL_{r_2}(\mathbb{A})$, right invariant by the standard maximal compacts everywhere, with central characters ω_1 and ω_2, necessarily unramified. Anticipating the behavior of corresponding genuine Eisenstein series, we require that f_1 and f_2 be eigenfunctions for all the spherical Hecke algebras, including the archimedean places. This includes an eigenfunction condition for invariant Laplacians. That is, f_1 and f_2 are *cuspforms* in the strong sense, beyond satisfaction of the Gelfand condition on vanishing of constant terms. The theory of the constant term [8.3] shows that cuspforms in this strong sense are *of rapid decay*. Then $f = f_1 \otimes f_2$ is a function on $GL_{r_1}(\mathbb{A}) \times GL_{r_2}(\mathbb{A}) \approx M^P_{\mathbb{A}}$. In the extreme cases where $r_1 = 1$ or $r_2 = 1$, the situation degenerates a little: there is no corresponding f_j, that is, the corresponding f_j is simply the identically-1 function. For a test function η on the ray $(0, \infty)$, define

$$\varphi(znmk) = \varphi_{\eta,f}(znmk) = \eta\left(\frac{|\det m_1|^{r_2}}{|\det m_2|^{r_1}}\right) \cdot f_1(m_1) \cdot f_2(m_2)$$

with $m = \begin{pmatrix} m_1 & 0 \\ 0 & m_2 \end{pmatrix} \in M^P_{\mathbb{A}}$, $z \in Z^+$, $n \in N_{\mathbb{A}}$, $k \in K_{\mathbb{A}}$. The possibly counter-intuitive exponents on the idele norms of the determinants make φ invariant under $Z_{\mathbb{A}}$. The corresponding pseudo-Eisenstein series is formed as expected,

$$\Psi^P_\varphi(g) = \sum_{\gamma \in P_k \backslash G_k} \varphi(\gamma \cdot g)$$

However, this sum is not locally finite, so convergence is subtler and needs properties of strong-sense-cuspform data. Convergence will follow from comparison to similarly formed genuine Eisenstein series in their range of absolute convergence as in [3.11.2].

[3.9.1] Remark: The argument of [3.11.3] for orthogonality of genuine Eisenstein series with cuspidal data attached to *nonassociate* parabolics applies to pseudo-Eisenstein series with cuspidal data as well, showing orthogonality of those attached to nonassociate parabolics. For associate parabolics P, Q, as for GL_2 in [2.13.5], spectral decompositions of pseudo-Eisenstein series will make clear [3.17.3] that pseudo-Eisenstein series $\Psi^P_{\eta,f}$ and $\Psi^Q_{\theta,f'}$ with test functions η, θ and cuspidal-data f, f', are orthogonal if $P = Q$ but $\langle f, f' \rangle = 0$, or if $M^P = wM^Q w^{-1}$ but $f^w \neq f'$.

3.10 Minimal-Parabolic Eisenstein Series

In the often-treated example of automorphic forms on GL_2, there are no Eisenstein series made from *cuspidal data*, because GL_2 is so small. In contrast, for

GL_n with $n > 2$, cuspidal-data Eisenstein series play an essential role. However, the *minimal*-parabolic Eisenstein series for GL_r involve no cuspidal data because the Levi component is a product of groups GL_1, where the cuspidal condition is vacuous. Further, especially in the everywhere-spherical case of right $K_{\mathbb{A}}$-invariant minimal-parabolic Eisenstein series, much of the behavior reduces to GL_2 via *Bochner's lemma* [3.B], as we will see. Hartogs' lemma on separate analyticity implying joint analyticity [15.C] removes several ambiguities and potential imprecisions in discussion of functions of one complex variable versus several.

With δ mapping $(0, \infty)$ to the archimedean factors of $\mathbb{J}$ so that $|\delta(t)| = t$, as earlier, describe Hecke characters $\widetilde{\chi}$ as

$$\widetilde{\chi}(\delta(t) \cdot t_1) = t^s \cdot \chi(t_1) \qquad (\text{with } t > 0, t_1 \in \mathbb{J}^1, s \in \mathbb{C})$$

Given an r-tuple of Hecke characters $\widetilde{\chi}_1, \ldots, \widetilde{\chi}_r$ with the relation $s_1 + \cdots + s_r = 0$ among the complex parameters $s = (s_1, \ldots, s_r)$, the right $K_{\mathbb{A}}$-invariant, $Z_{\mathbb{A}}$-invariant minimal-parabolic Eisenstein series $E_{s,\chi} = E^{\min}_{s,\chi}$ on GL_r is formed as usual:

$$E_{s,\chi}(g) = \sum_{\gamma \in P_k \backslash G_k} \varphi^o_{s,\chi}(\gamma \cdot g)$$

where

$$\varphi^o_{s,\chi}(nmk) = \widetilde{\chi}(m_1) \cdot \ldots \cdot \widetilde{\chi}_r(m_r)$$

for $n \in N^{\min}_{\mathbb{A}}$, $m = \begin{pmatrix} m_1 & & \\ & \ddots & \\ & & m_r \end{pmatrix}$, $k \in K_{\mathbb{A}}$. For Hecke characters all of the simplest form $\widetilde{\chi}_j(\delta(t) \cdot t_1) = t^{s_j}$, this is

$$\varphi^o_s(nmk) = \varphi^o_{s,1}(nmk) = |m_1|^{s_1} |m_2|^{s_2} \ldots |m_r|^{s_r}$$

That is, in terms of the parameter s, $E_{s,\chi}$ is a function-valued function of $r - 1$ complex variables, but the parameter space is the complex hyperplane $s_1 + \cdots + s_r = 0$ in $\mathbb{C}^r$, rather than $\mathbb{C}^{r-1}$. In terms of the positive simple roots $\alpha_i(m) = m_i/m_{i+1}$, using $s_1 + \cdots + s_r = 0$,

$$\varphi^o_{s,\chi}(nmk) = |\alpha_1(m)|^{s_1} \cdot |\alpha_2(m)|^{s_1+s_2} \cdot |\alpha_3(m)|^{s_1+s_2+s_3} \cdot \ldots \cdot |\alpha_{r-1}(m)|^{s_1+\cdots+s_{r-1}}$$

[3.10.1] Claim: *(In coordinates)* The minimal-parabolic Eisenstein series $E_{s,\chi}(g)$ on GL_n converges (absolutely and uniformly for g in compacts) for $\frac{\sigma_j - \sigma_{j+1}}{2} > 1$ for $j = 1, \ldots, r-1$, where $s = (s_1, \ldots, s_r) \in \mathbb{C}^r$ and $\sigma = (\mathrm{Re}(s_1), \ldots, \mathrm{Re}(s_r))$. *(Proof provided subsequently.)*

The inequalities describing the region of convergence can be rewritten in a more intrinsic form later relevant to functional equations, as follows. Let $\mathfrak{gl}_r(\mathbb{R})$ be the Lie algebra of $GL_r(\mathbb{R})$, that is, all r-by-r real matrices. Let $\mathfrak{a}$ be the Lie algebra of the diagonal matrices in $GL_r(\mathbb{R})$. The nonzero eigenvalues (roots) of $\mathfrak{a}$ on $\mathfrak{gl}_r(\mathbb{R})$ are functionals $a \to a_i - a_j$ in the dual space $\mathfrak{a}^*$. For $i \neq j$, the corresponding eigenspace (rootspace) is matrices with nonzero entry only at the ij^{th} entry. The standard *positive* roots and rootspaces are those with $i < j$. Write $\beta > 0$ for positive root β, and $\beta < 0$ when $-\beta > 0$. The standard simple positive roots are $a \to a_i - a_{i-1}$ The *half-sum* of positive roots is

$$\rho(a) = \sum_{i<j} (a_i - a_j) \qquad (\text{for } a \in \mathfrak{a})$$

There is a sort of logarithm map $M_{\mathbb{A}} \to \mathfrak{a}$ by

$$\log \left| \begin{pmatrix} m_1 & & \\ & \ddots & \\ & & m_r \end{pmatrix} \right| = \begin{pmatrix} \log |m_1| & & \\ & \ddots & \\ & & \log |m_r| \end{pmatrix}$$

and then for $m \in M_{\mathbb{A}}$ and $\alpha \in \mathfrak{a}^*$, write

$$m^{\alpha} = e^{\alpha(\log |m|)}$$

This enables interpretation of the parameter s as lying in the complexification $\mathfrak{a}^* \otimes_{\mathbb{R}} \mathbb{C}$ of the dual $\mathfrak{a}^*$ of $\mathfrak{a}$. Using $\langle x, y \rangle = \text{tr}(xy)$ on $\mathfrak{a}$, we can identify $\mathfrak{a}$ with $\mathfrak{a}^*$, and transport to $\mathfrak{a}^*$ the pairing $\langle , \rangle$.

[3.10.2] Corollary: *(Intrinsic/conceptual version)* The minimal-parabolic Eisenstein series $E_{s,\chi}(g)$ on GL_r converges (absolutely and uniformly for g in compacts) for $\langle \alpha, \sigma - 2\rho \rangle > 0$ for all positive simple roots α. *(Proof provided subsequently.)*

That is, the Eisenstein series $E_{s,\chi}$ converges absolutely for $\sigma \in \mathfrak{a}$ in the translate by 2ρ of

$$\text{positive Weyl chamber} = \{x \in \mathfrak{a}^* : \langle x, \alpha \rangle > 0, \text{ for all positive roots } \alpha\} \subset \mathfrak{a}^*$$

Proof: (of claim) For convergence, it suffices to treat Hecke characters only of the form $\widetilde{\chi}(y) = |y|^s$. With number field k, let h be the standard *height* function on a k-vectorspace with specified basis. Let $P = P^{\min}$ be the standard minimal parabolic of G. Let $e_1, \ldots, e_r$ be the standard basis of k^r. Any exterior power $\wedge^{\ell}(k^r)$ has (unordered) basis of wedges of the e_j, and an associated height function. Let $v_o = e_j \wedge \ldots \wedge e_r$, and

$$\eta_j(g) = \frac{h(v_o \cdot \wedge^{r-j+1} g)}{h(v_o)} \qquad (\text{for } g \in GL_r(\mathbb{A}))$$

where $\wedge^\ell g$ is the natural action of g on $\wedge^\ell k^r$. The spherical vector $\varphi_s = \varphi^o_{s,1}$, from which the sth minimal-parabolic Eisenstein series $E_s = E_{s,1}$ is made, is expressible as

$$\varphi^o_s = \eta_1^{s_1}\, \eta_2^{s_2-s_1}\, \eta_3^{s_3-s_1-s_2} \ldots \eta_r^{s_r-s_1-s_2-\ldots-s_{r-1}}$$

where $s = (s_1, \ldots, s_r)$. From *reduction theory*, given compact $C \subset Z_{\mathbb{A}}\backslash G_{\mathbb{A}}$, for some implied constants depending only on C,

$$h(v) \ll_C h(v \cdot g) \ll_C h(v) \qquad (\text{for all } 0 \neq v \in k^r \text{ and } g \in C)$$

and similarly for heights on $\wedge^\ell k^r$. Therefore, convergence of the series defining the Eisenstein series $E_s(g_o)$ is equivalent to convergence of

$$\int_C \sum_{\gamma \in Z_{\mathbb{A}} P_k \backslash G_k} \varphi^o_s(\gamma g)\, dg$$

Shrinking C sufficiently so that $\gamma \cdot C \cap C \neq \phi$ implies $\gamma = 1$,

$$\int_C \sum_{\gamma \in P_k \backslash G_k} \varphi^o_s(\gamma g)\, dg = \int_{Z_{\mathbb{A}} P_k \backslash G_k \cdot C} \varphi^o_s(g)\, dg$$

From reduction theory, the infimum μ of $h(v)$ over nonzero primitive v in $\wedge^\ell(\mathbb{A}^r)$ is attained, so is positive. In particular, $\mu \le h(v_o \gamma g)$ for all $g \in C$ and $\gamma \in G_k$. Thus, $G_k \cdot C$ is contained in a set

$$Y = \{g \in G_{\mathbb{A}} : 1 \ll_C \eta_j(g) \text{ for } j = 1, \ldots, r\}$$

Thus, convergence of the Eisenstein series is *implied* by convergence of

$$\int_{Z_{\mathbb{A}} P_k \backslash Y} \left|\varphi^o_s(g)\right| dg$$

The set Y is stable by right multiplication by the maximal compact subgroup $K_v \subset G_v$ at all places v, so by Iwasawa decomposition this integral is

$$\int_{Z_{\mathbb{A}} P_k \backslash (Y \cap P_{\mathbb{A}})} \left|\varphi^o_s(p)\right| dp \qquad (\text{left Haar measure on } Z_{\mathbb{A}}\backslash P_{\mathbb{A}})$$

With ρ the half-sum of positive roots, the left Haar measure on $Z_{\mathbb{A}} P_{\mathbb{A}}$ is $d(nm) = dn\, dm/m^{2\rho}$, where dn is Haar measure on the unipotent radical and dm is Haar measure on $Z_{\mathbb{A}}\backslash M_{\mathbb{A}}$. Since φ^o_s is left $N_{\mathbb{A}}$-invariant and $N_k\backslash N_{\mathbb{A}}$ is compact, convergence of the latter integral is equivalent to convergence of

$$\int_{Z_{\mathbb{A}} M_k \backslash (Y \cap M_{\mathbb{A}})} \left|\varphi^o_s(a)\right| \frac{dm}{m^{2\rho}} = \int_{Z_{\mathbb{A}} M_k \backslash (Y \cap M_{\mathbb{A}})} m^{\sigma - 2\rho}\, dm$$

where $\sigma = (\mathrm{Re}\, s_1, \ldots, \mathrm{Re}\, s_r)$. The quotient $k^\times \backslash \mathbb{J}^1$ of norm-one ideles $\mathbb{J}^1$ is *compact*, by [2.A], and the discrepancy between $Z_\mathbb{A} \backslash G_\mathbb{A}$ and $SL_r(\mathbb{A})$ is absorbed by $M_\mathbb{A} \cap \prod_v K_v$. Thus, convergence of the following integral suffices.

Parametrize a subgroup H of $SL_n(\mathbb{A})$ by $r-1$ maps from $(GL_1(\mathbb{A})$, namely,

$$h_j : t \longrightarrow \begin{pmatrix} 1 & & & & & & \\ & \ddots & & & & & \\ & & 1 & & & & \\ & & & t & & & \\ & & & & t^{-1} & & \\ & & & & & 1 & \\ & & & & & & \ddots \\ & & & & & & & 1 \end{pmatrix}$$

with t and t^{-1} at j^{th} and $(j+1)^{\text{th}}$ positions. From

$$\eta_i(h_j(t)) = \begin{cases} |t|^{-1} & (\text{for } i = j+1) \\ 1 & (\text{otherwise}) \end{cases}$$

we have

$$Y \cap M_\mathbb{A} \cap SL_n(\mathbb{A}) = \{\prod_j h_j(t_j) : t_j \in \mathbb{J} \text{ and } |t_j^{-1}| \gg 1\}$$

Again using compactness of $k^\times \backslash \mathbb{J}^1$, noting that $h_j(t)^{2\rho} = |t|^2$ for all j, convergence of the Eisenstein series is implied by convergence of the archimedean integral

$$\int_0^{\ll 1} t^{\sigma_j - \sigma_{j+1} - 2} \, \frac{dt}{t} \qquad (\text{for } j = 1, \ldots, r-1)$$

These integrals are absolutely convergent for $\sigma_i - \sigma_{i+1} - 2 > 0$ for all i. ///

Proof: (of corollary) The absolute convergence condition is $\langle \sigma - 2\rho, \alpha \rangle > 0$ for all simple roots α. ///

The general shape of the $P = P^{\min}$ constant term of the simplest P Eisenstein series is easily determined:

[3.10.3] Claim: In the region of convergence, for suitable holomorphic functions $s \to c_{w,s}$, with $c_{1,s} = 1$, the constant term is

$$c_P E^P_{\rho+s}(m) = m^{\rho+s} + \sum_{1 \neq w \in W} c_{w,s} \, m^{\rho + w \cdot s} \qquad (\text{with } m \in M^{\min}_\mathbb{A})$$

[3.10.4] Remark: We could explicitly compute the coefficients $c_{w,s}$ as part of the proof of this claim, but essentially the same computation occurs in the proof of the functional equations [3.12.1] and the corollary [3.12.3]. Quoting [3.12.3] and [3.12.6] to have a more complete statement here:

[3.10.5] Corollary: $c_{\tau,s} = \xi\langle s,\alpha\rangle/\xi(1+\langle s,\alpha\rangle)$ for reflections τ, and the cocycle relation $c_{w',w\cdot s}\cdot c_{w,s} = c_{ww',s}$ holds for $w, w' \in W$ and $s \in \mathfrak{a}^* \otimes_R \mathbb{C}$. We have

$$c_{w,s} = \prod_{\beta>0:w\cdot\beta<0} \frac{\xi\langle s,\beta\rangle}{\xi(\langle s,\beta\rangle+1)} \qquad ///$$

Proof: This begins with an archetypical unwinding argument, using the *disjointness* of $G_k = \bigsqcup_w P_k w N_k$ from the Bruhat decomposition [3.1.1].

$$\begin{aligned}
c_P E^P_{\rho+s}(m) &= \int_{N_k\backslash N_{\mathbb{A}}} E^P_{\rho+s}(n\cdot m)\,dn \\
&= \int_{N_k\backslash N_{\mathbb{A}}} \sum_{\gamma\in P_k\backslash G_k} \varphi^o_{\rho+s}(\gamma\cdot nm)\,dn \\
&= \sum_w \int_{N_k\backslash N_{\mathbb{A}}} \sum_{\gamma\in P_k\backslash P_k w N_k} \varphi^o_{\rho+s}(\gamma\cdot nm)\,dn \\
&= \sum_w \int_{N_k\backslash N_{\mathbb{A}}} \sum_{\beta\in(w^{-1}P_k w\cap N_k)\backslash N_k} \varphi^o_{\rho+s}(w\beta\cdot nm)\,dn \\
&= \sum_w \int_{(w^{-1}P_k w\cap N_k)\backslash N_{\mathbb{A}}} \varphi^o_{\rho+s}(w\cdot nm)\,dn \\
&= \sum_w \int_{(w^{-1}N_k w\cap N_k)\backslash N_{\mathbb{A}}} \varphi^o_{\rho+s}(w\cdot nm)\,dn
\end{aligned}$$

Since $\varphi^o_{\rho+s}$ is left $N_{\mathbb{A}}$-invariant, $g \to \varphi^o_{\rho+s}(wg)$ is still left invariant by $w^{-1}N_{\mathbb{A}}w \cap N_{\mathbb{A}}$. Thus, with the volume of $(w^{-1}N_k w\cap N_k)\backslash(w^{-1}N_{\mathbb{A}}w\cap N_{\mathbb{A}})$ normalized to 1, the constant term is

$$\sum_w \int_{(w^{-1}N_{\mathbb{A}} w\cap N_{\mathbb{A}})\backslash N_{\mathbb{A}}} \varphi^o_{\rho+s}(w\cdot nm)\,dn$$

The case $w = 1$ gives $\varphi^o_{\rho+s}(m) = m^{\rho+s}$. More generally, there is a convenient *complementary subgroup* N^w to $w^{-1}N_{\mathbb{A}}w \cap N_{\mathbb{A}}$ inside $N_{\mathbb{A}}$:

[3.10.6] Lemma: Let N^{opp} be *lower*-triangular matrices with 1s on the diagonal, and let $N^w = N \cap w^{-1}N^{\text{opp}}w$. Then

$$N^w \cap (w^{-1}Nw \cap N) = \{1\} \qquad \text{and} \qquad N^w \cdot (w^{-1}Nw\cap N) = N$$

Proof: (of Lemma) First, of course,

$$w^{-1}N^{\text{opp}}w \cap w^{-1}Nw = w^{-1}(N^{\text{opp}} \cap N)w = w^{-1}\{1\}w = \{1\}$$

For a root $\alpha(m) = m_i/m_j$ for $i \neq j$ and $m \in M = M^{\min}$, letting $\mathfrak{n}$ be the Lie algebra of N, and $x \to e^x$ is the usual matrix exponential, the corresponding *root subgroup* is

$$N^{\alpha} = \{n = e^x : x \in \mathfrak{n},\ mxm^{-1} = \alpha(m) \cdot x\}$$

Thus, N is generated by all the N_β for positive roots β, and N^{opp} is generated by the N_β with negative roots β. The action of W permutes roots, so it permutes root subgroups. Every root subgroup is inside either $w^{-1}Nw$ or $w^{-1}N^{\text{opp}}w$, so the intersections of these with N generate N. ///

Then

$$c_P E^P_{\rho+s}(m) = \sum_w \int_{N^w_{\mathbb{A}}} \varphi^o_{\rho+s}(w \cdot nm)\, dn$$

Each root subgroup is stable under conjugation by M, so any product that is a subgroup of N is stable by M. Thus, in the integral, replace n by mnm^{-1}: letting $\delta^w(m)$ be the change of measure $d(mnm^{-1})/dn$,

$$\begin{aligned} c_P E^P_{\rho+s}(m) &= \sum_w \delta^w(m) \int_{N^w_{\mathbb{A}}} \varphi^o_{\rho+s}(wmn)\, dn \\ &= \sum_w \delta^w(m) \int_{N^w_{\mathbb{A}}} \varphi^o_{\rho+s}(wmw^{-1} \cdot wn)\, dn \\ &= \sum_w \delta^w(m)\, (wmw^{-1})^{\rho+s} \int_{N^w_{\mathbb{A}}} \varphi^o_{\rho+s}(wn)\, dn \\ &= \sum_w \delta^w(m)\, m^{w^{-1}\cdot(\rho+s)} \int_{N^w_{\mathbb{A}}} \varphi^o_{\rho+s}(wn)\, dn \end{aligned}$$

As usual, the sign in the exponent of w in the latter expression is necessary for the action of W on $\mathfrak{a}^* \otimes_{\mathbb{R}} \mathbb{C}$ to be associative. Thus,

$$c_{w,s} = \int_{N^w_{\mathbb{A}}} \varphi^o_{\rho+s}(w^{-1}n)\, dn$$

Optimistically, to understand $\delta^{w^{-1}}(m)\, m^{w\cdot(\rho+s)} = \delta^{w^{-1}}(m)\, m^{w\cdot\rho} \cdot m^{w\cdot s}$, apparently

[3.10.7] Lemma: $\delta^{w^{-1}}(m)\, m^{w\cdot\rho} = m^{\rho}$ for $m \in M$.

Proof: (of Lemma) Write $\beta > 0$ or $\beta < 0$ as β is a positive or negative root. The character $m \to \delta^w(m)$ is the modular function of $N \cap w^{-1}N^{\text{opp}}w$, so

$\delta^w(m) = m^\gamma$ where

$$\gamma = \sum_{\beta<0:w^{-1}\beta>0} w^{-1}\cdot\beta = \sum_{\beta>0:w^{-1}\beta<0} w^{-1}\cdot(-\beta)$$
$$= -\sum_{\beta>0:w^{-1}\beta<0} w^{-1}\cdot\beta$$

Meanwhile,

$$w^{-1}\cdot 2\rho = \sum_{\beta>0} w^{-1}\cdot\beta = \sum_{\beta>0:w^{-1}\cdot\beta>0} w^{-1}\cdot\beta + \sum_{\beta>0:w^{-1}\cdot\beta<0} w^{-1}\cdot\beta$$
$$= \sum_{\beta>0:w^{-1}\cdot\beta>0} w^{-1}\cdot\beta - \sum_{\beta<0:w^{-1}\cdot\beta>0} w^{-1}\cdot\beta$$

Thus, $w^{-1}\cdot\rho + \gamma = \rho$. ///

Thus, we obtain an expression for $c_P E_s(m)$ of the asserted form. ///

3.11 Cuspidal-Data Eisenstein Series

To keep things relatively simple, our examples of cuspidal-data Eisenstein series for nonminimal proper parabolics will include only *maximal* proper parabolics. In fact, the general case is a combination of the features of the minimal-parabolic and maximal-proper parabolic.

Let f_1, f_2 be cuspforms on $GL_{r_1}(\mathbb{A})$ and $GL_{r_2}(\mathbb{A})$, right invariant by the standard maximal compacts everywhere, with trivial central characters. We require that f_1 and f_2 be eigenfunctions for all the spherical Hecke algebras, including the archimedean places. This includes an eigenfunction condition for invariant Laplacians. That is, f_1 and f_2 are *cuspforms* in a strong sense, beyond satisfaction of the Gelfand condition on vanishing of constant terms.

The corollary [7.3.19] of the discrete decomposition of cuspforms shows that cuspforms in this strong sense are *of rapid decay*. The cuspidal data $f = f_1 \otimes f_2$ is a function on $GL_{r_1}(\mathbb{A}) \times GL_{r_2}(\mathbb{A}) \approx M^P_{\mathbb{A}}$. In the extreme cases where $r_1 = 1$ or $r_2 = 1$, the situation degenerates: there is no corresponding f_j, that is, the corresponding f_j is simply the identically-1 function. Let

$$\varphi(znmk) = \varphi_{s,f}(znmk) = \left|\frac{(\det m_1)^{r_2}}{(\det m_2)^{r_1}}\right|^s \cdot f_1(m_1)\cdot f_2(m_2)$$

with $m = \begin{pmatrix} m_1 & 0 \\ 0 & m_2 \end{pmatrix} \in M^P_{\mathbb{A}}$, $z \in Z^+$, $n \in N_{\mathbb{A}}$, $k \in K_{\mathbb{A}}$. The exponents on the idele norms of the determinants make φ invariant under $Z_{\mathbb{A}}$. The corresponding genuine Eisenstein series is formed as expected:

$$E_{s,f}(g) = \sum_{\gamma \in P_k \backslash G_k} \varphi_{s,f}(\gamma \cdot g)$$

[3.11.1] Claim: The cuspidal-data Eisenstein series $E_{s,f}(g)$ converges (absolutely and uniformly for g in compacts) for $\mathrm{Re}\,(s) > 1$.

[3.11.2] Corollary: All cuspidal-data pseudo-Eisenstein series converge (absolutely and uniformly on compacts).

Proof: (of corollary) The genuine Eisenstein series with any $\mathrm{Re}\,(s) > 1$ and the same cuspidal data dominates every pseudo-Eisenstein series with that cuspidal data. ///

Proof: (of claim) As we have seen, hypotheses on the cuspform f ensure that it is *bounded*, so it suffices to prove the claim with f replaced by 1. Then the argument becomes a variant of that of [3.10.1] and [3.10.2], with $\varphi_{s,f}$ replaced by

$$\varphi_s(nmk) = \left| \frac{(\det m_1)^{r_2}}{(\det m_2)^{r_1}} \right|^s$$

The sum

$$E_s(g) = \sum_{\gamma \in P_k \backslash G_k} \varphi_s(\gamma \cdot g)$$

dominates that for $E_{s,f}$. This E_s is a *degenerate* Eisenstein series when either $r_1 + r_1 > 2$, so-called because it is missing the cuspidal data and does not play a direct role in spectral theory.

With $e_1, \ldots, e_r$ the standard basis of k^r, let h be the standard *height* function on k-vectorspace $\wedge^{r_2}(k^r)$ with basis consisting of r_2-fold exterior products of the e_i. Put

$$\eta(g) = \frac{h(e_{r_1+1} \wedge e_{r_1+2} \wedge \ldots \wedge e_r \cdot \wedge^{r_2} g)}{h(e_{r_1+1} \wedge e_{r_1+2} \wedge \ldots \wedge e_r)}$$

with g acting on $\wedge^{r_2}(k^r)$ by $\wedge^{r_2} g$. Note that η is right K_v-invariant at all v, left $N_{\mathbb{A}}$-invariant, and $\eta \begin{pmatrix} m_1 & * \\ 0 & m_2 \end{pmatrix} = |\det m_2|$. Thus,

$$|\varphi_s(g)| = \left(\frac{|\det g|^{r_2}}{\eta(g)^r} \right)^{\sigma} \qquad \text{(with } \sigma = \mathrm{Re}\,(s)\text{)}$$

From *reduction theory*, given compact $C \subset Z_{\mathbb{A}} \backslash G_{\mathbb{A}}$, $h(v) \ll_C h(v \cdot g) \ll_C h(v)$ for all $0 \neq v \in k^r$ and $g \in C$. Therefore, convergence of the series defining

$E_s(g_o)$ is equivalent to convergence of

$$\int_C \sum_{\gamma \in P_k \backslash G_k} \varphi_\sigma(\gamma g)\, dg$$

Shrinking C sufficiently so that $\gamma \cdot C \cap C \neq \phi$ implies $\gamma = 1$,

$$\int_C \sum_{\gamma \in P_k \backslash G_k} \varphi_\sigma(\gamma g)\, dg = \int_{Z_{\mathbb{A}} P_k \backslash G_k \cdot C} \varphi_\sigma(g)\, dg$$

Let μ be the infimum of $h(v)$ over nonzero primitive v in $\wedge^{r_2}(\mathbb{A}^r)$. From reduction theory, this infimum is attained, so it is $\mu > 0$, and $\mu \ll h(v_o \gamma g)$ for all $g \in C$ and $\gamma \in G_k$. Thus, $G_k \cdot C$ is contained in a set

$$Y = \{g \in G_{\mathbb{A}} : |\det g|^{r_2}/\eta(g)^r \ll_C 1\}$$

and convergence of the Eisenstein series is implied by convergence of

$$\int_{Z_{\mathbb{A}} P_k \backslash Y} \varphi_\sigma(g)\, dg$$

The set Y is stable by right multiplication by the maximal compact subgroup $K_v \subset G_v$ at all places v, so via the Iwasawa decomposition this integral is

$$\int_{Z_{\mathbb{A}} P_k \backslash (Y \cap P_{\mathbb{A}})} \varphi_\sigma(p)\, dp \qquad \text{(left Haar measure on } P)$$

The left Haar measure on $P_{\mathbb{A}}$ is

$$d(nm) = \frac{dn\, dm}{|\det m_1|^{r_2} \cdot |\det m_2|^{-r_1}} \qquad \left(\text{where } m = \begin{pmatrix} m_1 & 0 \\ 0 & m_2 \end{pmatrix}\right)$$

where dn is Haar measure on the unipotent radical and dm is Haar measure on $M^P_{\mathbb{A}}$. Since φ_σ is left $N^P_{\mathbb{A}}$-invariant and $N^P_k \backslash N^P_{\mathbb{A}}$ is compact, convergence of the latter integral is equivalent to convergence of

$$\int_{Z_{\mathbb{A}} M_k \backslash (Y \cap M_{\mathbb{A}})} \varphi_\sigma(m) \frac{dm}{|\det m_1|^{r_2} \cdot |\det m_2|^{-r_1}}$$
$$= \int_{Z_{\mathbb{A}} M_k \backslash (Y \cap M_{\mathbb{A}})} |\det m_1|^{r_2(\sigma-1)}\, |\det m_2|^{-r_1(\sigma-1)}\, dm$$

We have

$$\begin{aligned} Y \cap M_{\mathbb{A}} &= \{m \in M_{\mathbb{A}} : |\det m|^{r_2}/\eta(m)^r \ll_C 1\} \\ &= \{m \in M_{\mathbb{A}} : |\det m_1|^{r_2}/|\det m_2|^{r_1} \ll_C 1\} \end{aligned}$$

By reduction theory, for example, the quotients $GL_{r_i}(k) \backslash GL_{r_i}(\mathbb{A})^1$ have finite total measure, and $|\det m_1|^{r_2}/|\det m_2|^{r_1}$ is $Z_{\mathbb{A}} M^1$-invariant. Let M^1 be the copy of $GL_{r_1}(\mathbb{A})^1 \times GL_{r_2}(\mathbb{A})^1$ inside $M_\mathbb{A}$. It suffices to prove convergence of

$$\int_{Z_\mathbb{A} M^1\backslash(Y\cap M_\mathbb{A})} |\det m_1|^{r_2\sigma} \cdot |\det m_2|^{-r_1\sigma} \frac{dm}{|\det m_1|^{r_2} \cdot |\det m_2|^{-r_1}}$$

$$= \int_{Z_\mathbb{A} M^1\backslash(Y\cap M_\mathbb{A})} |\det m_1|^{r_2(\sigma-1)} \cdot |\det m_2|^{-r_1(\sigma-1)}\, dm$$

The map

$$\Delta(t) = \begin{pmatrix} t\cdot 1_{r_1} & 0 \\ 0 & 1_{r_2} \end{pmatrix} \qquad (\text{for } t > 0)$$

surjects to $Z_\mathbb{A} M^1\backslash M_\mathbb{A}$, so it suffices to prove convergence of

$$\int_0^{\ll 1} (\det(t\cdot 1_{r_1}))^{r_2(\sigma-1)}\, \frac{dt}{t} = \int_0^{\ll 1} t^{r_1 r_2(\sigma-1)}\, \frac{dt}{t}$$

Convergence is implied by $\sigma > 1$. ///

One benefit of cuspidal data for Eisenstein series is that many constant terms vanish for general reasons. For maximal proper parabolics, the outcome is especially clear. Continue to assume that f_1 and f_2 are everywhere spherical, for simplicity. The *vanishing* conclusion in the following does not need much beyond the Gelfand condition on the cuspforms f_1, f_2, and enough decay for convergence of the Eisenstein series in $\mathrm{Re}(s) > 1$. However, the explicit computation of constant terms in the nonvanishing case will need more.

[3.11.3] Theorem: Let $P = P^{r_1,r_2}$, and $f = f_1 \otimes f_2$ cuspform(s) on $M = M^P$. Let Q be another parabolic. Then $c_Q E^P_{s,f} = 0$ unless $Q = P$ or $Q = P^{r_2,r_1}$, the *associate* of P.

Proof: First, since we claim that it suffices to consider *maximal proper* Q, the underlying reason being that all standard parabolics are intersections of maximal proper standard parabolics, and for standard parabolics $N^{Q\cap Q'} = N^Q \cdot N^{Q'}$. Giving $X = (N_k^Q \cap N_k^{Q'})\backslash(N_\mathbb{A}^Q \cap N_\mathbb{A}^{Q'})$ measure 1, and noting that the constant-term integrals make sense for any left $P_k^{\min}$-invariant functions,

$$c_{Q\cap Q'} f(g) = \int_{N_k^{Q\cap Q'}\backslash N_\mathbb{A}^{Q\cap Q'}} f(ng)\, dn = \int_{N_k^{Q\cap Q'}\backslash N_\mathbb{A}^{Q\cap Q'}} \int_X f(nxg)\, dx\, dn$$

$$= \int_{N_k^Q\backslash N_\mathbb{A}^Q} \int_{N_k^{Q'}\backslash N_\mathbb{A}^{Q'}} \int_X f(n'ng)\, dx\, dn = c_Q(c_{Q'} f)(g)$$

Next, we show that $c_Q E^P_{s,f} = 0$ for maximal proper Q, unless $Q = P$ or its associate. Let $Q = P^{r_1',r_2'}$, and write $\varphi = \varphi_{s,f}$. Take $\mathrm{Re}(s) > 1$ for convergence of the series expression for $E^P_{s,f}$.

$$c_Q E^P_{s,f}(g) = \int_{N^Q_k \backslash N^Q_{\mathbb{A}}} E^P_{s,f}(ng)\,dn = \int_{N^Q_k \backslash N^Q_{\mathbb{A}}} \sum_{\gamma \in P_k \backslash G_k} \varphi(\gamma \cdot ng)\,dn$$

$$= \int_{N^Q_k \backslash N^Q_{\mathbb{A}}} \sum_{\delta \in P_k \backslash G_k / Q_k} \sum_{\gamma \in (\delta^{-1} P_k \delta \cap Q_k) \backslash Q_k} \varphi(\delta\gamma \cdot ng)\,dn$$

$$= \sum_{\delta \in P_k \backslash G_k / Q_k} \int_{N^Q_k \backslash N^Q_{\mathbb{A}}} \sum_{\gamma \in (\delta^{-1} P_k \delta \cap Q_k) \backslash Q_k} \varphi(\delta\gamma \cdot ng)\,dn$$

It certainly suffices to show that the integral vanishes for every δ. The idea is that enough of the unipotent radical $N^Q_{\mathbb{A}}$ conjugates across each $\delta\gamma$ so that the integral vanishes because of the Gelfand property of f. We need to understand $\delta^{-1} P_k \delta \cap Q_k$.

By the Bruhat decomposition [3.1,1], the Weyl group W gives a collection of representatives for $P_k \backslash G_k / Q_k$. Indeed, letting $W^P = P_k \cap W$ and $W^Q = Q_k \cap W$, a set of representatives for $W^P \backslash W / W^Q$ is a set of representatives for $P_k \backslash G_k / Q_k$. In fact, $W^P \backslash W / W^Q$ is in *bijection* with $P_k \backslash G_k / Q_k$ by $W^P w W^Q \longleftrightarrow P_k w Q_k$, although we only proved this for $P = Q = P^{\min}$ in [3.1]. We determine representatives for $W^P \backslash W / W^Q$. Write a permutation matrix $w \in W$ in blocks corresponding to the Levi components of P, Q:

$$w = \begin{pmatrix} a & b \\ c & d \end{pmatrix}$$

with $a = r_1 \times r_1'$, $b = r_1 \times r_2'$, $c = r_2 \times r_1'$, $d = r_2 \times r_2'$. The left action of the upper-left GL_{r_1} part of W^P inside $GL_{r_1} \times GL_{r_2} \approx M^P$, and the action of the upper-left $GL_{r_1'}$ part of W^Q inside $GL_{r_1'} \times GL_{r_2'} \approx M^Q$ adjust the matrix a to the form $a = \begin{pmatrix} 1_{t_1} & 0 \\ 0 & 0 \end{pmatrix}$ for some size t_1, so w becomes

$$w = \begin{pmatrix} 1_{t_1} & 0 & 0 \\ 0 & 0 & * \\ 0 & * & * \end{pmatrix}$$

The lower-right parts of W^P and W^Q further adjust the lower right block of w to be of the form $\begin{pmatrix} 0 & 0 \\ 0 & 1_{t_4} \end{pmatrix}$ for some t_4, putting the permutation matrix w into the form

$$w = \begin{pmatrix} 1_{t_1} & 0 & 0 & 0 \\ 0 & 0 & * & 0 \\ 0 & * & 0 & 0 \\ 0 & 0 & 0 & 1_{t_4} \end{pmatrix}$$

Necessarily the remaining entries can be adjusted to be identity matrices of suitable sizes. That is, $W^P \backslash W / W^Q$ has representatives of the form

$$w = \begin{pmatrix} 1_{t_1} & 0 & 0 & 0 \\ 0 & 0 & 1_{t_3} & 0 \\ 0 & 1_{t_2} & 0 & 0 \\ 0 & 0 & 0 & 1_{t_4} \end{pmatrix}$$

where $t_1 + t_3 = r_1$, $t_1 + t_2 = r_1'$, and so on. With suitable block sizes,

$$w^{-1}Pw \cap Q = \begin{pmatrix} 1 & 0 & 0 & 0 \\ 0 & 0 & 1 & 0 \\ 0 & 1 & 0 & 0 \\ 0 & 0 & 0 & 1 \end{pmatrix} \begin{pmatrix} * & * & * & * \\ * & * & * & * \\ 0 & 0 & * & * \\ 0 & 0 & * & * \end{pmatrix} \begin{pmatrix} 1 & 0 & 0 & 0 \\ 0 & 0 & 1 & 0 \\ 0 & 1 & 0 & 0 \\ 0 & 0 & 0 & 1 \end{pmatrix} \cap Q$$

$$= \begin{pmatrix} * & * & * & * \\ 0 & * & 0 & * \\ * & * & * & * \\ 0 & * & 0 & * \end{pmatrix} \cap Q = \begin{pmatrix} * & * & * & * \\ 0 & * & 0 & * \\ 0 & 0 & * & * \\ 0 & 0 & 0 & * \end{pmatrix}$$

Thus, writing the sum as an iterated sum and unwinding,

$$\int_{N_k^Q \backslash N_{\mathbb{A}}^Q} \sum_{\gamma \in (w^{-1}P_k w \cap Q_k) \backslash Q_k} \varphi(w\gamma \cdot ng)\, dn$$

$$= \int_{N_k^Q \backslash N_{\mathbb{A}}^Q} \sum_{\gamma \in (w^{-1}P_k w \cap M_k^Q) \backslash M_k^Q} \sum_{\nu \in ((w\gamma)^{-1}P_k w\gamma \cap N_k^Q) \backslash N_k^Q} \varphi(w\gamma\nu \cdot ng)\, dn$$

$$= \sum_{\gamma \in (w^{-1}P_k w \cap M_k^Q) \backslash M_k^Q} \int_{(w\gamma)^{-1}P_k w\gamma \cap N_k^Q) \backslash N_{\mathbb{A}}^Q} \varphi(w\gamma \cdot ng)\, dn$$

For fixed γ, replacing n by $\gamma^{-1}n\gamma$ gives

$$\int_{(w^{-1}P_k w \cap N_k^Q) \backslash N_{\mathbb{A}}^Q} \varphi(wn \cdot \gamma g)\, dn$$

$$= \int_{(w^{-1}P_k w \cap N_k^Q) \backslash N_{\mathbb{A}}^Q} \varphi(wnw^{-1} \cdot w\gamma g)\, dn$$

A similar computation to the foregoing shows that

$$wN^Q w^{-1} \cap P = \begin{pmatrix} 1_{t_1} & * & 0 & * \\ 0 & 1_{t_3} & 0 & 0 \\ 0 & 0 & 1_{t_2} & * \\ 0 & 0 & 0 & 1_{t_4} \end{pmatrix}$$

This contains the unipotent radical N' of the parabolic $P' = P^{t_1,t_3} \times P^{t_2,t_4}$ of the Levi component $M^P \approx GL_{r_1} \times GL_{r_2}$ of P. Unless $(r_1', r_2') = (r_1, r_2)$ or $(r_1', r_2') = (r_2, r_1)$, at least one of those parabolic subgroups of GL_{r_j} must be a *proper* parabolic of the corresponding GL_{r_j}. That is, for each fixed γ the integral over $(w^{-1}P_k w \cap N_k^Q)\backslash N_{\mathbb{A}}^Q$ has a subintegral over $N_k'\backslash N_{\mathbb{A}}'$, which computes the P' constant term of the cuspidal data f, giving 0.

This almost gives the vanishing assertion of the theorem. One anomalous case remains, namely, $P \cap Q$ when $P = P^{r_1,r_2}$ and $Q = P^{r_2,r_1}$ with $r_1 \neq r_2$. Still, use the fact that $c_{Q\cap P} = c_Q \circ c_P$. Compute the constant term along P, using the foregoing fact that only $w = 1$ gives a nonzero outcome. Thus, for non-self-associate proper maximal P and cuspidal data,

$$c_P E_{s,f}^P(g) = \int_{(P_k \cap N_k^Q)\backslash N_{\mathbb{A}}^Q} \varphi(ng)\, dn = \int_{N_k^Q\backslash N_{\mathbb{A}}^Q} \varphi_{s,f}(ng)\, dn$$

$$= \int_{N_k^Q\backslash N_{\mathbb{A}}^Q} \varphi_{s,f}(g)\, dn = \varphi_{s,f}(g) \cdot \int_{N_k^Q\backslash N_{\mathbb{A}}^Q} 1\, dn$$

Because $r_1 \neq r_2$, $N^Q \cap N^P$ contains a unipotent radical of some proper parabolic in M^P, so the cuspidality of f means $c_Q \varphi_{s,f} = 0$. Thus, $c_{Q\cap P} E_{s,f}^P = 0$. ///

[3.11.4] Remark: More generally, Eisenstein series with cuspidal data for parabolics $P = P^{r_1,\ldots,r_\ell}$ have constant term 0 along parabolics Q unless Q *contains* some associate of P, that is, contains some $P^{r_1',\ldots,r_\ell'}$ with the r_i''s a permutation of the r_i's.

The same arguments and vanishing conclusions apply to constant terms of *pseudo*-Eisenstein series with cuspidal data:

[3.11.5] Corollary: For maximal proper P and cuspidal data f on M^P, for another parabolic Q, $c_Q \Psi_{\eta,f}^P = 0$ unless Q is associate to P. ///

An optimist would have to hope that cuspidal-data Eisenstein series $E_{s,f}^P$ formed from spherical Hecke eigenfunction cuspforms $f = f_1 \otimes f_2$ would itself be a spherical Hecke eigenfunction and that this is so because $\varphi_{s,f}$ is a spherical Hecke eigenfunction for all s. Happily, this is nearly true, with a yet-stronger notion of *cuspform*, as follows. Fix a non-archimedean v and square-integrable right K_v-invariant cuspform f, and consider the space

$$\pi_v = \{\text{finite linear combinations of right translates } g \to f(gh) \text{ with } h \in G_v\}$$

generated by f under the action of G_v by right translation, suitably topologized. The most direct way to *begin* description of a suitable requirement[6] on f at v is that π_v be isomorphic as G_v representation to a G_v-*sub*representation of an *unramified principal series* attached to the standard minimal-parabolic B:

$$I_\chi^B = \{F \in C^\infty(G_v) : F(bg) = \chi(b) \cdot F(g) \text{ for all } b \in B_v,\ g \in G_v\}$$

for unramified χ on M_v^B. Since f is K_v-fixed, its image in I_χ^B contains the subspace of K_v-fixed vectors, which is one-dimensional by the Iwasawa decomposition $G_v = B_v \cdot K_v$. This is the right feature to prove

[3.11.6] Theorem: Fix a non-archimedean place v. With f_1 and f_2 as just described, that is, under right translation by $GL_{r_j}(k_v)$ generating representations isomorphic to subrepresentations of unramified principal series representations of $GL_{r_j}(k_v)$, with trivial central characters, the function $\varphi_{s,f}$ is a spherical Hecke-algebra eigenfunction for G_v. In the region of convergence, $E_{s,f}^P$ is a spherical Hecke-algebra eigenfunction with the same eigenvalues as $\varphi_{s,f}$. *(Proof in [8.5].)*

[3.11.7] Remark: Quantitative details about the spherical Hecke eigenvalues of $\varphi_{s,f}$ and $E_{s,f}^P$ for G_v in terms of the spherical Hecke eigenvalues of $f = f_1 \otimes f_2$ for M_v and $s \in \mathbb{C}$ are visible in the proof [8.5].

[3.11.8] Remark: In the cases of $c_Q E_{s,f}^P \neq 0$, the foregoing proof shows that nonvanishing occurs only in a few cases: $w = 1_r$ for $Q = P$ always gives the summand $\varphi_{s,f}^P$ of the constant term, and $w = \begin{pmatrix} 0 & 1_{r_1} \\ 1_{r_2} & 0 \end{pmatrix}$ for $Q = P^{r_2,r_1}$ for both $r_1 = r_2$ and $r_1 \neq r_2$. Happily, in both these cases, $w^{-1}P_k w \cap M_k^Q) = M_k^Q$,

[6] Our description of what is needed to have cuspidal-data Eisenstein series be Hecke eigenfunctions would usually be the *conclusion* of a highly nontrivial chain of reasoning. That is, we have directly described what is needed to *set up* the proof that Eisenstein series formed from spherical Hecke-algebra eigenfunctions on Levi components are Hecke eigenfunctions. A more usual characterization, inherited from the chaotic historical order of developments, would be to require that the local representation generated by the cuspform be *admissible* and *irreducible*. Admissibility is equivalent to K_v-finiteness of every vector in π_v, and irreducibility has the usual meaning of having no closed M_v^P-stable subspaces, with the representation space suitably topologized. The Borel-Casselman-Matsumoto theorem [Borel 1976], [Casselman 1980], [Matsumoto 1977] asserts that π_v *is* a subrepresentation of an unramified principal series. Given that, the key point is that *induction in stages* is legitimate, as in [6.9]. *Unitariness* of the representation, which follows from square-integrability of the cuspform, implies admissibility, from [Harish-Chandra 1970], for example. The global result, stated in [3.7] and proven in Chapter 7, on discrete decomposition of cuspforms, in fact proves there is an orthogonal basis for everywhere-locally spherical square-integrable cuspforms generating irreducible representations of the global spherical Hecke algebra $C_c^\infty(K_\mathbb{A}\backslash G_\mathbb{A}/K_\mathbb{A})$.

so the sum over $\gamma \in (w^{-1}P_k w \cap M_k^Q)\backslash M_k^Q$ is *trivial*. Thus, for $Q = P$ and $w = 1_r$, that part of the constant term is easily made explicit, as in the preceding proof:

$$c_P E_{s,f}^P(g) = \varphi_{s,f}(g) \cdot \int_{N_k^Q \backslash N_{\mathbb{A}}^Q} 1\, dn$$

The other part of the constant terms is significantly more complicated, as follows. With or without $r_1 = r_2$, when $Q = P^{r_2,r_1}$ and $w = \begin{pmatrix} 0 & 1_{r_1} \\ 1_{r_2} & 0 \end{pmatrix}$, that part of the constant term is *unwound completely*, since $w^{-1}P_k w \cap N_k^Q = \{1\}$, so

$$\int_{N_k^Q \backslash N_{\mathbb{A}}^Q} E_{s,f}^P(ng)\, dn = \int_{N_{\mathbb{A}}^Q} \varphi_{s,f}(ng)\, dn$$

Since we have supposed that f is right $K_{\mathbb{A}}$-invariant, the integral produces a left $Z^+ N_{\mathbb{A}} M_k^Q$-invariant, right $K_{\mathbb{A}}$-invariant function, so by Iwasawa is a function on $M_k^Q \backslash M_{\mathbb{A}}^Q$ and right $M_{\mathbb{A}} \cap K_{\mathbb{A}}$-invariant. The behavior under the center of $M_{\mathbb{A}}^Q$ is also easy to assess by changing variables in the integer. Thus, it is reasonable to imagine that it is of the form $\varphi_{1-s,f'}^Q$ for some cuspform(s) on M^Q. However, we would *not* want f' to depend on s, so the dependence on s should be somehow separate, and this integral should be expressible as $c_{s,f} \cdot \varphi_{1-s,f'}^Q(m)$ with cuspform(s) on M^Q independent of s, and $m \in M_{\mathbb{A}}^Q$.

This conclusion does hold, but only with substantial assumptions on $f \approx f_1 \otimes f_2$, as follows. Continue to assume that f is a spherical Hecke algebra eigenfunction on M^P for all non-archimedean M_v^P. The best further simplifying hypothesis[7] is a form of *strong multiplicity one*, that the only other cuspforms on $M^P \approx GL_{r_1} \times GL_{r_2}$ with the same spherical Hecke eigenvalues at all finite primes are scalar multiples of $f = f_1 \otimes f_2$. Let $f^w = (f_1 \otimes f_2)^w = f_2 \otimes f_1$.

[3.11.9] Theorem: In the nonvanishing cases, with maximal proper P, and $Q = P$ or its associate, with the *strong multiplicity one* assumption given earlier,

$$\begin{cases} c_P E_{s,f}^P &= \varphi_{s,f}^P \qquad\qquad\qquad\quad (\text{for } r_1 \neq r_2) \\ c_P E_{s,f}^P &= \varphi_{s,f}^P + c_{s,f}^P \varphi_{1-s,f^w}^P \quad (\text{for } r_1 = r_2) \\ c_Q E_{s,f}^P &= c_{s,f}^Q \cdot \varphi_{1-s,f^w}^Q \qquad\quad (\text{for } r_1 \neq r_2, Q = P^{r_2,r_1}) \end{cases}$$

with meromorphic $c_{s,f}^P$ and $c_{s,f}^Q$.

[7] The *strong multiplicity one* assumption convenient for GL_r is a theorem of [Shalika 1981]; see also [Piatetski-Shapiro 1979]. It apparently does not hold for most other groups. That is, in general, only more complicated conclusions can be reached about unwound integrals appearing in constant term computations.

Proof: From the proof of [3.11.3], due to the cuspidal data f, most summands of $c_Q E^P_{s,f}$ corresponding to double cosets $P_k w Q_k \in P_k\backslash G_k/Q_k$ are 0. In all cases, for $P = Q$, the small Bruhat cell $P = P \cdot 1 \cdot Q$ gives contribution

$$\int_{N^Q_k\backslash N^Q_{\mathbb{A}}} \sum_{\gamma \in (w^{-1}P_k w \cap Q_k)\backslash Q_k} \varphi(w\gamma \cdot ng)\, dn = \int_{N^P_k\backslash N^P_{\mathbb{A}}} \varphi(ng)\, dn$$

$$= \varphi(g) \int_{N^P_k\backslash N^P_{\mathbb{A}}} 1\, dn \qquad (\text{with } w = 1 \text{ and } P = Q)$$

since φ is left $N_{\mathbb{A}}$-invariant.

Now consider $P = P^{r_1,r_2}$ and $Q = P^{r_2,r_1}$. From the end of the proof of [3.11.3], a double coset PwQ can give a nonzero contribution to the constant term only if $wN^Q w^{-1} \cap P$ contains *no* unipotent radical of a proper parabolic of the Levi component M^P of P. As in the proof of [3.11.3], $P_k\backslash G_k/Q_k \approx W^P\backslash W/W^Q$ has representatives

$$w = \begin{pmatrix} 1_{t_1} & 0 & 0 & 0 \\ 0 & 0 & 1_{t_3} & 0 \\ 0 & 1_{t_2} & 0 & 0 \\ 0 & 0 & 0 & 1_{t_4} \end{pmatrix}$$

where $t_1 + t_3 = r_1$, $t_1 + t_2 = r_1$, and so on. The only case other than $w = 1$ meeting the condition is with $t_1 = 0 = t_4$ and $t_3 = r_1$, $t_2 = r_2$. This has the effect that $w^{-1}Pw \cap Q = M^Q$. This summand in the constant term *unwinds completely:*

$$\int_{N^Q_k\backslash N^Q_{\mathbb{A}}} \sum_{\gamma \in w^{-1}P_k w \cap Q_k\backslash Q_k} \varphi(w\gamma \cdot ng)\, dn = \int_{N^Q_k\backslash N^Q_{\mathbb{A}}} \sum_{\gamma \in M^Q_k\backslash Q_k} \varphi(w\gamma \cdot ng)\, dn$$

$$= \int_{N^Q_k\backslash N^Q_{\mathbb{A}}} \sum_{\gamma \in N^Q_k} \varphi(w\gamma \cdot ng)\, dn = \int_{N^Q_{\mathbb{A}}} \varphi(w \cdot ng)\, dn$$

Let $\varphi'(g)$ be the latter integral. With φ right K_v-invariant at all places, to understand $\varphi'(g)$ it suffices to take

$$g = m = \begin{pmatrix} m_2 & 0 \\ 0 & m_1 \end{pmatrix} = w^{-1} \begin{pmatrix} m_1 & 0 \\ 0 & m_2 \end{pmatrix} w$$

by the Iwasawa decomposition.

Certainly φ' is left $N^Q_{\mathbb{A}}$-invariant and invariant under the center. It is left M^Q_k-invariant, since for $\gamma \in M^Q_k$

$$\varphi'(\gamma m) = \int_{N^Q_{\mathbb{A}}} \varphi(w \cdot n\gamma g)\, dn = \int_{N^Q_{\mathbb{A}}} \varphi(w \cdot \gamma ng)\, dn$$

$$= \int_{N^Q_{\mathbb{A}}} \varphi(w\gamma w^{-1} \cdot wng)\, dn = \int_{N^Q_{\mathbb{A}}} \varphi(wng)\, dn = \varphi'(m)$$

by changing variables in the integral, and observing that the change-of-measure is 1, by the product formula. Since the right translation action commutes with the integration along $N^Q_{\mathbb{A}}$ on the left, $m_1 \times m_2 \to \varphi'(m)$ is a spherical Hecke eigenfunction on $GL(r_1) \times GL(r_2)$ with the same eigenvalues as φ.

To see the behavior of the s parameter, it suffices to consider left translation by

$$h = \begin{pmatrix} t \cdot 1_{r_2} & 0 \\ 0 & 1_{r_1} \end{pmatrix} = w^{-1} \begin{pmatrix} 1_{r_1} & 0 \\ 0 & t \cdot 1_{r_2} \end{pmatrix} w$$

with $t > 0$ imbedded diagonally at archimedean places. Then

$$\varphi'(hm) = \int_{N^Q_{\mathbb{A}}} \varphi(w \cdot n \cdot hm)\, dn = |\det(t \cdot 1_{r_2})|^{r_1} \int_{N^Q_{\mathbb{A}}} \varphi(w \cdot h \cdot nm)\, dn$$

$$= |\det(t \cdot 1_{r_2})|^{r_1} \int_{N^Q_{\mathbb{A}}} \varphi(whw^{-1} \cdot wnm)\, dn$$

by replacing n by hnh^{-1}, picking up the indicated change-of-measure. The left equivariance of φ under elements of the form whw^{-1} is

$$\varphi(whw^{-1} \cdot wnm) = \left|1/\det(t \cdot 1_{r_2})^{r_1}\right|^s \cdot \varphi(wnm)$$

Thus,

$$\varphi'(hm) = \left|\det(t \cdot 1_{r_2})/1\right|^{1-s} \cdot \varphi'(m)$$

as claimed in the assertion of the theorem.

Finally, the multiplicity-one assumption says that $m_1 \times m_2 \to \varphi'(m)$ must be a scalar multiple $c^P_{s,f}$ of $\varphi(m)$. Since $s \to E^P_{s,f}$ is a meromorphic smooth-function-valued function of s, composition with c^Q gives a meromorphic smooth-function-valued function of s. Since it differs by the scalar $c^Q_{s,f}$ from $\varphi_{s,f}$, this scalar must be meromorphic in s. ///

[3.11.10] Remark: The meromorphic functions $c_{s,f}$ have *Euler product expansions* attached to f_1 and f_2, but we do not have immediate need of this fact.[8]

In parallel with the spherical Hecke algebra behavior of $E^P_{s,f}$ at finite places, keeping the assumption that f_1, f_2 have trivial central character and are right $K_{\mathbb{A}}$-invariant,

[3.11.11] Theorem: For v archimedean, for $f = f_1 \otimes f_2$ with f_1, f_2 eigenfunctions for the invariant Laplacians on the factors $GL_{r_1}(k_v)$ and $GL_{r_2}(k_v)$ of M^P_v, the function $\varphi^P_{s,f}$ is an eigenfunction for the invariant Laplacian on G_v, and, thus, $E^P_{s,f}$ is also an eigenfunction. In particular, letting λ_j be the eigenvalue of f_j,

$$\Omega \cdot E^P_{s,f} = (r_1 r_2 (r_1 + r_2)(s^2 - s) + \lambda_1 + \lambda_2) \cdot E^P_{s,f}$$

In particular, the eigenvalue is invariant under $s \longrightarrow 1 - s$.

Proof: For simplicity, treat $G_v \approx GL_r(\mathbb{R})$. Accommodations for the complex case are illustrated in [4.6]. This is a purely local issue, and it suffices to consider arbitrary functions $f_1 \otimes f_2$ on $GL_{r_1}(\mathbb{R}) \times GL_{r_2}(\mathbb{R})$ with trivial central character. That is, the possibility that f_1, f_2 are automorphic forms of any sort is irrelevant. Similarly, we have a purely locally defined function

$$\varphi_s(\begin{pmatrix} a & * \\ 0 & d \end{pmatrix} k) = \left| \frac{(\det a)^{r_2}}{(\det d)^{r_1}} \right|^s \cdot f_1(a) \cdot f_2(d)$$

for $a \in GL_{r_1}(\mathbb{R})$, $d \in GL_{r_2}(\mathbb{R})$, $k \in O(r, \mathbb{R})$. As in [4.2] and [4.4], the invariant Laplacian on G_v/K_v is Casimir Ω on G_v descended to that quotient, and then to any further quotient. For any choice of basis $\{x_i\}$ of the Lie algebra $\mathfrak{g}$ of G_v, and and dual basis $\{x_i^*\}$ with respect to the pairing $\langle x, y \rangle = \mathrm{tr}(xy)$, Casimir is expressible as an element in the (center of the) universal enveloping algebra [4.2], [4.3] as $\Omega = \sum_i x_i x_i^*$. It is easy to exhibit a basis so that the summands separate into three pieces: Casimir Ω_1 of $GL_{r_1}(\mathbb{R})$ acting only on f_1, Casimir Ω_2 of $GL_{r_2}(\mathbb{R})$ acting only on f_2, and a leftover acting only on $|(\det a)^{r_2}/(\det d)^{r_1}|^s$.

Let h_i be the diagonal matrix with 1 at the ith place 0's otherwise. For $i < j$, let x_{ij} be the matrix with a unique nonzero entry, a 1, at the ij^{th} location, and for $i > j$ let y_{ij} be the matrix with a unique nonzero entry, a 1, at the ij^{th} location. The $\{x_{ij}\}$ and $\{y_{ji}\}$ are dual under $\langle , \rangle$. Thus, the Casimir operators for $GL_{r_1}(\mathbb{R})$

[8] [Langlands 1971] considers consequences of the appearance of Euler products in constant terms of Eisenstein series series. A part of that program is completed in [Shahidi 1978, 1985].

and $GL_{r_2}(\mathbb{R})$ are

$$\Omega_1 = \sum_{i=1}^{r_1} h_i^2 + \sum_{i<j\le r_1} (x_{ij}y_{ji} + y_{ji}x_{ij})$$

$$\Omega_2 = \sum_{i=r_1+1}^{r} h_i^2 + \sum_{r_1<j\le r} (x_{ij}y_{ji} + y_{ji}x_{ij})$$

and $\Omega = \Omega_1 + \Omega_2 + \Omega'$ with leftover

$$\Omega' = \sum_{1\le i\le r_1,\ r_1<j\le r} (x_{ij}y_{ji} + y_{ji}x_{ij})$$

Use the fact that Casimir commutes with conjugation by G_v, so we can let the associated differential operators [4.1] act on the *left* on left N_v^P-invariant functions such as φ_s, so that the Lie algebra $\mathfrak{n}$ of N^P annihilates such functions. There is a sign or order-of-operations issue: for a smooth function φ on G_v, the effect of Casimir acting on the *right* is

$$\Omega\varphi(g) = \sum_i \frac{\partial}{\partial t_1}\frac{\partial}{\partial t_2}\varphi(g \cdot e^{t_1 x_i} \cdot e^{t_2 x_i^*})$$

Invoking the invariance under conjugation by G_v, this is

$$\Omega\varphi(g) = \sum_i \frac{\partial}{\partial t_1}\frac{\partial}{\partial t_2}\varphi(e^{t_1 x_i} \cdot e^{t_2 x_i^*} g)$$

Thus, terms $x_{ij}y_{ji}$ with $1 \le i \le r_1$ and $r_1 < j \le r$ annihilate φ_s, because, after conjugating, x_{ij} acts first and is in $\mathfrak{n}$. For $1 \le i \le r_1$ and $r_1 < j \le r$, we can move toward invocation of this annihilation property by noting that $x_{ij}y_{ji} - y_{ij}x^{ij} = h_i - h_j$, so

$$\begin{aligned} x_{ij}y_{ji} + y_{ji}x_{ij} &= 2x_{ij}y_{ji} + y_{ji}x_{ij} - x_{ij}y_{ji} \\ &= 2x_{ij}y_{ji} - [x_{ij}, y_{ji}] = 2x_{ij}y_{ji} - h_i + h_j \end{aligned}$$

which acts just by $-h_i + h_j$ on left N_v^P-invariant functions:

$$\begin{aligned} \Omega' \cdot \varphi_s &= \sum_{1\le i\le r_1,\ r_1<j\le r} (x_{ij}y_{ji} + y_{ji}x_{ij}) \cdot \varphi_s = \sum_{1\le i\le r_1,\ r_1<j\le r} (-h_i + h_j) \cdot \varphi_s \\ &= -r_2 \sum_{1\le i\le r_1} h_i \cdot \varphi_s + r_1 \sum_{r_1<j\le r} h_j \cdot \varphi_s \end{aligned}$$

Thus, with $z_1 = \sum_{1\le i\le r_1} h_i$ and $z_2 = \sum_{r_1<j\le r} h_j$,

$$\Omega \cdot \varphi_s = (\Omega_1 - r_2 z_1) \cdot |\det m_1|^{r_2 s} f_1(m_1) + (\Omega_2 + r_1 z_2) \cdot |\det m_2|^{r_1 s} f_2(m_2)$$

Since z_1 is in the Lie algebra of the center of the GL_{r_1} factor of M^P, and f_1 has trivial central character, $z_1 \cdot f_1 = 0$, and

$$\begin{aligned}(\Omega_1 - r_2 z_1) \cdot |\det m_1|^{r_2 s} f_1(m_1) \\ = \Omega_1 \cdot (|\det m_1|^{r_2 s} f_1(m_1)) - (r_2 z_1 \cdot |\det m_1|^{r_2 s}) f_1(m_1)\end{aligned}$$

and similarly for $\Omega_2 + r_1 z_2$. The effect of z_1 on that power of determinant is straightforward:

$$\begin{aligned} z_1 \cdot |\det m_1|^{r_2 s} &= \frac{\partial}{\partial t}\Big|_{t=0} |\det(e^{tz_1} \cdot m_1)|^{r_2 s} \\ &= \frac{\partial}{\partial t}\Big|_{t=0} (e^t)^{r_1 r_2 s} \cdot |\det m_1|^{r_2 s} = r_1 r_2 s \cdot |\det m_1|^{r_2 s}\end{aligned}$$

Similarly, $z_2 \cdot |\det m_2|^{-r_1 s} = -r_1 r_2 s \cdot |\det m_2|^{-r_1 s}$. Thus, the $-r_2 z_1 + r_1 z_2$ terms combine to $-r_1 r_2 (r_1 + r_2) s \cdot \varphi_s$.

The effect of Ω_1 on $f_1(m_1)$ adjusted by a power of determinant is only slightly more complicated, using Leibniz' rule. The terms $x_{ij} y_{ji}$ and $y_{ji} x_{ij}$ annihilate the determinant, and $h_i \cdot |\det m_1|^{r_2 s} = r_2 s \cdot |\det m_1|^{r_2 s}$, so

$$\begin{aligned} &\Omega_1 \, (|\det m_1|^{r_2 s} \cdot f_1(m_1)) \\ &= \sum_{1 \le i \le r_1} h_1^2 \cdot (|\det m_1|^{r_2 s} \cdot f_1(m_1)) \\ &\quad + |\det m_1|^{r_2 s} \sum_{1 \le i < j \le r_1} (x_{ij} y_{ji} + y_{ji} x_{ij}) \cdot f_1(m_1) \\ &= \sum_{1 \le i \le r_1} \big(h_i^2 |\det m_1|^{r_2 s} \cdot f_1(m_1) + 2 h_i |\det m_1|^{r_2 s} \cdot h_i f_1(m_1) \big) \\ &\quad + |\det m_1|^{r_2 s} \Omega_1 f_1(m_1) \\ &= r_1 (r_2 s)^2 |\det m_1|^{r_2 s} \cdot f_1(m_1) + 2 r_2 s |\det m_1|^{r_2 s} \cdot \Big(\sum_{1 \le i \le r_1} h_i \Big) \cdot f_1(m_1) \\ &\quad + |\det m_1|^{r_2 s} \Omega_1 f_1(m_1) \\ &= r_1 (r_2 s)^2 |\det m_1|^{r_2 s} \cdot f_1(m_1) + 0 + |\det m_1|^{r_2 s} \Omega_1 f_1(m_1) \\ &= |\det m_1|^{r_2 s} \big(r_1 (r_2 s)^2 + \Omega_1 \big) \cdot f_1(m_1) \end{aligned}$$

since $\sum_i h_i$ annihilates f_1 due to the latter's trivial central character. A similar computation applies to Ω_2 and f_2. Letting $\Omega_1 f_1 = \lambda_1 \cdot f$ and $\Omega_2 f_2 = \lambda_2 \cdot f_2$, these computations give

$$\Omega \cdot \varphi_s = \big(r_1 r_2 (r_1 + r_2)(s^2 - s) + \lambda_1 + \lambda_2 \big) \cdot \varphi_s$$

That is, φ_s is an eigenfunction for Casimir on G_v, and by the invariance so is $E^P_{s,f}$. ///

[3.11.12] Remark: The argument can be recast as an application of *induction in stages*, in the archimedean case, analogous to the corresponding non-archimedean argument [6.9].[9]

3.12 Continuation of Minimal-Parabolic Eisenstein Series

We show that the meromorphic continuations of some simple types of *minimal-parabolic* Eisenstein series on GL_r follow from the GL_2 case [2.B] via Bochner's Lemma [3.B]. That determines the $r!$ functional equations corresponding to elements of the Weyl group W, the latter identified with permutation matrices in GL_r. We can also use this to compute the minimal-parabolic constant terms. To illustrate the points with minimal clutter, we consider just the simplest Eisenstein series

$$E_s(g) = \sum_{\gamma \in P_k \backslash G_k} \varphi_s^o(\gamma \cdot g)$$

where $\varphi_s^o(nmk) = |m_1|^{s_1}|m_2|^{s_2} \ldots |m_r|^{s_r}$ with $s_1 + \cdots + s_r = 0$ with $P = P^{\min}$, $n \in N_{\mathbb{A}} = N^{\min}$, $m \in M_{\mathbb{A}}^{\min}$, and $k \in K_{\mathbb{A}}$. Let $\xi(s)$ be the completed zeta function of the underlying number field. For $s \in \mathfrak{a}^* \otimes_{\mathbb{R}} \mathbb{C}$, write $w \cdot s$ for the action of $w \in W$, that is, $(wmw^{-1})^s = m^{w \cdot s}$. In the following, because the fixed point in $\mathfrak{a}^* \otimes_{\mathbb{R}} \mathbb{C}$ of all the functional equations turns out to be the half-sum

$$\rho = (\rho_1, \ldots, \rho_r) = \Big(\frac{r-1}{2}, \frac{r-3}{2}, \frac{r-5}{2}, \ldots, \frac{3-r}{2}, \frac{1-r}{2}\Big)$$

of positive roots, we will express the functional equation in terms of

$$E_{\rho+s}(g) = \sum_{\gamma \in P_k \backslash G_k} \varphi_{\rho+s}^o(\gamma \cdot g)$$

where $\varphi_{\rho+s}^o(nmk) = |m_1|^{\rho_1+s_1}|m_2|^{\rho_2+s_2} \ldots |m_r|^{\rho_r+s_r}$. We prove the following theorem and the corollary together.

[3.12.1] Theorem: (*Selberg, Langlands, et alia*) Minimal-parabolic Eisenstein series E_s have *meromorphic continuations* in $s \in \mathfrak{a}^* \otimes_{\mathbb{R}} \mathbb{C}$, with functional equations

$$E_{\rho+w\cdot s} = c_{w,s}^{-1} \cdot E_{\rho+s}$$

[3.12.2] Corollary: The meromorphic continuation of $E_{\rho+s}$ is holomorphic for s off the zero-sets of $\xi(\langle s, \beta\rangle + 1)$, $1 \pm \langle s, \beta\rangle$, and $\langle s, \beta\rangle$, for positive roots β. ///

[9] The archimedean analogue of the Borel-Casselman-Matsumoto result [Borel 1976], [Casselman 1980], [Matsumoto 1977] is the sharper *subrepresentation theorem* [Casselman-Miličić 1982], considerably improving the *subquotient theorem* of [Harish-Chandra 1954].

[3.12.3] Corollary: $c_{\tau,s} = \xi\langle s,\alpha\rangle/\xi(1+\langle s,\alpha\rangle)$ for reflections τ, and the cocycle relation $c_{w',w\cdot s}\cdot c_{w,s} = c_{ww',s}$ holds for $w, w' \in W$ and $s \in \mathfrak{a}^* \otimes_R \mathbb{C}$. ///

Proof: In brief, the idea is to view the minimal-parabolic Eisenstein series as an iterated object, variously as an Eisenstein series for all the next-to-minimal parabolics $Q^i = P^{1,\ldots,1,2,1,\ldots,1}$ with the 2 at the ith place, formed from data including a suitably normalized GL_2 Eisenstein series $\widetilde{E}$ on the Levi component factor GL_2 of Q^i, rather than cuspidal data on that GL_2. Phragmén-Lindelöf gives boundedness of the analytic continuation of $\widetilde{E}$ in vertical strips, yielding convergence of the Q^i Eisenstein series in a larger region

$$\Omega_i = \{s \in \mathbb{C}^n : \mathrm{Re}(s_j) - \mathrm{Re}(s_{j+1}) > 2 \text{ for } j \neq i\}$$

That is, in Ω_i there is no constraint on $\mathrm{Re}(s_i) - \mathrm{Re}(s_{i+1})$. This applies to all the 2-by-2 blocks along the diagonal, giving a meromorphic continuation of E_s to $\bigcup_i \Omega_i$. Then Bochner's lemma [3.B] analytically continues the whole Eisenstein series to the convex hull of $\bigcup_i \Omega_i$, namely, $\mathbb{C}^n$.

Partial analytic continuation: Let $P = P^{\min}$. For each fixed index $1 \le i < r$, there is the next-to-minimal standard parabolic $Q = Q_i$ with standard Levi components and unipotent radicals given by $Q = N^Q \cdot M^Q$ with

$$N^Q = \begin{pmatrix} 1 & * & * & * & * & * & * & * \\ & \ddots & * & * & * & * & * & * \\ & & 1 & * & * & * & * & * \\ & & & 1 & 0 & * & * & * \\ & & & 0 & 1 & * & * & * \\ & & & & & 1 & * & * \\ & & & & & & \ddots & * \\ & & & & & & & 1 \end{pmatrix}$$

$$M^Q = \begin{pmatrix} * & 0 & 0 & 0 & 0 & 0 & 0 & 0 \\ & \ddots & 0 & 0 & 0 & 0 & 0 & 0 \\ & & * & 0 & 0 & 0 & 0 & 0 \\ & & & * & * & 0 & 0 & 0 \\ & & & * & * & 0 & 0 & 0 \\ & & & & & * & 0 & 0 \\ & & & & & & \ddots & 0 \\ & & & & & & & * \end{pmatrix}$$

with the anomalous block at the (i,i), $(i,i+1)$, $(i+1,i)$, and $(i+1,i+1)$ positions. The minimal-parabolic Eisenstein series can be written as an iterated sum

$$E_s(g) = \sum_{\gamma \in P_k \backslash G_k} \varphi_s^o(\gamma g) = \sum_{\gamma \in Q_k \backslash G_k} \Big(\sum_{\delta \in P_k \backslash Q_k} \varphi_s^o(\delta \gamma g) \Big)$$

The quotient $P_k \backslash Q_k$ has representatives

$$M_k^P \backslash M_k^Q \approx \{\delta = \begin{pmatrix} 1 & 0 & 0 & 0 & 0 & 0 & 0 & 0 \\ & \ddots & 0 & 0 & 0 & 0 & 0 & 0 \\ & & 1 & 0 & 0 & 0 & 0 & 0 \\ & & & a & b & 0 & 0 & 0 \\ & & & c & d & 0 & 0 & 0 \\ & & & & & 1 & 0 & 0 \\ & & & & & & \ddots & 0 \\ & & & & & & & 1 \end{pmatrix} \}$$

with $\begin{pmatrix} a & b \\ c & d \end{pmatrix} \in P_k^{1,1} \backslash GL_2(k)\} \approx P_k^{1,1} \backslash GL_2(k)$, where $P^{1,1}$ is the standard upper-triangular parabolic in GL_2. Further,

$$\varphi_s^o \begin{pmatrix} a_1 & * & * & * & * & * & * & * \\ & \ddots & * & * & * & * & * & * \\ & & a_{i-1} & * & * & * & * & * \\ & & & a_i & * & * & * & * \\ & & & 0 & a_{i+1} & * & * & * \\ & & & & & a_{i+2} & * & * \\ & & & & & & \ddots & * \\ & & & & & & & a_r \end{pmatrix} = |a_1|^{s_1} \ldots |a_r|^{s_r}$$

$$= |a_1|^{s_1} \ldots |a_{i-1}|^{s_{i-1}} |a_i/a_{i+1}|^{\frac{s_i - s_{i+1}}{2}} |a_i a_{i+1}|^{\frac{s_i + s_{i+1}}{2}} |a_{i+2}|^{s_{i+2}} \ldots |a_r|^{s_r}$$

Thus, the inner sum in the expression for E_s is

$$\sum_{\delta \in P_k \backslash Q_k} \varphi_s^o\Big(\delta \cdot \begin{pmatrix} a_1 & * & * & * & * & * & * & * \\ & \ddots & * & * & * & * & * & * \\ & & a_{i-1} & * & * & * & * & * \\ & & & a & b & * & * & * \\ & & & c & d & * & * & * \\ & & & & & a_{i+2} & * & * \\ & & & & & & \ddots & * \\ & & & & & & & a_r \end{pmatrix}\Big)$$

$$= |a_1|^{s_1} \ldots |a_{i-1}|^{s_{i-1}} \cdot E^{1,1}_{\frac{s_i - s_{i+1}}{2}} \begin{pmatrix} a & b \\ c & d \end{pmatrix} \left| \det \begin{pmatrix} a & b \\ c & d \end{pmatrix} \right|^{\frac{s_i + s_{i+1}}{2}} \cdot |a_{i+2}|^{s_{i+2}} \ldots |a_r|^{s_r}$$

where $E^{1,1}$ is the usual GL_2 Eisenstein series with trivial central character. So let $g = nmk$ be an Iwasawa decomposition with $n \in N^Q$, $m \in M^Q$, and $k \in \prod_v K_v$ with m in the form just displayed, and put

$$\Phi_s^i(g) = |a_1|^{s_1} \dots |a_{i=1}|^{s_{i-1}} \cdot E^{1,1}_{\frac{s_i - s_{i+1}}{2}} \begin{pmatrix} a & b \\ c & d \end{pmatrix}$$
$$\times \left| \det \begin{pmatrix} a & b \\ c & d \end{pmatrix} \right|^{\frac{s_i + s_{i+1}}{2}} \cdot |a_{i+2}|^{s_{i+2}} \dots |a_r|^{s_r}$$

Then

$$E_s(g) = \sum_{\gamma \in Q_k \backslash G_k} \Phi_s^i(\gamma\, g) \qquad (\text{for } g \in GL_r)$$

This expresses the GL_r minimal-parabolic Eisenstein series as Q-Eisenstein series formed from the $P^{1,1}$ Eisenstein series on the GL_2 part of its Levi component.

The usual normalization of the GL_2 Eisenstein series to eliminate poles, for boundedness on vertical strips for g in compacts in $GL_2(\mathbb{A})$, and to be *invariant* under $s \to 1 - s$, is

$$\widetilde{E}_s(g) = s(1-s) \cdot \xi(2s) \cdot E_s^{1,1}(g) \qquad (\text{for } s \in \mathbb{C})$$

Thus, let

$$\widetilde{\Phi}_s^i = (\frac{s_i - s_{i+1}}{2})(1 - \frac{s_i - s_{i+1}}{2}) \cdot \xi(s_i - s_{i+1}) \cdot \Phi_s^i$$

An argument similar to [3.10.1] for convergence of the minimal-parabolic Eisenstein series E_s and [3.11.1] for maximal proper parabolic Eisenstein series will prove the absolute convergence of

$$(\frac{s_i - s_{i+1}}{2})(1 - \frac{s_i - s_{i+1}}{2}) \cdot \xi(s_i - s_{i+1}) \cdot E_s(g) = \sum_{\gamma \in P_k \backslash G_k} \widetilde{\Phi}_s^i(\gamma\, g)$$

for $\frac{\mathrm{Re}(s_j) - \mathrm{Re}(s_{j+1})}{2} > 1$ for $j \neq i$, with *no condition* on $s_i - s_{i+1}$, because we use the analytically continued Eisenstein series on GL_2 rather than the expression of it as a series.

The convergence argument is as follows. For g in a fixed compact and $s_i - s_{i+1}$ in a fixed vertical strip, $\Phi_s^i(g)$ is dominated by the function obtained by replacing $\widetilde{E}^{1,1}$ by a constant, namely, with $\sigma_j = \mathrm{Re}(s_j)$,

$$\theta(g) = |a_1|^{\sigma_1} \dots |a_{i-1}|^{\sigma_{i-1}} \cdot \left| \det \begin{pmatrix} a & b \\ c & d \end{pmatrix} \right|^{\frac{\sigma_i + \sigma_{i+1}}{2}} \cdot |a_{i+2}|^{\sigma_{i+2}} \dots |a_n|^{\sigma_n}$$

We prove the absolute convergence of the Eisenstein series $E(g) = \sum_{\gamma \in P_k \backslash G_k} \theta(\gamma g)$, which is *degenerate* in the same sense as the approximating Eisenstein series in the proof of [3.11.1]. As in the earlier convergence arguments, convergence is equivalent to convergence of an integrated form, namely

$$\int_{Z_{\mathbb{A}} \backslash C} \sum_{\gamma \in P_k \backslash G_k} \theta(\gamma g) \, dg$$

Shrinking C sufficiently so that $\gamma \cdot C \cap C \neq \phi$ implies $\gamma = 1$,

$$\int_{Z_{\mathbb{A}} \backslash C} \sum_{\gamma \in P_k \backslash G_k} \theta(\gamma g) \, dg = \int_{Z_{\mathbb{A}} P_k \backslash G_k \cdot C} \theta(g) \, dg$$

As in the earlier convergence arguments, letting η_j be the norm of the determinant of the lower right $n - j$ minor, $G_k \cdot C$ is contained in

$$Y = \{g \in G_{\mathbb{A}} : 1 \ll_C \eta_j(g) \ \text{ for } j = 1, \ldots, n\}$$

To compare with M^Q with $Q = Q^i$, drop the $(i+1)^{\text{th}}$ condition: $G_k \cdot C$ is contained in

$$Y' = \{g \in G_{\mathbb{A}} : 1 \ll_C \eta_j(g) \ \text{ for } j \neq i+1\}$$

Thus, convergence of the Eisenstein series is implied by convergence of

$$\int_{Z_{\mathbb{A}} Q_k \backslash Y'} \theta(g) \, dg$$

As Y' is stable by right multiplication by the maximal compact subgroup $K_v \subset G_v$ at all places v, by an Iwasawa decomposition this integral is

$$\int_{Z_{\mathbb{A}} Q_k \backslash (Y' \cap Q_{\mathbb{A}})} \theta(p) \, dp \qquad \text{(left Haar measure on } Q)$$

Let $\alpha = \alpha_i$ be the ith simple positive root, and ρ the half-sum of positive roots. The left Haar measure on $Q_{\mathbb{A}}$ is $d(nm) = dn \, dm / m^{2\rho - \alpha}$, where dn is Haar measure on N^Q and dm is Haar measure on the Levi component M^Q. Since θ is left $N^Q_{\mathbb{A}}$-invariant and $N^Q_k \backslash N^Q_{\mathbb{A}}$ is compact, convergence of the latter integral is equivalent to convergence of

$$\int_{Z_{\mathbb{A}} M^Q_k \backslash (Y' \cap M^Q_{\mathbb{A}})} \theta(m) \, \frac{dm}{m^{2\rho - \alpha}}$$

As in the earlier convergence argument, the compactness lemma [2.A] and right action of $M \cap \prod_v K_v$ reduce the convergence question to that of a simpler integral.

As in the proof of [3.10.1], parametrize a subgroup H of $SL_n(\mathbb{A})$ by $r-1$ maps from $(GL_1(\mathbb{A})$, namely,

$$h_j : t \longrightarrow \begin{pmatrix} 1 & & & & & & \\ & \ddots & & & & & \\ & & 1 & & & & \\ & & & t & & & \\ & & & & t^{-1} & & \\ & & & & & 1 & \\ & & & & & & \ddots & \\ & & & & & & & 1 \end{pmatrix}$$

with t and t^{-1} at the jth and $(j+1)^{\text{th}}$ positions, with the ith replaced by the obvious map from $SL_2(\mathbb{A})$, namely,

$$h_i' : \begin{pmatrix} a & b \\ c & d \end{pmatrix} \longrightarrow \begin{pmatrix} 1 & & & & & & \\ & \ddots & & & & & \\ & & 1 & & & & \\ & & & a & b & & \\ & & & c & d & & \\ & & & & & 1 & \\ & & & & & & \ddots & \\ & & & & & & & 1 \end{pmatrix}$$

at i^{th} and $(i+1)^{\text{th}}$ positions. Then

$$\begin{aligned} Y' \cap M^Q_{\mathbb{A}} \cap SL_n(\mathbb{A}) = \{&\prod_{j\neq i} h_j(t_j) : t_j \in \mathbb{J} \text{ and } |t_j^{-1}| \gg 1\} \\ &\times \{h_i'(T) : T \in SL_2(\mathbb{A}),\ |\det T| \gg 1\} \end{aligned}$$

Noting that $h_j(t)^{2\rho} = |t|^2$, convergence is implied by convergence of

$$\begin{cases} \displaystyle\int_0^{\ll 1} t^{\sigma_j - \sigma_{j+1} - 2} \, \frac{dt}{t} & \text{(for } j \neq i\text{)} \\ \displaystyle\int_{SL_2(k)\backslash SL_2(\mathbb{A})} 1 \, dt & \text{(right invariant measure } dt\text{)} \end{cases}$$

The GL_1 integrals are absolutely convergent for $\sigma_j - \sigma_{j+1} - 2 > 0$ for $j \neq i$. By reduction theory [2.2] and [3.3], for example, $SL_2(k)\backslash SL_2(\mathbb{A})$ has finite volume, so the SL_2 integral is convergent. This is the desired convergence conclusion: there is no constraint on $\sigma_i - \sigma_{i+1}$. Thus, the iterated expression for the Eisenstein series analytically continues as indicated.

Functional equations for reflections: In addition to a partial analytic continuation, the previous argument gives the functional equations for the *reflections* in W attached to the simple roots, as follows. The main issue is making the functional equations understandable. In the earlier iterated expression for E_s in terms of a GL_2 Eisenstein series, the functional equation $\widetilde{E}^{1,1}_{1-z} = \widetilde{E}^{1,1}_z$ of that GL_2 Eisenstein series gives

$$\begin{aligned}\widetilde{\Phi}^i_s(g) &= |a_1|^{s_1} \ldots |a_{i=1}|^{s_{i-1}} \cdot \widetilde{E}^{1,1}_{\frac{s_i - s_{i+1}}{2}} \begin{pmatrix} a & b \\ c & d \end{pmatrix} \\ &\quad \times \left|\det \begin{pmatrix} a & b \\ c & d \end{pmatrix}\right|^{\frac{s_i+s_{i+1}}{2}} \cdot |a_{i+2}|^{s_{i+2}} \ldots |a_r|^{s_r} \\ &= |a_1|^{s_1} \ldots |a_{i=1}|^{s_{i-1}} \cdot \widetilde{E}^{1,1}_{1-\frac{s_i - s_{i+1}}{2}} \begin{pmatrix} a & b \\ c & d \end{pmatrix} \\ &\quad \times \left|\det \begin{pmatrix} a & b \\ c & d \end{pmatrix}\right|^{\frac{s_i+s_{i+1}}{2}} \cdot |a_{i+2}|^{s_{i+2}} \ldots |a_r|^{s_r}\end{aligned}$$

This is not presented immediately in terms of $s = (s_1, \ldots, s_r)$, but, instead, says

$$\begin{aligned}&(s_1, \ldots, s_{i-1}, \frac{s_i - s_{i+1}}{2}, \frac{s_i + s_{i+1}}{2}, s_{i+2}, \ldots, s_r) \\ &\qquad \longrightarrow (s_1, \ldots, s_{i-1}, 1 - \frac{s_i - s_{i+1}}{2}, \frac{s_i + s_{i+1}}{2}, s_{i+2}, \ldots, s_r)\end{aligned}$$

We hope for clarification by identifying the simultaneous *fixed point(s)*, if any, of *all* these, for $i = 1, \ldots, r$, together with the condition $s_1 + \cdots + s_r = 0$: the ith transformation fixes all by the ith and $(i+1)^{\text{th}}$ coordinate, and in those two coordinates the fixed-point condition is

$$\frac{s_i - s_{i+1}}{2} = 1 - \frac{s_i - s_{i+1}}{2} \qquad \text{and} \qquad \frac{s_i + s_{i+1}}{2} = \frac{s_i + s_{i+1}}{2}$$

The second equation is a tautology, so the ith fixed-point condition is simply $s_i - s_{i+1} = 1$. These conditions for $i = 1, \ldots, r-1$ and $s_1 + \cdots + s_r = 0$, give a unique fixed point,

$$\text{fixed point} = \left(\frac{n-1}{2}, \frac{n-3}{2}, \frac{n-5}{2}, \ldots, \frac{3-n}{2}, \frac{1-n}{2}\right)$$

This is the half-sum ρ of positive roots

$$\rho = (\rho_1, \ldots, \rho_r) = \tfrac{1}{2} \sum_{i<j} (0, \ldots, 0, 1, 0, \ldots, 0, -1, 0, \ldots, 0)$$

at the ith and j^{th} places. Replacing s by $\rho + s$ replaces $s_i - s_{i+1}$ by $(\rho_i + s_i) - (\rho_{i+1} + s_{i+1}) = s_i - s_{i+1} + 1$, and the map $\frac{s_i - s_{i+1}}{2} \to 1 - \frac{s_i - s_{i+1}}{2}$ becomes

$$\frac{(\rho_i + s_i) - (\rho_{i+1} + s_{i+1})}{2} \to 1 - \frac{(\rho_i + s_i) - (\rho_{i+1} s_{i+1})}{2}$$

which simplifies to $s_i \longleftrightarrow s_{i+1}$. That is, in the $\rho + s$ coordinates, this functional equation is the interchange of s_i and s_{i+1}:

$$\begin{aligned} &\rho + (s_1, \ldots, s_{i-1}, s_i, s_{i+1}, s_{i+2}, \ldots, s_r) \\ &\qquad \longleftrightarrow \rho + (s_1, \ldots, s_{i-1}, s_{i+1}, s_i, s_{i+2}, \ldots, s_r) \end{aligned}$$

This is the same as the effect $m^s \longrightarrow (\tau_i m \tau_i^{-1})^s$ with

$$\tau_i = \begin{pmatrix} 1 & & & & & & \\ & \ddots & & & & & \\ & & 1 & & & & \\ & & & 0 & 1 & & & \\ & & & 1 & 0 & & & \\ & & & & & 1 & & \\ & & & & & & \ddots & \\ & & & & & & & 1 \end{pmatrix} \in W$$

at i^{th} and $(i+1)^{\text{th}}$ positions. This $\tau_i \in W$ is usually considered to be *attached to* the ith simple root $\alpha_i(m) = m_i - m_{i+1}$ on the Lie algebra, as it is characterized by interchanging $\pm\alpha_i$ and *permuting* the other *positive* roots $\beta_{j\ell}(m) = m_j - m_\ell$: for m in the Lie algebra of $M^{\min}$, *unless* $j = i$ and $\ell = i + 1$,

$$\beta_{j\ell}\left(\tau_i m \tau_i^{-1}\right) = \left(\tau_i m \tau_i^{-1}\right)_j - \left(\tau_i m \tau_i^{-1}\right)_\ell = m_{j'} - m_{\ell'}$$

for some $j' < \ell'$, thus giving some other positive root evaluated on m. When $j = i$ and $\ell = i + 1$, the effect is qualitatively different, reversing the sign, producing $-\alpha_i$.

To rewrite the foregoing in more geometric terms, use the pairing $\langle x, y \rangle = \operatorname{tr}(xy)$ on the Lie algebra $\mathfrak{g}$ of GL_r to identify $\mathfrak{a}$ and $\mathfrak{a}^*$ and give each of them a nondegenerate inner product. The Weyl group preserves this inner product, since

$$\begin{aligned} \langle wxw^{-1}, wyw^{-1} \rangle &= \operatorname{tr}(wxw^{-1} \cdot wyw^{-1}) = \operatorname{tr}(wxyw^{-1}) \\ &= \operatorname{tr}(xy) = \langle x, y \rangle \end{aligned}$$

by conjugation invariance of trace. The geometric characterization of the reflection $\tau = \tau_\alpha$ associated to a vector (here a simple positive root) α is that τ should fix the hyperplane orthogonal to α, and should send $\alpha \to -\alpha$: this is expressed by

$$\tau x = x - 2 \cdot \frac{\langle x, \alpha \rangle}{\langle \alpha, \alpha \rangle} \cdot \alpha \qquad \text{(for } x \in \mathfrak{a}^* \approx \text{diagonal matrices)}$$

Via this pairing, α_i is identified with

$$\alpha_i = \begin{pmatrix} 0 & & & & & & \\ & \ddots & & & & & \\ & & 0 & & & & \\ & & & \begin{matrix} 1 & 0 \\ 0 & -1 \end{matrix} & & & \\ & & & & 0 & & \\ & & & & & \ddots & \\ & & & & & & 0 \end{pmatrix}$$

at i^{th} and $(i+1)^{\text{th}}$ positions, because $\alpha_i(m) = \langle m, \alpha_i \rangle$. Similarly, $s_i - s_{i+1} = \langle s, \alpha_i \rangle$. It is immediate that $\tau \cdot \alpha = \tau \alpha \tau^{-1} = -\alpha$, as the reflection should. Since τ preserves $\langle , \rangle$, it preserves the orthogonal complement to α, so truly is the associated reflection. Since τ flips the sign on α and permutes the other positive roots, we can compute, using $\tau^{-1} = \tau$,

$$\langle 2\rho, \alpha \rangle = -\langle 2\rho, \tau \cdot \alpha \rangle = -\langle \tau \cdot 2\rho, \tau \cdot \alpha \rangle$$

and

$$\tau \cdot 2\rho = \tau \cdot \sum_{\beta>0} \beta = \tau \cdot \Big(\sum_{\beta>0,\, \beta\neq\alpha} \beta \Big) + \tau \cdot \alpha$$
$$= \Big(\sum_{\beta>0,\, \beta\neq\alpha} \beta \Big) - \alpha = 2\rho - 2\alpha$$

Thus, $\langle 2\rho, \alpha \rangle = -\langle 2\rho - 2\alpha, \alpha \rangle$, from which $\langle \rho, \alpha \rangle = \langle \alpha, \alpha \rangle / 2 = 1$. Thus, the αth functional equation, inherited from the GL_2 Eisenstein series, is

$$\xi \langle \rho + \tau \cdot s, \alpha \rangle \cdot E_{\rho + \tau \cdot s} = \xi \langle \rho + s, \alpha \rangle \cdot E_{\rho+s} \qquad \text{(reflection } \tau = \tau_\alpha)$$

Using $\langle \rho, \alpha \rangle = 1$, this is

$$\xi(1 + \langle \tau \cdot s, \alpha \rangle) \cdot E_{\rho+\tau\cdot s} = \xi(1 + \langle s, \alpha \rangle) \cdot E_{\rho+s} \qquad \text{(reflection } \tau = \tau_\alpha)$$

or, since $\langle \tau \cdot s, \alpha \rangle = \langle s, \tau \cdot \alpha \rangle = \langle s, -\alpha \rangle = -\langle s, \alpha \rangle$,

$$\begin{aligned} E_{\rho+\tau \cdot s} &= \frac{\xi(1+\langle s,\alpha\rangle)}{\xi(1+\langle \tau \cdot s,\alpha\rangle)} \cdot E_{\rho+s} = \frac{\xi(1+\langle s,\alpha\rangle)}{\xi(1-\langle s,\alpha\rangle)} \cdot E_{\rho+s} \\ &= \frac{\xi(1+\langle s,\alpha\rangle)}{\xi\langle s,\alpha\rangle} \cdot E_{\rho+s} \text{ (reflection } \tau = \tau_\alpha) \end{aligned}$$

using the functional equation $\xi(1-z) = \xi(z)$.

Application of Bochner's Lemma: The $n-1$ partial analytic continuations can be organized to allow application of Bochner's Lemma. Above, for $\alpha = \alpha_i$ the ith simple root, we showed that the function

$$\begin{aligned} E^{\alpha}_{\rho+s} &= \langle \rho+s,\alpha\rangle \cdot (1-\langle \rho+s,\alpha\rangle) \cdot \xi\langle \rho+s,\alpha\rangle \cdot E_{\rho+s} \\ &= (1+\langle s,\alpha\rangle) \cdot (1-\langle s,\alpha\rangle) \cdot \xi(1+\langle s,\alpha\rangle) \cdot E_{\rho+s} \end{aligned}$$

admits an analytic continuation in which $s_i - s_{i+1} = \langle s, \alpha \rangle$ is not constrained, and this normalized version of $E_{\rho+s}$ is *invariant* under $\rho + s \to \rho + \tau_\alpha \cdot s$ for the reflection τ_α. This *might* suggest adding normalization factors for *all* positive roots, to obtain an eventually W-invariant expression:

$$E_{\rho+s} \cdot \prod_{\beta>0} (1+\langle s,\beta\rangle) \cdot (1-\langle s,\beta\rangle) \cdot \xi(1+\langle s,\beta\rangle)$$

The intention is that $E^{\alpha}_{\rho+s}$ is invariant under the reflection τ_α for each simple root α, and the remaining factors should be permuted among themselves, since the other positive roots are permuted among themselves by τ_α.

A minor technical issue arises: to be sure to cancel the pole of $\xi(1+\langle s,\beta\rangle)$ at $1+\langle s,\beta\rangle = 1$, to most easily justify application of Bochner's lemma, add additional polynomial factors, squared to avoid disturbing the sign in functional equations: let

$$E^{\#}_{\rho+s} = E_{\rho+s} \cdot \prod_{\beta>0} (1+\langle s,\beta\rangle) \cdot (1-\langle s,\beta\rangle) \cdot \langle s,\beta\rangle^2 \cdot \xi(1+\langle s,\beta\rangle)$$

The exponential decay of the gamma factor in ξ is more than sufficient to preserve boundedness in vertical strips for real part s in compacts.

[3.12.4] Claim: $E^{\#}_{\rho+s}$ has an analytic continuation to a holomorphic function on $\mathbb{C}^r$, and is invariant under $s \to w \cdot s$ for all $w \in W$.

Proof: By the GL_2 discussion and the above adaptations, $E^{\#}_{\rho+s}$ has an analytic continuation to the *tube domain* $\Omega = \{z \in \mathbb{C}^r : \text{Re}(z) \in \Omega_o\}$ over $\Omega_o \subset \mathbb{R}^r$

given by

$$\Omega_o = \{\sigma \in \mathbb{R}^r : \langle \rho + \sigma, \alpha \rangle > 1 \text{ for all but possibly a single simple root } \alpha\}$$

In Ω, for $\mathrm{Re}(s)$ in compacts, $E^{\#}_{\rho+s}$ is *bounded*, so certainly has sufficiently modest growth for application of Bochner's Lemma, and $E^{\#}_{\rho+s}$ has an analytic continuation to the convex hull of Ω, which is $\mathbb{C}^r$. It is invariant under all reflections attached to simple roots, and these generate W. This proves the claim. ///

Returning to the proof of the theorem, this last claim gives the meromorphic continuation of $E_{\rho+w}$, and the first corollary.

Given the meromorphic continuation of $E_{\rho+s}$, the functional equations

$$E_{\rho+\tau\cdot s} = \frac{\xi(1+\langle s,\alpha\rangle)}{\xi\langle s,\alpha\rangle} \cdot E_{\rho+s} \quad (\text{reflection } \tau = \tau_\alpha)$$

of $E_{\rho+s}$ proven earlier for *reflections* $\tau = \tau_\alpha$ attached to simple roots α can be *iterated*. Taking constant terms gives, by the general form of the constant term [3.10.3],

$$\sum_w c_{w,\tau\cdot s}\, m^{\rho+w\cdot\tau\cdot s} = c_P E_{\rho+\tau\cdot s} = \frac{\xi(1+\langle s,\alpha\rangle)}{\xi\langle s,\alpha\rangle} \cdot c_P E_{\rho+s}$$

$$= \frac{\xi(1+\langle s,\alpha\rangle)}{\xi\langle s,\alpha\rangle} \sum_w c_{w,s}\, m^{\rho+w\cdot s}$$

For generic $s \in \mathfrak{a}^* \otimes_{\mathbb{R}} \mathbb{C}$, the coefficients of the various characters of $m \in M_{\mathbb{A}}$ must be equal, so by the identity principle are equal for all s. In particular, equating the coefficient of $m^{\rho+\tau\cdot s}$ gives

$$1 = c_{1,\tau\cdot s} = \frac{\xi(1+\langle s,\alpha\rangle)}{\xi\langle s,\alpha\rangle} \cdot c_{\tau,s}$$

That is, $c_{\tau,s} = \xi\langle s,\alpha\rangle/\xi(1+\langle s,\alpha\rangle)$, and

$$E_{\rho+\tau\cdot s} = \frac{\xi(1+\langle s,\alpha\rangle)}{\xi\langle s,\alpha\rangle} \cdot E_{\rho+s} = \frac{1}{c_{\tau,s}} \cdot E_{\rho+s} \qquad (\text{for reflection } \tau)$$

Equating the coefficients of $m^{\rho+w\cdot\tau\cdot s}$ gives

$$c_{w,\tau\cdot s} = \frac{\xi(1+\langle s,\alpha\rangle)}{\xi\langle s,\alpha\rangle} \cdot c_{w\tau,s} = \frac{1}{c_{\tau,s}} \cdot c_{w\tau,s}$$

from which $c_{w,\tau\cdot s} \cdot c_{\tau,s} = c_{w\tau,s}$. For two reflections σ, τ,

$$c_{w\sigma\tau,s} = c_{(w\sigma)\tau,s} = c_{w\sigma,\tau\cdot s} \cdot c_{\tau,s} = c_{w,\sigma\tau\cdot s} \cdot c_{\sigma,\tau\cdot s} \cdot c_{\tau,s} = c_{w,\sigma\tau\cdot s} \cdot c_{\sigma\tau,s}$$

Induction gives $c_{ww',s} = c_{w,w'\cdot s} \cdot c_{w',s}$. Then

$$E_{\rho+\sigma\tau\cdot s} = E_{\rho+\sigma\cdot(\tau\cdot s)} = \frac{1}{c_{\sigma,\tau\cdot s}} \cdot E_{\rho+\tau\cdot s}$$
$$= \frac{1}{c_{\sigma,\tau\cdot s}} \cdot \frac{1}{c_{\tau,s}} \cdot E_{\rho+s} = \frac{1}{c_{\sigma\tau,s}} \cdot E_{\rho+s}$$

and a similar induction on the length of $w \in W$ gives the general functional equation. Qualitatively, the number of factors in both numerator and denominator of $c_w(s)$ is the *length* of w. This proves the theorem and corollary. ///

[3.12.5] Example: For $G = GL_3$ there are two simple positive roots,

$$\langle x, \alpha\rangle = x_1 - x_2 \qquad \langle x, \beta\rangle = x_2 - x_3$$

for $x \in \mathfrak{a}$ with diagonal entries x_i. The other positive root is $\alpha + \beta$, so $\rho = \frac{1}{2}(\alpha + \beta + (\alpha_\beta)) = \alpha + \beta$. Let σ, τ be the reflections corresponding to α, β, respectively. The whole Weyl group is $W = \{1, \sigma, \tau, \sigma\tau, \tau\sigma, \sigma\tau\sigma\}$ and $\sigma\tau\sigma = \tau\sigma\tau$. From the GL_2 computation,

$$c_{\sigma,\rho+s} = \frac{\xi\langle s, \alpha\rangle}{\xi(\langle s, \alpha\rangle + 1)} \qquad c_{\tau,s} = \frac{\xi\langle s, \beta\rangle}{\xi(\langle s, \beta\rangle + 1)}$$

By the cocycle relation $c_{wr,s} = c_{w,r\cdot s} \cdot c_{r,s}$ for reflection $r \in W$ and $w \in W$, we have

$$c_{\sigma\tau,s} = c_{\sigma,\tau\cdot s} \cdot c_{\tau,s} = \frac{\xi\langle \tau\cdot s, \alpha\rangle}{\xi(\langle \tau\cdot s, \alpha\rangle + 1)} \cdot \frac{\xi\langle s, \beta\rangle}{\xi(\langle s, \beta\rangle + 1)}$$

Since $\langle \tau x, \alpha\rangle = \langle x, \tau\alpha\rangle = \langle x, \alpha + \beta\rangle$,

$$c_{\sigma\tau,s} = \frac{\xi\langle s, \alpha+\beta\rangle}{\xi(\langle s, \alpha+\beta\rangle + 1)} \cdot \frac{\xi\langle s, \beta\rangle}{\xi(\langle s, \beta\rangle + 1)}$$

Similarly,

$$c_{\tau\sigma,s} = \frac{\xi\langle s, \alpha+\beta\rangle}{\xi(\langle s, \alpha+\beta\rangle + 1)} \cdot \frac{\xi\langle s, \alpha\rangle}{\xi(\langle s, \alpha\rangle + 1)}$$

Finally,

$$c_{\tau\sigma\tau,s} = c_{\sigma\tau\sigma,s} = c_{\sigma\tau,\sigma\cdot s} \cdot c_{\sigma,s}$$
$$= \frac{\xi\langle \sigma\cdot s, \alpha+\beta\rangle}{\xi(\langle \sigma\cdot s, \alpha+\beta\rangle + 1)} \cdot \frac{\xi\langle \sigma\cdot s, \beta\rangle}{\xi(\langle \sigma\cdot s, \beta\rangle + 1)} \cdot \frac{\xi\langle s, \alpha\rangle}{\xi(\langle s, \alpha\rangle + 1)}$$

Using $\sigma\beta = \alpha + \beta$ and $\sigma(\alpha + \beta) = \beta$, this is

$$c_{\tau\sigma\tau,s} = c_{\sigma\tau\sigma,s} = \frac{\xi\langle s, \beta\rangle}{\xi(\langle s, \beta\rangle + 1)} \cdot \frac{\xi\langle s, \alpha+\beta\rangle}{\xi(\langle s, \alpha+\beta\rangle + 1)} \cdot \frac{\xi\langle s, \alpha\rangle}{\xi(\langle s, \alpha\rangle + 1)}$$

The latter example suggests that more can be said about $c_{w,s}$:

[3.12.6] Claim:

$$c_{w,s} = \prod_{\beta>0:w\cdot\beta<0} \frac{\xi\langle s,\beta\rangle}{\xi(\langle s,\beta\rangle+1)}$$

Proof: Induction on the length of w in the generating reflections associated to simple roots. With $\tau = \tau_\alpha$ for simple root α the cocycle relation gives

$$c_{w\tau,s} = c_{w,\tau\cdot s}\cdot c_{\tau,s} = \prod_{\beta>0:w\cdot\beta<0} \frac{\xi\langle \tau\cdot s,\beta\rangle}{\xi(\langle \tau\cdot s,\beta\rangle+1)}\cdot\frac{\xi\langle s,\alpha\rangle}{\xi(\langle s,\alpha\rangle+1)}$$

$$= \prod_{\beta>0:w\cdot\beta<0} \frac{\xi\langle s,\tau\cdot\beta\rangle}{\xi(\langle s,\tau\cdot\beta\rangle+1)}\cdot\frac{\xi\langle s,\alpha\rangle}{\xi(\langle s,\alpha\rangle+1)}$$

The effect of $\tau = \tau_\alpha$ on roots is to interchange $\pm\alpha$, permute the *other* positive roots, and permute the *other* negative roots. There are two cases.

First, if $w\cdot\alpha < 0$, then α itself appears in the product, and $(w\tau_\alpha)\cdot\alpha = w(-\alpha) = -w\cdot\alpha > 0$. So α will *not* appear in the corresponding product for $w\tau$. Using the functional equation $\xi(1-z) = \xi(z)$,

$$\frac{\xi\langle s,\tau\cdot\alpha\rangle}{\xi(\langle s,\tau\cdot\alpha\rangle+1)}\cdot\frac{\xi\langle s,\alpha\rangle}{\xi(\langle s,\alpha\rangle+1)} = \frac{\xi\langle s,-\alpha\rangle}{\xi(\langle s,-\alpha\rangle+1)}\cdot\frac{\xi\langle s,\alpha\rangle}{\xi(\langle s,\alpha\rangle+1)}$$

$$= \frac{\xi(1-(\langle s,\alpha\rangle+1))}{\xi(1-\langle s,\alpha\rangle)}\cdot\frac{\xi\langle s,\alpha\rangle}{\xi(\langle s,\alpha\rangle+1)}$$

$$= \frac{\xi(\langle s,\alpha\rangle+1)}{\xi\langle s,\alpha\rangle}\cdot\frac{\xi\langle s,\alpha\rangle}{\xi(\langle s,\alpha\rangle+1)} = 1$$

Thus, the leftover factor from the product for w cancels the new factor from the cocycle relation, and the desired relation holds for $w\tau_\alpha$ in the case that $w\cdot\alpha < 0$.

Second, similarly but oppositely, suppose $w\cdot\alpha > 0$. Then α does *not* appear in the product for w. But $(w\tau_\alpha)\alpha = w(-\alpha) < 0$, so α *should* appear in the product for $w\tau_\alpha$. The extra term provides this, proving the relation in this case. ///

3.13 Continuation of Cuspidal-Data Eisenstein Series

The functional equations of Eisenstein series attached to nonminimal parabolics $P^{d_1,\ldots,d_\ell}$ involve *all* the parabolics $P^{d'_1,\ldots,d'_\ell}$ with $d'_1,\ldots,d'_\ell$ a permutation of $d_1,\ldots,d_\ell$, called the *associates* of $P^{d_1,\ldots,d_\ell}$. This is so even with the simplifying assumption of cuspidal data, without which the situation is messier. Then, the expression of pseudo-Eisenstein series for such parabolics in terms of genuine Eisenstein series, even with the corresponding assumption of cuspidal data, involves all these. Thus, as in [3.11], we consider only maximal proper parabolics $P = P^{r_1,r_2}$, right $M^P_{\mathbb{A}} \cap K_{\mathbb{A}}$-invariant *cuspidal data* $f = f_1 \otimes f_2$ on $M^P_{\mathbb{A}}$ with trivial central character. Assume f_1, f_2 are spherical Hecke eigenfunctions at all finite places, so by [3.11.6] the Eisenstein series $E^P_{s,f}$ is a spherical Hecke eigenfunction at all finite places. Similarly, at archimedean places v, we assume (at least) that f_1 and f_2 are eigenfunctions for the invariant Laplacians on the factors of the Levi component, so by [3.11.11] $E^P_{s,f}$ is an eigenfunction for the invariant Laplacian on G_v.

Assume *strong multiplicity one* for $f_1 \otimes f_2$, as in [3.11.9], so that the constant terms of $E^P_{s,f}$ are as simple as possible. With $P = P^{r_1,r_2}$, let $Q = P^{r_2,r_1}$, so $Q = P$ for self-associate P and otherwise is the unique other associate of P. Thus, the following is special, but perhaps more palatable than the general case. Write $f^w = (f_1 \otimes f_2)^w = f_2 \otimes f_1$.

[3.13.1] Theorem: (*Langlands, Bernstein, Wong, et alia*) With the constant-term conventions as in [3.11.9] such Eisenstein series $E^P_{s,f}$ have meromorphic continuations in s, with functional equation

$$E^P_{1-s,f} = (c^P_{s,f^w})^{-1} \cdot E^Q_{s,f^w} \qquad \text{and} \qquad c^Q_{1-s,f} \cdot c^P_{s,f^w} = 1$$

(Proofs in [11.10], [11.12].) ///

Although the proof of meromorphic continuation is postponed to [11.10], [11.12], if we *grant* meromorphic continuation then the form of the *functional equation* is determined by the constant term, using the theory of the constant term [8.3], as follows. As in [3.11.9], the constant terms of $E^P_{s,f}$ and E^Q_{1-s,f^w} are explicit. First, for P self-associate,

$$c_P E^P_{s,f^w} = \varphi^P_{s,f^w} + c_{s,f^w} \varphi^P_{1-s,f}$$

and

$$c_P E^P_{1-s,f} = \varphi^P_{1-s,f} + c_{1-s,f} \varphi^P_{s,f^w}$$

and all other constant terms are 0. Thus,

$$c_P \left(E^P_{1-s,f} - c^{-1}_{s,f^w} \cdot E^P_{s,f^w} \right) = \left(c_{1-s,f} - c^{-1}_{s,f^w} \right) \cdot \varphi^P_{s,f^w}$$

and all other constant terms are 0. The functions f_1 and f_2 are cuspforms in a strong sense, so by [8.2] are *bounded*. Thus,

$$|\varphi^P_{s,f}(nmk)| \le \left| \frac{(\det m_1)^{r_2}}{(\det m_2)^{r_1}} \right|^{\mathrm{Re}\,(s)}$$

is *bounded* in standard Siegel sets for $\mathrm{Re}\,(s)$ sufficiently negative. By the theory of the constant term [8.3], $E^P_{1-s,f^w} - c^{-1}_{s,f^w} \cdot E^P_{s,f}$ is bounded in Siegel sets. Thus, this difference is in $L^2(Z^+G_k\backslash G_{\mathbb{A}})$. However, from [3.11.11], both E^P_{1-s,f^w} and $E^P_{s,f}$ have the same eigenvalues for invariant Laplacians at archimedean places, namely, $r_1r_2(r_1+r_2)(s^2-s)+\lambda_1+\lambda_2$ where λ_j is the eigenvalue for f_j on the corresponding archimedean factor GL_{r_j} of M^P. Thus, the difference has that eigenvalue. There are many choices of s with $\mathrm{Re}(s)<0$ that make this eigenvalue nonreal, however, which is impossible for an L^2 eigenfunction other than 0, as in the proof of [1.10.5]. Thus, this difference must be 0, and have constant term 0. This gives the functional equation, assuming the meromorphic continuation, in the self-associate case.

For the non-self-associate case, for P and its other associate Q, both constant terms must be considered to invoke the theory of the constant term. Starting from

$$c_P E^P_{1-s,f} = \varphi^P_{1-s,f} \qquad c_Q E^P_{1-s,f} = c^Q_{1-s,f}\varphi^Q_{s,f^w}$$

$$c_P E^Q_{s,f^w} = c^P_{s,f^w}\varphi^P_{1-s,f} \qquad c^Q E^Q_{s,f^w} = \varphi^Q_{s,f^w}$$

we have

$$\begin{cases} c_P\big(E^P_{1-s,f} - (c^P_{s,f^w})^{-1}\cdot E^Q_{s,f^w}\big) &= 0 \\ c_Q\big(E^P_{1-s,f} - (c^P_{s,f^w})^{-1}\cdot E^Q_{s,f^w}\big) &= \big(c^Q_{1-s,f} - (c^P_{s,f^w})^{-1}\big)\cdot\varphi^Q_{s,f^w}\end{cases}$$

and all other constant terms are 0. As usual, the cuspforms are bounded by [8.3], so, for $\mathrm{Re}(s)$ sufficiently negative, the constant term along Q is square-integrable on Siegel sets. By the theory of the constant term [8.3], the difference $E^P_{1-s,f} - (c^P_{s,f^w})^{-1}\cdot E^Q_{s,f^w}$ is square-integrable. However, again by [3.11.11], this difference is an eigenfunction for invariant Laplacians at archimedean places, with nonreal eigenvalues for many choices of s. Thus, it is identically 0. ///

3.14 Truncation and Maaß-Selberg Relations

First, we make precise a notion of *truncation* of automorphic forms, relative to a choice of parabolic subgroup, especially maximal proper parabolics. For the self-associate maximal proper parabolic $P^{r,r}$ in GL_{2r}, the computation of inner products of truncations of $P^{r,r}$ Eisenstein series with cuspidal data is parallel to the computation for GL_2. As in [1.11] and [2.10], corollaries give information about possible poles of Eisenstein series, and square-integrability of residues of Eisenstein series.

This bears on the occurrence of nontrivial *residual* square-integrable automorphic forms coming from cuspforms on smaller groups, anticipating that such automorphic forms occur as residues of Eisenstein series. For example, there is no nonconstant noncuspidal discrete spectrum for $GL_2(\mathbb{Z})$ nor for $GL_3(\mathbb{Z})$, but only for $GL_4(\mathbb{Z})$ and larger. Namely, the Eisenstein series on GL_3 with GL_2 cuspidal data have no poles in the right half-plane, as follows immediately from the following Maaß-Selberg relations.

The simplest nontrivial examples of Maaß-Selberg relations and corollaries concern spherical Eisenstein series on GL_n associated to *cuspidal data* on the Levi component of maximal (proper) parabolics $P = P^{r_1,r_2}$. For simplicity, we continue to consider only right $K_{\mathbb{A}}$-invariant Eisenstein series $E^P_{s,f_1\otimes f_2}$, where f_1, f_2 are cuspforms in the strong sense of being spherical Hecke eigenfunctions everywhere, with trivial central characters, allowing the simple outcomes of computations of constant terms as in [3.11.8] and [3.11.9]. When $r_1 \neq r_2$, that is, when P is not self-associate, let $Q = P^{r_2,r_1}$ be its other associate parabolic.

Let δ^P be the modular function of $P_{\mathbb{A}}$

$$\delta\begin{pmatrix} m_1 & 0 \\ 0 & m_2 \end{pmatrix} = \left|\frac{(\det m_1)^{r_2}}{(\det m_2)^{r_1}}\right|$$

and extend this to a *height function* aligned with P, by making it right $K_{\mathbb{A}}$-invariant: $h^P(nmk) = \delta^P(nm) = \delta^P(m)$ for $n \in N^P_{\mathbb{A}}$, $m \in M^P_{\mathbb{A}}$, and $k \in K_{\mathbb{A}}$. For fixed large real T, the T-*tail* of the P-constant term of an automorphic form F is

$$c^T_P F(g) = \begin{cases} c_P F(g) & (\text{for } h^P(g) \geq T) \\ 0 & (\text{for } h^P(g) < T) \end{cases}$$

Similarly, the T-tail of the Q-constant term is

$$c^T_Q F(g) = \begin{cases} c_Q F(g) & (\text{for } h^Q(g) \geq T) \\ 0 & (\text{for } h^Q(g) < T) \end{cases}$$

Suitable truncations of these cuspidal-data Eisenstein series should be square integrable (potentially accomplished a number of ways), and their inner products calculable in explicit, straightforward terms. There should be no obstacle to meromorphic continuation of the *tail* in the truncation. These requirements are at odds with each other. Writing $\Psi^P(\varphi) = \Psi^P_\varphi$ for the pseudo-Eisenstein series attached to data φ, the *truncation* at height T of the Eisenstein series is

$$\wedge^T E^P_{s,f} = \begin{cases} E^P_{s,f} - \Psi^P(c^T_P E^P_{s,f}) & (\text{for } n_1 = n_2) \\ E^P_{s,f} - \Psi^P(c^T_P E^P_{s,f}) - \Psi^Q(c^T_Q E^P_{s,f}) & (\text{for } n_1 \neq n_2) \end{cases}$$

[3.14.1] Proposition: The truncated Eisenstein series $\wedge^T E^P_{s,f}$ is of rapid decay in Siegel sets.

Proof: The argument is simpler in the self-associate case, which we carry out first. From the computations in [3.11.8] and [3.11.9] of constant terms of such Eisenstein series, for self-associate maximal proper P in GL_r all such constant terms are 0 except that along P itself. By the theory of the constant term, on standard Siegel sets $E^P_{s,f} - c_P E^P_{s,f}$ is of rapid decay. Thus, $E^P_{s,f} - c^T_P E^P_{s,f}$ is of rapid decay on standard Siegel sets, and then the automorphic form

$$\wedge^T E^P_{s,f} = E^P_{s,f} - \Psi^P\left(c^T_P E^P_{s,f}\right)$$

is of rapid decay on *all* Siegel sets.

As in the discussion immediately before [3.10.2], for a root α of G, for $a \in M^{\min}_{\mathbb{A}}$, let $a^\alpha = e^{\alpha(\log|a|)}$. For $g \in G_{\mathbb{A}}$, in an Iwasawa decomposition let $g \in N^{\min} \cdot a_g \cdot K_{\mathbb{A}}$ with $a_g \in M^{\min}_{\mathbb{A}}$, so we can consider the functions $g \longrightarrow a^\alpha_g$. The ambiguity of a by $M^{\min}_{\mathbb{A}} \cap K_{\mathbb{A}}$ does not affect the value of this function. In the non-self-associate, case, let α, β be the simple positive roots corresponding to P and Q, in the sense that N^P contains the α root subgroup

$$N^\alpha = \{n = e^x : x \in \mathfrak{n},\ axa^{-1} = a^\alpha \cdot x\}$$

where $\mathfrak{n}$ is the Lie algebra of $N^{\min}$, and N^Q contains the β root subgroup N^β. Because f is a cuspform and P is not self-associate, only a single Bruhat cell contributes to $c_P E^P_{s,f}$, and $c_P E^P_{s,f} = \varphi^P_{s,f}$, which is rapidly decreasing on standard Siegel sets as $a^\gamma_g \to +\infty$ for any simple positive root $\gamma \neq \alpha$, because f is a cuspform in a strong sense. Similarly, only a single Bruhat cell (corresponding to the longest Weyl element) contributes to the constant term $c_Q E^P_{s,f}$, which similarly is rapidly decreasing on standard Siegel sets as $a^\gamma_g \to +\infty$ for any simple (positive) root $\gamma \neq \beta$. Thus, the truncation

$$\wedge^T E^P_\varphi = E^P_\varphi - \Psi^P\left(c^T_P E^P_\varphi\right) - \Psi^Q\left(c^T_Q E^P_\varphi\right)$$

has decay properties as follows. If $a^\gamma_g \to +\infty$ for γ other than α, β, then all three terms on the right-hand side are of rapid decay in standard Siegel sets. If $\alpha \to +\infty$, then each of the two expressions $E^P_\varphi - \Psi^P(c^T_P E^P_\varphi)$ and $\Psi^Q(c^T_Q E^P_\varphi)$ is of rapid decay. If $\beta \to +\infty$, then each of the two expressions $E^P_\varphi - \Psi^Q(c^T_Q E^P_\varphi)$ and $\Psi^P(c^T_P E^P_\varphi)$ is of rapid decay. Thus, as the value of any simple positive root goes to $+\infty$ in a standard Siegel set, the truncation goes rapidly to zero. ///

Let $h = h_1 \otimes h_2$ be a another cuspform on $M = M^P$. Let $\langle f, h\rangle_{M^P}$ be the inner product on the quotient $Z^M_{\mathbb{A}} M_k \backslash M_{\mathbb{A}}$. For brevity, write $f^w = (f_1 \otimes f_2)^w = f_2 \otimes f_1$.

[3.14.2] Theorem: *(Maaß-Selberg relations)* The Hermitian inner product $\langle \wedge^T E^P_{s,f}, \wedge^T E^P_{r,h}\rangle$ of truncations of two cuspidal-data Eisenstein series for maximal proper parabolic P is given as follows. For P *self-associate*,

$$\begin{aligned}\big\langle \wedge^T E^P_{s,f}, \wedge^T E^P_{r,h}\big\rangle &= \langle f, h\rangle_M \cdot \frac{T^{s+\overline{r}-1}}{s+\overline{r}-1} + \langle f, h^w\rangle_M \cdot \overline{c_{r,h}}\frac{T^{s+(1-\overline{r})-1}}{s+(1-\overline{r})-1} \\ &\quad + \langle f^w, h\rangle_M \cdot c_{s,f}\frac{T^{(1-s)+\overline{r}-1}}{(1-s)+\overline{r}-1} \\ &\quad + \langle f^w, h^w\rangle_M \cdot c_{s,f}\overline{c_{r,h}}\frac{T^{(1-s)+(1-\overline{r})-1}}{(1-s)+(1-\overline{r})-1}\end{aligned}$$

For P *not self-associate*, that is, for $r_1 \neq r_2$,

$$\begin{aligned}\big\langle \wedge^T E^P_{s,f}, \wedge^T E^P_{r,h}\big\rangle &= \langle f, h\rangle_{M^P} \cdot \frac{T^{s+\overline{r}-1}}{s+\overline{r}-1} \\ &\quad + \langle f^w, h^w\rangle_{M^Q} \cdot c^Q_{s,f}\overline{c^Q_{r,h}}\frac{T^{(1-s)+(1-\overline{r})-1}}{(1-s)+(1-\overline{r})-1}\end{aligned}$$

[3.14.3] Remark: The expression for the not-self-associate case is that of the self-associate case with the middle two terms missing. In the non-self-associate case, the inner products $\langle f^w, h\rangle$ and $\langle f, h^w\rangle$ would not make sense because in that case $wM^Pw^{-1} \neq M^P$, so the two functions live on different groups.

[3.14.4] Corollary: For maximal proper parabolics P in GL_r, on the half-plane $\mathrm{Re}(s) \geq 1/2$ an Eisenstein series $E^P_{s,f}$ with cuspidal data f has no poles if P is *not* self-associate. If P is self-associate, the only possible poles are on the real line, and only occur if $\langle f, f^w\rangle_M \neq 0$. In that case, any pole is simple, and the residue is square-integrable. In particular, taking $f = f_o \otimes f_o$

$$\big\langle \mathrm{Res}_{s_o} E^P_{s,f}, \mathrm{Res}_{s_o} E^P_{s,f}\big\rangle = \langle f_o, f_o\rangle^2_M \cdot \mathrm{Res}_{s_o} c_{s,f}$$

Proof: (of theorem) The self-associate case admits a simpler argument because in this case the truncated Eisenstein series $\wedge^T E^P_{s,f}$ is itself a pseudo-Eisenstein series

$$\wedge^T E^P_{s,f} = \Psi^P(\varphi_{s,f}) - \Psi^P\big(c^T_P E^P_{s,f}\big) = \Psi^P\big(\varphi_{s,f} - c^T_P E^P_{s,f}\big)$$

As in smaller cases [1.11] and [2.10], the pseudo-Eisenstein series, made from the tail of the constant term integrates to zero against the truncated Eisenstein series: fortunately, for cuspidal-data Eisenstein series, this fact need not refer to subtle reduction theory but only needs the $r_1 = r_2$ instance of the following:

[3.14.5] Lemma: With $P = P^{r_1,r_2}$, $Q = P^{r_2,r_2}$, and $w = \begin{pmatrix} 0 & 1_{r_2} \\ 1_{r_1} & 0 \end{pmatrix}$,

$$h^Q(wn \cdot m) \le h^P(m)^{-1} \qquad \text{(for all } n \in N^P_{\mathbb{A}},\ m \in M^P_{\mathbb{A}})$$

Proof: This lemma observes a qualitative aspect of Iwasawa decompositions of special elements. First,

$$h^Q(wnm) = h^Q(wmw^{-1} \cdot wm^{-1}nm) = \delta^Q(wmw^{-1}) \cdot h^Q(wm^{-1}nm)$$
$$= \delta^P(m)^{-1} \cdot h^Q(wm^{-1}nm)$$

since w conjugates M^P to M^Q, and inverts the modular function. M^P normalizes N^Q, and thus it suffices to prove $h^Q(wn) \le 1$ for all $n \in N^P_{\mathbb{A}}$. Because h^Q is right $K_{\mathbb{A}}$-invariant, this is equivalent to

$$1 \ge h^Q(wnw^{-1}) = h^Q\begin{pmatrix} 1_{r_2} & 0 \\ x & 1_{r_1} \end{pmatrix} \qquad \text{(with } x\ r_1\text{-by-}r_2)$$

In fact, we claim that the same inequality holds *locally* at every place v with the local analogues h^Q_v and h^P_v. At all places v, right multiplication of wnw^{-1} by K_v produces $q = \begin{pmatrix} a & b \\ 0 & d \end{pmatrix}$ with a, d upper-triangular.

For finite v, we imagine achieving the Iwasawa decomposition in stages, first putting the bottom row into the correct shape, then the next-to-bottom, and so on. To begin, right multiply by $k_1 \in K_v$ to put the *gcd* of the bottom row of x and of 1 into the far right entry of the bottom row of the whole, and to replace the bottom row of x by 0's. This can be done without disturbing the left r_1-by-$(r_1 - 1)$ part of the lower right block of wnw^{-1}. Next, without disturbing the adjusted bottom row, right multiply by $k_2 \in K_v$ to put the *gcd* of the next-to-bottom row of (the new) x and 1 in the next-to-last entry of the next-to-bottom row of the whole, and to replace the next-to-bottom row of (the new) x by 0's. Continuing, every diagonal entry of d will be a *gcd* of some v-adic numbers and 1, so *not* divisible by the local parameter ϖ. Thus, $|\det d|_v \ge 1$. At the same time, the entries of a are among the entries of an element of K_v, so are all v-integral, and $|\det a|_v \le 1$. Thus, $h^Q_v(wn) = h^Q_v(wnw^{-1}) \le 1$.

Somewhat analogously, for archimedean v, the ith diagonal entry of d is *lengths* of vectors with entries including the diagonal 1's in the lower-right block of wnw^{-1}. Thus, all the diagonal entries of d will be at least 1 in size, and certainly $|\det d| \ge 1$. At the same time, the rows of a are fragments of rows of a matrix in K_v, so have length at most 1. The absolute value of the determinant of a is the volume of the parallelopiped spanned by those rows, so is at most 1. ///

Returning to the computation of the inner product in the self-associate case, the integral of $\Psi^P(\varphi_{s,f} - c_P^T E_{s,f}^P)$ against $\Psi^P(c_P^T E_{r,h}^P)$ unwinds to the integral of $\varphi_{s,f} - c_P^T E_{s,f}^P$ against $\Psi^P(c_P^T E_{r,h}^P)$, which is then the integral of $\varphi_{s,f} - c_P^T E_{s,f}^P$ against $c_P(\Psi^P(c_P^T E_{r,h}^P))$. By construction, $(\varphi_{s,f} - c_P^T E_{s,f}^P))(m)$ is supported where $h^P(m) \geq T$, for $m \in M^P$. The proof of [3.11.3] and remarks in [3.11.8] apply as well to pseudo-Eisenstein series, so

$$c_P\big(\Psi^P\big(c_P^T E_{r,h}^P\big)\big)(m) = c_P^T E_{r,h}^P + \int_{N_{\mathbb{A}}^P} \big(c_P^T E_{r,h}^P\big)(wn \cdot m)\, dn$$

for $m \in M^P$. By definition of the truncation, the integrand is 0 unless $h^P(wn \cdot m) \geq T$. The lemma gives $h^P(wn \cdot m) \leq h^P(m)^{-1}$. Thus, for $T > 1$, there is no overlap of supports of $\varphi_{s,f} - c_P^T E_{s,f}^P$ and the second part of the constant term. That is,

$$\begin{aligned}\langle \wedge^T E_{s,f}^P, \wedge^T E_{r,h}^P\rangle &= \int \Psi^P\big(\varphi_{s,f} - c_P^T E_{s,f}^P\big) \cdot \overline{\Psi^P\big(c_P^T E_{r,h}^P\big)} \\ &= \int \big(\varphi_{s,f} - c_P^T E_{s,f}^P\big) \cdot \overline{c_P^T E_{r,h}^P} \\ &= \int \Psi^P\big(\varphi_{s,f} - c_P^T E_{s,f}^P\big) \cdot \overline{E_{r,h}^P} = \langle \wedge^T E_{s,f}^P, E_{r,h}^P\rangle\end{aligned}$$

Unwind the truncated Eisenstein series:

$$\begin{aligned}\langle \wedge^T E_{s,f}^P, E_{r,h}^P\rangle &= \int_{Z^+ G_k\backslash G_{\mathbb{A}}} \Psi^P\big(\varphi_{s,f} - c_P^T E_{s,f}^P\big)\, \overline{E_{r,h}^P} \\ &= \int_{Z^+ \cdot P_k\backslash G_{\mathbb{A}}} \big(\varphi_{s,f} - c_P^T E_{s,f}^P\big)\, \overline{E_{r,h}^P} \\ &= \int_{Z^+ \cdot P_k\backslash G_{\mathbb{A}}} \begin{Bmatrix} -c_{s,f}\varphi_{1-s,f^w} & (\text{for } h^P \geq T) \\ \varphi_{s,f} & (\text{for } h^P < T)\end{Bmatrix} \cdot \overline{E_{r,h}^P}\end{aligned}$$

This is

$$\begin{aligned}&\int_{Z^+ N_{\mathbb{A}} M_k\backslash G_{\mathbb{A}}} \begin{Bmatrix} -c_{s,f}\varphi_{1-s,f^w} & (\text{for } h^P \geq T) \\ \varphi_{s,f} & (\text{for } h^P < T)\end{Bmatrix} \cdot \overline{c_P E_{r,h}^P} \\ &= \int_{Z^+ N_{\mathbb{A}} M_k\backslash G_{\mathbb{A}}} \begin{Bmatrix} -c_{s,f}\varphi_{1-s,f^w} & (h^P \geq T) \\ \varphi_{s,f} & (h^P < T)\end{Bmatrix} \cdot \overline{\big(\varphi_{r,h} + c_{1-r,h^w}\, \varphi_{1-r,h^w}\big)}\end{aligned}$$

Since the integrand is now left $N_{\mathbb{A}}$-invariant, $Z_{\mathbb{A}}$-invariant, and right $K_{\mathbb{A}}$-invariant, this integral may be computed as an integral over the Levi component M^P, using the Iwasawa decomposition, noting that the Haar integral on $G_{\mathbb{A}}$ in such coordinates is

$$\int_{G_{\mathbb{A}}} f(g)\,dg = \int_{N_{\mathbb{A}}}\int_{M_{\mathbb{A}}}\int_{K_{\mathbb{A}}} f(nmk)\,dn\,\frac{dm}{\delta^P(m)}\,dk$$

Then

$$\langle \wedge^T E^P_{s,f},\ \wedge^T E^P_{r,h}\rangle$$

$$= \int\limits_{Z_{\mathbb{A}}M_k\backslash M_{\mathbb{A}}} \left\{ \begin{matrix} -c_{s,f}\varphi_{1-s,f^w} & (h^P \ge T) \\ \varphi_{s,f} & (h^P < T) \end{matrix} \right\} \cdot \overline{\left(\varphi_{r,h} + c_{1-r,h^w}\,\varphi_{1-r,h^w}\right)}\,\frac{dm}{\delta^P(m)}$$

This gives the four terms of the theorem for the self-associate case. We evaluate one in detail, as follows.

Use $Z^+M_k\backslash M_{\mathbb{A}} \approx Z^+\backslash A_P^+ \times M_k\backslash M^1$ and the A_P^+-invariance of f, parametrizing $Z^+\backslash A_P^+$ by

$$t \longrightarrow a_t = \begin{pmatrix} \delta(t)^{1/r_2}\cdot 1_{r_1} & 0 \\ 0 & 1_{r_2} \end{pmatrix} \qquad (\text{for } t > 0)$$

with the diagonal imbedding of the ray $(0, \infty)$ into the archimedean part of the ideles, so that $\delta^P(a_t) = t$. Then, for example,

$$\int\limits_{m\in Z_{\mathbb{A}}M_k\backslash M_{\mathbb{A}}:h^P(m)<T} \varphi_{s,f}\cdot\overline{\varphi_{1-r,h^w}}\,\frac{dm}{\delta^P(m)}$$

$$= \int\limits_{0<t<T,\ m\in Z_{\mathbb{A}}M_k\backslash M^1} t^s\cdot f(a_t m_1)\cdot\overline{t^{1-r}\cdot h^w(a_t m_1)}\,\frac{dt}{t}\,dm_1$$

$$= \int_0^T t^{s+(1-r)}\,\frac{dt}{t}\ \cdot \int\limits_{Z_{\mathbb{A}}M_k\backslash M^1} f(m_1)\cdot\overline{h^w(m_1)}\,dm_1 = \frac{T^{s+(1-r)}}{s+(1-r)}\cdot\langle f, h^w\rangle$$

The other three integrals are evaluated in identical fashion.

In the non-self-associate case, invoking the Lemma in similar fashion,

$$\begin{cases} \langle E^P_{s,f} - \Psi^P(c^T_P E^P_{s,f}),\ \Psi^P(c^T_P E^P_{s,h})\rangle = 0 \\ \langle E^P_{s,f} - \Psi^Q(c^T_Q E^P_{s,f}),\ \Psi^Q(c^T_Q E^P_{s,h})\rangle = 0 \\ \langle \Psi^P(c^T_P E^P_{s,f}),\ \Psi^Q(c^T_Q E^P_{s,h})\rangle = 0 \end{cases}$$

so the inner product of the truncated Eisenstein series is

$$\langle \wedge^T E^P_{s,f},\ \wedge^T E^P_{r,h}\rangle = \langle E^P_{s,f} - \Psi^P(c^T_P E^P_{s,f}), E^P_{r,h}\rangle + \langle \Psi^Q(c^T_Q E^P_{s,f}),\ \Psi^Q(c^T_Q E^P_{r,h})\rangle$$

The pairings unwind. First,

$$\langle E^P_{s,f} - \Psi^P(c^T_P E^P_{s,f}),\, E^P_{r,h}\rangle = \langle \Psi^P(c_P E^P_{s,f} - c^T_P E^P_{s,f}),\, E^P_{r,h}\rangle$$
$$= \int_{Z^+ P_k\backslash G_{\mathbb{A}}} \begin{Bmatrix} 0 & (\text{for } h^P \geq T) \\ \varphi^P_{s,f} & (\text{for } h^P < T) \end{Bmatrix} \cdot \overline{E^P_{r,h}}$$

Because of the left $N_{\mathbb{A}}$-invariance of the first part of the integral, this is

$$\int_{Z^+ N_{\mathbb{A}} M_k\backslash G_{\mathbb{A}}} \begin{Bmatrix} 0 & (\text{for } h^P \geq T) \\ \varphi^P_{s,f} & (\text{for } h^P < T) \end{Bmatrix} \cdot \overline{c_P E^P_{r,h}}$$
$$= \int_{Z^+ N_{\mathbb{A}} M_k\backslash G_{\mathbb{A}}} \begin{Bmatrix} 0 & (\text{for } h^P \geq T) \\ \varphi^P_{s,f} & (\text{for } h^P < T) \end{Bmatrix} \cdot \overline{\varphi^P_{r,h}}$$

Again, the integrand is left $N_{\mathbb{A}}$-invariant and right $K_{\mathbb{A}}$-invariant, so may be computed as an integral over the Levi component using the Iwasawa decomposition: it is

$$\int_{Z^M_{\mathbb{A}} M_k\backslash M_{\mathbb{A}}} \begin{Bmatrix} 0 & (\text{for } h^P \geq T) \\ \varphi^P_{s,f} & (\text{for } h^P < T) \end{Bmatrix} \cdot \overline{\varphi^P_{r,h}}\, \frac{dm}{\delta^P(m)}$$

giving one term in the non-self-associate case. The other pairing unwinds similarly, and becomes

$$\langle \Psi^Q(c^T_Q E^P_{s,f}),\ \Psi^Q(c^T_Q E^P_{r,h})\rangle$$
$$= \int_{Z_{\mathbb{A}} M^Q_k\backslash M^Q_{\mathbb{A}}} \begin{Bmatrix} c^Q_{s,f}\,\varphi^Q_{1-s,f^w} & (\text{for } h^Q \geq T) \\ 0 & (\text{for } h^Q < T) \end{Bmatrix} \cdot \overline{c^Q_{r,h}\,\varphi^Q_{1-r,h^w}}\, \frac{dm}{\delta^Q(m)}$$

giving the second term of the theorem for the not self-associate case. ///

Proof: (*of corollary*) From the theory of the constant term, the only possible poles of the Eisenstein series are at poles of the constant terms, which in this case means a pole of $c_{s,f}$. Invoke the Maaß-Selberg relation with $r = s$ and $h = f$. In the non-self-associate case, this is

$$\langle \wedge^T E^P_{s,f}, \wedge^T E^P_{s,f}\rangle = \langle f, f\rangle \frac{T^{2\sigma-1}}{2\sigma - 1} + \langle f^w, h^w\rangle\, |c_s|^2\, \frac{T^{1-2\sigma}}{1-2\sigma}$$

with $\sigma = \mathrm{Re}(s)$. The non-self-associate case is slightly unlike the simple case of GL_2, in that the inner product of truncated Eisenstein series is missing the two middle terms which for GL_2 made a pole possible. Specifically, in the non-self-associate case, let $s_o = \sigma_o + it_o$ be an alleged pole s_o of c_s of order ℓ in that half-plane. Letting $s = \sigma_o + it$ approach s_o vertically the left-hand side of the relation is asymptotic to a *positive* multiple of $t^{-2\ell}$, while on right-hand side only the second of the two terms blows up at all. In particular, that expression

$$|c_s|^2 \cdot \langle f^w, f^w \rangle \cdot \frac{T^{1-2\sigma}}{1-2\sigma}$$

is asymptotic to a *negative* multiple of $t^{-2\ell}$, since $\sigma = \mathrm{Re}(s) > \frac{1}{2}$. Thus, there is no pole in that half-plane.

Similarly, in the self-associate case, for there to be any pole in the right half-plane, the two middle terms on the right-hand side of the relation must not vanish, or the same contradiction occurs, so $\langle f, f^w \rangle$ must be nonzero, and the alleged pole must be on the real axis and must also be simple: if any of these conditions fail, the middle terms cannot keep up with the negative value of the fourth term. Letting $f = f_o \otimes f_o$ with real-valued f_o, we have $f^w = f$ and $\langle f, f \rangle = \langle f_o, f_o \rangle \cdot \langle f_o, f_o \rangle$. Letting $s = \sigma + it$,

$$\langle \wedge^T E^P_{s,f}, \wedge^T E^P_{s,f} \rangle = \langle f_o, f_o \rangle^2 \frac{T^{2\sigma-1}}{2\sigma-1} + \langle f_o, f_o \rangle^2 \overline{c_{s,f}} \frac{T^{2it}}{2it}$$
$$+ \langle f_o, f_o \rangle^2 c_{s,f} \frac{T^{-2it}}{-2it} + \langle f_o, f_o \rangle^2 |c_{s,f}|^2 \frac{T^{1-2\sigma}}{1-2\sigma}$$

Multiplying through by $t^2 = (it)(-it)$ and taking the limit as $t \to 0$ gives

$$\langle \mathrm{Res}_\sigma \wedge^T E^P_{s,f}, \mathrm{Res}_\sigma \wedge^T E^P_{s,f} \rangle = \langle f_o, f_o \rangle^2 \overline{\mathrm{Res}_\sigma c_{s,f}} \cdot \frac{1}{2} + \langle f_o, f_o \rangle^2 \mathrm{Res}_\sigma c_{s,f} \cdot \frac{1}{2}$$
$$+ \langle f_o, f_o \rangle^2 |\mathrm{Res}_\sigma c_{s,f}|^2 \frac{T^{1-2\sigma}}{1-2\sigma}$$

Letting $T \to +\infty$ causes the last term to go to zero, and yields the indicated finite limit in the self-associate case, since $c_{\bar{s},f} = \overline{c_{s,f}}$ and the supposed pole is on the real axis. ///

[3.14.6] Claim: All residues of Eisenstein series $E^P_{s,f}$ are orthogonal to cuspforms.

Proof: There exists an orthogonal basis for cuspforms consisting of strong-sense cuspforms [3.7.3]. Thus, by the theory of the constant term [8.3], that basis consists of functions *of rapid decay* in Siegel sets. Eisenstein series are *of moderate growth* even when analytically continued, so the integral for $\langle E_{s,f}, F \rangle$ with strong-sense cuspform F is absolutely convergent, and is 0. By properties of vector-valued integrals [14.1] and holomorphic vector-valued functions [15.2], taking residues commutes with the integral, so the integral of any residue against a cuspform is 0, whether or not that residue is square-integrable. ///

3.15 Minimal-Parabolic Decomposition

The harmonic analysis required to express pseudo-Eisenstein series in terms of genuine Eisenstein series reduces to Fourier transform on Euclidean spaces. Here we treat the extreme case $P = P^{\min}$, where no cuspidal data enters. We consider minimal-parabolic pseudo-Eisenstein series Ψ_φ with test function data φ. All the $r!$ functional equations [3.12.1] of the genuine Eisenstein series are needed to obtain the expression of pseudo-Eisenstein series as integrals of Eisenstein series.

Let A^+ be the *archimedean split component* of $M = M^{\min} = M^P$, that is, the image of r copies of $(0, +\infty)$ imbedded diagonally on the archimedean factors $M_\infty = \prod_{v|\infty} M_v$ of $M_\mathbb{A}$. With M^1 the subgroup of $M = M^{\min}_\mathbb{A}$ with diagonal entries m_i all satisfying $|m_i| = 1$, we have $M^{\min}_\mathbb{A} = A^+ \cdot M^1$. Via the exponential $\mathbb{R} \to (0, +\infty)$, we have $A^+ \approx \mathbb{R}^r$ and $Z^+ \backslash A^+ \approx \mathbb{R}^{r-1}$. Spectral decomposition along the Euclidean space $Z^+\backslash A^+$ and the functional equations of the minimal-parabolic Eisenstein series $E_s = E^P_s$ yield the spectral decomposition of minimal-parabolic pseudo-Eisenstein series. Let $\langle , \rangle$ be the invariant pairing on the Lie algebra $\mathfrak{q}$ of $Z^+\backslash A^+$, as in [3.10.2], where it was shown that E_s converges nicely in the cone

$$\{s \in \mathfrak{q} \otimes_\mathbb{R} \mathbb{C} : \langle \alpha, \mathrm{Re}(s) - 2\rho \rangle > 0, \text{ for all simple positive roots } \alpha\}$$

For simplicity, we only consider right $K_\mathbb{A}$-invariant Ψ_φ with trivial central character formed from $\varphi \in \mathcal{D}(Z^+\backslash A^+)$. Further, suppose that the pseudo-Eisenstein series Ψ_φ is orthogonal to all residues of $E_{\rho+s}$ in the cone

$$\{s \in \mathfrak{q} \otimes_\mathbb{R} \mathbb{C} : \langle \alpha, \mathrm{Re}(s) \rangle > 0, \text{ for all simple positive roots } \alpha\}$$

[3.15.1] Theorem: Ψ_φ is an integral of Eisenstein series:

$$\Psi_\varphi = \frac{1}{r!\,(2\pi i)^{r-1}} \int_{i\mathfrak{a}^*} \langle \Psi_\varphi, E_{\rho+s} \rangle \cdot E_{\rho+s}\, ds$$

[3.15.2] Remark: From [3.8.1], the pseudo-Eisenstein series Ψ_φ is compactly supported on $Z^+G_k\backslash G_\mathbb{A}$, and $E_{\rho+s}$ is of moderate growth, so the integral

$$\langle \Psi_\varphi, E_{\rho+s} \rangle = \int_{Z^+G_k\backslash G_\mathbb{A}} \Psi_\varphi \cdot \overline{E_{\rho+s}}$$

implied by $\langle \Psi_\varphi, E_{\rho+s} \rangle$, while *not* an inner product, converges well.

Proof: To decompose right $K_\mathbb{A}$-invariant pseudo-Eisenstein series as integrals of minimal-parabolic Eisenstein series, begin with Fourier transform on the Lie algebra $\mathfrak{q} \approx \mathbb{R}^{r-1}$ of $Z^+\backslash A^+$. Let $\langle , \rangle : \mathfrak{q}^* \times \mathfrak{q} \to \mathbb{R}$ be the $\mathbb{R}$-bilinear pairing of $\mathfrak{q}$ with its $\mathbb{R}$-linear dual $\mathfrak{q}^*$. For $f \in \mathcal{D}(\mathfrak{q})$, the Fourier transform and inversion are

$$\widehat{f}(\xi) = \int_{\mathfrak{q}} e^{-i\langle x,\xi\rangle} f(x)\,dx \qquad\qquad f(x) = \frac{1}{(2\pi)^{r-1}} \int_{\mathfrak{q}^*} e^{i\langle x,\xi\rangle}\,\widehat{f}(\xi)\,d\xi$$

Let $\exp : \mathfrak{q} \to Z^+\backslash A^+$ be the Lie algebra exponential, and $\log : Z^+\backslash A^+ \to \mathfrak{q}$ the inverse. Given $\varphi \in \mathcal{D}(Z^+\backslash A^+)$, let $f = \varphi \circ \exp$ be the corresponding function in $\mathcal{D}(\mathfrak{q})$. The *Mellin transform* $\mathcal{M}\varphi$ of φ is the Fourier transform of f:

$$\mathcal{M}\varphi(i\xi) = \widehat{f}(\xi)$$

Mellin inversion is Fourier inversion in these coordinates:

$$\varphi(\exp x) = f(x) = \frac{1}{(2\pi)^{r-1}} \int_{\mathfrak{q}^*} e^{i\langle\xi,x\rangle}\,\widehat{f}(\xi)\,d\xi$$

$$= \frac{1}{(2\pi)^{r-1}} \int_{\mathfrak{q}^*} e^{i\langle\xi,x\rangle}\,\mathcal{M}\varphi(i\xi)\,d\xi$$

Extend the pairing $\langle,\rangle$ on $\mathfrak{q}^* \times \mathfrak{q}$ to a $\mathbb{C}$-bilinear pairing on the complexification. Use the convention

$$(\exp x)^{i\xi} = e^{i\langle\xi,x\rangle} = e^{\langle i\xi,x\rangle}$$

With $a = \exp x \in Z^+\backslash A^+$, Mellin inversion is

$$\varphi(a) = \frac{1}{(2\pi)^{r-1}} \int_{\mathfrak{q}^*} a^{i\xi}\,\mathcal{M}\varphi(i\xi)\,d\xi = \frac{1}{(2\pi i)^{r-1}} \int_{i\mathfrak{q}^*} a^{s}\,\mathcal{M}\varphi(s)\,ds$$

with $a \in Z^+\backslash A^+$ and $s = i\xi$. With this notation, the Mellin transform itself is

$$\mathcal{M}\varphi(s) = \int_{Z^+\backslash A^+} a^{-s}\,\varphi(a)\,da \qquad (\text{with } s \in i\mathfrak{q}^*)$$

Since φ is a test function, its Fourier-Mellin transform is *entire* on $\mathfrak{q}^* \otimes_{\mathbb{R}} \mathbb{C}$. (It is in the Paley-Wiener space.) Thus, for any $\sigma \in \mathfrak{q}^*$, Mellin inversion can be written

$$\varphi(a) = \frac{1}{(2\pi i)^{r-1}} \int_{\sigma+i\mathfrak{q}^*} a^{s}\,\mathcal{M}\varphi(s)\,ds$$

Via Iwasawa, identify $Z^+N_{\mathbb{A}}M^1\backslash G_{\mathbb{A}}/K_{\mathbb{A}} \approx A^+$, and let $g \to a(g)$ be the function that picks out the A^+ component. For $\sigma \in \mathfrak{q}^*$ suitable for convergence

[3.10.1], the following rearrangement is legitimate:

$$\begin{aligned}\Psi_\varphi(g) &= \sum_{\gamma\in P_k\backslash G_k} \varphi(a(\gamma \circ g)) \\ &= \sum_{\gamma\in P_k\backslash G_k} \frac{1}{(2\pi i)^{r-1}} \int_{\sigma+i\mathfrak{q}^*} \mathcal{M}\varphi(s)\, a(\gamma g)^s\, ds \\ &= \frac{1}{(2\pi i)^{r-1}} \int_{\sigma+i\mathfrak{q}^*} \mathcal{M}\varphi(s) \Big(\sum_{\gamma\in P_k\backslash G_k} a(\gamma g)^s \Big)\, ds \\ &= \frac{1}{(2\pi i)^{r-1}} \int_{\sigma+i\mathfrak{q}^*} \mathcal{M}\varphi(s) \cdot E_s(g)\, ds\end{aligned}$$

Anticipating the invocation of the functional equations, using the rapid vertical decay of $\mathcal{M}\varphi(s)$, we can move the $(r-1)$-fold integration to $\rho + i\mathfrak{q}^*$. *For simplicity, we assume Ψ_φ is orthogonal to any (multi-) residues:*

$$\Psi_\varphi = \frac{1}{(2\pi i)^{r-1}} \int_{i\mathfrak{q}^*} \mathcal{M}\varphi(\rho+s) \cdot E_{\rho+s}\, ds$$

This does express the pseudo-Eisenstein series as a superposition of Eisenstein series. However, the coefficients $\mathcal{M}\varphi$ are not expressed in terms of Ψ_φ itself. This is rectified using the functional equations of $E_{\rho+s}$, as follows.

Since $dn\, dm\, da\, dk/a^{2\rho}$ with $n \in N_\mathbb{A}, m \in M^1, a \in A^+, k \in K_\mathbb{A}$ is a Haar measure on $G_\mathbb{A}$, $da\, dk/a^{2\rho}$ is a right G-invariant measure on $N_\mathbb{A} M^1\backslash G_\mathbb{A}$, and $da/a^{2\rho}$ is the associated measure on $N_A M^1\backslash G_\mathbb{A}/K_\mathbb{A} \approx A^+$ and it descends to $Z^+\backslash A^+$. In the region of convergence, for $\varphi \in \mathcal{D}(Z^+ G_k\backslash G_\mathbb{A})$,

$$\begin{aligned}\int_{Z^+G_k\backslash G_\mathbb{A}} f \cdot E_{\rho+s} &= \int_{Z^+P_k\backslash G_\mathbb{A}} f \cdot a^{\rho+s} \\ &= \int_{Z^+N_\mathbb{A}M_k\backslash G_\mathbb{A}} \int_{N_k\cap N_\mathbb{A}} f(ng)\, a(ng)^{\rho+s}\, dn\, dg \\ &= \int_{Z^+N_\mathbb{A}M_k\backslash G_\mathbb{A}} c_P f \cdot a^{\rho+s} = \int_{Z^+\backslash A^+} c_P f(a) \cdot a^{\rho+s}\, \frac{da}{a^{2\rho}} \\ &= \int_{Z^+\backslash A^+} c_P f(a) \cdot a^{-(\rho-s)}\, da = \mathcal{M} c_P f(\rho - s)\end{aligned}$$

That is, with $f = \Psi_\varphi$,

$$\int_{Z^+G_k\backslash G_\mathbb{A}} \Psi_\varphi \cdot E_{\rho+s} = \mathcal{M} c_P \Psi_\varphi(\rho - s)$$

On the other hand, a similar unwinding of the pseudo-Eisenstein series, and recollection of the constant term $c_P E_{\rho+s}$ from [3.10.3], gives

$$\int_{Z^+ G_k \backslash G_{\mathbb{A}}} \Psi_\varphi \cdot E_{\rho+s} = \int_{Z^+ \backslash A^+} \varphi(a) \cdot c_P E_{\rho+s}(a) \, \frac{da}{a^{2\rho}}$$
$$= \int_{Z^+ \backslash A^+} \varphi(a) \cdot \sum_w c_{w,s} \, a^{\rho + w\cdot s} \, \frac{da}{a^{2\rho}}$$
$$= \sum_w c_{w,s} \int_{A^+} \varphi(a) \, a^{-(\rho - w\cdot s)} \, da = \sum_w c_{w,s} \mathcal{M}\varphi(\rho - w \cdot s)$$

Combining these,

$$\mathcal{M} c_P \Psi_\varphi(\rho - s) = \int_{Z^+ G_k \backslash G_{\mathbb{A}}} \Psi_\varphi \cdot E_{\rho+s} = \sum_w c_{w,s} \, \mathcal{M}\varphi(\rho - w \cdot s)$$

Replacing s by $-s$,

$$\mathcal{M} c_P \Psi_\varphi(\rho + s) = \int_{Z^+ G_k \backslash G_{\mathbb{A}}} \Psi_\varphi \cdot E_{\rho-s} = \sum_w c_{w,-s} \, \mathcal{M}\varphi(\rho + w \cdot s)$$

The Eisenstein series E_s behaves reasonably under complex conjugation: $\overline{E_s} = E_{\bar{s}}$. This is visible in the region of convergence, and persists under analytic continuation, since $\overline{E_{\bar{s}}} = E_s$ is an equality of meromorphic functions. Thus, the previous equality becomes

$$\mathcal{M} c_P \Psi_\varphi(\rho + s) = \int_{Z^+ G_k \backslash G_{\mathbb{A}}} \Psi_\varphi \cdot \overline{E_{\rho+s}} = \sum_w c_{w,-s} \, \mathcal{M}\varphi(\rho + w \cdot s)$$

Behavior under complex conjugation is inherited by the constant term along P:

$$\sum_w \overline{c_{w,s}} \cdot \overline{a^{\rho+w\cdot s}} = c_P \overline{E_{\rho+s}} = c_P E_{\rho+\bar{s}} = \sum_w c_{w,\bar{s}} \cdot a^{\rho + w\cdot \bar{s}}$$

Since $\overline{a^{\rho+w\cdot s}} = a^{\rho+w\cdot\bar{s}}$, this gives $\overline{c_{w,s}} = c_{w,\bar{s}}$. For $\rho + s$ on the unitary hyperplane $\rho + i\mathfrak{a}^*$, conveniently $\bar{s} = -s$, and $c_{w,-s} = \overline{c_{w,s}}$, so

$$\mathcal{M} c_P \Psi_\varphi(\rho + s) = \int_{Z^+ G_k \backslash G_{\mathbb{A}}} \Psi_\varphi \cdot \overline{E_{\rho+s}} = \sum_w \overline{c_{w,s}} \, \mathcal{M}\varphi(\rho + w \cdot s)$$

With these points in hand, average the relation

$$\Psi_\varphi = \frac{1}{(2\pi i)^{r-1}} \int_{i\mathfrak{q}^*} \mathcal{M}\varphi(\rho + s) \cdot E_{\rho+s} \, ds$$

over $w \in W$ to convert it a W-symmetric expression, thereby to obtain an expression in terms of $c_P\Psi_\varphi$, using the functional equations:

$$\Psi_\varphi = \frac{1}{|W|}\sum_w \frac{1}{(2\pi i)^{r-1}}\int_{i\mathfrak{a}^*} \mathcal{M}\varphi(\rho + w\cdot s)\cdot E_{\rho+w\cdot s}\,ds$$

$$= \frac{1}{|W|}\frac{1}{(2\pi i)^{r-1}}\int_{i\mathfrak{a}^*}\left(\sum_w \frac{1}{c_{w,s}}\mathcal{M}\varphi(\rho + w\cdot s)\right)\cdot E_{\rho+s}\,ds$$

Fortunately, from [3.12.6], $|c_{w,s}| = 1$ for $s \in i\mathfrak{a}^*$, so this becomes

$$\Psi_\varphi = \frac{1}{|W|}\frac{1}{(2\pi i)^{r-1}}\int_{i\mathfrak{a}^*}\left(\sum_w \overline{c_{w,s}}\,\mathcal{M}\varphi(\rho + w\cdot s)\right)\cdot E_{\rho+s}\,ds$$

$$= \frac{1}{|W|}\frac{1}{(2\pi i)^{r-1}}\int_{i\mathfrak{a}^*}\langle \Psi_\varphi, E_{\rho+s}\rangle\cdot E_{\rho+s}\,ds$$

The cardinality of the Weyl group W is r-factorial. ///

3.16 Cuspidal-Data Decomposition

Now we treat the opposite extreme, the case of maximal proper parabolics $P = P^{r_1,r_2}$, where *cuspforms* on Levi components unavoidably enter. With $\delta : (0,\infty) \longrightarrow \mathbb{J}$ the diagonal imbedding at archimedean places, the *split component* of P is

$$A_P^+ = \{\begin{pmatrix} \delta(t_1)\cdot 1_{r_1} & 0 \\ 0 & \delta(t_2)\cdot 1_{r_2}\end{pmatrix} : t_1 > 0,\ t_2 > 0\}$$

Let

$$M^1 = \{m = \begin{pmatrix} m_1 & 0 \\ 0 & m_2\end{pmatrix} \in M_{\mathbb{A}}^P : |\det m_1| = 1 = |\det m_2|\}$$

The family of pseudo-Eisenstein series Ψ_φ with fixed *cuspidal* data $f = f_1 \otimes f_2$ on M^1 with test-function data just on the quotient $Z^+\backslash A_P^+ \approx (0,\infty)$ of split components, as in [3.9], constitute the smallest natural vector spaces of functions expressible as integrals of genuine Eisenstein series. In contrast, the pseudo-Eisenstein series with test-function data on $Z_{\mathbb{A}}M_k\backslash M_{\mathbb{A}}$, as in [3.8], are smeared out across these smaller spaces of functions.

For simplicity, we only consider *everywhere spherical* automorphic forms with trivial central character, that is, right $K_{\mathbb{A}}$-invariant and left $Z_{\mathbb{A}}$-invariant functions. Thus, via Iwasawa decomposition, constant terms $c_P f$ are identifiable with functions on the quotient of the Levi component of P, allowing easier description of the cuspidal data, as follows. Let f_1, f_2 be cuspforms on

$GL_{r_1}(\mathbb{A})$ and $GL_{r_2}(\mathbb{A})$, right invariant by the standard maximal compacts everywhere, themselves with trivial central characters. We require that f_1 and f_2 be eigenfunctions for all the spherical Hecke algebras, including the archimedean places. That is, f_1 and f_2 are cuspforms in a *strong* sense, beyond vanishing of constant terms. The theory of the constant term [8.3] shows that cuspforms in this strong sense are *of rapid decay*. Then $f = f_1 \otimes f_2$ is a function on $GL_{r_1}(\mathbb{A}) \times GL_{r_2}(\mathbb{A}) \approx M^P_{\mathbb{A}}$. For a test function η on the ray $(0, \infty)$, define

$$\varphi(znmk) = \varphi_{\eta,f}(znmk) = \eta\left(\frac{|\det m_1|^{r_2}}{|\det m_2|^{r_1}}\right) \cdot f_1(m_1) \cdot f_2(m_2)$$

with $m = \begin{pmatrix} m_1 & 0 \\ 0 & m_2 \end{pmatrix} \in M^P_{\mathbb{A}}$, $z \in Z^+$, $n \in N_{\mathbb{A}}$, $k \in K_{\mathbb{A}}$, with corresponding pseudo-Eisenstein series

$$\Psi^P_\varphi(g) = \Psi^P_{\eta,f} = \sum_{\gamma \in P_k \backslash G_k} \varphi(\gamma \cdot g)$$

Convergence follows from comparison to similarly formed genuine Eisenstein series in their range of absolute convergence in [3.11.2]. The decomposition of such pseudo-Eisenstein series in terms of the analogous genuine Eisenstein series reduces to Fourier inversion on $\mathbb{R}$ together with the functional equation (and analytic continuation) of the genuine Eisenstein series.

Without loss of generality, normalize so that $\int_{M_k \backslash M^1} |f|^2 = 1$, and f is real-valued. In the self-associate case, we can assume that either $f_1 = f_2$ or they are *orthogonal*.

[3.16.1] Theorem:

$$\Psi^P_{\eta,f} = \frac{1}{4\pi i} \int_{\frac{1}{2}-i\infty}^{\frac{1}{2}+i\infty} \langle \Psi^P_{\eta,f}, E^P_{s,f} \rangle \cdot E^P_{s,f} \, ds + \sum_{s_o} \langle \Psi^P_{\eta,f}, \operatorname{Res}_{s_o} E^P_{s,f} \rangle \cdot \operatorname{Res}_{s_o} E^P_{s,f}$$

The residual part is nonzero only for self-associate P and $f^w = f$, in which case there are at most finitely many residues, all in L^2.

[3.16.2] Remark: The argument literally only proves the previous equality *pointwise*. In fact, a natural extension of the argument shows that the integral converges as a vector-valued integral, stemming from corresponding convergence of Euclidean Fourier inversion, one instance of the latter proven in [14.3], and already exploited in [1.12] and [2.11–12] to prove Plancherel theorems for fragments of the spectrum. The current form of the issue is addressed in [3.17].

Proof: Euclidean Fourier-Mellin inversion as in [1.12] expresses $\eta \in \mathcal{D}(0, \infty)$ as

$$\eta(y) = \frac{1}{2\pi} \int_{-\infty}^{\infty} \mathcal{M}\eta(\sigma + it)\, y^{\sigma+it}\, dt \qquad (\text{for any } \sigma \in \mathbb{R})$$

Thus,

$$\begin{aligned} &\eta\left(\frac{|\det m_1|^{r_2}}{|\det m_2|^{r_1}}\right) \cdot f_1(m_1) \cdot f_2(m_2) \\ &= \frac{1}{2\pi} \int_{-\infty}^{\infty} \mathcal{M}\eta(\sigma + it)\; f(m) \left|\frac{(\det m_1)^{r_2}}{(\det m_2)^{r_1}}\right|^{\sigma+it} dt \\ &= \frac{1}{2\pi i} \int_{\sigma-i\infty}^{\sigma+i\infty} \mathcal{M}\eta(s)\; f(m) \left|\frac{(\det m_1)^{r_2}}{(\det m_2)^{r_1}}\right|^{s} ds \end{aligned}$$

As usual, to see how a genuine Eisenstein series arises, let

$$\varphi_{s,f}(znmk) = \left|\frac{(\det m_1)^{r_2}}{(\det m_2)^{r_1}}\right|^{s} \cdot f(m)$$

with $z \in Z_{\mathbb{A}}$, $n \in N^P_{\mathbb{A}}$, $m \in M^P_{\mathbb{A}}$, and $k \in K_{\mathbb{A}}$. Moving σ to $\mathrm{Re}\,(s) > 1$ for convergence of the sum, *wind up* to

$$\begin{aligned} \Psi^P_{\eta,f} &= \frac{1}{2\pi i} \int_{\sigma-i\infty}^{\sigma+i\infty} \mathcal{M}\eta(s) \sum_{\gamma \in P_k \backslash G_k} \varphi_{f,s}(\gamma g)\, ds \\ &= \frac{1}{2\pi i} \int_{\sigma-i\infty}^{\sigma+i\infty} \mathcal{M}\eta(s) \cdot E^P_{s,f}(g)\, ds \end{aligned}$$

with the genuine Eisenstein series

$$E^P_{s,f}(g) = \sum_{\gamma \in P_k \backslash G_k} \varphi_{s,f}(\gamma g) \qquad (\text{for } \mathrm{Re}\,(s) > 1)$$

Expression of $\Psi^P_{\eta,f}$ in terms of η should be replaced by an intrinsic expression in terms of $\Psi^P_{\eta,f}$. The non-self-associate and self-associate cases are somewhat different from each other, due to the different behavior of the constant terms of Eisenstein series in those two cases. We treat the non-self-associate case first.

In the non-self-associate case, from [3.14.4], $E^P_{f,s}$ has *no poles* in $\mathrm{Re}\, s \geq \frac{1}{2}$, and has reasonable vertical behavior. Meanwhile, being essentially the Fourier transform of a compactly supported smooth function, $\mathcal{M}\eta(s)$ is in the Paley-Wiener space, so is entire with rapid decay on vertical lines. Thus, we can shift the vertical integral to the line $\sigma = \frac{1}{2}$ without picking up any residues:

$$\Psi^P_{\eta,f} = \frac{1}{2\pi i} \int_{\frac{1}{2}-i\infty}^{\frac{1}{2}+i\infty} \mathcal{M}\eta(s) \cdot E^P_{s,f}(g)\, ds$$

To obtain an intrinsic expression for $\mathcal{M}\eta(s)$: unwind the pseudo-Eisenstein series: using an Iwasawa decomposition, spherical-ness, and trivial central character, and the fact that $c_P E^P_{s,f}$ is just $\varphi_{s,f}$ in the non-self-associate case: with $\delta^P(m) = |\det m_1|^{r_1}/|\det m_2|^{r_1}$ the modular function of P,

$$\int_{Z^+G_k\backslash G_{\mathbb{A}}} \Psi^P_{f,\eta} \cdot \overline{E^P_{s,f}}$$

$$= \int_{Z^+M_k\backslash M_{\mathbb{A}}} f(m)\,\eta\left(\frac{|\det m_1|^{r_2}}{|\det m_2|^{r_1}}\right) \cdot \overline{c_P E^P_{s,f}(m)}\ \frac{dm}{\delta^P(m)}$$

$$= \int_{Z^+M_k\backslash M_{\mathbb{A}}} f(m)\,\eta\left(\frac{|\det m_1|^{r_2}}{|\det m_2|^{r_1}}\right) \cdot \overline{f}(m) \left|\frac{(\det m_1)^{r_2}}{(\det m_2)^{r_1}}\right|^{1-s} \frac{dm}{\delta^P(m)}$$

From

$$Z^+M_k\backslash M_{\mathbb{A}} \approx Z^+\backslash A^+_P \times M_k\backslash M^1$$

the pairing of pseudo-Eisenstein series against Eisenstein series becomes

$$\int_{Z^+G_k\backslash G_{\mathbb{A}}} \Psi^P_{\eta,f} \cdot \overline{E^P_{s,f}} = \int_{Z^+\backslash A^+_P\times M_k\backslash M^1} |f(m)| \cdot \eta(\delta(a)) \cdot \delta(a)^{-s}\,dm\,da$$

$$= \int_{M_k\backslash M^1} |f|^2 \cdot \int_0^\infty \eta(r)\cdot r^s\ \frac{1}{r}\,\frac{dr}{r} = \langle f, f\rangle \cdot \mathcal{M}\eta(s)$$

yielding an intrinsic expression for $\mathcal{M}\eta$,

$$\mathcal{M}\eta(s) = \frac{1}{\langle f, f\rangle} \int_{Z^+G_k\backslash G_{\mathbb{A}}} \Psi^P_{\eta,f} \cdot \overline{E^P_{s,f}}$$

This computation incidentally demonstrates the absolute convergence of the integral. Unlike GL_2 and self-associate cases, the previous computation of the integral of a pseudo-Eisenstein series against an Eisenstein series *already* gives an intrinsic expression for the coefficient $\mathcal{M}\eta(s)$ in the spectral decomposition, with no immediate reason to use functional equations to symmetrize the integral. With the normalization $\langle f, f\rangle = 1$, the *spectral decomposition* is

$$\Psi^P_{\eta,f} = \frac{1}{2\pi i}\int_{\frac{1}{2}-i\infty}^{\frac{1}{2}+i\infty} \mathcal{M}\eta(s)\cdot E^P_{s,f}\,ds$$

$$= \frac{1}{2\pi i}\int_{\frac{1}{2}-i\infty}^{\frac{1}{2}+i\infty} \langle \Psi^P_{\eta,f}, E^P_{s,f}\rangle \cdot E^P_{s,f}\,ds$$

where, as usual, the pairing $\langle , \rangle$ cannot be the L^2 pairing, because $E^P_{s,f}$ is not in L^2, but the implied integral converges absolutely, as the foregoing unwinding argument demonstrates.

In the self-associate case, the subcase where f and f^w are orthogonal is similar to the non-self-associate case, as follows. First,

$$\int_{Z^+G_k\backslash G_{\mathbb{A}}} \Psi^P_{f,\eta} \cdot \overline{E^P_{s,f}}$$

$$= \int_{Z^+M_k\backslash M_{\mathbb{A}}} f(m)\,\eta(\delta(m)) \cdot \overline{c_P E^P_{s,f}(m)}\, \frac{dm}{\delta^P(m)}$$

$$= \int_{Z^+M_k\backslash M_{\mathbb{A}}} f(m)\,\eta(\delta(m)) \cdot \overline{\Big(f(m)\cdot\delta(m)^s + c^P_{s,f} f^w(m)\,\delta(m)^{1-s}\Big)}\, \frac{dm}{\delta^P(m)}$$

Since f and f^w are orthogonal, the second summand in the constant term of the Eisenstein series integrates to 0 against the unwound pseudo-Eisenstein series. From [3.14.4], the Eisenstein series has no poles in $\mathrm{Re}(s) \ge \frac{1}{2}$, so we can move the contour to that line without picking up any residues. Again from

$$Z^+M_k\backslash M_{\mathbb{A}} \;\approx\; Z^+\backslash A^+_P \times M_k\backslash M^1$$

the pairing of pseudo-Eisenstein series against Eisenstein series with $\mathrm{Re}(s) = \frac{1}{2}$ becomes

$$\int_{Z^+G_k\backslash G_{\mathbb{A}}} \Psi^P_{\eta,f} \cdot \overline{E^P_{s,f}} = \int_{M_k\backslash M_{\mathbb{A}}} |f|^2 \cdot \int_0^\infty \eta(r)\cdot r^{1-s}\,\frac{1}{r}\,\frac{dr}{r}$$

$$= \langle f, f\rangle \cdot \mathcal{M}\eta(s)$$

yielding the same decomposition

$$\Psi^P_{\eta,f} = \frac{1}{2\pi i}\int_{\frac{1}{2}-i\infty}^{\frac{1}{2}+i\infty} \mathcal{M}\eta(s)\cdot E^P_{s,f}\, ds$$

$$= \frac{1}{2\pi i}\int_{\frac{1}{2}-i\infty}^{\frac{1}{2}+i\infty} \langle \Psi^P_{\eta,f}, E^P_{s,f}\rangle \cdot E^P_{s,f}\, ds$$

in the self-associate case with f orthogonal to f^w.

In the subcase of the self-associate case where $f = f^w$, moving the contour from $\mathrm{Re}(s) > 1$ to $\mathrm{Re}(s) = \frac{1}{2}$ may pick up finitely many residues of $E^P_{s,f}$, which by [3.14.4] are in L^2. Thus,

$$\Psi^P_{\eta,f} - (\text{residual part}) = \frac{1}{2\pi i}\int_{\frac{1}{2}-i\infty}^{\frac{1}{2}+i\infty} \mathcal{M}\eta(s)\cdot E^P_{s,f}\, ds$$

As suggested by both the GL_2 case [2.11] and the minimal-parabolic GL_n case [3.15], average the original expression of $\Psi^P_{\eta,f}$ with its image under replacing s by $1-s$, to symmetrize:

$$\Psi^P_{\eta,f} - (\text{residual part})$$

$$= \frac{1}{4\pi i} \int_{\frac{1}{2}-i\infty}^{\frac{1}{2}+i\infty} \left(\mathcal{M}\eta(s) \cdot E^P_{s,f} + \mathcal{M}\eta(1-s) \cdot E^P_{1-s,f}\right) ds$$

$$= \frac{1}{4\pi i} \int_{\frac{1}{2}-i\infty}^{\frac{1}{2}+i\infty} \left(\mathcal{M}\eta(s) + \mathcal{M}\eta(1-s)\frac{1}{c_{s,f}}\right) \cdot E^P_{s,f}\, ds$$

Because $f^w = f$, applying the functional equation twice gives $c_{s,f} \cdot c_{1-s,f} = 1$, and $|c_{s,f}| = 1$ on $\mathrm{Re}(s) = \frac{1}{2}$. Unwind and use Iwasawa:

$$\langle \Psi^P_{\eta,f}, E^P_{s,f} \rangle = \int_{Z^+ N_{\mathbb{A}} M_k \backslash G_{\mathbb{A}}} \eta(\delta(m)) \cdot \overline{c_P E^P_{s,f}(m)}\, \frac{dm}{\delta^P(m)}$$

$$= \int_{Z^+ \backslash A^+_P \times M_k \backslash M^1} \eta(\delta(a)) \cdot f(m)$$

$$\cdot \overline{\left(f(m) \cdot \delta(a)^s + c_{s,f} f(m) \cdot \delta(a)^{1-s}\right)} \cdot \delta(a)^{-1} dm\, da$$

$$= \int_0^\infty \eta(t) \cdot \left(t^{-s} + \overline{c_{s,f}} t^s\right) \frac{dt}{t} = \int_0^\infty \eta(t) \cdot \left(t^{-s} + \frac{1}{c_{s,f}} t^s\right) \frac{dt}{t}$$

$$= \mathcal{M}\eta(s) + \mathcal{M}\eta(1-s)\frac{1}{c_{s,f}}$$

so once again

$$\Psi^P_{\eta,f} - (\text{residual part}) = \frac{1}{4\pi i} \int_{\frac{1}{2}-i\infty}^{\frac{1}{2}+i\infty} \langle \Psi^P_{\eta,f}, E^P_{s,f} \rangle \cdot E^P_{s,f}\, ds$$

To address the finitely many residues of $E^P_{s,f}$ in the self-associate situation with $f^w = f$, recall that the poles s_o of $E_{s,f}$ in $\mathrm{Re}(s) > \frac{1}{2}$ are poles of $c_{s,f}$ of the same order. Since $c_{s,f} \cdot c_{1-s,f} = 1$, necessarily $c_{1-s,f}$ has a *zero* at s_o.

Thus, from

$$\mathcal{M}c_P\Psi^P_{\eta,f}(s) = \mathcal{M}\eta(s)\, f(m) + c_{1-s,f}\mathcal{M}\eta(1-s)\, f(m)$$

at a pole s_o of E_s

$$\begin{aligned}\mathcal{M}c_P\Psi_{\eta,f}(s_o) &= \big(\mathcal{M}\eta(s_o) + c_{1-s_o,f}\mathcal{M}\eta(1-s_o)\big)\cdot f(m)\\ &= (\mathcal{M}\eta(s_o) + 0\cdot\mathcal{M}\eta(1-s_o))\cdot f(m) = \mathcal{M}\eta(s_o)\, f(m)\end{aligned}$$

That is, the value $\mathcal{M}c_P\Psi_{\eta,f}$ at s_o is just the value of $\mathcal{M}\eta(s_o)\, f(m)$, so the coefficients appearing in the decomposition of $\Psi_{\eta,f}$ *are intrinsic*. Thus, the preceding decomposition has the form as in the statement of the theorem. ///

3.17 Plancherel for Pseudo-Eisenstein Series

For a fixed cuspform $f = f_1 \otimes f_2$ on the Levi component M^P of a maximal proper parabolic P, we show that the map from pseudo-Eisenstein series $\Psi^P_{\eta,f}$ attached to test functions η on the ray $Z^+\backslash A^+_P$ and that fixed cuspidal data f to decomposition coefficients $\langle\Psi^P_{\eta,f}, E^P_{s,f}\rangle$ against genuine Eisenstein series *with the same cuspidal data* is an *isometry* to the image. This gives a Plancherel theorem for the fragment of L^2 given by (the closure of) the span of $\{\Psi^P_{\eta,f}$: test function $\eta\}$. Similarly, we show that the map from minimal-parabolic pseudo-Eisenstein series to their decomposition coefficients against minimal-parabolic Eisenstein series is an *isometry to its image*, giving Plancherel for this part of the spectrum. In both cases, retain the simplifying hypotheses of the last sections: trivial central character and right $K_{\mathbb{A}}$-invariance.

Often, legitimization of natural procedures requires a linear map to commute with an integral. The best assurance of commuting is when the integral converges as a *vector-valued* integral, as in [14.1]. Thus, we will see that we want the spectral decomposition integrals to converge not merely *pointwise*, but as *vector-valued integrals* in spaces of functions on $Z^+G_k\backslash G_{\mathbb{A}}$ on which integration against cuspidal-data pseudo-Eisenstein series are continuous functionals. In earlier examples [1.12] and [2.11–12], the pseudo-Eisenstein series were *compactly supported*, so convergence of spectral integrals in $C^\infty(Z^+G_k\backslash G_{\mathbb{A}})$, without growth or support constraints, was sufficient, and this was inherited from the analogous argument [14.3] for Fourier series on $\mathbb{R}$. The present discussion requires somewhat more, since cuspidal-data pseudo-Eisenstein series do not have compact support but are of rapid decay for strong-sense cuspidal data, as in [3.9] and [3.16].

Fix a strong-sense cuspform $f = f_1 \otimes f_2$ on M^P with P maximal proper. The Plancherel theorem for associated pseudo-Eisenstein series comes from

[3.17.1] Theorem: With $|f|^2 = 1$, letting s_o run over poles of $E^P_{s,f}$ in $\mathrm{Re}(s) \geq \frac{1}{2}$, with η, θ test functions on $Z^+\backslash A_P^+) \approx (0, \infty)$,

$$\langle \Psi^P_{\eta,f}, \Psi^P_{\theta,f} \rangle = \frac{1}{2\pi i} \int_{\frac{1}{2}-i\infty}^{\frac{1}{2}+i\infty} \langle \Psi^P_{\eta,f}, E^P_{s,f} \rangle \cdot \overline{\langle \Psi^P_{\theta,f}, E^P_{s,f} \rangle}\, ds$$
$$+ \sum_{s_o} \langle \Psi^P_{\eta,f}, \mathrm{Res}_{s_o} E^P_{s,f} \rangle \cdot \overline{\langle \Psi^P_{\eta,f}, \mathrm{Res}_{s_o} E^P_{s,f} \rangle}$$

[3.17.2] Remark: There are no residues in $\mathrm{Re}(s) \geq \frac{1}{2}$ unless P is self-associate and $\langle f, f^w \rangle \neq 0$, by [3.14.4].

Proof: Grant that the decomposition integral for $\Psi^P_{\eta,f}$ converges well enough to pass integration against $\Psi^P_{\theta,f}$ inside it: a vector-valued convergence certainly is sufficient, and more pedestrian arguments are also possible. Then

$$\langle \Psi^P_{\eta,f}, \Psi^P_{\theta,f} \rangle = \Big\langle \frac{1}{2\pi i} \int_{\frac{1}{2}-i\infty}^{\frac{1}{2}+i\infty} \langle \Psi^P_{\eta,f}, E^P_{s,f} \rangle \cdot E^P_{f,s}\, ds$$
$$+ \sum_{s_o} \langle \Psi^P_{\eta,f}, \mathrm{Res}_{s_o} E^P_{s,f} \rangle \cdot \mathrm{Res}_{s_o} E^P_{s,f},\ \Psi^P_{\theta,f} \Big\rangle$$
$$= \frac{1}{2\pi i} \int_{\frac{1}{2}-i\infty}^{\frac{1}{2}+i\infty} \langle \Psi^P_{\eta,f}, E^P_{s,f} \rangle \cdot \overline{\langle \Psi^P_{\theta,f}, E^P_{s,f} \rangle}\, ds$$
$$+ \sum_{s_o} \langle \Psi^P_{\eta,f}, \mathrm{Res}_{s_o} E^P_{s,f} \rangle \cdot \overline{\langle \Psi^P_{\eta,f}, \mathrm{Res}_{s_o} E^P_{s,f} \rangle}$$

as claimed. ///

This spectral decomposition facilitates demonstration of the orthogonality of pseudo-Eisenstein series remarked upon in [3.9.1], beyond the non-self-associate case from [3.11.5]:

[3.17.3] Corollary: Pseudo-Eisenstein series $\Psi^P_{\eta,f}$ and $\Psi^Q_{\theta,f'}$ for maximal proper parabolics P, Q, with test functions η, θ and cuspidal data f, f', are mostly *mutually orthogonal*: they are orthogonal if P, Q are not associate, or if $P = Q$ but $\langle f, f' \rangle = 0$, or if $M^P = wM^Q w^{-1}$ but $\langle f^w, f' \rangle = 0$.

Proof: In all cases, the first part of the proof of the decomposition of pseudo-Eisenstein series in terms of genuine Eisenstein series yields a vector-valued integral

$$\Psi^P_{\eta,f} = \frac{1}{2\pi i} \int_{\frac{1}{2}-i\infty}^{\frac{1}{2}+i\infty} \mathcal{M}\eta(s) \cdot E^P_{s,f}\, ds + \sum_{s_o} \mathcal{M}\eta(s_o) \cdot \mathrm{Res}_{s_o} E^P_{s,f}$$

converging in the $C^\infty(Z^+G_k\backslash G_\mathbb{A})$ topology, from the discussions just above. By [14.1], the inner product functional against $\Psi^Q_{\theta,f'}$ can pass through the integral, and through the operation of residue as well [15.2]. Thus,

$$\langle \Psi^P_{\eta,f}, \Psi^Q_{\theta,f'}\rangle = \frac{1}{2\pi i}\int_{\frac{1}{2}-i\infty}^{\frac{1}{2}+i\infty} \mathcal{M}\varphi(s)\cdot \langle E^P_{s,f}, \Psi^Q_{\theta,f'}\rangle\, ds$$
$$+ \sum_{s_o} \mathcal{M}\varphi(s_o)\cdot \mathrm{Res}_{s_o}\langle E^P_{s,f}, \Psi^Q_{\theta,f'}\rangle$$

The integrals of $\Psi^Q_{\theta,f'}$ against genuine Eisenstein series *unwind* to integrals of $\varphi^Q_{\theta,f'}$ against the Q-constant terms of the Eisenstein series, computed in [3.11.9]. If Q is not associate to P, this is 0. If $Q = P$, or if P is not self-associate and Q is the other associate to P, the integral includes an inner integral of f against f' or f^w against f', as in the proof of Maaß-Selberg relations [3.14.2], and these are 0 by assumption. ///

These formulas suggest the form of a Plancherel theorem for this fragment of L^2. For P self-associate and $f^w = f$, let $s_1, \ldots, s_n$ be the poles of $E^P_{s,f}$ in $\mathrm{Re}(s) \geq \frac{1}{2}$, and let

$$\mathcal{F}^P_f : \Psi^P_{\eta,f} \longrightarrow \left(t \to \langle \Psi^P_{\eta,f}, E^P_{\frac{1}{2}+it,f}\rangle\right) \oplus \left(\ldots, \langle \Psi^P_{\eta,f}, \mathrm{Res}_{s_o} E^P_{s,f}\rangle, \ldots\right)$$

with the latter in $L^2(\frac{1}{2}+i\mathbb{R}) \oplus \mathbb{C}^n$, be the spectral coefficient map. It is necessary to identify the *image* of $\mathcal{F}^P_f$. For P self-associate and $f^w = f$, let

$$V = \left\{F \in L^2\left(\tfrac{1}{2}+i\mathbb{R}\right) : F(1-s) = c_{s,f}F(s)\right\} \oplus \mathbb{C}^n$$

for P not self-associate or $\langle f, f^w\rangle = 0$, let

$$V = L^2\left(\tfrac{1}{2}+i\mathbb{R}\right)$$

[3.17.4] Claim: The spectral coefficient map $\mathcal{F}^P_f$ is an isometry to its image in V, and that image is *dense* in V.

Proof: The fact that it is an isometry to its image is [3.17.1]. The map $\eta \to \mathcal{M}\eta$ is essentially Fourier transform, so sends the dense subset of test functions inside the Schwartz space to the dense subset of Paley-Wiener functions inside the Schwartz space. The Schwartz space is dense in L^2, so the functions $\mathcal{M}\eta$ are dense in $L^2(\frac{1}{2}+i\mathbb{R})$.

For P not self-associate, this is all we need, since there are no residues, and no folding-up of the spectral integral, since the functional equation of the Eisenstein series does not relate it to itself. The case of P self-associate but $\langle f, f^w\rangle = 0$ is similar.

Now consider P self-associate and $f^w = f$. The residues of $E_{s,f}$ in $\mathrm{Re}(s) \geq \frac{1}{2}$ are orthogonal to cuspforms, by [3.14.6]. These residues have Q-constant terms 0 unless Q is associate to P, by [3.11.3], since taking residues commutes with evaluation of constant terms, from generalities about vector-valued integrals [14.1] and holomorphic or meromorphic vector-valued functions [15.2]. With Q associate to P, for cuspidal data f' with $\langle f', f\rangle = 0 = \langle f', f^w\rangle$, we have orthogonality by [3.17.3]. Thus, by a process of elimination, these residues must in the closure of the space of pseudo-cuspforms $\Psi^P_{\eta,f}$ with test-function data η.

The decomposition integral gets folded up via the functional equation of $E_{s,f}$ to obtain coefficients $\langle \Psi_{\eta,f}, E_{s,f}\rangle = \mathcal{M}\eta(s) + c^{-1}_{s,f}\mathcal{M}\eta(1-s)$. The map $F \to F(s) + c_{1-s,f}F(1-s)$ is a continuous map of L^2 to the subspace in V, and is continuous because $|c_{s,f}|$ is constant on $\mathrm{Re}(s) = \frac{1}{2}$. Since the residues are in the closure, the integral part of the spectral decomposition is in the closure. This proves that the spectral map has dense image. ///

[3.17.5] Remark: Continuing in this vein, the L^2 closure of the image of $\Psi^P_{\eta,f}$ for fixed P, for *all* cuspforms f, and for all test functions η, is the collection of functions orthogonal to cuspforms on G and with all constant terms vanishing *except* c_P and c_Q.

In the opposite case, with minimal-parabolic $P = P^{\min}$, in [3.10.1] the Eisenstein series E^P_s was shown convergent for $s \in 2\rho + C$, with cone

$$C = \{s \in \mathfrak{q} \otimes_{\mathbb{R}} \mathbb{C} : \langle \alpha, \mathrm{Re}(s)\rangle > 0, \text{ for all simple positive roots } \alpha\}$$

The argument for the spectral decomposition of pseudo-Eisenstein series for $P = P^{\min}$ in [3.15] showed that only (multi-) residues of E^P_s for in $s \in \rho + C$ are relevant to the spectral decomposition. Let $\mathfrak{q}$ be the Lie algebra of $Z^+\backslash A^+_P$, as in [3.15].

[3.17.6] Theorem: Let η, θ be test functions on $Z^+\backslash A^+_P \approx (0, \infty)^{r-1}$ such that Ψ^P_φ is orthogonal to all residues of $E^P_{\rho+s}$ with $s \in C$. Then

$$\langle \Psi^P_\eta, \Psi^P_\theta\rangle = \frac{1}{r!}\frac{1}{(2\pi i)^{r-1}}\int_{i\mathfrak{q}^*} \langle \Psi_\eta, E_{\rho+s}\rangle \cdot \overline{\langle \Psi_\theta, E_{\rho+s}\rangle}\, ds$$

Proof: In this example, since the pseudo-Eisenstein series have only test-function data, they are compactly supported on $Z^+G_k\backslash G_{\mathbb{A}}$, by [3.8.1]. Thus, integration against Ψ^P_θ in the following is justified by observing that the spectral expansion of Ψ^P_η converges in the $C^\infty(Z^+G_k\backslash G_{\mathbb{A}})$ topology. The latter follows from the corresponding assertion for Fourier transform on $\mathbb{R}$ and $\mathbb{R}^n$, proven in [14.3], simply augmenting the argument for the spectral decomposition [3.15] to retain that aspect, rather than mere pointwise convergence. More pedestrian arguments are possible, as usual. Granting that, compute directly:

$$\langle \Psi_\eta^P, \Psi_\theta^P \rangle = \left\langle \frac{1}{r!} \frac{1}{(2\pi i)^{r-1}} \int_{iq^*} \langle \Psi_\eta^P, E_{\rho+s}^P \rangle \cdot E_{\rho+s}^P \, ds, \ \Psi_\theta^P \right\rangle$$

$$= \frac{1}{r!} \frac{1}{(2\pi i)^{r-1}} \int_{iq^*} \langle \Psi_\varphi^P, E_{\rho+s}^P \rangle \cdot \overline{\langle \Psi_\theta^P, E_{\rho+s}^P \rangle} \, ds$$

as claimed. ///

Toward Plancherel: *Any* $F \in L^2(iq^*)$ satisfying $F(w \cdot s) = c_{w,s} \cdot F(s)$ for all $w \in W$ is in the closure of the image of the map $\Psi_\eta^P \to \langle \Psi_\eta^P, E_{\rho+s} \rangle$ ranging over all test functions η on iq^*. Indeed, the map $\eta \to \mathcal{M}\eta$ is essentially Fourier transform, and as in other examples maps test functions to the Paley-Wiener space, dense in L^2. The averaging map

$$F \longrightarrow \sum_{w \in W} \frac{1}{c_{w,s}} F(w \cdot s)$$

surjects $L^2(iq^*)$ to its subspace where $F(w \cdot s) = c_{w,s} \cdot F(s)$ for all $w \in W$, since $|c_{w,s}| = 1$ on iq^*, by [3.12.6]. However, we have not identified the (multi-) residues of E_s that appear when moving contours and cannot immediately distinguish the subspace of pseudo-Eisenstein series orthogonal to these residues.

Further, we would need to be able to argue that these multi-residues are entirely inside the closure of the images of the pseudo-Eisenstein series. For the latter, it seems necessary to invoke the *complete* spectral decomposition of $L^2(\Gamma \backslash G/K)$, that *cuspforms* and *cuspidal data* Eisenstein series attached to *non-minimal* parabolics, and their L^2 residues, as well as the minimal-parabolic pseudo-Eisenstein series, span $L^2(\Gamma \backslash G/K)$. Only then is the orthogonality of integrals of minimal-parabolic Eisenstein series to all the other spectral components clear.

Thus, although we did prove that the map from the space of pseudo-Eisenstein series to integrals of Eisenstein series is an isometry to its image, we did not quite identify that image.

3.18 Automorphic Spectral Expansions

We would like to express $L^2(Z^+ G_k \backslash G_\mathbb{A} / K_\mathbb{A})$ as the closure of subspaces consisting of eigenfunctions for invariant differential operators and for spherical Hecke operators. Analogous decomposition of $L^2(Z^+ G_k \backslash G_\mathbb{A})$ needs more general integral operators.

First, we can decompose by *central characters* into pieces $L^2(Z^+G_k\backslash G_{\mathbb{A}}, \omega)$, for general reasons [3.6].

Then the general pattern is that there are *cuspforms*, and we are left to sift through their orthogonal complement. That orthogonal complement is spanned by pseudo-Eisenstein series with cuspidal data attached to the various parabolics. The cuspidal-data pseudo-Eisenstein series themselves are *not* eigenfunctions for invariant differential operators or Hecke operators but are essentially integrals of *genuine* Eisenstein series, and the latter *are* eigenfunctions [3.11.6] and [3.11.11]. The functional equations of genuine Eisenstein series show that parabolics P, Q that are *associate*, in the sense that their Levi components are conjugate, produce the same functions on the group. Thus, a rough indexing of parts of L^2 is by associate-class of parabolics.

In general, *residues* of cuspidal-data Eisenstein series also enter the expression of pseudo-Eisenstein series. The relevant residues are square-integrable and inherit eigenfunction properties from the genuine Eisenstein series. For GL_2, these residues are relatively uninteresting, as in [2.B]. For GL_3, the $P^{2,1}$ and $P^{1,2}$ parabolics' Eisenstein series have no relevant residues [3.14.4], so the only residues are those from $P^{1,1,1} = P^{\min}$, *which turn out to be constants*. Granting the latter, fact, for example, with trivial central character, over groundfield $\mathbb{Q}$, since there are no unramified Hecke characters, functions Φ in $L^2(Z_{\mathbb{A}}GL_3(\mathbb{Q})\backslash GL_3(\mathbb{A})/GL_3(\mathbb{A}))$ have L^2 decompositions

$$\Phi = \sum_{GL_3 \text{ cfm } F} \langle \Phi, F\rangle \cdot F + \sum_{GL_2 \text{ cfm } f} \frac{1}{2\pi i} \int_{\frac{1}{2}-i\infty}^{\frac{1}{2}+i\infty} \langle \Phi, E^{2,1}_{s,f}\rangle \cdot E^{2,1}_{s,f}\, ds$$

$$+ \frac{1}{3!\cdot 2\pi i} \int_{i\mathfrak{a}^*} \langle \Phi, E^{\min}_{\rho+s}\rangle \cdot E^{\min}_{\rho+s}\, ds + \frac{\langle \Phi, 1\rangle \cdot 1}{\langle 1, 1\rangle}$$

where the first sum is over an orthonormal basis of spherical cuspforms for $GL_3(\mathbb{Z})$ with trivial central character, and the second sum is over an orthonormal basis for spherical cuspforms for $GL_2(\mathbb{Z})$ with trivial central character. The right-hand side is only promised to converge in an L^2 sense, and the explicit and implicit integrals involving Eisenstein series are merely isometric extensions of the corresponding literal integrals.

Nontrivial residual spectrum for GL_4: This is the smallest GL_n in which some Eisenstein series $E^{2,2}_{s,f}$ with real-valued $f = f^w$ have non-constant residues.[10]

[10] Nonconstant residues for $P^{n,n}$ are called *Speh forms*, since for $GL_4(\mathbb{R})$ the relevant unitary representations appear in [Speh 1981/1982]. The general pattern for residual spectrum for GL_n was conjectured in [Jacquet 1982/1983] and proven in [Moeglin-Waldspurger 1989].

The Maaß-Selberg relations do not *exclude* the possibility of a pole of such an Eisenstein series. A computation of the constant term of that Eisenstein series shows that it is a ratio of values of the Rankin-Selberg L-function attached to $f \times \overline{f}$, which definitely has a pole in $\mathrm{Re}(s) > \frac{1}{2}$, yielding a square-integrable residue. Granting this, for example, over $\mathbb{Q}$, spherical Φ in $L^2(Z_{\mathbb{A}}GL_4(\mathbb{Q})\backslash GL_4(\mathbb{A})$, that is, in $L^2(Z_{\mathbb{A}}GL_4(\mathbb{Q})\backslash GL_4(\mathbb{A})/K_{\mathbb{A}})$, have L^2 decompositions described as follows. For a modicum of coherence, let Ξ_n be a fixed orthonormal basis for spherical cuspforms with trivial central character for $GL_n(\mathbb{Z})$, consisting of spherical Hecke eigenfunctions, etc. We grant that there is a unique relevant residue F_f of $E^{2,2}_{s,f\otimes\overline{f}}$ for cuspforms f on GL_2. Then

$$\Phi = \sum_{f\in\Xi_4} \langle \Phi, F\rangle \cdot F + \sum_{f\in\Xi_3} \frac{1}{2\pi i} \int_{\frac{1}{2}-i\infty}^{\frac{1}{2}+i\infty} \langle \Phi, E^{3,1}_{s,f}\rangle \cdot E^{3,1}_{s,f}\, ds$$

$$+ \sum_{f_1,f_2\in\Xi_2,\ f_1\neq\overline{f}_2} \frac{1}{2\pi i} \int_{\frac{1}{2}-i\infty}^{\frac{1}{2}+i\infty} \langle \Phi, E^{2,2}_{s,f_1\otimes f_2}\rangle \cdot E^{2,2}_{s,f_1\otimes f_2}\, ds$$

$$+ \sum_{f\in\Xi_2} \frac{1}{4\pi i} \int_{\frac{1}{2}-i\infty}^{\frac{1}{2}+i\infty} \langle \Phi, E^{2,2}_{s,f\otimes\overline{f}}\rangle \cdot E^{2,2}_{s,f\otimes\overline{f}}\, ds + \sum_{f\in\Xi_2} \langle \Phi, F_f\rangle \cdot F_f$$

$$+ \sum_{f\in\Xi_2} \frac{1}{4\pi i} \int_{\frac{1}{2}-i\infty}^{\frac{1}{2}+i\infty} \langle \Phi, E^{2,1,1}_{s,f}\rangle \cdot E^{2,1,1}_{s,f}\, ds$$

$$+ \frac{1}{4!\cdot 2\pi i} \int_{i\mathfrak{q}^*} \langle \Phi, E^{1,1,1,1}_{\rho+s}\rangle \cdot E^{1,1,1,1}_{\rho+s}\, ds + \frac{\langle \Phi, 1\rangle \cdot 1}{\langle 1, 1\rangle}$$

Again, the explicit and implicit integrals involving Eisenstein series are in fact isometric extensions of the literal integrals.

3.A Appendix: Bochner's Lemma

Bochner's Lemma is a one-of-a-kind device for meromorphic continuation in two or more complex variables.[11] Let Ω_o be a nonempty, connected, open set

[11] I first saw Bochner's lemma in the appendix of [Langlands 1967/1976] treating minimal-parabolic Eisenstein series for $SL_n(\mathfrak{o})$ for rings of integers $\mathfrak{o}$. General accounts of several complex variables are [Bochner-Martin 1948] and [Hörmander 1973].

in $\mathbb{R}^n$ with $n > 1$. The *tube domain* Ω over Ω_o is $\Omega = \Omega_o + i\mathbb{R}^n$, that is, the collection of $z \in \mathbb{C}^n$ with real part in Ω_o. Let f be a holomorphic $\mathbb{C}$-valued function on Ω, of not-too-awful vertical growth, in the sense that, for x in fixed compact $C \subset \Omega_o$, there is $1 \le N \in \mathbb{Z}$ such that

$$|f(x+iy)| \ll_C e^{|y|^N} \left(\text{with } |(y_1, \ldots, y_n)|^2 = y_1^2 + \cdots + y_n^2\right)$$

[3.A.1] Claim: f extends to a holomorphic function on the convex hull of Ω.

Proof: First, let x, ξ be two points in Ω_o, such that the line segment connecting them lies entirely within Ω_o. We will specify a rectangle inside Ω with x, ξ the midpoints of opposite sides. Let $\gamma = \gamma_{x,\xi,R}$ parametrize the rectangle with sides individually parametrized by

$$\begin{cases} \text{side through } x\text{: } x + it(x-\xi), \quad -R \le t \le R \\ \text{top: } (1-t)(x + iR(x-\xi)) + t(\xi + iR(x-\xi)), \; 0 \le t \le 1 \\ \text{side through } \xi\text{: } \xi - it(x-\xi), \quad -R \le t \le R \\ \text{bottom: } (1-t)(\xi - iR(\xi)) + t(x - iR(x-\xi)), \; 0 \le t \le 1 \end{cases}$$

The expressions for the top and bottom simplify to

$$\begin{cases} \text{top:} & (1-t)x + t\xi + iR(x-\xi) \quad (\text{with } 0 \le t \le 1) \\ \text{bottom:} & (1-t)\xi + tx - iR(x-\xi) \quad (\text{with } 0 \le t \le 1) \end{cases}$$

This rectangle lies inside $Z = x + \mathbb{C} \cdot (x-\xi) \approx \mathbb{C}$ and is contractible in Ω. Let $j(\zeta) = x + \zeta \cdot (x-\xi)$. In Z, Cauchy's formula in one variable is

$$f \circ j(\zeta_o) = \frac{1}{2\pi i} \int_\gamma \frac{f \circ j(\zeta)\; d\zeta}{\zeta - \zeta_o}$$

To legitimately push the top and bottom of the rectangle to infinity, use the growth assumption on f, and the modified integral expression

$$f \circ j(\zeta_o) = e^{-\zeta_o^{2N}} \frac{1}{2\pi i} \int_\gamma \frac{e^{\zeta^{2N}} \cdot f \circ j(\zeta)\, d\zeta}{\zeta - \zeta_o}$$

Thus, taking the limit $R \to +\infty$,

$$e^{\zeta_o^{2N}} \cdot f(\zeta_o \cdot (x-\xi)) = \frac{1}{2\pi} \int_{-\infty}^{+\infty} \frac{e^{(x+it(x-\xi))^{2N}} f(x + it(x-\xi))\, dt}{it - \zeta_o}$$
$$+ \frac{1}{2\pi} \int_{-\infty}^{+\infty} \frac{e^{(\xi - it(x-\xi))^{2N}} f(\xi - it(x-\xi))\, dt}{-1 - it - \zeta_o}$$

The right-hand side makes sense for any $x, \xi \in \Omega_o$, whether or not the line segment connecting them lies in Ω_o. Further, the right-hand side is holomorphic in $x, \xi \in \Omega$. Thus, the left-hand side is holomorphic, and gives the extension to the convex hull of Ω. ///

3.B Appendix: Phragmén-Lindelöf Theorem

This is from [Lindelöf 1908] and [Phragmén-Lindelöf 1908].

The *maximum modulus principle* can easily be misapplied on *unbounded* open sets. That is, while for an open set $U \subset \mathbb{C}$ with *bounded* closure $\overline{U}$, it *does* follow that the sup of a holomorphic function f on U extending continuously to $\overline{U}$ occurs on the boundary ∂U of U, holomorphic functions on an *unbounded* set can be bounded by 1 on the edges but be violently unbounded in the interior.

The usual simple example is $f(z) = e^{e^z}$:

$$|e^{e^{x+iy}}| = e^{\mathrm{Re}\,(e^{x+iy})} = e^{e^x \cdot \cos y}$$

On one hand, for fixed $y = \mathrm{Im}\, z$ with $\cos\, y > 0$, the function blows up as $x = \mathrm{Re}\, z \to +\infty$. On the other hand, for $\cos y = 0$ the function is *bounded*. Thus, on the strip $-\frac{\pi}{2} \le y \le \frac{\pi}{2}$, the function e^{e^z} is bounded on the edges but blows up as $x \to +\infty$.

This example suggests *growth conditions* under which a bound of 1 on the edges implies the same bound throughout the strip. In fact, the suggested bound is essentially sharp, in light of the example. For a *half-strip*, the theorem is

[3.B.1] Theorem: For f a holomorphic function on the horizontal half-strip

$$\{z : -\frac{\pi}{2} \le y \le \frac{\pi}{2} \ \text{ and } \ 0 \le x\}$$

satisfying

$$|f(z)| \ll e^{e^{C \cdot \mathrm{Re}\, z}} \qquad \text{(for some constant } 0 \le C < 1)$$

$|f(z)| \le 1$ on the edges of the half-strip implies $|f(z)| \le 1$ in the interior, as well.

Proof: Unsurprisingly, the proof is a reduction to the usual maximum modulus principle. Take any fixed D in the range

$$C < D < 1$$

The function

$$F_\varepsilon(z) = f(z)/e^{\varepsilon e^{D \cdot z}} \qquad \text{(for } \varepsilon > 0)$$

is bounded by 1 on the edges of the half-strip, and in the interior goes to 0 uniformly in y as $x \to +\infty$, for fixed $\varepsilon > 0$, exploiting the modification with

D. Thus, on a rectangle

$$R_T = \{z : -\frac{\pi}{2} \le y \le \frac{\pi}{2}, \text{ and } 0 \le x \le T\}$$

for sufficiently large $T > 0$ depending on ε, the function F_ε is bounded by 1 on the edge. The usual maximum modulus principle implies that F_ε is bounded by 1 throughout. That is, *for each fixed* z_o in the half-strip,

$$|f(z_o)| \le e^{\varepsilon \cdot e^{D \mathrm{Re}\, z_o}} \qquad \text{(for all } \varepsilon > 0)$$

Let $\varepsilon \to 0^+$, giving $|f(z_o)| \le 1$. ///

[3.B.2] Remark: Analogous theorems on strips of other widths follow by using $e^{c \cdot e^z}$ with suitable constants c.

The theorem on a full strip follows by using $e^{\cosh z}$ in place of e^{e^z}, as follows.

[3.B.3] Theorem: For f a holomorphic function on the horizontal strip

$$\{z : -\frac{\pi}{2} \le \mathrm{Im}\, z \le \frac{\pi}{2}\}$$

satisfying

$$|f(z)| \ll e^{\cosh C \cdot \mathrm{Re}\, z} \qquad \text{(for some constant } 0 \le C < 1)$$

$|f(z)| \le 1$ on the edges of the strip implies $|f(z)| \le 1$ in the interior, as well.

Proof: Again, reduce to the maximum modulus principle. Fix D in the range $C < D < 1$. The function

$$F_\varepsilon(z) = f(z)/e^{\varepsilon \cosh Dz} \qquad \text{(for } \varepsilon > 0)$$

is bounded by 1 on the edges of the strip, and in the interior goes to 0 uniformly in y as $x \to \pm\infty$, for fixed $\varepsilon > 0$. Thus, on a rectangle

$$R_T = \{z : -\frac{\pi}{2} \le y \le \frac{\pi}{2}, \text{ and } -T \le x \le T\}$$

for large $T > 0$, depending upon ε, the function F_ε is bounded by 1 on the edge. The usual maximum modulus principle implies that F_ε is bounded by 1 throughout. That is, *for each fixed* z_o in the half-strip,

$$|f(z_o)| \le e^{\varepsilon \cosh D \mathrm{Re}\, z_o} \qquad \text{(for all } \varepsilon > 0)$$

We can let $\varepsilon \to 0^+$, giving $|f(z_o)| \le 1$. ///

4

Invariant Differential Operators

We want a method to determine natural Laplacian-like differential operators invariant under group actions in *coordinate-free* terms, and also exhibit the operators in convenient coordinate systems. That is, we do *not* want to specify operators in coordinate systems and *check* invariance, but, rather, know invariance a priori. The first example is the well-known operator

$$\Delta = y^2 \left(\frac{\partial^2}{\partial x^2} + \frac{\partial^2}{\partial y^2} \right) \qquad \text{(coordinate(s) } z = x + iy \text{ on } \mathfrak{H})$$

Although, once exhibited, this operator is certifiably invariant under the linear fractional action of $SL_2(\mathbb{R})$, it is oppressive and unenlightening to do this checking. Worse, it is misguided to think in terms of such verification. The relevant issue is the coordinate-independent *origin* of the operator, expressed subsequently in coordinates. *No prior acquaintance with Lie groups or Lie algebras is assumed.*

4.1 Derivatives of Group Actions: Lie Algebras

Let G be a subgroup of $GL_n(\mathbb{R})$ or $GL_n(\mathbb{C})$ or $GL_n(\mathbb{H})$ acting *differentiably*[1] on the right on a smooth manifold[2] thereby acting on *functions* f on M by

$$(g \cdot f)(m) = f(mg)$$

Our operational definition of the (real) *Lie algebra* $\mathfrak{g}$ of G is

$$\mathfrak{g} = \{\text{real } n\text{-by-}n \text{ real matrices } x \ : \ e^{tx} \in G \ \text{ for all real } t\}$$

where the matrix exponential is

$$\exp(x) = e^x = 1 + x + \frac{x^2}{2!} + \frac{x^3}{3!} + \dots$$

This definition makes clear that $\mathfrak{g}$ is closed under scalar multiplication but not that it is closed under *addition*. When x and y are n-by-n real or complex matrices with $xy = yx$, then $e^{x+y} = e^x \cdot e^y$, but this does *not* hold more generally, so we cannot easily conclude closed-ness under addition. We will prove closure under addition as a side effect of proof in [4.A] that such a Lie algebra is closed under *Lie brackets*

$$x \times y \longrightarrow [x, y] = xy - yx$$

for $x, y \in \mathfrak{g}$, with $x \times y \to xy$ matrix multiplication. In any particular example, the vector space property is readily verified, as below. However, this binary operation $x \times y \to [x, y]$ is not similar to more elementary ring or algebra multiplications, as it is *not associative*, is *anti-commutative*, and $[x, x] = 0$.

For each $x \in \mathfrak{g}$ we have a *differentiation* X_x of functions f on M in the *direction* x, by

$$(X_x f)(m) = \frac{d}{dt}\bigg|_{t=0} f(m \cdot e^{tx})$$

This applies uniformly to *any* space M on which G acts (differentiably).[3] The differential operators X_x for $x \in \mathfrak{g}$ do *not* typically commute with the action of

1 When the group G and the set M are subsets of Euclidean spaces defined as zero sets or level sets of differentiable functions, *differentiability* of the action can be posed in the ambient Euclidean coordinates and the Implicit Function Theorem. In any particular example, even less is usually required to make sense of this requirement.

2 As in other instances of a group acting transitively on a set with additional structure, under modest hypotheses, M is a quotient $G_o \backslash G$ of G by the isotropy group G_o of a chosen point in M.

3 The action of the Lie algebra by differentiating the action of the Lie group also applies *abstractly* to certain vectors v in vectorspaces V on which G acts, namely, those v such that $g \to g \cdot v$ is a differentiable V-valued function on G. Under mild hypotheses, smooth vectors are *dense* [14.6].

$g \in G$, although the relation between the two is reasonable:

$$(g \circ X_x \circ g^{-1})f(h) = \frac{\partial}{\partial t}\bigg|_{t=0} f(h \cdot g e^{tx} g^{-1})$$
$$= \frac{d}{dt}\bigg|_{t=0} f(h \cdot e^{t \cdot gxg^{-1}}) = X_{gxg^{-1}} f(h)$$

[4.1.1] Example: The condition $e^{tx} \in SL_n(\mathbb{R})$ for all real t is that $\det(e^{tx}) = 1$. To see what this requires of x, observe that for n-by-n (real or complex) matrices x

$$\det(e^x) = e^{\operatorname{tr} x} \qquad \text{(where tr is trace)}$$

To see why, both determinant and trace are invariant under conjugation $x \to gxg^{-1}$, so without loss of generality x is *upper-triangular*. Then e^x is upper-triangular, with diagonal entries $e^{x_{ii}}$, with diagonal entries x_{ii} of x. Thus,

$$\det(e^x) = e^{x_{11}} \cdots e^{x_{nn}} = e^{x_{11}+\ldots+x_{nn}} = e^{\operatorname{tr} x}$$

Using this, the determinant-one condition is

$$1 = \det(e^{tx}) = e^{t \cdot \operatorname{tr} x} = 1 + t \cdot \operatorname{tr} x + \frac{(t \cdot \operatorname{tr} x)^2}{2!} + \cdots$$

Taking the derivative with respect to t and setting $t = 0$ gives $0 = \operatorname{tr} x$. Looking at the right-hand side of the expanded $1 = \det(e^{tx})$, this condition is also *sufficient* for $\det(e^{tx}) = 1$. Thus,

$$\text{Lie algebra } \mathfrak{sl}_n(\mathbb{R}) \text{ of } SL_n(\mathbb{R}) = \{\, n\text{-by-}n \text{ real } x : \operatorname{tr} x = 0\}$$

[4.1.2] Example: Similarly,

$$\text{Lie algebra } \mathfrak{sl}_n(\mathbb{C}) \text{ of } SL_n(\mathbb{C}) = \{\, n\text{-by-}n \text{ complex } x : \operatorname{tr} x = 0\}$$

[4.1.3] Example: From $\det(e^x) = e^{\operatorname{tr} x}$, any matrix e^x is invertible, so

$$\text{Lie algebra } \mathfrak{sl}_n(\mathbb{R}) \text{ of } GL_n(\mathbb{R}) = \{\textit{all} \text{ real } n\text{-by-}n \text{ matrices}\}$$

[4.1.4] Example: For the simplest *real orthogonal group* $G = O(n, \mathbb{R}) = \{g \in GL_n(\mathbb{R}) : g^\top \cdot g = 1_n\}$, using $(e^{tx})^\top = e^{tx^\top}$,

$$1 = (e^{tx})^\top \cdot e^{tx} = (1 + tx^\top + \cdots) \cdot (1 + tx + \cdots) = 1 + t(x + x^\top) + \cdots$$

Thus, necessarily $x^\top + x = 0$. On the other hand, when $x^\top + x = 0$ we have $x^\top = -x$, so

$$(e^{tx})^\top \cdot e^{tx} = e^{-tx} \cdot e^{tx} = (e^{tx})^{-1} \cdot e^{tx} = 1$$

This shows that the Lie algebra of $O(n, \mathbb{R})$ is skew-symmetric matrices.

[4.1.5] Example: Exponentiation of matrices with quaternion entries is similar:

$$\text{Lie algebra } \mathfrak{sl}_n(\mathbb{H}) \text{ of } GL_n(\mathbb{H}) = \{\textit{all} \text{ quaternion } n\text{-by-}n \text{ matrices}\}$$

Slightly more subtly,

$$\begin{aligned}&\text{Lie algebra } \mathfrak{sl}_n(\mathbb{H}) \text{ of } SL_n(\mathbb{H})\\ &= \{n\text{-by-}n \text{ quaternionic } x : \sum_i \operatorname{tr} x_{ii} = 0\} \quad \text{(quaternion trace)}\end{aligned}$$

[4.1.6] Example: For $G = Sp^*_{1,1} \subset GL_2(\mathbb{H})$, let $S = \begin{pmatrix} 0 & 1 \\ 1 & 0 \end{pmatrix}$ and let σ be quaternion-conjugate-transpose. The defining condition $S = (e^{tx})^\sigma S(e^{tx})$ is $S = e^{tx^\sigma} Se^{tx}$. Differentiating with respect to $t \in \mathbb{R}$ gives $0 = x^\sigma e^{tx^\sigma} Se^{tx} + e^{tx^\sigma} Sxe^{tx}$. Setting $t = 0$ gives $x^\sigma S + Sx = 0$, which is $Sx^\sigma S^{-1} = -x$. Multiplying out, this condition is

$$\begin{pmatrix} -a & -b \\ -c & -d \end{pmatrix} = S\begin{pmatrix} a & b \\ c & d \end{pmatrix}^\sigma S^{-1} = S\begin{pmatrix} \overline{a} & \overline{c} \\ \overline{b} & \overline{d} \end{pmatrix} S^{-1} = \begin{pmatrix} \overline{d} & \overline{b} \\ \overline{c} & \overline{a} \end{pmatrix}$$

The *sufficiency* of this necessary condition is seen by exponentiating, noting that exponentiation respects conjugation: first,

$$\begin{aligned}(e^{tx})^\sigma S(e^{tx}) &= S \cdot S(e^{tx})^\sigma S \cdot (e^{tx})S \cdot Se^{tx^\sigma} Se^{tx}\\ &= S \cdot e^{t \cdot Sx^\sigma S} e^{tx} = S \cdot e^{t\cdot(-x)} e^{tx} = S \cdot e^{-tx} e^{tx} = S\end{aligned}$$

Thus,

$$\begin{aligned}&\text{Lie algebra } \mathfrak{sp}^*_{1,1}(\mathbb{H}) \text{ of } Sp^*_{1,1}\\ &= \{\text{2-by-2 quaternionic } \begin{pmatrix} a & b \\ c & -\overline{a} \end{pmatrix} : b = -\overline{b},\ c = -\overline{c}\}\end{aligned}$$

[4.1.7] Example: In the simple case that the space M is G itself, there is a second action of G on itself in addition to right multiplication, namely, *left* multiplication. The *right* differentiation by elements of $\mathfrak{g}$ *does* commute with the *left* multiplication by G, for the simple reason that

$$F(h \cdot (ge^{tx})) = F((h \cdot g) \cdot e^{tx}) \qquad (\text{for } g, h \in G, x \in \mathfrak{g})$$

That is, $\mathfrak{g}$ gives left G-invariant differential operators on G.

[4.1.8] Claim: The conjugation action of G on $\mathfrak{g}$ stabilizes $\mathfrak{g}$, and $g \cdot X_x \cdot g^{-1} = X_{gxg^{-1}}$ for $g \in G$ and $x \in \mathfrak{g}$.

Proof: For smooth f on M,

$$(g \cdot X_x \cdot g^{-1} \cdot f)(m) = (g(X_x(g^{-1}f)))(m) = (X_x(g^{-1}f))(mg)$$
$$= \frac{d}{dt}\bigg|_{t=0} (g^{-1}f)(m\,g\,e^{tx}) = \frac{d}{dt}\bigg|_{t=0} f(m\,g\,e^{tx}\,g^{-1})$$

Again, conjugation and exponentiation interact well:

$$g\,e^{tx}\,g^{-1} = g\left(1 + tx + \frac{(tx)^2}{2!} + \ldots\right)g^{-1}$$
$$= 1 + tgxg^{-1} + \frac{(tgxg^{-1})^2}{2!} + \ldots = e^{tgxg^{-1}}$$

Thus,

$$(g \cdot X_x \cdot g^{-1} \cdot f)(m) = \frac{d}{dt}\bigg|_{t=0} f(m\,g\,e^{tx}\,g^{-1}) = \frac{d}{dt}\bigg|_{t=0} f(m\,e^{tgxg^{-1}})$$
$$= (X_{gxg^{-1}}\,f)(m)$$

as claimed, and $gxg^{-1} \in \mathfrak{g}$. ///

The commutant expression $[x, y] = xy - yx$, the *Lie bracket*, arises naturally:

[4.1.9] Claim: Since G is *non-abelian* in many cases of interest, typically $e^x \cdot e^y \neq e^y \cdot e^x$ for $x, y \in \mathfrak{g}$. Specifically,

$$e^{tx}\,e^{ty}\,e^{-tx}\,e^{-ty} = 1 + t^2[x, y] + \text{(higher-order terms)}$$

where $[x, y] = xy - yx$.

Proof: This is a direct computation, easy if we drop cubic and higher-order terms.

$$e^{tx}\,e^{ty}\,e^{-tx}\,e^{-ty}$$
$$= (1 + tx + t^2x^2/2)(1 + ty + t^2y^2/2)(1 - tx + t^2x^2/2)(1 - ty + t^2y^2/2)$$
$$= (1 + t(x+y) + \frac{t^2}{2}(x^2 + 2xy + y^2))\;(1 - t(x+y) + \frac{t^2}{2}(x^2 + 2xy + y^2))$$
$$= 1 + t^2\left(x^2 + 2xy + y^2 - (x+y)(x+y)\right)$$
$$= \left(1 + t^2(2xy - xy - yx)\right) = 1 + t^2[x, y]$$

as claimed. ///

[4.1.10] Claim: The conjugation/adjoint action of G on $\mathfrak{g}$ respects brackets:

$$[gxg^{-1},\, gyg^{-1}] = g[x, y]g^{-1} \qquad \text{(for } x, y \in \mathfrak{g} \text{ and } g \in G)$$

Proof: For Lie brackets expressed in terms of matrix operations, this is straightforward:

$$[gxg^{-1}, gyg^{-1}] = gxg^{-1}\, gyg^{-1} - gyg^{-1}\, gxg^{-1}$$
$$= gxyg^{-1} - gyxg^{-1} = g(xy - yx)g^{-1} = g[x, y]g^{-1}$$

as claimed. ///

Composition of the derivatives X_x operators mirrors the bracket in the Lie algebra:

[4.1.11] Theorem: The map $x \to X_x$ is a *Lie algebra homomorphism*, meaning it respects these commutants (brackets): $X_x \circ X_y - X_y \circ X_x = X_{[x,y]}$. *(Proof: see [4.A].)*

4.2 Laplacians and Casimir Operators

As in the last theorem, *commutants* of differential operators coming from Lie algebras $\mathfrak{g}$ are again differential operators coming from the Lie algebra, namely[4]

$$X_x \circ X_y - X_y \circ X_x = [X_x, X_y] = X_{[x,y]} = X_{xy-yx}$$

However, the *composition* of differential operators has no analogue inside the Lie algebra. That is, typically,

$$X_x \circ X_y \neq X_\varepsilon \qquad (\text{for any } \varepsilon \in \mathfrak{g})$$

We want an object associated to the Lie algebra that allows this composition.[5] That is, we want an *associative* algebra $U\mathfrak{g}$ *universal* in the sense that any linear map $\varphi : \mathfrak{g} \to B$ to an associative algebra B respecting brackets

$$\varphi([x, y]) = \varphi(x)\,\varphi(y) - \varphi(y)\,\varphi(x) \qquad (\text{for } x, y \in \mathfrak{g})$$

should give a unique *associative algebra* homomorphism $\Phi : U\mathfrak{g} \longrightarrow B$. There must be a connection to the original $\varphi : \mathfrak{g} \to B$, so we require existence of a fixed map $i : \mathfrak{g} \to U\mathfrak{g}$ *respecting brackets* and commutativity of a diagram

[4] For matrix groups with Lie bracket described via matrix multiplication $[x, y] = xy - yx$, properties otherwise needing explicit declaration, such as the *Jacobi identity* $[x, [y, z]] - [y, [x, z]] = [[x, y], z]$, can be verified directly by expanding the brackets. The *content* of the Jacobi identity is that the map $\mathrm{ad} : \mathfrak{g} \to \mathrm{End}(\mathfrak{g})$ by $(\mathrm{ad}x)(y) = [x, y]$ is a Lie algebra *homomorphism*. That is, $[\mathrm{ad}x, \mathrm{ad}y] = \mathrm{ad}[x, y]$.

[5] For Lie algebras $\mathfrak{g}$ such as $\mathfrak{so}(n)$, $\mathfrak{sl}_n$, or $\mathfrak{gl}_n$ lying inside matrix rings, typically $X_x \circ X_y \neq X_{xy}$. That is, multiplication of *matrices* is definitely *not* multiplication in any sense that will match multiplication (composition) of *differential operators*.

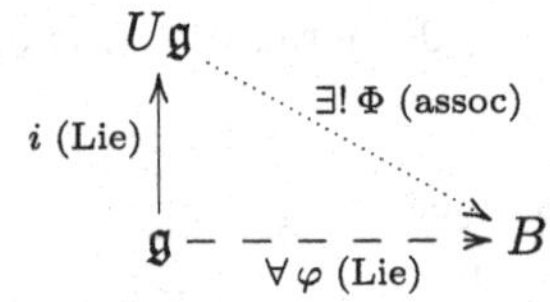

where the labels tell the *type* of the maps.

In what follows, we see that $U\mathfrak{g}$ is a canonical quotient of the *universal associative algebra* AV of a vector space V over a field k, very often called the *tensor algebra* and denoted $\bigotimes^\bullet V$, although, unhelpfully, this name refers to details of a specific *construction*, rather than to the *characterizing property* of the algebra. The characterizing property of the universal associative algebra AV is that there is a fixed linear $j : V \to AV$, and any linear map $V \to B$ to an (associative) algebra B extends to a unique associative algebra map $AV \to B$. That is, there is a commutative diagram

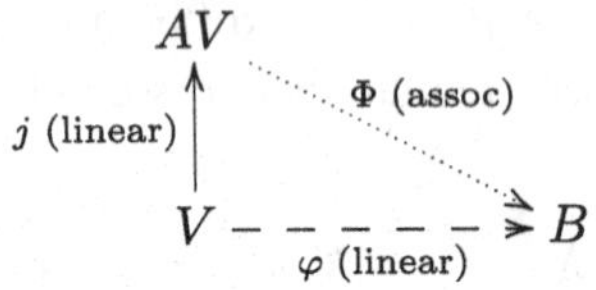

Since the universal associative algebra $j : \mathfrak{g} \to A\mathfrak{g}$ is universal with respect to maps $\mathfrak{g} \to B$ that are merely *linear*, not necessarily preserving the Lie brackets, there is a (unique) natural (quotient) map $q : A\mathfrak{g} \to U\mathfrak{g}$.

The conjugation (Adjoint) action $x \to gxg^{-1}$ of G on $\mathfrak{g}$ should extend to an action of G on $U\mathfrak{g}$ (which we may still write as conjugation) compatible with the multiplication in $U\mathfrak{g}$. That is, we expect

$$\begin{cases} g(\alpha) = g\alpha g^{-1} & (\text{for } \alpha \in \mathfrak{g} \text{ and } g \in G) \\ g(\alpha\,\beta) = g(\alpha) \cdot g(\beta) & (\text{for } \alpha, \beta \in U\mathfrak{g} \text{ and } g \in G) \end{cases}$$

The action of G on $\mathfrak{g}$ should extend to $A\mathfrak{g}$, too, and the quotient map $q : A\mathfrak{g} \to U\mathfrak{g}$ should respect that action. We also *assume* for the moment that we have a nondegenerate symmetric bilinear form $\langle , \rangle$ on $\mathfrak{g}$, and that this form is G-invariant: $\langle gxg^{-1}, gyg^{-1} \rangle = \langle x, y \rangle$ for $x, y \in \mathfrak{g}$ and $g \in G$.

Granting these things, we can *intrinsically* describe the simplest nontrivial G-invariant element in $U\mathfrak{g}$, the *Casimir element* Ω. In any action of G, the Casimir element gives rise to a G-invariant differential operator, the corresponding *Casimir operator*. In many situations the Casimir operator is the suitable notion of *invariant Laplacian*. Map $\zeta : \mathrm{End}_{\mathbb{C}}(\mathfrak{g}) \to U\mathfrak{g}$ by

$$\mathrm{End}_{\mathbb{C}}(\mathfrak{g}) \xrightarrow{\approx} \mathfrak{g}\otimes\mathfrak{g}^* \xrightarrow{\approx} \mathfrak{g}\otimes\mathfrak{g} \xrightarrow{\text{inc}} A\mathfrak{g} \xrightarrow{\text{quot}} U\mathfrak{g} \qquad (\text{composite: } \zeta)$$

where the first map is an instance of the inverse of the isomorphism $V \otimes V^* \to \mathrm{End}\, V$ for finite-dimensional vectorspaces V, the second map uses the inverse of the isomorphism $V \to V^*$ given by $v \to \langle -, v\rangle$ for a nondegenerate bilinear form $\langle , \rangle$ on V. The action of G respects all the maps. An obvious endomorphism of $\mathfrak{g}$ commuting with the action of G on $\mathfrak{g}$ is the *identity map* $\mathrm{id}_{\mathfrak{g}}$. Thus,

[4.2.1] Claim: The *Casimir element* $\Omega = \zeta(\mathrm{id}_{\mathfrak{g}})$ is a G-invariant element of $U\mathfrak{g}$.

Proof: Since ζ is G-equivariant by construction,

$$g\zeta(\mathrm{id}_{\mathfrak{g}})g^{-1} = \zeta(g\,\mathrm{id}_{\mathfrak{g}}\,g^{-1}) = \zeta(gg^{-1}\,\mathrm{id}_{\mathfrak{g}}) = \zeta(\mathrm{id}_{\mathfrak{g}})$$

since id_g commutes with *any* endomorphism of $\mathfrak{g}$. Thus, $\zeta(\mathrm{id}_{\mathfrak{g}})$ is a G-invariant element of $U\mathfrak{g}$. ///

The possible hazard is that $\zeta(\mathrm{id}_{\mathfrak{g}})$ is accidentally 0. This non-vanishing can be proven by demonstrating at least one associative algebra B and $\mathfrak{g} \to B$ so that the induced image of Casimir is nonzero in B.[6]

The preceding prescription *does* implicitly tell how to express the Casimir element $\Omega = \zeta(\mathrm{id}_{\mathfrak{g}})$ in *various* coordinates. Namely, for any basis $x_1, \ldots, x_n$ of $\mathfrak{g}$, let $\lambda_1, \ldots, \lambda_n$ be the corresponding dual *dual basis* of the dual $\mathfrak{g}^*$: $\lambda_i(x_j)$ is 0 or 1 as $i = j$ or not. Let $x_1^*, \ldots, x_n^*$ be the corresponding dual basis for $\mathfrak{g}$ in terms of $\langle , \rangle$, namely, $\langle x_i, x_j^*\rangle$ is 0 or 1 as $i = j$ or not. Then in

$$\mathrm{End}_{\mathbb{C}}(\mathfrak{g}) \xrightarrow{\approx} \mathfrak{g}\otimes\mathfrak{g}^* \xrightarrow{\approx \text{ via } \langle,\rangle} \mathfrak{g}\otimes\mathfrak{g} \xrightarrow{\text{inc}} A\mathfrak{g} \xrightarrow{\text{quot}} U\mathfrak{g} \qquad (\text{composite: } \zeta)$$

we have

$$\mathrm{id}_{\mathfrak{g}} \to \sum_i x_i \otimes \lambda_i \to \sum_i x_i \otimes x_i^* \to \sum_i x_i \otimes x_i^* \to \sum_i x_i x_i^* = \Omega$$

This *intrinsic* description of the Casimir element as $\zeta(\mathrm{id}_{\mathfrak{g}})$ shows that it does not depend on the choice of basis $x_1, \ldots, x_n$.[7]

[6] The nonvanishing is also a corollary of the *Poincaré-Birkhoff-Witt* theorem, but we need not invoke it.

[7] Some sources *define* the Casimir element as $\sum_i x_i x_i^*$ in the universal enveloping algebra, show by computation that it is G-invariant, and show by change-of-basis that the defined object is

4.3 Details about Universal Algebras

We fill in details about $U\mathfrak{g}$ and $A\mathfrak{g}$, including constructions. Again, we want an *associative* algebra $U\mathfrak{g}$ such that any Lie algebra map $\varphi : \mathfrak{g} \to B$ to an associative algebra B with the property

$$\varphi([x, y]) = \varphi(x)\,\varphi(y) - \varphi(y)\,\varphi(x) \qquad (\text{for } x, y \in \mathfrak{g})$$

gives a unique *associative algebra* homomorphism $\Phi : U\mathfrak{g} \longrightarrow B$ fitting into a commutative diagram

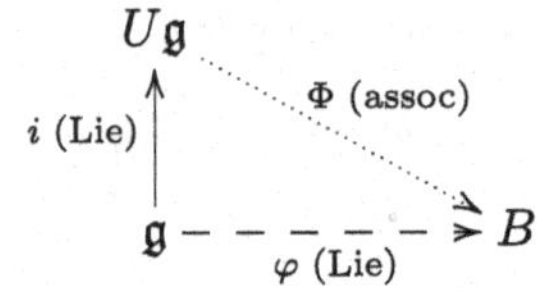

Similarly, we want a *universal associative algebra* AV of a vector space V over a field k, with a specified linear $j : V \to AV$, such that any linear map $V \to B$ to an associative algebra B extends to a unique associative algebra map $AV \to B$ fitting into a commutative diagram

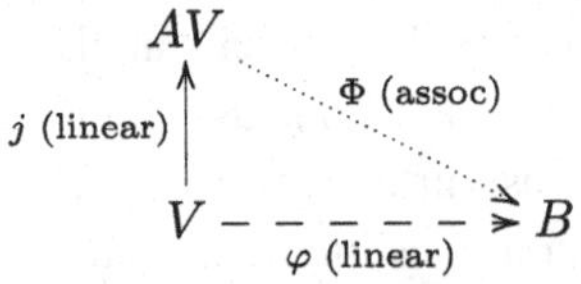

Granting for a moment the *existence* of $A\mathfrak{g}$, construct $U\mathfrak{g}$ as the quotient of $A\mathfrak{g}$ by the two-sided ideal generated by all elements

$$\big(jx \otimes jy - jy \otimes jx\big) - j[x, y] \qquad (\text{where } x, y \in \mathfrak{g})$$

The map $i : \mathfrak{g} \to U\mathfrak{g}$ is the obvious composite $q \circ j$. Given a Lie algebra map $\varphi : \mathfrak{g} \to B$ from $\mathfrak{g}$ to an associative algebra, we show that the induced map $\Phi : A\mathfrak{g} \longrightarrow B$ factors through $q : A\mathfrak{g} \longrightarrow U\mathfrak{g}$. Diagrammatically, we claim the existence of an arrow to fill in a commutative diagram

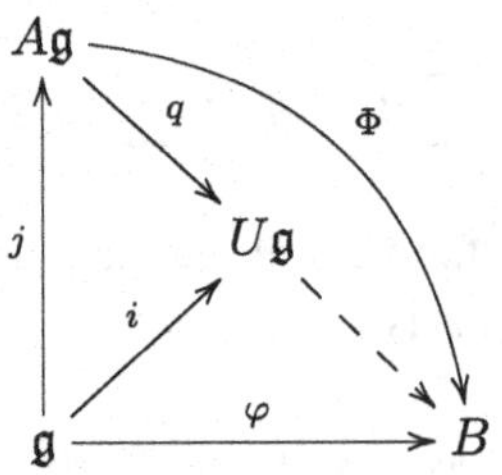

independent of the choice of basis. That element $\sum_i x_i x_i^*$ is of course the image in $U\mathfrak{g}$ of the tensor $\sum_i x_i \otimes x_i^*$ (discussed here) which is simply the image of $\mathrm{id}_{\mathfrak{g}}$ in coordinates.

Indeed, the the Lie algebra homomorphism property $\varphi(x)\varphi(y) - \varphi(y)\varphi(x) - \varphi[x, y] = 0$ and the commutativity imply that

$$\Phi\Big(jx \otimes jy - jy \otimes jx) - j[x, y]\Big) = 0 \qquad \text{(for all } x, y \in \mathfrak{g})$$

That is, Φ vanishes on the kernel of the quotient map $q : A\mathfrak{g} \to U\mathfrak{g}$, so factors through this quotient map. This proves the existence of $U\mathfrak{g}$ in terms of $A\mathfrak{g}$.

The conjugation (Adjoint) action $x \to gxg^{-1}$ of G on $\mathfrak{g}$ should extend to an action of G on $U\mathfrak{g}$ (which we may still write as conjugation) compatible with the multiplication in $U\mathfrak{g}$. That is, we expect

$$\begin{cases} g(\alpha) = g\alpha g^{-1} & \text{(for } \alpha \in \mathfrak{g} \text{ and } g \in G) \\ \\ g(\alpha\,\beta) = g(\alpha) \cdot g(\beta) & \text{(for } \alpha, \beta \in U\mathfrak{g} \text{ and } g \in G) \end{cases}$$

The action of G on $\mathfrak{g}$ should extend to $A\mathfrak{g}$, too, and the quotient map $q : A\mathfrak{g} \to U\mathfrak{g}$ should respect that action. Fulfillment of this requirement, or the observation that it is automatically fulfilled, is best understood from further details about $A\mathfrak{g}$, just below.

[4.3.1] Construction of Universal Associative Algebras: The tensor *construction* of $A\mathfrak{g}$ gives enough further information so that we can see that it inherits an action of G from $\mathfrak{g}$, and that this action is inherited by $U\mathfrak{g}$. The *construction* of AV in terms of tensors is

$$AV = k \oplus V \oplus (V \otimes V) \oplus (V \otimes V \otimes V) \oplus \cdots$$

with multiplication given by (the bilinear extension of) the obvious

$$(v_1 \otimes \cdots \otimes v_m) \cdot (w_1 \otimes \cdots \otimes w_n) = v_1 \otimes \cdots \otimes v_m \otimes w_1 \otimes \cdots \otimes w_n$$

The well-definedness of the multiplication follows from noting that there is a unique linear map $\bigotimes^m V \otimes \bigotimes^n V \longrightarrow \bigotimes^{m+n} V$ induced from the bilinear map

$$(v_1 \otimes \cdots \otimes v_m) \times (w_1 \otimes \cdots \otimes w_n) \longrightarrow v_1 \otimes \cdots \otimes v_m \otimes w_1 \otimes \cdots \otimes w_n$$

Distributivity of multiplication over addition follows from the fact that the multiplication maps are induced from bilinear maps. The map $V \to AV$ is to the summand $V \subset AV$, which shows that this map is *injective*. It is also true that $\mathfrak{g} \to U\mathfrak{g}$ is injective, but the latter fact is considerably less trivial to prove.

To verify that this constructed object has the requisite universal property, let $\varphi : V \to B$ be a linear map to an associative algebra. Then the linear map $\Phi_n : \bigotimes^n V \to B$ defined by

$$\Phi(v_1 \otimes \cdots \otimes v_n) = \varphi(v_1) \cdots \varphi(v_n)$$

with the latter being multiplication in B, is *well-defined*, being induced from the n-multilinear map

$$\underbrace{V \times \cdots \times V}_{n} \longrightarrow B \qquad \text{by} \qquad v_1 \times \cdots \times v_n \longrightarrow \varphi(v_1) \cdots \varphi(v_n)$$

Letting k be the underlying field (probably either $\mathbb{C}$ or $\mathbb{R}$), there is also the map $\Phi_0 : k \to B$ by $a \to 1_B$. The collection of maps Φ_n gives a linear map $\Phi : AV \to B$. It also obviously preserves multiplication. This proves that the tensor construction yields the universal associative algebra.

[4.3.2] G-action on $A\mathfrak{g}$ and $U\mathfrak{g}$: The notationally obvious G-action on $A\mathfrak{g}$ is

$$g(x_1 \otimes \cdots \otimes x_m)g^{-1} = gx_1g^{-1} \otimes \cdots \otimes gx_mg^{-1}$$

This gives a well-defined linear map of each $\bigotimes^n \mathfrak{g}$ to itself, because it is the unique map induced by the multilinear map

$$\underbrace{\mathfrak{g} \times \cdots \times \mathfrak{g}}_{n} \longrightarrow \bigotimes\nolimits^n \mathfrak{g}$$

by

$$v_1 \times \cdots \times v_n \longrightarrow gv_1g^{-1} \otimes \cdots \otimes gv_ng^{-1}$$

The map is visibly compatible with multiplication. Since $\mathfrak{g}$ injects to $A\mathfrak{g}$, we can safely suppress the map j in this discussion. The G-action stabilizes the kernel of the kernel of $q : A\mathfrak{g} \to U\mathfrak{g}$, since

$$\begin{aligned} & g\Big((x \otimes y - y \otimes x) - [x, y]\Big)g^{-1} \\ & \quad = g(x \otimes y)g^{-1} - g(y \otimes x)g^{-1} - g[x, y]g^{-1} \\ & \quad = gxg^{-1} \otimes gyg^{-1} - gyg^{-1} \otimes gxg^{-1} - [gxg^{-1}, gyg^{-1}] \end{aligned}$$

This gives a natural action of G on $U\mathfrak{g}$, respecting the quotient $q : A\mathfrak{g} \to U\mathfrak{g}$, and, therefore, respecting the map $\mathfrak{g} \to U\mathfrak{g}$. The universal associative algebra $A\mathfrak{g}$ is sufficiently large that, roughly, *it has no nontrivial relations*. Thus, the notationally obvious apparent definition of the G-action on $A\mathfrak{g}$ is *well defined*. Then the G-action descends to $U\mathfrak{g}$.

[4.3.3] Killing's Bilinear Form: The last necessary item is more special and not possessed by all Lie algebras. We want a *nondegenerate* symmetric $\mathbb{R}$-bilinear map

$$\langle , \rangle : \mathfrak{g} \times \mathfrak{g} \longrightarrow \mathbb{R}$$

G-equivariant in the sense that

$$\langle gxg^{-1}, gyg^{-1}\rangle = \langle x, y\rangle$$

Happily, for $\mathfrak{so}(n)$, $\mathfrak{sl}_n(\mathbb{R})$, and $\mathfrak{gl}_n(\mathbb{R})$, the obvious *trace form*

$$\langle x, y\rangle = \operatorname{tr}(xy)$$

suffices. For G described as a subgroup of $GL_n(\mathbb{C})$, take $\langle x, y\rangle = \operatorname{Re}(\operatorname{tr} xy)$. For G described as a subgroup of $GL_n(\mathbb{H})$, take $\langle x, y\rangle = \sum_{ij}(x_{ij}y_{ji} + \overline{x_{ij}y_{ji}})$. For notational simplicity, we write out the arguments only for $G \subset GL_n(\mathbb{R})$. The behavior under the action of G is clear:

$$\begin{aligned}\langle gxg^{-1}, gyg^{-1}\rangle &= \operatorname{tr}\big(gxg^{-1}\cdot gyg^{-1}\big) = \operatorname{tr}\big(gxyg^{-1}\big)\\ &= \operatorname{tr}(xy) = \langle x, y\rangle\end{aligned}$$

The nondegeneracy and G-equivariance of $\langle,\rangle$ give a natural G-equivariant isomorphism $\mathfrak{g} \to \mathfrak{g}^*$ by

$$x \longrightarrow \lambda_x \quad \text{by} \quad \lambda_x(y) = \langle x, y\rangle \qquad (\text{for } x, y \in \mathfrak{g})$$

When G acts on a vector space V the action on the dual V^* is by $(g\cdot\lambda)(v) = \lambda(g^{-1}\cdot v)$ for $v \in V$ and $\lambda \in V^*$. As usual, the inverse appears to preserve *associativity*. The equivariance of $\langle,\rangle$ gives

$$\begin{aligned}\lambda_{g\cdot x}(y) &= \lambda_{gxg^{-1}}(y) = \langle gxg^{-1}, y\rangle = \langle x, g^{-1}yg\rangle\\ &= \lambda_x(g^{-1}yg) = \lambda_x(g^{-1}\cdot y) = (g\cdot\lambda_x)(y)\end{aligned}$$

proving that the map $x \to \lambda_x$ is a G-isomorphism.

Finally, recall that the natural isomorphism $V \otimes_k V^* \longrightarrow \operatorname{End}_k V$ for V a finite-dimensional vector space over a field k is given by the k-linear extension of the map $v \times \lambda \longrightarrow (w \to \lambda(w)\cdot v)$ for $v, w \in V$ and $\lambda \in V^*$. The fact that the map is an isomorphism follows by dimension counting, using the finite dimensionality.

4.4 Descending to G/K

Now we see how the Casimir operator Ω on G gives G-invariant Laplacian-like differential operators on quotients G/K. Let $\mathfrak{k} \subset \mathfrak{g}$ be the Lie algebra of a maximal compact $K \subset G$. Again, the action of $x \in \mathfrak{g}$ on the *right* on functions f on G, by

$$(x\cdot f)(g) = \frac{d}{dt}\bigg|_{t=0} f(g\,e^{tx})$$

is *left* G-invariant for the straightforward reason that

$$f(h \cdot (g\, e^{tx})) = f((h \cdot g) \cdot e^{tx})) \qquad (\text{for } g, h \in G, x \in \mathfrak{g})$$

For a (closed) subgroup K of G let $q : G \to G/K$ be the quotient map. A function f on G/K gives the right K-invariant function $F = f \circ q$ on G. Given $x \in \mathfrak{g}$, the differentiation

$$(x \cdot (f \circ q))(g) = \frac{d}{dt}\bigg|_{t=0} (f \circ q)(g\, e^{tx})$$

makes sense. *However*, $x \cdot (f \circ q)$ is *not* usually right K-invariant. Indeed, the condition for right K-invariance is

$$\frac{d}{dt}\bigg|_{t=0} F(g\, e^{tx}) = (x \cdot F)(g) = (x \cdot F)(gk) = \frac{d}{dt}\bigg|_{t=0} F(gk\, e^{tx})$$

for $k \in K$. Using the right K-invariance of $F = f \circ q$,

$$F(gk\, e^{tx}) = F(g\, k e^{tx} k^{-1}\, k) = F(g\, e^{t \cdot kxk^{-1}})$$

Thus, unless $kxk^{-1} = x$ for all $k \in K$, it is unlikely that $x \cdot F$ is still right K-invariant. That is, the left G-invariant differential operators coming from $\mathfrak{g}$ usually do *not* descend to differential operators on G/K.

The differential operators in the G-invariant subalgebra

$$\mathfrak{z} = \{\alpha \in U\mathfrak{g} : g\alpha g^{-1}\}$$

do descend to G/K, exactly because of the commutation property, as follows. For any function φ on G let $(k \cdot \varphi)(g) = \varphi(gk)$. For F right K-invariant on G, for $\alpha \in Z(\mathfrak{g})$ compute directly

$$k \cdot (\alpha \cdot F) = \alpha \cdot (k \cdot F) = \alpha \cdot F$$

showing the right K-invariance of $\alpha \cdot F$. Thus, $\alpha \cdot F$ gives a well-defined function on G/K.

4.5 Example Computation: $SL_2(\mathbb{R})$ and $\mathfrak{H}$

Let $\mathfrak{g} = \mathfrak{sl}_2(\mathbb{R})$, the Lie algebra of $G = SL_2(\mathbb{R})$. A typical choice of basis for $\mathfrak{g}$ is[8]

$$H = \begin{pmatrix} 1 & 0 \\ 0 & -1 \end{pmatrix} \quad X = \begin{pmatrix} 0 & 1 \\ 0 & 0 \end{pmatrix} \quad Y = \begin{pmatrix} 0 & 0 \\ 1 & 0 \end{pmatrix}$$

[8] Yes, the notation is somewhat in conflict with previous use of X_x to denote the differential operator attached to $x \in \mathfrak{g}$.

These have the easily verified relations

$$[H, X] = HX - XH = 2X \quad [H, Y] = HY - YH = -2Y$$
$$[X, Y] = XY - YX = H$$

To see that the pairing $\langle x, y\rangle = \text{tr}(xy)$ is *nondegenerate*, use the stability of $\mathfrak{g}$ under *transpose* $v \to v^\top$, and then

$$\langle v, v^\top\rangle = \text{tr}(vv^\top) = 2a^2 + b^2 + c^2 \qquad (\text{for } v = \begin{pmatrix} a & b \\ c & -a \end{pmatrix})$$

We easily compute that

$$\langle H, H\rangle = 2 \quad \langle H, X\rangle = 0 \quad \langle H, Y\rangle = 0 \quad \langle X, Y\rangle = 1$$

Thus, for the basis H, X, Y we have *dual* basis $H^* = H/2, X^* = Y$, and $Y^* = X$, and in these coordinates the Casimir operator is

$$\Omega = HH^* + XX^* + YY^* = \tfrac{1}{2}H^2 + XY + YX \qquad (\text{now inside } U\mathfrak{g})$$

Since $XY - YX = H$,[9] the expression for Ω can be rewritten in various useful forms, such as

$$\Omega = \tfrac{1}{2}H^2 + XY + YX = \tfrac{1}{2}H^2 + XY - YX + 2YX = \tfrac{1}{2}H^2 + H + 2YX$$

and, similarly,

$$\begin{aligned}\Omega &= \tfrac{1}{2}H^2 + XY + YX = \tfrac{1}{2}H^2 + XY - (-YX) \\ &= \tfrac{1}{2}H^2 + 2XY - (XY - YX) = \tfrac{1}{2}H^2 + 2XY - H\end{aligned}$$

To obtain a G-invariant differential operator on the upper half-plane $\mathfrak{H}$ from Ω, use the G-space isomorphism $\mathfrak{H} \approx G/K$ where $K = SO_2(\mathbb{R})$ is the isotropy group of the point $i \in \mathfrak{H}$. Let $q : G \to G/K$ be the quotient map $q(g) = gK \longleftrightarrow g(i)$. A function f on $\mathfrak{H}$ naturally yields the right K-invariant function $f \circ q$

$$(f \circ q)(g) = f(g(i)) \qquad (\text{for } g \in G)$$

As earlier, for any $z \in \mathfrak{g}$ there is the corresponding left G-invariant differential operator on a function F on G by

$$(z \cdot F)(g) = \frac{d}{dt}\bigg|_{t=0} F(g e^{tz})$$

but these linear operators should not be expected to descend to operators on G/K. Nevertheless, G-invariant elements such as the Casimir operator Ω in $Z(\mathfrak{g})$ *do* descend.

9 The identity $XY - YX = H$ holds *both* in the universal enveloping algebra *and* as matrices.

The computation of Ω on $f \circ q$ can be simplified by using the right K-invariance of $f \circ q$, which implies that $f \circ q$ is annihilated by

$$\begin{aligned} \mathfrak{so}_2(\mathbb{R}) &= \text{Lie algebra of } SO_2(\mathbb{R}) \\ &= \text{skew-symmetric 2-by-2 real matrices} = \{\begin{pmatrix} 0 & t \\ -t & 0 \end{pmatrix} : t \in \mathbb{R}\} \end{aligned}$$

Thus, in terms of the basis H, X, Y above, $X - Y$ annihilates $f \circ q$.

Among other possibilities, a point $z = x + iy \in \mathfrak{H}$ is the image $x + iy = (n \cdot m)(i)$, where

$$n_x = \begin{pmatrix} 1 & x \\ 0 & 1 \end{pmatrix} \qquad m_y = \begin{pmatrix} \sqrt{y} & 0 \\ 0 & \frac{1}{\sqrt{y}} \end{pmatrix}$$

These are convenient group elements because they match the exponentiated Lie algebra elements:

$$e^{tX} = n_t \qquad e^{tH} = m_{e^{2t}}$$

In contrast, the exponentiated Y has a more complicated action on $\mathfrak{H}$. This suggests invocation of the fact that $X - Y$ acts trivially on right K-invariant functions on G. *Order matters*: application of a differential operator typically disrupts right K-invariance. For right K-invariant F on G,

$$\begin{aligned} (\Omega F)(n_x m_y) &= (\tfrac{1}{2}H^2 + XY + YX)F(n_x m_y) \\ &= (\tfrac{1}{2}H^2 + 2XY - XY + YX)F(n_x m_y) \\ &= (\tfrac{1}{2}H^2 + 2X^2 + 2X(Y - X) - [X, Y])F(n_x m_y) \\ &= (\tfrac{1}{2}H^2 + 2X^2 - H)F(n_x m_y) \end{aligned}$$

Compute the pieces separately. First, using the normalization identity

$$\begin{aligned} m_y n_x &= (m_y n_x m_y^{-1})\, m_y = \begin{pmatrix} \sqrt{y} & 0 \\ 0 & \frac{1}{\sqrt{y}} \end{pmatrix} \begin{pmatrix} 1 & x \\ 0 & 1 \end{pmatrix} \begin{pmatrix} \sqrt{y} & 0 \\ 0 & \frac{1}{\sqrt{y}} \end{pmatrix}^{-1} m_y \\ &= n_{yx}\, m_y \end{aligned}$$

we compute the effect of X

$$\begin{aligned} (X \cdot F)(n_x m_y) &= \frac{d}{dt}\Big|_{t=0} F(n_x m_y n_t) = \frac{d}{dt}\Big|_{t=0} F(n_x n_{yt} m_y) \\ &= \frac{d}{dt}\Big|_{t=0} F(n_{x+yt} m_y) = y\frac{\partial}{\partial x}F(n_x m_y) \end{aligned}$$

Thus,

$$2X^2 \longrightarrow 2(y\frac{\partial}{\partial x})^2 = 2y^2(\frac{\partial}{\partial x})^2$$

The action of H is

$$(H \cdot F)(n_x m_y) = \frac{d}{dt}\Big|_{t=0} F(n_x m_y m_{e^{2t}})$$
$$= \frac{d}{dt}\Big|_{t=0} F(n_x m_{ye^{2t}}) = 2y\frac{\partial}{\partial y}F(n_x m_y)$$

Then

$$\tfrac{1}{2}H^2 - H = \tfrac{1}{2}(2y\frac{\partial}{\partial y})^2 - (2y\frac{\partial}{\partial y}) = 2y^2(\frac{\partial}{\partial y})^2 + 2y\frac{\partial}{\partial y} - 2y\frac{\partial}{\partial y}$$
$$= 2y^2(\frac{\partial}{\partial y})^2$$

Altogether, on right K-invariant functions F,

$$(\Omega F)(n_x m_y) = 2y^2\left((\frac{\partial}{\partial x})^2 + (\frac{\partial}{\partial y})^2\right) F(m_x n_y)$$

That is, in the usual coordinates $z = x + iy$ on $\mathfrak{H}$, discarding the leading constant,

$$\text{(image of) } \Omega = y^2\left(\frac{\partial^2}{\partial x^2} + \frac{\partial^2}{\partial y^2}\right)$$

4.6 Example Computation: $SL_2(\mathbb{C})$

Let $\mathfrak{g} = \mathfrak{sl}_2(\mathbb{C})$, the Lie algebra of $G = SL_2(\mathbb{C})$, with basis

$$H = \begin{pmatrix} 1 & 0 \\ 0 & -1 \end{pmatrix} \quad H' = \begin{pmatrix} i & 0 \\ 0 & -i \end{pmatrix} \quad X = \begin{pmatrix} 0 & 1 \\ 0 & 0 \end{pmatrix}$$
$$X' = \begin{pmatrix} 0 & i \\ 0 & 0 \end{pmatrix} \quad Y = \begin{pmatrix} 0 & 0 \\ 1 & 0 \end{pmatrix} \quad Y' = \begin{pmatrix} 0 & 0 \\ -i & 0 \end{pmatrix}$$

with $[X, Y] = H$ and $[X', Y'] = H$, and so on. To see that the pairing $\langle x, y\rangle = \mathrm{Re}\,\mathrm{tr}(xy)$ is *nondegenerate*, use the stability of $\mathfrak{g}$ under *conjugate-transpose* $v \to v^* = \overline{v}^\top$, and then

$$\langle v, v^*\rangle = \mathrm{Re}\,\mathrm{tr}(vv^*) = 2|a|^2 + |b|^2 + |c|^2 \qquad (\text{for } v = \begin{pmatrix} a & b \\ c & -a \end{pmatrix})$$

We easily compute that

$$\langle H, H\rangle = 2 \quad \langle H', H'\rangle = -2 \quad \langle X, Y\rangle = 1 \quad \langle X', Y'\rangle = 1$$

and all other pairings give 0. Thus, for the basis H, X, Y, H', X', Y', we have *dual* basis $H^* = H/2$, $X^* = Y$, $Y^* = X$, $H'^* = -H'/2$, $X'^* = Y'$, $Y'^* = X'$. In these coordinates the Casimir operator is

$$\begin{aligned}\Omega &= HH^* + XX^* + YY^* + H'H'^* + X'X'^* + Y'Y'^* \\ &= \tfrac{1}{2}H^2 + XY + YX - \tfrac{1}{2}H'^2 + X'Y' + Y'X' \qquad (\text{in } U\mathfrak{g})\end{aligned}$$

Let $q : G \to G/K$ be the quotient map and use Iwasawa coordinates

$$n_x a_y k = \begin{pmatrix} 1 & x \\ 0 & 1 \end{pmatrix} \begin{pmatrix} \sqrt{y} & 0 \\ 0 & \frac{1}{\sqrt{y}} \end{pmatrix} \cdot k$$

with $x \in \mathbb{C}$, $y > 0$, and $k \in K = SU_2$. For any $z \in \mathfrak{g}$ there is the corresponding *left* G-invariant differential operator on a function F on G by

$$(z \cdot F)(g) = \frac{d}{dt}\Big|_{t=0} F(g e^{tz})$$

but these linear operators generally do not descend to operators on G/K, that is, are not *right* K-invariant. Nevertheless, G-invariant elements such Casimir *do* descend.

Differentiating the condition $1_2 = (e^{tx})^*(e^{tx})$ with respect to t gives

$$0 = \frac{d}{dt} 1_2 = x^*(e^{tx})^*(e^{tx}) + (e^{tx})^*(e^{tx})x$$

and setting $t = 0$ gives a necessary condition for x to be in the Lie algebra $\mathfrak{k}$ of $K = SU_2$, namely, $0 = x^* + x$. Conversely, for $x^* = -x$, exponentiation gives $1_2 = (e^{tx})^*(e^{tx})$. The determinant-one condition gives $\operatorname{tr} x = 0$. Thus, the Lie algebra $\mathfrak{k} = \mathfrak{su}_2$ of $K = SU_2$ is

$$\begin{aligned}\mathfrak{k} &= \text{skew-Hermitian 2-by-2 trace 0 complex matrices} \\ &= \{\begin{pmatrix} it & \tau \\ -\overline{\tau} & -it \end{pmatrix} : t \in \mathbb{R},\ \tau \in \mathbb{C}\}\end{aligned}$$

The computation of Ω on G/K is simplified by using the right K-invariance: the right action of $\mathfrak{k}$ *annihilates* right K-invariant functions on G. In terms of the basis H, X, Y, H', X', Y', $H', X - Y$ and $X' + Y'$ are all in $\mathfrak{k}$, so annihilate right K-invariant functions. *Order matters*: application of a differential operator typically disrupts right K-invariance. First, rearrange Ω in anticipation of application to right K-invariant f on G:

$$\begin{aligned}\Omega &= \tfrac{1}{2}H^2 + XY + YX - \tfrac{1}{2}H'^2 + X'Y' + Y'X' \\ &= \tfrac{1}{2}H^2 + 2XY - H - \tfrac{1}{2}H'^2 + 2X'Y' - H \\ &= \tfrac{1}{2}H^2 - 2H + 2X^2 + 2X(Y - X) - \tfrac{1}{2}H'^2 - 2X'^2 + 2X'(Y' + X')\end{aligned}$$

Since $H', X - Y$, and $X' + Y'$ annihilate right K-invariant functions f, this gives

$$\Omega f = (\tfrac{1}{2}H^2 - 2H + 2X^2 - 2X'^2)f$$

Compute the pieces separately. Use coordinates $x = x_1 + ix_2 \in \mathbb{C}$. Using $m_y\, n_x = n_{yt}\, m_y$, the effects of X and X' are

$$\begin{aligned}(X \cdot f)(n_x\, m_y) &= \frac{d}{dt}\Big|_{t=0} f(n_x\, m_y\, n_t) = \frac{d}{dt}\Big|_{t=0} f(n_x\, n_{yt}\, m_y)\\ &= \frac{d}{dt}\Big|_{t=0} f(n_{x+yt}\, m_y) = y\frac{\partial}{\partial x_1} f(n_x\, m_y)\\ (X' \cdot f)(n_x\, m_y) &= \frac{d}{dt}\Big|_{t=0} f(n_x\, m_y\, n_{it}) = \frac{d}{dt}\Big|_{t=0} f(n_x\, n_{iyt}\, m_y)\\ &= \frac{d}{dt}\Big|_{t=0} f(n_{x+iyt}\, m_y) = iy\frac{\partial}{\partial x_2} f(n_x\, m_y)\end{aligned}$$

Thus,

$$2X^2 \longrightarrow 2(y\frac{\partial}{\partial x_1})^2 = 2y^2(\frac{\partial}{\partial x_1})^2$$

and

$$-2X'^2 \longrightarrow -2(iy\frac{\partial}{\partial x_2})^2 = 2y^2(\frac{\partial}{\partial x_2})^2$$

As for $SL_2(\mathbb{R})$, the action of H is

$$\begin{aligned}(H \cdot f)(n_x\, m_y) &= \frac{d}{dt}\Big|_{t=0} f(n_x\, m_y\, m_{e^{2t}})\\ &= \frac{d}{dt}\Big|_{t=0} f(n_x\, m_{ye^{2t}}) = 2y\frac{\partial}{\partial y} f(n_x\, m_y)\end{aligned}$$

so

$$\begin{aligned}\tfrac{1}{2}H^2 - 2H &= \tfrac{1}{2}(2y\frac{\partial}{\partial y})^2 - 2(2y\frac{\partial}{\partial y}) = 2y^2(\frac{\partial}{\partial y})^2 + 2y\frac{\partial}{\partial y} - 4y\frac{\partial}{\partial y}\\ &= 2y^2(\frac{\partial}{\partial y})^2 - 2y\frac{\partial}{\partial y}\end{aligned}$$

Altogether, on right K-invariant functions f, discarding the irrelevant leading constant 2,

$$\Omega \longrightarrow y^2\left(\frac{\partial^2}{\partial x_1^2} + \frac{\partial^2}{\partial x_2^2} + \frac{\partial^2}{\partial y^2}\right) - y\frac{\partial}{\partial y}$$

In contrast to the situation for $SL_2(\mathbb{R})$ and $\mathfrak{H}$, this is *not* merely a multiple of the Euclidean Laplacian.

4.7 Example Computation: $Sp^*_{1,1}$

Let $\mathfrak{g} = \mathfrak{sp}^*_{1,1}$, the Lie algebra of $G = Sp^*_{1,1}$, with 10-element basis

$$H = \begin{pmatrix} 1 & 0 \\ 0 & -1 \end{pmatrix} \quad H_\ell = \begin{pmatrix} \ell & 0 \\ 0 & -\ell \end{pmatrix} \quad X_\ell = \begin{pmatrix} 0 & \ell \\ 0 & 0 \end{pmatrix} \quad Y_\ell = \begin{pmatrix} 0 & 0 \\ -\ell & 0 \end{pmatrix}$$

with $\ell = i, j, k$. Note that

$$[X_i, Y_i] = [X_j, Y_j] = [X_k, Y_k] = H$$

With $\mathrm{tr}(a + bi + cj + dk) = a$, use the pairing

$$\left\langle \begin{pmatrix} \alpha & \beta \\ \gamma & \delta \end{pmatrix}, \begin{pmatrix} \alpha' & \beta' \\ \gamma' & \delta' \end{pmatrix} \right\rangle = \mathrm{tr}\Big(\alpha \cdot \alpha' + \beta \cdot \gamma' + \gamma \cdot \beta' + \delta \cdot \delta'\Big)$$

The nondegeneracy follows from the stability of $\mathfrak{g}$ under *conjugate-transpose* $v \to v^*$, and then noting that $v \to \langle v, v^* \rangle$ is positive-definite. Compute that

$$\langle H, H \rangle = 2 \qquad \langle H_\ell, H_\ell \rangle = -2 \qquad \langle X_\ell, Y_\ell \rangle = 1$$

and all other pairings give 0. Thus, we have *dual* basis $H^* = H/2$, $H_\ell^* = -H_\ell/2$, $X_\ell^* = Y_\ell$, $Y_\ell^* = X_\ell$. In these coordinates the Casimir operator is

$$\begin{aligned}\Omega &= HH^* + \sum_{\ell=i,j,k} (H_\ell H_\ell^* + X_\ell X_\ell^* + Y_\ell Y_\ell^*) \\ &= \tfrac{1}{2}H^2 + \sum_{\ell=i,j,k} (-\tfrac{1}{2}H_\ell^2 + X_\ell Y_\ell + Y_\ell X_\ell)\end{aligned}$$

in $U\mathfrak{g}$. Use Iwasawa coordinates

$$n_x a_y k = \begin{pmatrix} 1 & x \\ 0 & 1 \end{pmatrix} \begin{pmatrix} \sqrt{y} & 0 \\ 0 & \frac{1}{\sqrt{y}} \end{pmatrix} \cdot k$$

with $x = ix_1 + jx_2 + kx_3 \in \mathbb{H}$, $y > 0$, and $k \in K$. The Lie algebra $\mathfrak{k}$ of the compact subgroup $K \approx Sp_1^* \times Sp_1^*$ of G can be identified by observing two copies of the Lie algebra of Sp_1^*, as follows. By differentiating,

$$\begin{aligned} 0 = \frac{d}{dt}1 &= \frac{d}{dt}\overline{(e^t x)}(e^t x) = \frac{d}{dt}(e^t \overline{x})(e^t x) \\ &= \overline{x} \cdot (e^t \overline{x})(e^t x) + (e^t \overline{x})(e^t x)x \end{aligned}$$

and at $t = 0$ this is $\overline{x} + x = 0$. We can observe two suitable copies of this inside $\mathfrak{sp}^*_{1,1}$, by

$$\mathfrak{k} = \{\begin{pmatrix} \alpha & 0 \\ 0 & -\alpha \end{pmatrix} : \overline{\alpha} = -\alpha\} \oplus \{\begin{pmatrix} 0 & \beta \\ -\beta & 0 \end{pmatrix} : \overline{\beta} = -\beta\}$$

The computation of Ω on G/K is simplified by using the right K-invariance, which entails annihilation by $\mathfrak{k}$. In terms of the basis above, H_ℓ and $X_\ell + Y_\ell$, for $\ell = i, j, k$, are all in $\mathfrak{k}$. Rearrange Ω in anticipation of application to right K-invariant f on G:

$$\begin{aligned}
\Omega &= \tfrac{1}{2}H^2 + \sum_{\ell=i,j,k} (-\tfrac{1}{2}H_\ell^2 + X_\ell Y_\ell + Y_\ell X_\ell) \\
&= \tfrac{1}{2}H^2 + \sum_{\ell=i,j,k} (-\tfrac{1}{2}H_\ell^2 + 2X_\ell Y_\ell - X_\ell Y_\ell + Y_\ell X_\ell) \\
&= \tfrac{1}{2}H^2 + \sum_{\ell=i,j,k} (-\tfrac{1}{2}H_\ell^2 + 2X_\ell Y_\ell - [X_\ell, Y_\ell]) \\
&= \tfrac{1}{2}H^2 - 3H + \sum_{\ell=i,j,k} (-\tfrac{1}{2}H_\ell^2 - 2X_\ell^2 + X_\ell(X_\ell + Y_\ell))
\end{aligned}$$

Since the elements H_ℓ and $X_\ell + Y_\ell$ annihilate right K-invariant functions f, this gives

$$\Omega f = \Big(\tfrac{1}{2}H^2 - 3H - \sum_{\ell=i,j,k} 2X_\ell^2\Big) f$$

Compute the pieces separately, with coordinates $x = ix_1 + jx_2 + kx_3 \in \mathbb{H}$. Using $m_y\, n_x = n_{yt}\, m_y$, the effects of X_i, X_j, X_k are

$$\begin{aligned}
(X_i \cdot f)(n_x\, m_y) &= \frac{d}{dt}\Big|_{t=0} f(n_x\, m_y\, n_{it}) = \frac{d}{dt}\Big|_{t=0} f(n_x\, n_{iyt}\, m_y) \\
&= \frac{d}{dt}\Big|_{t=0} f(n_{x+iyt}\, m_y) = y\frac{\partial}{\partial x_1} f(n_x\, m_y) \\
(X_j \cdot f)(n_x\, m_y) &= \frac{d}{dt}\Big|_{t=0} f(n_x\, m_y\, n_{jt}) = \frac{d}{dt}\Big|_{t=0} f(n_x\, n_{jyt}\, m_y) \\
&= \frac{d}{dt}\Big|_{t=0} f(n_{x+jyt}\, m_y) = jy\frac{\partial}{\partial x_2} f(n_x\, m_y) \\
(X_k \cdot f)(n_x\, m_y) &= \frac{d}{dt}\Big|_{t=0} f(n_x\, m_y\, n_{kt}) = \frac{d}{dt}\Big|_{t=0} f(n_x\, n_{kyt}\, m_y) \\
&= \frac{d}{dt}\Big|_{t=0} f(n_{x+kyt}\, m_y) = ky\frac{\partial}{\partial x_3} f(n_x\, m_y)
\end{aligned}$$

Thus,

$$\begin{aligned}
-2X_i^2 - 2X_j^2 - 2X_k^2 &\longrightarrow -2(iy\frac{\partial}{\partial x_1})^2 - 2(jy\frac{\partial}{\partial x_2})^2 - 2(ky\frac{\partial}{\partial x_3})^2 \\
&= 2y^2\left(\frac{\partial^2}{\partial x_1^2} + \frac{\partial^2}{\partial x_2^2} + \frac{\partial^2}{\partial x_3^2}\right)
\end{aligned}$$

As for $SL_2(\mathbb{R})$ and $SL_2(\mathbb{C})$, the action of H is

$$(H \cdot f)(n_x m_y) = \frac{d}{dt}\Big|_{t=0} f(n_x m_y m_{e^{2t}}) = \frac{d}{dt}\Big|_{t=0} f(n_x m_{ye^{2t}}) = 2y\frac{\partial}{\partial y} f(n_x m_y)$$

so

$$\tfrac{1}{2}H^2 - 3H = \tfrac{1}{2}(2y\frac{\partial}{\partial y})^2 - 3(2y\frac{\partial}{\partial y}) = 2y^2(\frac{\partial}{\partial y})^2 + 2y\frac{\partial}{\partial y} - 6y\frac{\partial}{\partial y} = 2y^2(\frac{\partial}{\partial y})^2 - 4y\frac{\partial}{\partial y}$$

Altogether, on right K-invariant functions f, discarding the irrelevant leading constant 2,

$$\Omega \longrightarrow y^2\left(\frac{\partial^2}{\partial x_1^2} + \frac{\partial^2}{\partial x_2^2} + \frac{\partial^2}{\partial x_3^2} + \frac{\partial^2}{\partial y^2}\right) - 2y\frac{\partial}{\partial y}$$

Again, as with $SL_2(\mathbb{C})$, in contrast to the situation for $SL_2(\mathbb{R})$ and $\mathfrak{H}$, this is *not* merely a multiple of the Euclidean Laplacian.

4.8 Example Computation: $SL_2(\mathbb{H})$

Let $\mathfrak{g} = \mathfrak{sl}_2(\mathbb{H})$, the Lie algebra of $G = SL_2(\mathbb{H})$. Letting ℓ run over i, j, k, take basis

$$H = \begin{pmatrix} 1 & 0 \\ 0 & -1 \end{pmatrix} \quad H_\ell = \begin{pmatrix} \ell & 0 \\ 0 & -\ell \end{pmatrix} \quad H'_\ell = \begin{pmatrix} 0 & 0 \\ 0 & \ell \end{pmatrix}$$

$$X = \begin{pmatrix} 0 & 1 \\ 0 & 0 \end{pmatrix} \quad X_\ell = \begin{pmatrix} 0 & \ell \\ 0 & 0 \end{pmatrix} \quad Y = \begin{pmatrix} 0 & 0 \\ 1 & 0 \end{pmatrix} \quad Y_\ell = \begin{pmatrix} 0 & 0 \\ -\ell & 0 \end{pmatrix}$$

Note that

$$[X, Y] = [X_i, Y_i] = [X_j, Y_j] = [X_k, Y_k] = H$$

With $\mathrm{tr}(a + bi + cj + dk) = a$, as with $Sp^*_{1,1}$, use the pairing

$$\left\langle \begin{pmatrix} \alpha & \beta \\ \gamma & \delta \end{pmatrix}, \begin{pmatrix} \alpha' & \beta' \\ \gamma' & \delta' \end{pmatrix} \right\rangle = \mathrm{tr}\Big(\alpha \cdot \alpha' + \beta \cdot \gamma' + \gamma \cdot \beta' + \delta \cdot \delta'\Big)$$

Compute that

$$\langle H, H\rangle = 2 \quad \langle H_\ell, H_\ell\rangle = -2 \qquad \langle H'_\ell, H'_\ell\rangle = -1 \quad \langle X_\ell, Y_\ell\rangle = 1$$

and all other pairings give 0. Thus, we have *dual* basis $H^* = H/2$, $H_\ell^* = -H_\ell/2$, $H_\ell'^* = -H_\ell'$, $X_\ell^* = Y_\ell$, $Y_\ell^* = X_\ell$. In these coordinates the Casimir operator is

$$\begin{aligned}\Omega &= HH^* + XX^* + YY^* + \sum_{\ell=i,j,k} (H_\ell H_\ell^* + H_\ell' H_\ell'^* + X_\ell X_\ell^* + Y_\ell Y_\ell^*)\\ &= \tfrac{1}{2}H^2 + XY + YX + \sum_{\ell=i,j,k} (-\tfrac{1}{2}H_\ell^2 - H_\ell' H_\ell' + X_\ell Y_\ell + Y_\ell X_\ell)\end{aligned}$$

in $U\mathfrak{g}$. Use Iwasawa coordinates

$$n_x a_y k = \begin{pmatrix} 1 & x \\ 0 & 1 \end{pmatrix} \begin{pmatrix} \sqrt{y} & 0 \\ 0 & \frac{1}{\sqrt{y}} \end{pmatrix} \cdot k$$

with $x = x_1 + ix_2 + jx_3 + kx_4 \in \mathbb{H}$, $y > 0$, and $k \in K$. To determine the Lie algebra $\mathfrak{k}$ of the compact subgroup $K = Sp_2^*$ of G, differentiate

$$0 = \frac{d}{dt} 1_2 = \frac{d}{dt}\Big((e^{tx})^*(e^{tx})\Big) = x^*(e^{tx})^*(e^{tx}) + (e^{tx})^*(e^{tx})x$$

and at $t = 0$ obtain $x^* + x = 0$. As usual, the converse follows by exponentiating. Thus,

$$\mathfrak{k} = \mathfrak{sp}_2^* = \{\begin{pmatrix} a & b \\ -\bar{b} & d \end{pmatrix} : \bar{a} = -a,\ \bar{d} = -d,\ a, b, d \in \mathbb{H}\}$$

The computation of Ω on G/K can be simplified by using the right K-invariance, which entails annihilation by $\mathfrak{k}$. The basis elements H_i, H_j, H_k and H_i', H_j', H_k' are in $\mathfrak{k}$, as are $X - Y$ and $X_i + Y_i$, $X_j + Y_j$, and $X_k + Y_k$. Rearrange Ω anticipating application to right K-invariant f on G:

$$\begin{aligned}\Omega &= \tfrac{1}{2}H^2 + XY + YX\\ &\quad + \sum_{\ell=i,j,k} (-\tfrac{1}{2}H_\ell^2 - H_\ell' H_\ell' + X_\ell Y_\ell + Y_\ell X_\ell)\\ &= \tfrac{1}{2}H^2 + 2XY + YX - XY\\ &\quad + \sum_{\ell=i,j,k} (-\tfrac{1}{2}H_\ell^2 - H_\ell' H_\ell' + 2X_\ell Y_\ell - X_\ell Y_\ell + Y_\ell X_\ell)\\ &= \tfrac{1}{2}H^2 + 2XY - H\\ &\quad + \sum_{\ell=i,j,k} (-\tfrac{1}{2}H_\ell^2 - H_\ell' H_\ell' + 2X_\ell Y_\ell - H)\\ &= \tfrac{1}{2}H^2 - 4H + 2X^2 + 2X(-X + Y)\\ &\quad + \sum_{\ell=i,j,k} (-\tfrac{1}{2}H_\ell^2 - H_\ell' H_\ell' - 2X_\ell^2 + 2X_\ell(X_\ell + Y_\ell))\end{aligned}$$

Since the elements H_ℓ, H'_ℓ, $X - Y$, and $X_\ell + Y_\ell$ annihilate right K-invariant functions f, this gives

$$\Omega f = \left(\tfrac{1}{2}H^2 - 4H - \sum_{\ell=i,j,k} 2X_\ell^2\right)f$$

The individual terms are computed as in the previous three cases. For example,

$$\begin{aligned} X^2 - X_i^2 - X_j^2 - X_k^2 \longrightarrow& (y\frac{\partial}{\partial x_1})^2 - (iy\frac{\partial}{\partial x_2})^2 - (jy\frac{\partial}{\partial x_3})^2 - (ky\frac{\partial}{\partial x_4})^2 \\ =& \; y^2\left(\frac{\partial^2}{\partial x_1^2} + \frac{\partial^2}{\partial x_2^2} + \frac{\partial^2}{\partial x_3^2} + \frac{\partial^2}{\partial x_4^2}\right) \end{aligned}$$

and

$$\begin{aligned} \tfrac{1}{2}H^2 - 4H =& \tfrac{1}{2}(2y\frac{\partial}{\partial y})^2 - 4(2y\frac{\partial}{\partial y}) = 2y^2(\frac{\partial}{\partial y})^2 + 2y\frac{\partial}{\partial y} - 8y\frac{\partial}{\partial y} \\ =& 2y^2(\frac{\partial}{\partial y})^2 - 6y\frac{\partial}{\partial y} \end{aligned}$$

Altogether, on right K-invariant functions f, discarding the irrelevant leading constant 2,

$$\Omega \longrightarrow y^2\left(\frac{\partial^2}{\partial x_1^2} + \frac{\partial^2}{\partial x_2^2} + \frac{\partial^2}{\partial x_3^2} + \frac{\partial^2}{\partial x_4^2} + \frac{\partial^2}{\partial y^2}\right) - 3y\frac{\partial}{\partial y}$$

Again, as with $SL_2(\mathbb{C})$ and $Sp^*_{1,1}$, in contrast to $SL_2(\mathbb{R})$, this is *not* merely a multiple of the Euclidean Laplacian.

4.A Appendix: Brackets

Here we prove the basic result about intrinsic derivatives. Let G act on itself by *right translations*, and on functions on G by $(g \cdot f)(h) = f(hg)$, for $g, h \in G$. For $x \in \mathfrak{g}$, the corresponding differential operator X_x on smooth functions f on G is $(X_x f)(h) = \frac{d}{dt}\big|_{t=0} f(h \cdot e^{tx})$.

[4.A.1] Theorem: $[X_x, X_y] = X_{x,y}$, that is, $X_x \circ X_y - X_y \circ, X_x = X_{[x,y]}$ for $x, y \in \mathfrak{g}$.

Proof: First, *recharacterize* the Lie algebra $\mathfrak{g}$ less formulaically. The *tangent space* $T_m M$ to a smooth manifold M at a point $m \in M$ is *intended* to be the collection of first-order (homogeneous) differential operators, on functions near m, followed by *evaluation* of the resulting functions at the point m.

One way to make the description of the tangent space precise is as follows. Let $\mathcal{O}$ be the ring of germs[10] of smooth functions at m. Let $e_m : f \to f(m)$ be the evaluation-at-m map $\mathcal{O} \to \mathbb{R}$ on (germs of) functions in $\mathcal{O}$. Since evaluation is a ring homomorphism, (and $\mathbb{R}$ is a field) the kernel $\mathfrak{m}$ of e_m is a *maximal* ideal in $\mathcal{O}$. A first-order homogeneous differential operator D might be characterized by the *Leibniz rule*

$$D(f \cdot F) = Df \cdot F + f \cdot DF$$

Then $e_m \circ D$ vanishes on $\mathfrak{m}^2$, since

$$\begin{aligned}(e_m \circ D)(f \cdot F) &= f(m) \cdot DF(m) + Df(m) \cdot F(m) \\ &= 0 \cdot DF(m) + Df(m) \cdot 0 = 0 \qquad (\text{for } f, F \in \mathfrak{m})\end{aligned}$$

Thus, D gives a linear functional on $\mathfrak{m}$ that factors through $\mathfrak{m}/\mathfrak{m}^2$. *Define*

$$\text{tangent space to } M \text{ at } m = T_m M = (\mathfrak{m}/\mathfrak{m}^2)^* = \mathrm{Hom}_{\mathbb{R}}(\mathfrak{m}/\mathfrak{m}^2, \mathbb{R})$$

To see that we have included exactly what we want, and nothing more, use the defining fact (for *manifold*) that m has a neighborhood U and a homeomorphism-to-image $\varphi : U \to \mathbb{R}^n$.[11] The precise definition of *smoothness* of a function f near m is that $f \circ \varphi^{-1}$ be smooth on some subset of $\varphi(U)$.[12] In brief, the nature of $\mathfrak{m}/\mathfrak{m}^2$ and $(\mathfrak{m}/\mathfrak{m}^2)^*$ can be immediately transported to an open subset of $\mathbb{R}^n$. From Maclaurin-Taylor expansions, the pairing

$$v \times f \longrightarrow (\nabla f)(m) \cdot v \qquad (\text{for } v \in \mathbb{R}^n \text{ and } f \text{ smooth at } m \in \mathbb{R}^n)$$

induces an isomorphism $\mathbb{R}^n \to (\mathfrak{m}/\mathfrak{m}^2)^*$. Thus, $(\mathfrak{m}/\mathfrak{m}^2)^*$ is a good notion of *tangent space*.

10 The *germ* of a smooth function f near a point x_o on a smooth manifold M is the equivalence class of f under the equivalence relation $\sim$, where $f \sim g$ if f, g are smooth functions defined on some neighborhoods of x_o, and which *agree* on *some* neighborhood of x_o. This is a *construction*, which does admit a more functional reformulation. That is, for each neighborhood U of x_o, let $\mathcal{O}(\mathcal{U})$ be the ring of smooth functions on U, and for $U \supset V$ neighborhoods of x_o let $\rho_{UV} : \mathcal{O}(\mathcal{U}) \to \mathcal{O}(\mathcal{V})$ be the restriction map. Then the *colimit* $\mathrm{colim}_U \mathcal{O}(\mathcal{U})$ is exactly the ring of germs of smooth functions at x_o.

11 This map φ is presumably part of an *atlas*, meaning a maximal family of *charts* (homeomorphisms-to-image) φ_i of opens U_i in M to subsets of a fixed $\mathbb{R}^n$, with the *smooth* manifold property that on *overlaps* things fit together smoothly, in the sense that

$$\varphi_i \circ \varphi_j^{-1} : \varphi_j(U_i \cap U_j) \longrightarrow U_i \cap U_j \longrightarrow \varphi_i(U_i \cap U_j)$$

is a *smooth* map from the subset $\varphi_j(U_i \cap U_j)$ of $\mathbb{R}^n$ to the subset $\varphi_i(U_i \cap U_j)$.

12 The well-definedness of this definition depends on the *maximality* property of an *atlas*.

[4.A.2] Claim: The Lie algebra $\mathfrak{g}$ of G is naturally identifiable with the tangent space to G at 1, via

$$x \times f \longrightarrow \left.\frac{d}{dt}\right|_{t=0} f(e^{tx}) \qquad \text{(for } x \in \mathfrak{g} \text{ and } f \text{ smooth near 1)}$$

Proof: The exponential map is a diffeomorphism of the Lie algebra $\mathfrak{g}$ to its image, and the image is a neighborhood of the identity in G. For *linear* Lie groups, the invertibility is immediate from existence of an explicit local inverse to the exponential near 1, given by the usual logarithm. ///

Left translation action of G on functions on G is $(L_g f)(h) = f(g^{-1}h)$ with $g, h \in G$, with the inverse for associativity, as usual.

[4.A.3] Claim: The map $x \longrightarrow X_x$ gives an $\mathbb{R}$-linear isomorphism

$$\mathfrak{g} \longrightarrow \text{left } G\text{-invariant vector fields on } G$$

Proof: (of claim) On one hand, since the action of x is on the *right*, it is not surprising that X_x is invariant under the *left* action of G, namely,

$$(X_x \circ L_g)f(h) = X_x f(g^{-1}h) = \left.\frac{d}{dt}\right|_{t=0} f(g^{-1}he^{tx})$$
$$= L_g \left.\frac{d}{dt}\right|_{t=0} f(he^{tx}) = (L_g \circ X_x)f(h)$$

On the other hand, for a left-invariant vector field X,

$$(Xf)(h) = (L_h^{-1} \circ X)f(1) = (X \circ L_h^{-1})f(1) = X(L_h^{-1}f)(1)$$

That is, X is completely determined by what it does to functions at 1.

Let $\mathfrak{m}$ be the maximal ideal of functions vanishing at 1, in the ring $\mathcal{O}$ of germs of smooth functions at 1 on G. The first-order nature of *vector fields* is captured by the Leibniz rule $X(f \cdot F) = f \cdot XF + Xf \cdot F$. As above, the Leibniz rule implies that $e_1 \circ X$ *vanishes* on $\mathfrak{m}^2$. Thus, we can identify $e_1 \circ X$ with an element of

$$(\mathfrak{m}/\mathfrak{m}^2)^* = \text{Hom}_{\mathbb{R}}(\mathfrak{m}/\mathfrak{m}^2, \mathbb{R}) = \text{tangent space to } G \text{ at } 1 = \mathfrak{g}$$

Thus, the map $x \to X_x$ is an isomorphism from $\mathfrak{g}$ to left invariant vector fields, proving the claim. ///

Now use the recharacterized $\mathfrak{g}$ to prove $[X_x, X_y] = X_z$ for *some* $z \in \mathfrak{g}$. Consider $[X_x, X_y]$ for $x, y \in \mathfrak{g}$. That this differential operator is *left* G-invariant is clear, since it is a difference of composites of such. It is *less* clear that it satisfies Leibniz's rule (and thus is *first order*). But, indeed, for *any* two vector fields X, Y,

$$\begin{aligned}[X,Y](fF) &= XY(fF) - YX(Ff) \\ &= X(Yf\cdot F + f\cdot YF) - Y(Xf\cdot F + f\cdot XF) \\ &= (XYf\cdot F + Yf\cdot XF + Xf\cdot YF + f\cdot XYF) \\ &\quad - (YXf\cdot F + Xf\cdot YF + Yf\cdot XF + f\cdot YXF) \\ &= [X,Y]f\cdot F + f\cdot [X,Y]F\end{aligned}$$

so $[X, Y]$ *does* satisfy the Leibniz rule. In particular, $[X_x, X_y]$ is again a left-G-invariant vector field, so is of the form $[X_x, X_y] = X_z$ for *some* $z \in \mathfrak{g}$.

In fact, the relation $[X_x, X_y] = X_z$ is the intrinsic *definition* of the Lie bracket on $\mathfrak{g}$, since we could *define* the element $z = [x, y]$ by the relation $[X_x, X_y] = X_{[x,y]}$. However, we are burdened by having the *ad hoc* but convenient definition $[x, y] = xy - yx$ in terms of matrix multiplication. However, our assumption that G is a subgroup of some $GL_n(\mathbb{R})$ or $GL_n(\mathbb{C})$ allows us to use the explicit exponential and a local logarithm inverse to it, to determine the bracket $[X_x, X_y]$ somewhat more intrinsically, as follows.

Consider *linear* functions on $\mathfrak{g}$, *locally* transported to G via locally inverting the exponential near $1 \in G$. Thus, for $\lambda \in \mathfrak{g}^*$, near $1 \in G$, define

$$f(e^x) = \lambda(x)$$

Then

$$[X_x, X_y]f_\lambda(1) = \frac{d}{dt}\Big|_{t=0}\frac{d}{ds}\Big|_{s=0}\Big(\lambda\big(\log(e^{sx}e^{ty})\big) - \lambda(\log(e^{ty}e^{sx}))\Big)$$

Dropping $O(s^2)$ and $O(t^2)$ terms, this is

$$\begin{aligned}&= \frac{d}{dt}\Big|_{t=0}\frac{d}{ds}\Big|_{s=0}\Big(\lambda\big(\log(1+sx)(1+ty)\big) - \lambda\big(\log(1+ty)(1+sx)\big)\Big) \\ &= \frac{d}{dt}\Big|_{t=0}\frac{d}{ds}\Big|_{s=0}\lambda\big(\log(1+sx+ty+stxy) - \log(1+ty+sx+styx)\big) \\ &= \frac{d}{dt}\Big|_{t=0}\frac{d}{ds}\Big|_{s=0}\lambda\Big((sx+ty+stxy-\tfrac{1}{2}(sx+ty)^2) \\ &\quad - (ty+sx+styx-\tfrac{1}{2}(ty+sx)^2)\Big) \\ &= \frac{d}{dt}\Big|_{t=0}\frac{d}{ds}\Big|_{s=0}\lambda\Big((stxy-\tfrac{1}{2}stxy-\tfrac{1}{2}styx) \\ &\quad - (styx-\tfrac{1}{2}stxy-\tfrac{1}{2}styx)\Big) \\ &= \frac{d}{dt}\Big|_{t=0}\frac{d}{ds}\Big|_{s=0} st\cdot\lambda(xy-yx) = \lambda(xy-yx)\end{aligned}$$

where the multiplication and commutator $xy - yx$ is in the ring of matrices. Thus, since $\mathfrak{g}^*$ separates points on $\mathfrak{g}$, we have the equality $[X_x, X_y] = X_{[x,y]}$ with the ad hoc definition of $[x, y]$. ///

4.B Appendix: Existence and Uniqueness

The characterization of the Lie algebra of a subgroup G of $GL_n(F)$ with $F = \mathbb{R}, \mathbb{C}$, or $\mathbb{H}$, as $\mathfrak{g} = \{x : e^{tx} \in G, \text{ for all } t \in \mathbb{R}\}$ produces a set $\mathfrak{g}$ closed under scalar multiplication but not obviously closed under *addition*, although in the foregoing explicit examples, this latter property is clear.

One way to prove closure under addition is verification that the matrix exponentiation characterization of $\mathfrak{g}$ really does produce the tangent space to G at 1, since the tangent space is a vectorspace. For matrix x such that $e^{tx} \in G$ for small real t, the map $t \to e^{tx}$ is a curve inside G, and in an extrinsic-geometry sense x is a tangent vector to G at 1. The converse is not as elementary, namely, given a tangent vector x to G at 1, show that $e^{tx} \in G$ for all $t \in \mathbb{R}$.

The curve $u(t) = e^{tx}$ certainly lies in $GL_n(F)$ and satisfies the differential equation $\frac{d}{dt}u = u \cdot x$ with initial condition $u(0) = 1_n$. This differential equation can be viewed as a differential equation on G, and also as a differential equation on $GL_n(F)$, and $u(t) = e^{tx}$ a solution in $GL_n(F)$. To prove that in fact $e^{tx} \in G$ for small t would follow from *uniqueness* of solutions to this differential equation on $GL_n(F)$, and from *existence* of a solution (for small t) on G. Then because G is a group, $e^{tx} \in G$ for all $t \in \mathbb{R}$.

[4.B.1] Theorem: For smooth F on an open subset Ω of $\mathbb{R}^2$, and for $(x_o, y_o) \in \Omega$, the equation $\frac{df}{dx} = F(x, f(x))$ has a unique differentiable solution f on a neighborhood of (x_o, y_o) with $f(x_o) = y_o$. This solution is smooth.

Proof: Picard iteration converts the differential equation to an equivalent *integral* equation to prove *existence*. Uniqueness of fixed points of *contractive mappings* proves uniqueness. Assuming $f' = df/dx$ exists as a pointwise-valued function and f is continuous, the relation $df/dx = F(x, f(x))$ shows that f' is continuous. Thus, by the fundamental theorem of calculus,

$$f(x) = f(x_o) + \int_{x_o}^{x} f'(t)\, dt = y_o + \int_{x_o}^{x} F(t, f(t))\, dt$$

That is, f satisfies the integral equation

$$f(x) = y_o + \int_{x_o}^{x} F(t, f(t))\, dt$$

Conversely, for continuous f satisfying this integral equation, by the fundamental theorem of calculus f is differentiable and

$$f'(x) = F(x, f(x)) \qquad \text{(and } f \text{ satisfies } f(x_o) = y_o\text{)}$$

Thus, for continuous f, the integral equation is equivalent to the differential equation and initial value. Without loss of generality, $x_o = y_o = 0$. Picard's iteration scheme is to take $f_o(x) = 0$, and iterate:

$$f_{n+1}(x) = \int_0^x F(t, f_n(t))\, dt$$

These are continuous functions. The claim is that, on a sufficiently small neighborhood of $x = 0$, these f_n approach a solution to the integral equation on that interval, proving *existence*.

We should check that, with x restricted to a small-enough interval $|x| \le \delta$, the pairs $(x, f_n(x))$ stay inside Ω. By smoothness of F, given a finite rectangle

$$R = \{|x| \le \delta,\ |y| \le \eta\} \subset \Omega$$

there is a constant B such that $|F(x, y)| \le B$ for all $(x, y) \in R$. Shrink δ so that $0 < \delta < B^{-1}$. Assuming the pairs $(x, f_n(x))$ are inside R,

$$|f_{n+1}(x)| \ \le\ \int_0^x |F(t, f_n(t))|\, dt \ \le\ \int_0^x B\, dt \ \le\ \delta \cdot B$$

Thus, further shrinking δ so that $\delta \cdot B \le \eta$, the restriction $|x| \le \delta$ produces functions f_n with $(x, f_n(x))$ staying inside $R \subset \Omega$ of F.

We show that, possibly further shrinking δ, for $|x| \le \delta$ the sequence of functions f_n converges in sup norm to a solution of the integral equation. The natural estimate succeeds: first,

$$\sup_{|x|\le\delta} |f_{n+1}(x) - f_n(x)| \ \le\ \int_0^\delta |F(t, f_n(t)) - F(t, f_{n-1}(t))|\, dt$$

Since F is smooth, for a fixed compact-closure neighborhood U of $(0, 0)$ there is a constant C such that

$$|F(x, y) - F(x, y')| \ \le\ C \cdot |y - y'|$$

for $(x, y) \in U$ and $(x, y') \in U$. Thus,

$$\sup_{|x|\le\delta} |f_{n+1}(x) - f_n(x)| \ \le\ |\delta| \cdot C \cdot \sup_{|x|\le\delta} |f_n(x) - f_{n-1}(x)|$$

Shrinking δ so that $\delta \cdot C \le \frac{1}{2}$, for example, gives convergence in sup norm to a continuous function f. This further shrinking of δ occurs just once, *not* for each n.

To show that f is a solution of the integral equation, given $\varepsilon > 0$, take N large enough so that the sup norm of $f_m - f_n$ is less than ε for all $m, n \geq N$. Then the sup norm of $f - f_{N+1}$ is $\leq \varepsilon$, and

$$\left|f(x) - \int_0^x F(t, f(t))\, dt\right|$$
$$\leq |f(x) - f_{N+1}(x)| + \int_0^x |F(t, f_N(t)) - F(t, f(t))|\, dt$$
$$\leq \varepsilon + \int_0^x C \cdot |f_N(t) - f(t)|\, dt \;\leq\; \varepsilon + \delta \cdot C \cdot \varepsilon$$

This holds for every $\varepsilon > 0$, so we have equality, proving f is a solution of the integral equation.

Toward *uniqueness*, we showed above that the mapping

$$Tg(x) = \int_0^x F(t, g(t))\, dt$$

maps continuous functions g on $|x| \leq \delta$ with bounds $|g(x)| \leq \eta$ to continuous functions with the same bound, for sufficiently small δ and η. Let X be the set of such functions. With metric given by sup norm, X is *complete*. Shrinking δ if necessary, as above we have the *contractive mapping* property

$$\sup_x |Tg(x) - Th(x)| \;\leq\; \tfrac{1}{2} \cdot \sup_x |g(x) - h(x)|$$

Given two solutions g, h to the integral equation,

$$\sup_x |g(x) - h(x)| \;\leq\; \sup_x |Tg(x) - Th(x)| \leq \tfrac{1}{2} \cdot \sup_x |g(x) - h(x)|$$

proving that $g(x) = h(x)$, giving *uniqueness* of solution to the integral equation.

Smoothness follows from the differential equation: granting $f \in C^k$, the relation $df/dx = F(x, f(x))$ exhibits the derivative as C^k. By induction, $f \in C^\infty$. ///

5

Integration on Quotients

The simplest case of *unwinding* is for $f \in C_c^o(\mathbb{R})$:

$$\int_{\mathbb{R}/\mathbb{Z}} \Big(\sum_{n\in\mathbb{Z}} f(x+n) \Big) dx = \int_{\mathbb{R}} f(x)\, dx$$

In fact, the integral on the quotient $\mathbb{R}/\mathbb{Z}$ is unequivocally *characterized*,[1] by this relation, once we know that the averaged functions $\sum_n f(x+n)$ are at least *dense* in $C^o(\mathbb{R}/\mathbb{Z})$. As corollary, for $F \in C^o(\mathbb{R}/\mathbb{Z})$, since $F \cdot f \in C_c^o(\mathbb{R})$,

$$\begin{aligned} \int_{\mathbb{R}/\mathbb{Z}} F(x) \Big(\sum_{n\in\mathbb{Z}} f(x+n) \Big) dx &= \int_{\mathbb{R}/\mathbb{Z}} \Big(\sum_{n\in\mathbb{Z}} F(x)\, f(x+n) \Big) dx \\ &= \int_{\mathbb{R}/\mathbb{Z}} \Big(\sum_{n\in\mathbb{Z}} F(x+n)\, f(x+n) \Big) dx \\ &= \int_{\mathbb{R}} F(x)\, f(x)\, dx \end{aligned}$$

We need analogous assertions with less elementary group actions and less transparent representatives for the quotients. For example, with G, K, Γ as in

[1] The Riesz-Markov-Kakutani theorem asserts that every (continuous) functional on compactly supported continuous functions on a reasonable topological space X is $f \to \int_X f(x)\, d\mu(x)$ for some measure μ. Relying on this, specification of a functional (integration) on $C_c^o(X)$ specifies a measure. In fact, we care more about the integral than about the measure.

our examples, integration on $\Gamma\backslash G$ is characterized by requiring, for all $f \in C_c^o(G)$,

$$\int_{\Gamma\backslash G} \Big(\sum_{\gamma\in\Gamma} f(\gamma g)\Big)\, dg = \int_G f(g)\, dg$$

once we know that the averages $\sum_{\gamma\in\Gamma} f(\gamma z)$ are at least *dense* in $C_c^o(\Gamma\backslash G)$. In fact, such averaging maps are universally *surjective* on compactly supported continuous functions, as demonstrated below.

An important variant[2] uses $f \in C_c^o(\Gamma_\infty\backslash G)$ for a *subgroup* Γ_∞ of Γ, by the surjectivity of averaging maps, take $\varphi \in C_c^o(G)$ such that

$$\sum_{\beta\in\Gamma_\infty} \varphi\circ\beta = f$$

so then

$$\begin{aligned}\int_{\Gamma\backslash G}\Big(\sum_{\gamma\in\Gamma_\infty\backslash\Gamma} f\circ\gamma\Big) &= \int_{\Gamma\backslash G}\Big(\sum_{\gamma\in\Gamma_\infty\backslash\Gamma}\Big(\sum_{\beta\in\Gamma_\infty}(\varphi\circ\beta)\circ\gamma\Big)\Big)\\ &= \int_{\Gamma\backslash G}\Big(\sum_{\gamma\in\Gamma}\varphi\circ\gamma\Big) = \int_G \varphi\\ &= \int_{\Gamma_\infty\backslash G}\Big(\sum_{\beta\in\Gamma_\infty}\varphi\circ\beta\Big) = \int_{\Gamma_\infty\backslash G} f\end{aligned}$$

The corollary with $F \in C^o(\Gamma\backslash G)$ and $f \in C_c^o(\Gamma_\infty\backslash G)$ is

$$\int_{\Gamma\backslash G} F\cdot\Big(\sum_{\gamma\in\Gamma_\infty\backslash\Gamma} f\circ\gamma\Big) = \int_{\Gamma\backslash G}\Big(\sum_{\gamma\in\Gamma_\infty\backslash\Gamma}(F\cdot f)\circ\gamma\Big) = \int_{\Gamma_\infty\backslash G} F\cdot f$$

5.1 Surjectivity of Averaging Maps

By convention, a *topological group* is a *locally compact, Hausdorff* topological space G with a *continuous* group operation $G\times G \to G$, and continuous *inversion* map $g \to g^{-1}$. To avoid pathologies with regard to measures on products, we require that topological groups have a *countable basis*.

[2] This variant of *unwinding* arose most prominently in the Rankin-Selberg method, where $\int_{\Gamma\backslash\mathfrak{H}} |f|^2\cdot E_s$ for cuspform f and Eisenstein series E_s is unwound using the definition of E_s as wound *up* from y^s. This theme is pervasive in the theory of automorphic forms.

Let dg be a right G-invariant measure on G, meaning that for $f \in C_c^o(G)$, for all $h \in G$,

$$\int_G f(gh)\, dg = \int_G f(g)\, d(gh^{-1}) = \int_G f(g)\, dg$$

Let $\delta_G : G \to (0, +\infty)$ be the *modular function* of G, gauging the discrepancy between left and right invariant measures, in the sense that meas $(gE) = \delta_G(g) \cdot$ meas (E) for a measurable set $E \subset G$. It is immediate that δ_G is a group homomorphism $\delta_G : G \to (0, +\infty)$. It is *continuous*. And $\delta_G^{-1}(g)\, dg$ is a *left* invariant measure. A group with $\delta = 1$ is *unimodular*.

[5.1.1] Claim: Every *discrete*, *compact*, or *abelian* group is *unimodular*.

Proof: For an abelian group, $d(hg) = d(gh)$. For a discrete group, one uses *counting* measure, which is both left and right invariant. For a *compact* groups, observe that the modular function is *continuous*, and has image in the multiplicative group of positive reals. The only compact subgroup is $\{1\}$. ///

Let H be a closed subgroup of G, with right H-invariant measure dh. The quotient $H\backslash G$ has the quotient topology [5.B]. See also [5A].

[5.1.2] Claim: The *averaging map* $\alpha : C_c^o(G) \to C_c^o(H\backslash G)$ by

$$(\alpha F)(g) = \int_H F(hg)\, dh \qquad (\text{for } F \in C_c^o(G))$$

is *surjective*.

Proof: By design, the image consists of left H-invariant functions on G:

$$(\alpha F)(hg) = \int_H F(h'(hg))\, dh' = \int_H F((h'h)g))\, dh' = \int_H F(h'g))\, dh'$$

by replacing h' by $h'h^{-1}$. The surjectivity is much less trivial. Let $q : G \to H\backslash G$ be the quotient map. Let U be a neighborhood of $1 \in G$ having compact closure $\overline{U}$. For each $g \in G$, gU is a neighborhood of g. The images $q(gU)$ are open, by the characterization of the quotient topology. Given $f \in C_c^o(H\backslash G)$, the support spt($f$) of f is covered by the opens $q(gU)$, and admits a finite subcover $q(g_1U), \ldots, q(g_nU)$. The set

$$C = \text{spt}(f \circ q) \cap (g_1\overline{U} \cup \ldots \cup g_n\overline{U}) \subset G$$

is compact, and $q(C) = \text{spt}(f) \subset H\backslash G$. By Urysohn's lemma [9.E.2], let $\varphi \in C_c^o(G)$ be identically 1 on C, and nonnegative real-valued everywhere.

Let $F = \varphi \cdot (f \circ q)$. Since $\alpha\varphi$ is strictly positive on a neighborhood of the (compact) support of F, the quotient $F/\alpha\varphi$ is in $C_c^o(G)$. Since f is already left H-invariant,

$$\alpha F(g) = \int_H \varphi(hg) \cdot f(hg)\, dh = \int_H \varphi(hg) \cdot f(g)\, dh$$

$$= \int_H \varphi(hg)\, dh \;\cdot\; f(g) = \alpha\varphi(g) \cdot f(g)$$

Because f and $\alpha\varphi$ are left H-invariant,

$$\alpha\Big(\frac{F}{\alpha\varphi}\Big)(g) = \int_H \frac{\varphi(hg) \cdot f(hg)}{\alpha\varphi(hg)}\, dh = \int_H \frac{\varphi(hg) \cdot f(g)}{\alpha\varphi(g)}\, dh$$

$$= \int_H \varphi(hg)\, dh \;\cdot\; \frac{f(g)}{\alpha\varphi(g)} = \alpha\varphi(g) \cdot \frac{f(g)}{\alpha\varphi(g)} = f$$

giving the surjectivity. ///

5.2 Invariant Measures and Integrals on Quotients $H\backslash G$

Temporarily, for clarity in the proofs of this section, we may let $\dot{g}$ denote an element of the quotient $H\backslash G$, where H is a closed subgroup of G. Let H have modular function δ_H.

[5.2.1] Theorem: The quotient $H\backslash G$ has a right G-invariant measure if and only if $\delta_G\big|_H = \delta_H$. In that case, the integral is unique up to scalars, and is characterized as follows. For given right Haar measure dh on H and for given right Haar measure dg on G, there is a unique invariant measure $d\dot{g}$ on $H\backslash G$ such that for $f \in C_c^o(G)$

$$\int_{H\backslash G} \Big(\int_H f(h\dot{g})\, dh\Big)\, d\dot{g} = \int_G f(g)\, dg \qquad (\text{for } f \in C_c^o(G))$$

Proof: Proof: First, prove the *necessity* of the condition on the modular functions. Suppose that there is a right G-invariant measure on $H\backslash G$. Let α be the averaging map $f \to \int_H f(hg)\, dh$. For $f \in C_c^o(G)$ the map

$$f \longrightarrow \int_{H\backslash G} \alpha f(\dot{g})\, d\dot{g}$$

emphasizing the coordinate $\dot{g}$ on the quotient, is a right G-invariant functional (with the continuity property as above), so by uniqueness of right invariant measure on G must be a constant multiple of the Haar integral

$$f \longrightarrow \int_G f(g)\, dg$$

The averaging map behaves in a straightforward manner under left translation $L_h f(g) = f(h^{-1}g)$ for $h \in H$: for $f \in C_c^o(G)$ and for $h \in H$

$$\alpha(L_h f)(g) = \int_H f(h^{-1}xg)\, dx = \delta_H(h) \int_H f(xg)\, dx$$

by replacing x by hx. Then

$$\begin{aligned}\int_G f(g)\, dg &= \int_{H\backslash G} \alpha(f)(g)\, d\dot{g} \\ &= \delta(h)^{-1} \int_{H\backslash G} \alpha(L_h f)(g)\, d\dot{g} = \delta(h)^{-1} \int_G f(h^{-1}g)\, dg\end{aligned}$$

by comparing the iterated integral to the single integral. Replacing g by hg in the integral gives

$$\int_G f(g)\, dg = \delta(h)^{-1}\delta_G(h) \int_G f(g)\, dg$$

Choosing f such that the integral is not 0 implies the stated condition on the modular functions.

Proof of *sufficiency* starts from *existence* of Haar measures on G and on H. For simplicity, first suppose that both groups are *unimodular*. As expected, attempt to define an integral on $C_c^o(H\backslash G)$ by

$$\int_{H\backslash G} \alpha f(\dot{g})\, d\dot{g} = \int_G f(g)\, dg$$

invoking the fact that the averaging map α from $C_c^o(G)$ to $C_c^o(H\backslash G)$ is surjective. The potential problem is *well-definedness*. It suffices to prove that $\int_G f(g)\, dg = 0$ for $\alpha f = 0$. Indeed, for $\alpha f = 0$, for all $F \in C_c^o(G)$, the integral of F against αf is certainly 0. Rearrange

$$\begin{aligned}0 &= \int_G F(g)\, \alpha f(g)\, dg = \int_G \int_H F(g)\, f(hg)\, dh\, dg \\ &= \int_H \int_G F(h^{-1}g)\, f(g)\, dg\, dh\end{aligned}$$

by replacing g by $h^{-1}g$. Replace h by h^{-1}, so

$$0 = \int_G \alpha F(g)\, f(g)\, dg$$

Surjectivity of α shows that F can be chosen so that αF is identically 1 on the support of f. Then the integral of f is 0, as claimed, proving the well-definedness for unimodular H and G.

For not-necessarily-unimodular H and G, in the previous argument the left translation by h^{-1} produces a factor of $\delta_G(h^{-1})$. Then replacing h by h^{-1} converts right Haar measure to left Haar measure, so produces a factor of $\delta_H(h)^{-1}$, and the other factor becomes $\delta_G(h)$. If $\delta_G(h) \cdot \delta_H(h)^{-1} = 1$, then the product of these two factors is 1, and the same argument goes through, proving well-definedness. ///

[5.2.2] Corollary: Let G be a group, with closed subgroups $\Theta \subset H \subset G$. Suppose that $\delta_G|_H = \delta_H$ and $\delta_H|_\Theta = \delta_\Theta$. Given any two of: right H-invariant measure on $\Theta\backslash H$, right G-invariant measure on $\Theta\backslash G$, or right G-invariant measure on $H\backslash G$, the other one of the three is uniquely determined so that, for all $f \in C_c^o(\Theta\backslash G)$,

$$\int_{\Theta\backslash G} f(g)\,dg = \int_{H\backslash G} \left(\int_{\Theta\backslash H} f(hg)\,dh \right) dg$$

Proof: With the surjectivity of averaging maps in hand, this proof is just an iterative application of the previous theorem. Namely, given $f \in C_c^o(\Theta\backslash G)$, let $F \in C_c^o(G)$ map to f by the averaging-over-Θ map, so, by the theorem,

$$\int_{\Theta\backslash G} f = \int_{\Theta\backslash G} \int_\Theta F(\theta g)\,d\theta\,dg = \int_G F$$

Again by the theorem,

$$\int_G F = \int_{H\backslash G} \int_H F(hg)\,dh\,dg$$

At the same time, applying the theorem to $h \to F(hg)$,

$$\int_H F(hg)\,dh = \int_{\Theta\backslash H} \int_\Theta F(\theta hg)\,d\theta\,dh$$

The inner integral in the latter is $f(hg)$, giving the claim. ///

5.A Appendix: Apocryphal Lemma $X \approx G/G_x$

We prove that under mild hypotheses a topological space X acted on transitively by a *topological* group G is homeomorphic to the quotient G/G_x, where G_x is the isotropy group of a chosen point x in X. Ignoring the topology, the bijection $G/G_x \approx X$ by $g \cdot G_x \leftrightarrow gx$ is easy to see. In contrast, the *topological* aspects are not trivial, but are very general.

[5.A.1] Proposition: Let G be a locally compact, Hausdorff topological group and X a locally compact Hausdorff topological space with a continuous transitive action of G upon X. Suppose that G has a *countable basis*. Fix any $x \in X$, and let G_x be the isotropy group $G_x = \{g \in G : gx = x\}$. Then we have a *homeomorphism* $G/G_x \longrightarrow X$ given by the natural $gG_x \longrightarrow gx$.

Proof: A little systematic development of topological groups will allow a coherent argument.

[5.A.2] Claim: In a locally compact Hausdorff space X, given an open neighborhood U of a point x, there is a neighborhood V of x with compact closure $\overline{V}$ and $\overline{V} \subset U$.

Proof: By local compactness, x has a neighborhood W with compact closure. Intersect U with W if necessary so that U has compact closure $\overline{U}$. Note that the compactness of $\overline{U}$ implies that the *boundary* ∂U of U is compact. Using the Hausdorff-ness, for each $y \in \partial U$ let W_y be an open neighborhood of y and V_y an open neighborhood of x such that $W_y \cap V_y = \phi$. By compactness of ∂U, there is a finite list $y_1, \ldots, y_n$ of points on ∂U such that the sets U_{y_i} cover ∂U. Then $V = \bigcap_i V_{y_i}$ is open and contains x. Its closure is contained in $\overline{U}$ and in the complement of the open set $\bigcup_i W_{y_i}$, the latter containing ∂U. Thus, the closure $\overline{V}$ of V is contained in U. ///

[5.A.3] Claim: The map $gG_x \to gx$ is a continuous bijection of G/G_x to X.

Proof: First, $G \times X \to X$ by $g \times y \to gy$ is continuous by definition of the continuity of the action. Thus, with fixed $x \in X$, the restriction to $G \times \{x\} \to X$ is still continuous, so $G \to X$ by $g \to gx$ is continuous. The quotient topology on G/G_x is the unique topology on the *set* (of cosets) G/G_x such that any continuous $G \to Z$ constant on G_x cosets factors through the quotient map $G \to G/G_x$. That is, we have a commutative diagram

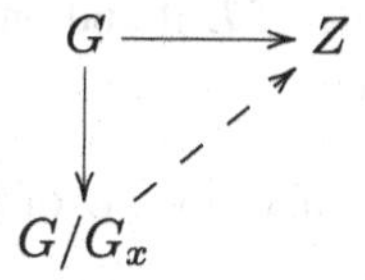

Thus, the induced map $G/G_x \to X$ by $gG_x \to gx$ is continuous. ///

We need to show that $gG_x \to gx$ is *open* to prove that it is a homeomorphism.

[5.A.4] Claim: For a given point $g \in G$, every neighborhood of g is of the form gV for some neighborhood V of 1.

Proof: First, again, $G \times G \to G$ by $g \times g \to gh$ is continuous, by assumption. Then, for fixed $g \in G$, the map $h \to gh$ is continuous on G, by restriction. And this map has a continuous inverse $h \to g^{-1}h$. Thus, $h \to gh$ is a homeomorphism of G to itself. In particular, since $1 \to g \cdot 1 = g$, neighborhoods of 1 are carried to neighborhoods of g, as claimed. ///

[5.A.5] Claim: Given an open neighborhood U of 1 in G, there is an open neighborhood V of 1 such that $V^2 \subset U$, where $V^2 = \{gh : g, h \in V\}$.

Proof: From the continuity of multiplication $G \times G \to G$, given the neighborhood U of 1, the inverse image W of U under the multiplication $G \times G \to G$ is open. Since $G \times G$ has the product topology, W contains an open of the form $V_1 \times V_2$ for opens V_i containing 1. With $V = V_1 \cap V_2$, we have $V^2 \subset V_1 \cdot V_2 \subset U$. ///

Similarly, but more simply, inversion $g \to g^{-1}$ is continuous and is its own (continuous) inverse, so the image $V^{-1} = \{g^{-1} : g \in V\}$ of an open V is open. For example, given a neighborhood V of 1, replacing V by $V \cap V^{-1}$ replaces V by a smaller *symmetric* neighborhood: the new V satisfies $V^{-1} = V$.

The following result is not strictly necessary but sheds some light on the nature of topological groups. It has an analogue for topological vector spaces.

[5.A.6] Claim: The *closure* of $E \subset G$ is $\bigcap_U E \cdot U$, where U runs over open neighborhoods of 1.

Proof: A point $g \in G$ is in the closure of E if and only if every neighborhood of g meets E. That is, from just above, every set gU meets E, for U an open neighborhood of 1. That is, $g \in E \cdot U^{-1}$ for every neighborhood U of 1. We have noted that inversion is a homeomorphism of G to itself (and sends 1 to 1), so the map $U \to U^{-1}$ is a bijection of the collection of neighborhoods of 1 to itself. Thus, g is in the closure of E if and only if $g \in E \cdot U$ for every open neighborhood U of 1, as claimed. ///

[5.A.7] Corollary: Given a neighborhood U of 1 in G, there is a neighborhood V of 1 such that $\overline{V} \subset U$.

Proof: From the continuity of $G \times G \to G$, there is V such that $V \cdot V \subset U$. From the previous claim, $\overline{V} \subset V \cdot V$, so $\overline{V} \subset V \cdot V \subset U$, as claimed. ///

We can improve the conclusion of the previous remark using the local compactness of G, as follows. Given a neighborhood U of 1 in G, there is a neighborhood V of 1 such that $\overline{V} \subset U$ and $\overline{V}$ is *compact*. Indeed, local compactness means exactly that there is a local basis at 1 consisting of opens with compact closures. Thus, given V as in the previous remark, shrink V if necessary to have the compact closure property, and still $\overline{V} \subset V \cdot V \subset U$, as claimed.

[5.A.8] Corollary: For an open subset U of G, given $g \in U$, there is a compact neighborhood V of $1 \in G$ such that $gV^2 \subset U$.

Proof: The set $g^{-1}U$ is an open containing 1, so there is an open $W \ni 1$ such that $W^2 \subset g^{-1}U$. Using the previous claim and remark, there is a compact neighborhood V of 1 such that $V \subset W$. Then $V^2 \subset W^2 \subset g^{-1}U$, so $gV^2 \subset U$ as desired. ///

[5.A.9] Claim: Given an open neighborhood V of 1, there is a countable list $g_1, g_2, \ldots$ of elements of G such that $G = \bigcup_i g_i V$.

Proof: To see this, first let $U_1, U_2, \ldots$ be a countable basis. For $g \in G$, by definition of a basis,

$$gV = \bigcup_{U_i \subset gV} U_i$$

Thus, for each $g \in G$, there is an index $j(g)$ such that $g \in U_{j(g)} \subset gV$. Do note that there are only countably many such indices. For each index i appearing as $j(g)$, let g_i be an element of G such that $j(g_i) = i$, that is,

$$g_i \in U_{j(g_i)} \subset g_i \cdot V$$

Then, for every $g \in G$ there is an index i such that

$$g \in U_{j(g)} = U_{j(g_i)} \subset g_i \cdot V$$

This shows that the union of these $g_i \cdot V$ is all of G. ///

Now we can prove that $G/G_x \approx X$:

Given an open set U in G and $g \in U$, let V be a compact neighborhood of 1 such that $gV^2 \subset U$. Let $g_1, g_2, \ldots$ be a countable set of points such that $G = \bigcup_i g_i V$. Let $W_n = g_n V x \subset X$. By the transitivity, $X = \bigcup_i W_i$.

We observed at the beginning of this discussion that $G \to X$ by $g \to gx$ is continuous, so W_n is compact, being a continuous image of the compact set $g_n V$. So W_n is closed since it is a compact subset of the Hausdorff space X. By the Baire category theorem [15.A] for locally compact Hausdorff spaces, some $W_m = g_m V x$ contains a nonempty open set S of X. For $h \in V$ so that $g_m h x \in S$,

$$gx = g(g_m h)^{-1}(g_m h)x \in gh^{-1}g_m^{-1}S$$

Every group element $y \in G$ acts by homeomorphisms of X to itself, since the continuous inverse is given by y^{-1}. Thus, the image $gh^{-1}g_m^{-1}S$ of the open set S is open in X. Continuing,

$$gh^{-1}g_m^{-1}S \subset gh^{-1}g_m^{-1}g_m V x \subset gh^{-1}Vx \subset gV^{-1} \cdot Vx \subset Ux$$

Therefore, gx is an interior point of Ux, for all $g \in U$. ///

5.B Appendix: Topology on Quotients $H\backslash G$ or G/H

As always, G is a *topological group*, which requires that G be locally compactn and Hausdorff, with the group operation and inversion *continuous*.

Let H be a *closed* subgroup of G. As *sets*, the left-quotient $H\backslash G$ is the set of cosets Hg with $g \in G$, and G/H is the set of cosets gH with $g \in G$. Let $q : G \to H\backslash G$ be the quotient map $q(g) = Hg$. A subset $U \subset H\backslash G$ is *open* when the inverse image $q^{-1}(U) = \{g \in G : Hg \subset U\}$ is open in G. Similarly for G/H.

[5.B.1] Claim: The quotient maps are *open* maps, meaning that they take open sets to open sets.

Proof: For open $U \subset G$,

$$q^{-1}(q(U)) = \{g \in G : Hg \subset \bigcup_{u \in U} Hu\} = \{g \in G : g \in H \cdot U\}$$
$$= H \cdot U = \bigcup_{h \in H} h \cdot U$$

Since the group operation and inverse are continuous, for every $h \in H$, the map $g \to h \cdot g$ is a homeomorphism of G to itself. Thus, every set $h \cdot U$ is open. An arbitrary union of opens is open. ///

6

Action of G on Function Spaces on G

The function spaces here are more complicated versions of the very concrete examples of Chapter 12, where various spaces of functions on the real line were given metrics or topologies so that they would be *complete* or *quasi-complete.* In some of those concrete examples of spaces of functions on $\mathbb{R}$, the *translation action* of $\mathbb{R}$ on functions plays a role. Spaces of automorphic forms are less visualizable examples. Fortunately, most of the specifics of the concrete examples are irrelevant to proofs.

6.1 Action of G on $L^2(\Gamma\backslash G)$

For this section, G need merely be a topological group, *unimodular* in the sense that its right-invariant measure is left-invariant. Let Γ be a discrete subgroup, and K a compact subgroup. This includes the assumptions that G is locally compact, Hausdorff, and countably based. This applies to both classical situations and adelic, such as $G = SL_n(\mathbb{R})$ and $\Gamma = SL_n(\mathbb{Z})$, and also to $G = Z^+GL_n(\mathbb{A})$ and $\Gamma = GL_n(k)$ for number fields k.

Identify functions on $\Gamma\backslash G/K$ with right K-invariant functions on the overlying space $\Gamma\backslash G$ by composition with the quotient map. Unlike $\Gamma\backslash G/K$, the space $\Gamma\backslash G$ admits an action of G by *right translation*. A right G-invariant measure dg on G (Haar measure) specifies a uniquc normalization for a right G-invariant measure on $\Gamma\backslash G$ by the *unwinding* characterization [5.2]

$$\int_{\Gamma\backslash G}\Big(\sum_{\gamma\in\Gamma} f(\gamma g)\Big)dg = \int_G f(g)\,dg \qquad (\text{for } f \in C_c^o(G))$$

Uniqueness of the Haar measure on G up to scalars, and uniqueness of a measure on $\Gamma\backslash G$ compatible with unwinding, are special cases of *uniqueness of invariant distributions*, as in [14.4]. In many examples, *existence* is not an issue because it is established by reduction to simpler cases. We have an isometry:

[6.1.1] Claim: For $f \in L^2(\Gamma\backslash G)$ and $g \in G$, the right translate $(g\cdot f)(x) = f(xg)$ is still in $L^2(\Gamma\backslash G)$, and $|g\cdot f|_{L^2} = |f|_{L^2}$.

Proof: Directly computing,

$$\int_{\Gamma\backslash G} |(g\cdot f)(x)|^2\,dx = \int_{\Gamma\backslash G} |f(xg)|^2\,dx = \int_{\Gamma\backslash G} |f(x)|^2\,dx$$

by replacing x by xg^{-1}, using the invariance of the measure. ///

The overriding point is the continuity of the group action:

[6.1.2] Theorem: $G \times L^2(\Gamma\backslash G) \longrightarrow L^2(\Gamma\backslash G)$ by $g \times f \longrightarrow \big(x \to f(xg)\big)$ for $x, g \in G$ is (jointly) *continuous*. That is, $L^2(\Gamma\backslash G)$ is a *unitary representation space* for G.

Proof: For the moment, write $|\cdot|$ for the L^2 norm. The crux of the matter is that L^2 functions can be approximated by continuous, compactly supported functions, and the latter are uniformly continuous. Granting that approximation property for a moment, given $\varepsilon > 0$, take $\varphi \in C_c^o(\Gamma\backslash G)$ such that $|f - \varphi| < \varepsilon$. Being compactly supported, φ is *uniformly* continuous, and the topology of $\Gamma\backslash G$ descends from that of G, so for every $\varepsilon_1 > 0$ there is a neighborhood U of 1 in G such that $|\varphi(xu) - \varphi(x)| < \varepsilon_1$ for all $x \in \Gamma\backslash G$ and for all $u \in U$. For every $F \in L^2(\Gamma\backslash G)$ such that $|f - F| < \varepsilon$, for $h = gu \in gU$, the triangle inequality breaks things into atomic issues:

$$\begin{aligned}|g\cdot f - h\cdot F| &= |f - g^{-1}h\cdot F| = |f - u\cdot F|\\ &\le |f-\varphi| + |\varphi - u\cdot\varphi| + |u\cdot\varphi - u\cdot f| + |u\cdot f - u\cdot F|\\ &= |f-\varphi| + |\varphi - u\cdot\varphi| + |\varphi - f| + |f - F| < 3\varepsilon + |u\cdot\varphi - \varphi|\end{aligned}$$

The support S of φ has finite measure μ, so for U a small enough neighborhood of 1 such that $\sup_x |\varphi(xu) - \varphi(x)| < \varepsilon/2\mu$ for $u \in U$,

$$\left(\int_{\Gamma\backslash G} |\varphi(xu) - \varphi(x)|^2 \, dx\right)^{\frac{1}{2}} < \varepsilon$$

and $|g \cdot f - h \cdot F| < 4\varepsilon$, proving joint continuity.

Density of continuous compactly supported functions in L^2 follows from general principles, but its importance justifies some attention to it. Now write $|\cdot|_{L^2}$ to distinguish this from the absolute value on real or complex numbers, and $|\cdot|_{L^1}$ for the L^1 norm. First, it suffices to approximate four pieces of f, the positive and negative parts of its real and imaginary parts, separately, by the triangle inequality. So without loss of generality, suppose f is real-valued and nonnegative. For given $\varepsilon > 0$ there is a bound M such that a truncated form of f, with maximum value replaced by M, satisfies

$$\int_{\Gamma\backslash G} |f - \min(f, M)|^2 < \varepsilon$$

Indeed, the sum of integrals of $|f|^2$ over the sets $\{M \leq |f| \leq M + 1\}$ converges, so the tails must go to zero. Replace f by that truncation.

For μ a (positive, regular, Borel) measure on a locally compact Hausdorff space, a *simple* function is a finite real-linear combination $s = \sum_i c_i \cdot \chi_{E_i}$ of characteristic functions χ_{E_i} of μ-measurable sets E_i. The *integral* of s is $\int s = \sum_i c_i \cdot \mu(E_i)$. The integral of a non-negative real-valued (measurable) function f is the sup of the integrals of *simple* functions s such that $0 \leq s(x) \leq f$ for all x. Since $0 \leq f \leq M$, for such s

$$\int |f - s|^2 = \int |f - s| \cdot |f - s| \leq \int 2M \cdot |f - s| = 2M \cdot \left(\int f - \int s\right)$$

Thus, with s such that $\int s$ is within $\varepsilon/2M$ of $\int f$, we have $|f - s|_{L^2}^2 \leq 2M \cdot \varepsilon/2M = \varepsilon$. By the triangle inequality, it suffices to approximate characteristic functions of measurable sets. *Regularity* of μ is that the measure of a set is the sup of the measures of compacts inside it, and is the inf of the opens containing it. Let $K \subset E \subset U$ with compact K and open U such that $\mu(U) - \mu(K) < \varepsilon$. Urysohn's lemma [9.E.2] yields a continuous function φ with values in the range $[0, 1]$ and 1 on K and 0 outside U, and

$$\int |\varphi - \chi_E|^2 = \int_K |\varphi - \chi_E|^2 + \int_{U-K} |\varphi - \chi_E|^2 \leq 0 + \int_{U-K} 1 < \varepsilon$$

Thus, continuous functions approximate simple functions, which are dense in L^2. ///

As in [13.13], the *strong* operator topology on the continuous linear endomorphisms $\mathrm{End}^o_{\mathbb{C}}(L^2(\Gamma\backslash G))$ is given by the collection of *seminorms*

$$T \longrightarrow |Tv|_{L^2} \qquad (\text{for } v \in L^2(\Gamma\backslash G))$$

The strong operator topology is weaker than the *(uniform) operator norm* topology $\sup_{|v|\le 1} |Tv|$. The strong operator topology is *not* complete-metrizable but is *quasi-complete* as a special case of [13.12], and immediately *locally convex* since the topology is given by seminorms [13.11].

[6.1.3] Corollary: The map $G \to \mathrm{End}^o_{\mathbb{C}}(L^2(\Gamma\backslash G))$ by right translation is *continuous* when $\mathrm{End}^o_{\mathbb{C}}(L^2(\Gamma\backslash G))$ has the strong operator topology.

Proof: This is a paraphrase of the joint continuity of $G \times L^2(\Gamma\backslash G) \to L^2(\Gamma\backslash G)$. ///

[6.1.4] Remark: As mentioned in [13.13], $G \to \mathrm{End}^o_{\mathbb{C}}(L^2(\Gamma\backslash G))$ is *not* continuous when $\mathrm{End}^o_{\mathbb{C}}(L^2(\Gamma\backslash G))$ has the operator norm topology. To see this, for each neighborhood N of $1 \in G$, we claim there is an L^2 function f such that $|f - g \cdot f|^2_L \ge 1$, so certainly cannot be made arbitrarily small. Indeed, shrink N if necessary so that it *injects* to the quotient $\Gamma\backslash G$, by the discreteness of Γ. Replace N by $N \cap N^{-1}$, so that it is closed under inverses. Take $1 \neq g \in N$, and let $U \subset N$ be a small-enough neighborhood of 1 such that $g \notin U$, by Hausdorff-ness. By continuity of multiplication, there is an open $V \ni 1$ such that $V \cdot V \subset U$. Replace V by $V \cap V^{-1}$ so that V is stable under inverses. Then $g \notin V^2$ gives $g^{-1} \notin V^2$, and $g^{-1}V \cap V = \phi$. For an L^2 function f of norm 1 and supported in V, the supports of f and $g \cdot f$, namely, V and $g^{-1}V$, are disjoint, so $|f - g \cdot f|_{L^2} = \sqrt{2}$. This quantity cannot be made arbitrarily small, so the representation is *not* continuous for the uniform operator norm topology.

The continuity property of the theorem gives a precise and useful sense to certain *integral operators*:

[6.1.5] Corollary: For $\varphi \in C^o_c(G)$, the integral operator $f \to \varphi \cdot f$ defined on $f \in L^2(\Gamma\backslash G)$ by a convergent vector-valued integral

$$\varphi \cdot f = \int_G \varphi(g)\, g \cdot f \, dg$$

is a continuous linear map $L^2(\Gamma\backslash G) \longrightarrow L^2(\Gamma\backslash G)$, with the natural property that for $F \in L^2(\Gamma\backslash G)$

$$\langle \varphi \cdot f,\ F\rangle = \int_G \varphi(g)\, \langle g \cdot f, F\rangle \, dg$$

In fact, letting $T_g f = g \cdot f$ and T_φ for the expected operator, we have a vector-valued integral convergent in the strong operator topology:

$$T_\varphi = \int_G \varphi(g)\, T_g\, dg$$

with the property

$$T_\varphi f = \int_G \varphi(g)\, T_g f\, dg$$

Proof: This is a special case of properties of Gelfand-Pettis vector-valued integrals [14.1], and using the quasi-completeness of the strong operator topology [13.12], [13.13]. First, because $G \times L^2(\Gamma\backslash G) \to L^2(\Gamma\backslash G)$ is continuous, the function $g \to g \cdot f$ is a continuous $L^2(\Gamma\backslash G)$-valued function on G. Then $g \to \varphi(g)\, g \cdot f$ is a compactly supported continuous $L^2(\Gamma\backslash G)$-valued functions, so by [14.8] the integral purporting to define $\varphi \cdot f$ converges in $L^2(\Gamma\backslash G)$ and enjoys the properties [14.1] of Gelfand-Pettis integrals. In particular, $f \to \langle f, F\rangle$ is a continuous linear functional on $L^2(\Gamma\backslash G)$, so by [14.1]

$$\langle \varphi \cdot f,\ F\rangle = \left\langle \int_G \varphi(g)\, g \cdot f\, dg,\ F \right\rangle = \int_G \langle \varphi(g)\, g \cdot f, F\rangle\, dg$$
$$= \int_G \varphi(g)\, \langle g \cdot f, F\rangle\, dg$$

Similarly, the continuity of the action of G on $L^2(\Gamma\backslash G)$ actually gives the stronger assertion that $g \to T_g$ is a continuous $\mathrm{End}^o_{\mathbb{C}}(L^2(\Gamma\backslash G))$-valued function on G. Multiplying by φ makes the function compactly supported, and again a Gelfand-Pettis integral exists, by [14.8]. The map $T \to Tf$ is a continuous $L^2(\Gamma\backslash G)$-valued function on $\mathrm{End}^o_{\mathbb{C}}(L^2(\Gamma\backslash G))$, so commutes with the integral, by [14.1]. ///

6.2 Action of G on $C_c^o(\Gamma\backslash G)$

The following instance of a general result is also a warm-up to the analogous result for *test functions* $C_c^\infty(\Gamma\backslash G)$. For this section, it still suffices that G is a topological group and Γ a discrete subgroup. Although any locally convex topological vector space topology *can* be given by a separating family of seminorms [13.11], this need not be the way the topology *arises*. The topology on $C_c^o(\Gamma\backslash G)$ most naturally arises from the expression of $C_c^o(\Gamma\backslash G)$ as an *ascending union* of subspaces

$$C_E^o(\Gamma\backslash G) = \{f \in C_c^o(\Gamma\backslash G) : \mathrm{spt} f \subset E\}$$

where E varies over compact subsets of $\Gamma\backslash G$. Recall the usual

[6.2.1] Lemma: Each $C_E^o(\Gamma\backslash G)$ is a Banach space, with sup norm

$$|f|_{C^o} = \sup_{x\in E} |f(x)| \qquad (\text{for } f \in C_E^o(\Gamma\backslash G))$$

Proof: First, as proven in [13.1.1], the space $C^o(E)$ of all continuous functions on a compact subset E of $\Gamma\backslash G$ is a Banach space. Then $C_E^o(\Gamma\backslash G)$ is a closed subspace defined by pointwise vanishing on the topological boundary of E in $\Gamma\backslash G$: each evaluation map $f \to f(x_o)$ is certainly *continuous* with the sup norm on $C^o(E)$, so the kernels are *closed*, and the intersection of all these closed subspaces for x_o on the boundary of E is a closed subspace of a Banach space, so is Banach. ///

Then $C_c^o(\Gamma\backslash G)$ is the ascending union, a *colimit*, as discussed in [13.8] and [13.9]:

$$C_c^o(\Gamma\backslash G) = \bigcup\nolimits_E C_E^o(\Gamma\backslash G) = \text{colim}_E\, C_E^o(\Gamma\backslash G)$$

There is a *countable cofinal* colimit over a countable collection of compact subsets $E_1 \subset E_2 \subset \cdots$ of G whose union is G. We can take the E_i to be closures of a nested family $U_1 \subset U_2 \subset \cdots$ of opens whose union is G. Cofinal colimits are isomorphic, for general reasons, so

$$C_c^o(\Gamma\backslash G) = \bigcup\nolimits_{E_i} C_{E_i}^o(\Gamma\backslash G) = \text{colim}_{E_i} C_{E_i}^o(\Gamma\backslash G)$$

Each inclusion $C_{E_i}^o(\Gamma\backslash G) \subset C_{E_{i+1}}^o(\Gamma\backslash G)$ is a homeomorphism to its image, and its image is closed, defined by vanishing at all $x \in E_{i+1} - E_i$. A countable colimit of such restricted inclusions is a *strict colimit*.[1] A strict colimit of Hilbert, Banach, or Fréchet spaces is an *LF-space* [13.8]. As in [13.12.4], LF-spaces are rarely *complete* in the strongest sense but are *quasi-complete* [13.8.5], and this is sufficient for use.

[6.2.2] Claim: G acts continuously on $C_c^o(\Gamma\backslash G)$ by right translation: this is the joint continuity of

$$G \times C_c^o(\Gamma\backslash G) \longrightarrow C_c^o(\Gamma\backslash G)$$

by

$$g \times f \longrightarrow \big(x \to f(xg)\big)$$

Proof: Of course, right translation by $g \in G$ does not stabilize any single $C_E^o(\Gamma\backslash G)$, only the colimit. Let $\nu_E(f)$ be $\sup_{x\in E} |f(x)|$. On the other hand, we

[1] Slightly older terminology is that a strict colimit is a *strict inductive limit*.

have the tautological

$$\nu_E(g \cdot f) = \nu_{Eg}(f) \qquad \text{and} \qquad g \cdot C_E^o(\Gamma\backslash G) = C_{Eg^{-1}}^o(\Gamma\backslash G)$$

Fix $f \in C_E^o(\Gamma\backslash G)$, $\varepsilon > 0$, and $g \in G$. By the uniform continuity of f there is a small-enough neighborhood U of $1 \in G$ such that $|f(x) - f(xu)| < \varepsilon$ for all $u \in U$ and for all $x \in \Gamma\backslash G$. Without loss of generality, U has *compact closure* V, and then $EV^{-1}g^{-1}$ is compact. For all $h = gu \in gU$,

$$h \cdot C_E^o(\Gamma\backslash G) \subset gU \cdot C_E^o(\Gamma\backslash G) \subset C_{EV^{-1}g^{-1}}^o(\Gamma\backslash G)$$

where $h = gu \in gU$. For all $F \in C_E^o(\Gamma\backslash G)$ with $\nu_E(f - F) < \varepsilon$, with $h = gu \in gU$

$$\begin{aligned}\nu_{EV^{-1}g^{-1}}(g \cdot f - h \cdot F) &= \nu_{EV^{-1}}(f - g^{-1}h \cdot F) = \nu_{EV^{-1}}(f - u \cdot F)\\ &\le \nu_{EV^{-1}}(f - u \cdot f) + \nu_{EV^{-1}}(u \cdot f - u \cdot F)\\ &= \nu_{EV^{-1}}(f - u \cdot f) + \nu_E(f - F) < \varepsilon + \varepsilon\end{aligned}$$

proving joint continuity at $g \times f$ of

$$gU \times C_E^o(\Gamma\backslash G) \longrightarrow C_{EV^{-1}g^{-1}}^o(\Gamma\backslash G)$$

The inclusion $C_{EV^{-1}g^{-1}}^o(\Gamma\backslash G) \to C_c^o(\Gamma\backslash G)$ is continuous, so $gU \times C_E^o(\Gamma\backslash G) \to C_c^o(\Gamma\backslash G)$ is continuous at $\{g\} \times \{f\}$. Since the colimit is stable under the action of G, now it makes sense to say that $G \times C_E^o(\Gamma\backslash G) \to C_c^o(\Gamma\backslash G)$ is continuous at $g \times f$. Since $g \in G$ and $f \in C_E^o(\Gamma\backslash G)$ were arbitrary, this shows that $G \times C_E^o(\Gamma\backslash G) \to C_c^o(\Gamma\backslash G)$ is continuous.

Maps *from* a colimit $X = \mathrm{colim}_i X_i$ *to* another object Y are exactly *compatible* families of maps $X_i \to Y$ from the limitands to Y, as in [13.8], [13.9]. Using the countable cofinal family $E_1 \subset \cdots \subset E_i \subset E_{i+1} \subset \cdots$ for notational convenience, the compatible family of jointly continuous maps (suppressing reference to $\Gamma\backslash G$),

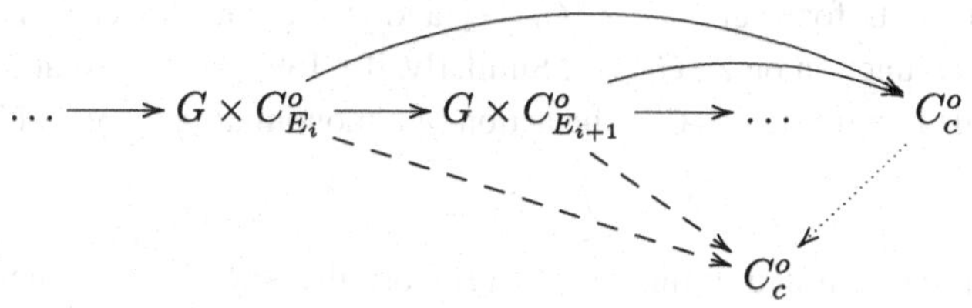

is the joint continuity of $G \times C_c^o(\Gamma\backslash G) \longrightarrow C_c^o(\Gamma\backslash G)$ for all U, as claimed. ///

[6.2.3] Corollary: The right translation action $G \times C_c^\infty(G) \to C_c^\infty(G)$ is jointly continuous.

Proof: Take $\Gamma = \{1\}$ in the previous. ///

6.3 Test Functions on $Z^+G_k\backslash G_{\mathbb{A}}$

As a preamble, we could consider $SL_n(\mathbb{Z})\backslash SL_n(\mathbb{R})$, a smooth manifold due to the discreteness of $SL_n(\mathbb{Z})$, on which the notion of *test function* as compactly supported smooth function has a general sense. The simple example $C_c^\infty(\mathbb{R})$ is in [13.9]. Edging toward generality, we could similarly consider $SL_n(\mathbb{Z}[\frac{1}{p}])\backslash(SL_n(\mathbb{R}) \times SL_n(\mathbb{Q}_p))$, where as usual $\mathbb{Z}[\frac{1}{p}]$ is $\mathbb{Z}$ with prime p inverted. Again, $SL_n(\mathbb{Z}[\frac{1}{p}])$ is demonstrably a discrete subgroup, basically because $\mathbb{Z}[\frac{1}{p}]$ is discrete in $\mathbb{R} \times \mathbb{Q}_p$. Certainly $SL_n(\mathbb{R}) \times SL_n(\mathbb{Q}_p)$ is not locally homeomorphic to any $\mathbb{R}^N$, but to $(\mathbb{R} \times \mathbb{Z}_p)^N$ for suitable N. Nevertheless, for a compact open subgroup K_p of $SL_n(\mathbb{Q}_p)$, such as $K_p = SL_n(\mathbb{Z}_p)$ or any congruence subgroup, the quotient $SL_n(\mathbb{Q}_p)/K_p$ is *discrete*, so $SL_n(\mathbb{Z}[\frac{1}{p}])\backslash SL_n(\mathbb{R}) \times SL_n(\mathbb{Q}_p)/K_p$ *is* a smooth manifold. But we object that $SL_n(\mathbb{Q}_p)$ *no longer acts on this quotient, nor on functions on it.*

To overcome this objection, and in anticipation of examination of the action of $G_{\mathbb{A}}$ on functions on $Z^+G_k\backslash G_{\mathbb{A}}$ in the next section, we can characterize differentiability slightly *indirectly*, taking advantage of the additional structure. At the same time, the appropriate notion of *smoothness* of functions f on totally disconnected groups such as $SL_n(\mathbb{Q}_p)$ is that there should exist an open subgroup K' such that f is right K'-invariant. We address these simultaneously. Let $G = GL_n(\mathbb{A})$ and $\Gamma = Z^+GL_n(k)$.[2] Let $\mathfrak{g}$ be the Lie algebra [4.1] of $G_\infty = \prod_{v|\infty} G_v$, and $U\mathfrak{g}$ its universal enveloping algebra [4.3]. Each $\gamma \in \mathfrak{g}$ gives difference quotients for functions on $Z^+G_k\backslash G_{\mathbb{A}}$:

$$X_\gamma f(x) = \lim_{t \to 0} \frac{f(xe^{t\gamma}) - f(x)}{t}$$

for $x \in Z^+G_k\backslash G_{\mathbb{A}}$ and $t \in \mathbb{R}$. The limit may or may not exist, depending on f and $x \in Z^+G_k\backslash G_{\mathbb{A}}$ and $\gamma \in \mathfrak{g}$. Given f and γ, when the limit does exist for every x, it gives a compactly supported function on $Z^+G_k\backslash G_{\mathbb{A}}$. Say f is C^1 if this limit exists for every $x \in Z^+G_k\backslash G_{\mathbb{A}}$ and $\gamma \in \mathfrak{g}$, and for each $\gamma \in \mathfrak{g}$ is a *continuous* function on $Z^+G_k\backslash G_{\mathbb{A}}$. Similarly, if ℓ-fold limits exist and produce continuous functions, f is C^ℓ. The action of a monomial $\gamma_1 \ldots \gamma_\ell \in U\mathfrak{g}$ is

$$X_{\gamma_1 \ldots \gamma_\ell} f = X_{\gamma_1} \circ \cdots \circ X_{\gamma_\ell} f$$

when all the implied limits exist. Temporarily, say that a function $f \in C_c^o(Z^+G_k\backslash G_{\mathbb{A}})$ such that these limits exist for all elements of $U\mathfrak{g}$ is *archimedean-smooth*. Also temporarily, say $f \in C_c^o(Z^+G_k\backslash G_{\mathbb{A}})$ is *non-archimedean smooth* when f is right K'-invariant for some open subgroup

[2] The same ideas apply to G a product $G = G_\infty \times G_o$ of a real Lie group G_∞ and a totally disconnected group G_o, with Γ a discrete subgroup.

$K' \subset G_{\text{fin}} = \prod_{v<\infty} G_v$.[3] A function $f \in C_c^o(Z^+G_k\backslash G_{\mathbb{A}})$ is *smooth* when it is both archimedean-smooth and non-archimedean smooth. Compactly supported smooth functions are *test functions*, denoted $C_c^\infty(Z^+G_k\backslash G_{\mathbb{A}})$.

The LF-space topology on $V = C_c^\infty(Z^+G_k\backslash G_{\mathbb{A}})$ is described much as $C_c^\infty(\mathbb{R})$ in [13.9] and as the colimit $\mathbb{C}^\infty$ of finite-dimensional spaces $\mathbb{C}^n$ in [13.8]. For compact $E \subset Z^+G_k\backslash G_{\mathbb{A}}$ and compact-open subgroup $K \subset G_{\text{fin}}$, let $C_E^\infty(Z^+G_k\backslash G_{\mathbb{A}})^K$ be the collection of right K-invariant test functions on $Z^+G_k\backslash G_{\mathbb{A}}$ with support in E. Below, we see that each $C_E^\infty(Z^+G_k\backslash G_{\mathbb{A}})^K$ is a Fréchet space, a (projective) limit of a countable collection of spaces $C_E^\ell(Z^+G_k\backslash G_{\mathbb{A}})^K$ of right K-invariant C^ℓ functions with supports on E, suitably topologized. Then $C_c^\infty(Z^+G_k\backslash G_{\mathbb{A}})$ is the *colimit* of the spaces $C_E^\infty(Z^+G_k\backslash G_{\mathbb{A}})^K$. Details are as follows.

As always, the C^o seminorm on $C_E^o(Z^+G_k\backslash G_{\mathbb{A}})$ is the sup norm, and this is a Banach space. Since evaluation at points is a continuous linear functional, a requirement of right K-invariance for open subgroup $K \subset G_{\text{fin}}$ is a collection of *closed* conditions, so defines a closed subspace, giving a Banach space. The C^ℓ seminorm on $C_E^\ell(Z^+G_k\backslash G_{\mathbb{A}})$ should be something like the sup of all derivatives of orders at most ℓ, with derivatives specifically given by $\mathfrak{g}$ and $U\mathfrak{g}$. In contrast to $\mathbb{R}$, the action of the group here does not generally commute with the natural differential operators. To topologize $C_E^1(Z^+G_k\backslash G_{\mathbb{A}})$ to behave well under the action of G_∞ requires examination of the interaction of the right translation action with these right derivatives. Of course, G_{fin} *does* commute with the action of $\mathfrak{g}$. For G_∞, the interaction is by the conjugation action[4] on $\mathfrak{g}$:

$$g \cdot e^{tY} \cdot f = ge^{tY}g^{-1} \cdot g \cdot f \qquad (\text{for } g \in G_\infty)$$

Conjugation interacts well with exponentiation:

$$ge^h g^{-1} = g\Big(\sum_{n\geq 0} \frac{h^n}{n!}\Big)g^{-1} = \sum_{n\geq 0} g\frac{h^n}{n!}g^{-1}$$
$$= \sum_{n\geq 0} \frac{(ghg^{-1})^n}{n!} = e^{ghg^{-1}}$$

so

$$g \cdot e^{tY} \cdot f = ge^{tgYg^{-1}} \cdot g \cdot f$$

[3] For f not necessarily compactly supported, the non-archimedean notion of smoothness would be *local*, allowing K' to vary. The compact support of f *implies* uniform non-archimedean smoothness, so we may as well give the simpler definition.

[4] This is an instance of an *Adjoint* action of a Lie group on its Lie algebra.

and by differentiating

$$g \cdot \gamma \cdot f = g\gamma g^{-1} \cdot g \cdot f$$

The right translation action of G_∞ on functions does not stabilize any individual differential operator coming from $\mathfrak{g}$, nor any finite subset, and does not stabilize the individual spaces $C^\ell_E(Z^+G_k\backslash G_{\mathbb{A}})$, since it does not stabilize supports. Nevertheless, for any open subgroup $K \subset G_{\text{fin}}$, the condition of right K-invariance is a collection of *closed* conditions, so defines a Banach space $C^\ell_E(Z^+G_k\backslash G_{\mathbb{A}})^K$.

One approach to a suitable topology is as follows. For each *bounded neighborhood* b of 0 in $\mathfrak{g}$, and for each compact $E \subset Z^+G_k\backslash G_{\mathbb{A}}$, define a semi-norm

$$\nu_{b,E}(f) = \sup_{\beta\in b}\sup_{x\in E} |(\beta \cdot f)(x)|$$

The *collection* of these has the desirable stability property that

$$\nu_{b,E}(f) = \nu_{gbg^{-1},Eg^{-1}}(g \cdot f)$$

Because $\mathfrak{g}$ is finite-dimensional, for every pair of bounded neighborhoods b, b' of $0 \in B$, there are constants $0 < c < C < \infty$ such that $c \cdot b' \subset b \subset C \cdot b'$, so

$$c \cdot \nu_{b',E}(f) \leq \nu_{b,E}(f) \leq C \cdot \nu_{b',E}(f) \qquad \text{(for all } f\text{)}$$

That is, *the topologies are the same,* for all bounded neighborhoods b of 0, for fixed E. That is, we can *topologize* each space $C^1_E(Z^+G_k\backslash G_{\mathbb{A}})$ by any one of the topologically equivalent seminorms $\nu_{b,E}$. As in the simplest Euclidean case in [13.1], for a fixed choice of $b \in B$ each $C^1_E(Z^+G_k\backslash G_{\mathbb{A}})$ is *complete* with respect to the (semi-) norm $\nu_{b,E}$, so is a Banach space. However, here there is no *canonical* Banach space structure, only a canonical *topology*, given by any one of the topologically equivalent Banach-space structures. In this topology, pointwise evaluation is continuous, so for open subgroup $K \subset G_{\text{fin}}$ the requirement of right K-invariance is a collection of closed conditions, so $C^1_E(Z^+G_k\backslash G_{\mathbb{A}})^K$ is a closed subspace of any of these Banach spaces.

Similarly, to *topologize* $C^\ell_E(Z^+G_k\backslash G_{\mathbb{A}})$, let B be the collection of bounded neighborhoods of 0 in the graded piece $U\mathfrak{g}^{\leq \ell}$ of elements of degree $\leq \ell$ in $U\mathfrak{g}$. Each $b \in B$ and compact E give a seminorm

$$\nu_{b,E}(f) = \sup_{\alpha\in b}\sup_{x\in E} |\alpha f(x)|$$

on ℓ-times differentiable functions supported on E. Since $U\mathfrak{g}^{\leq \ell}$ is finite-dimensional, these seminorms for varying bounded neighborhoods b of $0 \in U\mathfrak{g}^{\leq k}$ are all *comparable*, giving the same topology on $C^\ell_E(Z^+G_k\backslash G_{\mathbb{A}})$. The collection of such seminorms is stabilized by the right action of G, by the extension of the conjugation (Adjoint) action, written as conjugation:

$$\nu_{b,E}(f) = \nu_{gbg^{-1},Eg^{-1}}(g \cdot f)$$

As for C^1, each space $C^{\ell}_E(Z^+G_k\backslash G_{\mathbb{A}})$ is *complete*, although there is no canonical Banach-space structure. Again, for open subgroup K of G_{fin}, the K-fixed functions $C^{\ell}_E(Z^+G_k\backslash G_{\mathbb{A}})^K$ constitute a closed subspace, hence *complete*.

As in the simplest case [13.2], $C^{\infty}_E(Z^+G_k\backslash G_{\mathbb{A}})$ is a (projective) limit of topological vector spaces

$$C^{\infty}_E(Z^+G_k\backslash G_{\mathbb{A}}) = \bigcap_{\ell} C^{\ell}_E(Z^+G_k\backslash G_{\mathbb{A}}) = \lim_{\ell} C^{\ell}_E(Z^+G_k\backslash G_{\mathbb{A}})$$

This is equivalent to characterizing the topology on $C^{\infty}_E(Z^+G_k\backslash G_{\mathbb{A}})$ by the seminorms $\nu_{b,E}$ with compact $b \subset U\mathfrak{g}^{\leq \ell}$ for *all* ℓ and E. The completeness of the limitands implies completeness of the limit, for general reasons, as in [13.2] and other elementary examples in Chapter 12. For each open subgroup $K \subset G_{\text{fin}}$, taking K-invariant subspaces commutes with the projective limit, for elementary reasons: the evaluation maps $f \to f(x_o)$ are continuous and commute with the restriction maps $C^{\infty}_E(Z^+G_k\backslash G_{\mathbb{A}}) \longrightarrow C^{\infty}_{E'}(Z^+G_k\backslash G_{\mathbb{A}})$ for $E \supset E'$. Thus, we can unambiguously write

$$C^{\infty}_E(Z^+G_k\backslash G_{\mathbb{A}})^K = \bigcap_{\ell} C^{\ell}_E(Z^+G_k\backslash G_{\mathbb{A}})^K = \lim_{\ell} C^{\ell}_E(Z^+G_k\backslash G_{\mathbb{A}})^K$$

Then $C^{\infty}_E(Z^+G_k\backslash G_{\mathbb{A}})$ is a *(strict) colimit*

$$\begin{aligned} C^{\infty}_c(Z^+G_k\backslash G_{\mathbb{A}}) &= \bigcup_{E,K} C^{\infty}_E(Z^+G_k\backslash G_{\mathbb{A}})^K \\ &= \text{colim}_{E,K}\, C^{\infty}_E(Z^+G_k\backslash G_{\mathbb{A}})^K \end{aligned}$$

The *strictness* property resides first in the fact that there is the *countable* cofinal collection $E_1, E_2, \ldots$, and a countable local basis $K_1, K_2, \ldots$ for G_{fin}. For example, take E_i to be closures of a nested family $U_1 \subset U_2 \subset \cdots$ of opens whose union is $Z^+G_k\backslash G_A$. Second, the strictness resides in the fact that the inclusion maps are *isomorphisms to their images*, which are *closed* subspaces. Thus, as in the more elementary examples [13.8] and [13.9], this colimit is an *LF-space* and is *quasi-complete* [13.8.5].

The space of *distributions* $C^{\infty}_c(Z^+G_k\backslash G_{\mathbb{A}})^*$ is the *dual* to $C^{\infty}_c(Z^+G_k\backslash G_{\mathbb{A}})$, with the *weak-dual* (also called *weak-*$*$) topology, as in [13.14].

6.4 Action of $G_{\mathbb{A}}$ on $C_c^\infty(Z^+G_k\backslash G_{\mathbb{A}})$

[6.4.1] Theorem: $G_{\mathbb{A}}$ acts continuously on $C_c^\infty(Z^+G_k\backslash G_{\mathbb{A}})$ by right translation. This is the joint continuity of

$$G_{\mathbb{A}} \times C_c^\infty(Z^+G_k\backslash G_{\mathbb{A}}) \longrightarrow C_c^\infty(Z^+G_k\backslash G_{\mathbb{A}})$$

by

$$g \times f \longrightarrow (x \to f(xg))$$

Proof: As in the case of $C_c^o(\Gamma\backslash G)$ in the proof of [6.2.2], the *collection* of seminorms $\nu_{b,E}$ behaves reasonably under right translation by G_∞:

$$\nu_{b,E}(f) = \nu_{gbg^{-1},Eg^{-1}}(g \cdot f)$$

with compact neighborhood b of $0 \in U\mathfrak{g}^{\le k}$, and the *collection* of spaces $C_E^\infty(Z^+G_k\backslash G_{\mathbb{A}})^K$ behaves reasonably:

$$g \cdot C_E^\infty(Z^+G_k\backslash G_{\mathbb{A}})^K = C_{Eg^{-1}}^\infty(Z^+G_k\backslash G_{\mathbb{A}})^K$$

for $g \in G_\infty$, fixed $K \subset G_{\text{fin}}$, although G_∞ does not stabilize any *individual* $C_E^\infty(Z^+G_k\backslash G_{\mathbb{A}})$. Similarly, although right translation by G_{fin} does not preserve right K-invariance for any *individual* open subgroup $K \subset G_{\text{fin}}$, for $g \in G_{\text{fin}}$, for fixed open subgroup $K \subset G_{\text{fin}}$, and for right K-invariant f, the translate $g \cdot f$ is gKg^{-1}-invariant: for $h \in K$,

$$(g \cdot f)(x(ghg^{-1})) = f(xghg^{-1}g) = f(xgh) = f(xg) = (g \cdot f)(x)$$

Still gKg^{-1} is an open subgroup of G_{fin}, so

$$g \cdot C_E^\ell(Z^+G_k\backslash G_{\mathbb{A}})^K = C_{Eg^{-1}}^\ell(Z^+G_k\backslash G_{\mathbb{A}})^{gKg^{-1}}$$

for $g \in G_\infty$, fixed $K \subset G_{\text{fin}}$. As in the proof of [6.2.2], for joint continuity, we further need comparisons for g in a small open set containing a given $g_o \in G_{\mathbb{A}}$.

This uniformity is easiest to see for G_{fin}:

[6.4.2] Claim: Given $g \in G_{\text{fin}}$, a compact neighborhood C of g, and compact open subgroup K of G_{fin}, $\bigcap_{h \in C} hKh^{-1}$ is still an open subgroup of G_{fin}.

Proof (of claim): Since inversion and multiplication are continuous, $U = g^{-1}C$ is a compact neighborhood of 1. We may as well enlarge it to $C \cdot K$. Since K is compact, $C \cdot K$ is still compact. Thus, $C \cdot K$ consists of finitely many cosets

$c_1K, \ldots, c_\ell K$, and any $h \in C$ is $h = gc_ik$ for some i and some $k \in K$, and

$$hKh^{-1} = (gc_ik) \cdot K \cdot (gc_ik)^{-1} = g(c_i \cdot K \cdot c_i^{-1})g^{-1}$$

Thus, the indicated intersection is actually a *finite* intersection of open subgroups, so is open. ///

The uniformity at archimedean places is slightly more complicated, but is parallel to [6.2.2], with further details. Fix $f \in C_E^\infty(Z^+G_k\backslash G_{\mathbb{A}})$, $\varepsilon > 0$, $g \in G_\infty$, and $1 \le \ell \in \mathbb{Z}$. Given a compact neighborhood b of $0 \in U\mathfrak{g}^{\le \ell}$, by the *uniform* continuity of f and its derivatives βf for $\beta \in b$, there is a small-enough neighborhood U of $1 \in G_\infty$ such that $|\beta f(x) - \beta f(xu)| < \varepsilon$ for all $u \in U$, for all $x \in Z^+G_k\backslash G_{\mathbb{A}}$, and for all $\beta \in b$. Without loss of generality, U has *compact closure* V. Certainly $E' = EV^{-1}g^{-1}$ is compact. Being the continuous image of compact $V \times b$, the set $\bigcup_{v\in V} vbv^{-1}$ is itself a compact neighborhood of $0 \in U\mathfrak{g}^{\le\ell}$, so the seminorms $\nu_{g^{-1}bg,E'}$ are *uniformly* comparable to $\nu_{b,E'}$. For all $h = gu \in gU \subset G_\infty$,

$$\begin{aligned} h \cdot C_E^\infty(Z^+G_k\backslash G_{\mathbb{A}})^K &\subset gU \cdot C_E^\infty(Z^+G_k\backslash G_{\mathbb{A}})^K \\ &\subset C_{EV^{-1}g^{-1}}^\infty(Z^+G_k\backslash G_{\mathbb{A}})^K \qquad (h = gu \in gU) \end{aligned}$$

For all $F \in C_E^\infty(Z^+G_k\backslash G_{\mathbb{A}})^K$ with $\nu_{b,EV^{-1}}(f - F) < \varepsilon$, with $h = gu \in gU$, using the uniform comparability of these seminorms.

$$\begin{aligned} \nu_{b,EV^{-1}g^{-1}}(g \cdot f - h \cdot F) &= \nu_{g^{-1}bg,EV^{-1}}(f - g^{-1}h \cdot F) \\ &= \nu_{g^{-1}bg,EV^{-1}}(f - u \cdot F) \ll \nu_{b,EV^{-1}}(f - u \cdot F) \\ &\le \nu_{b,EV^{-1}}(f - u \cdot f) + \nu_{b,EV^{-1}}(u \cdot f - u \cdot F) \\ &= \nu_{b,EV^{-1}}(f - u \cdot f) + \nu_{b,E}(f - F) < \varepsilon + \varepsilon \end{aligned}$$

proving *archimedean* joint continuity at $\{g\} \times \{f\}$ of

$$gU \times C_E^\infty(Z^+G_k\backslash G_{\mathbb{A}})^K \longrightarrow C_{EV^{-1}g^{-1}}^\infty(Z^+G_k\backslash G_{\mathbb{A}})^K$$

The inclusion $C_{EV^{-1}g^{-1}}^\infty(Z^+G_k\backslash G_{\mathbb{A}}) \to C_c^\infty(Z^+G_k\backslash G_{\mathbb{A}})$ is continuous, so $gU \times C_E^\infty(Z^+G_k\backslash G_{\mathbb{A}}) \to C_c^\infty(Z^+G_k\backslash G_{\mathbb{A}})$ is continuous at $\{g\} \times \{f\}$. Since the colimit is stable under the action of G, now it makes sense to say that $G \times C_E^\infty(Z^+G_k\backslash G_{\mathbb{A}}) \to C_c^\infty(Z^+G_k\backslash G_{\mathbb{A}})$ is continuous at $\{g\} \times \{f\}$. Since $g \in G$ and $f \in C_E^\infty(Z^+G_k\backslash G_{\mathbb{A}})$ were arbitrary, this shows that $G \times C_E^\infty(Z^+G_k\backslash G_{\mathbb{A}}) \to C_c^\infty(Z^+G_k\backslash G_{\mathbb{A}})$ is continuous.

Maps *from* a colimit $X = \lim_i X_i$ *to* another object Y are exactly *compatible* families of maps $X_i \to Y$ from the limitands to Y, as in [13.8], [13.9], and

[13.10]. Thus, the compatible family of continuous maps (suppressing reference to the quotient $Z^+G_k\backslash G_{\mathbb{A}}$),

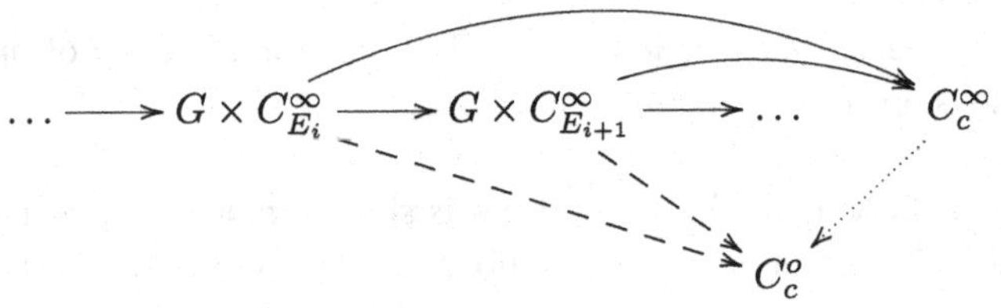

immediately gives the joint continuity of

$$G \times C_c^\infty(Z^+G_k\backslash G_{\mathbb{A}}) \longrightarrow C_c^\infty(Z^+G_k\backslash G_{\mathbb{A}}) \qquad \text{(for all } U\text{)}$$

as claimed. ///

[6.4.3] Corollary: The right translation action $G \times C_c^\infty(G) \to C_c^\infty(G)$ is jointly continuous.

Proof: The proof did not use specific features of Z^+G_k other than its discreteness in $Z^+\backslash G$, so we could as well replace Z^+G_k by $\{1\}$. ///

[6.4.4] Corollary: The *contragredient* or *dual* action of G on distributions $C_c^\infty(Z^+G_k\backslash G_{\mathbb{A}})^*$ defined by $(g \cdot u)(f) = u(g^{-1} \cdot f)$ for all $f \in C_c^\infty(Z^+G_k\backslash G_{\mathbb{A}})$ gives jointly continuous

$$G \times C_c^\infty(Z^+G_k\backslash G_{\mathbb{A}})^* \longrightarrow C_c^\infty(Z^+G_k\backslash G_{\mathbb{A}})^*$$

Proof: This is a special case of the continuity of $G \times V \to V$ giving that of $G \times V^* \to V^*$, where V^* is the dual space of V and is given the weak dual topology [13.11]. The group acts on the dual by $(g \cdot \lambda)(v) = \lambda(g^{-1} \cdot v)$. Given $g \in G, \lambda \in V^*, v \in V$, and $\varepsilon > 0$, we want a neighborhood of gU and a neighborhood $\lambda + N$ of λ such that for $h = gu \in gU$ and $\mu = \lambda + \nu \in \lambda + N$, $|(h\mu - g\lambda)(v)| < \varepsilon$. This is

$$\begin{aligned} \varepsilon > |(h\mu - g\lambda)(v)| &= \left|\mu(h^{-1}v) - \lambda(g^{-1}v)\right| \\ &= \left|\nu(u^{-1}g^{-1}v) + \lambda(u^{-1}g^{-1}v - g^{-1}v)\right| \end{aligned}$$

By continuity of $G \times V \to V$, we can take U small enough so that $u^{-1}g^{-1}v - g^{-1}v$ is in a small enough neighborhood E of 0 in V so that $|\lambda(E)| < \frac{\varepsilon}{2}$. Then we can take N small enough so that $|\nu(U^{-1}g^{-1}v)| < \frac{\varepsilon}{2}$. ///

6.5 Symmetry of Invariant Laplacians

Just as the density of $C_c^o(Z^+G_k\backslash G_{\mathbb{A}})$ in $L^2(Z^+G_k\backslash G_{\mathbb{A}})$ was used to examine the representation of G on $L^2(Z^+G_k\backslash G_{\mathbb{A}})$, we need the density of test functions in $L^2(Z^+G_k\backslash G_{\mathbb{A}})$ to prove things about the invariant Laplacians on right K_∞-invariant functions. As in [4.2], the invariant Laplacians Δ_v on G_v/K_v with v archimedean are the restrictions of the corresponding Casimir elements Ω_v to right K_v-invariant functions.

[6.5.1] Claim: In the four simplest examples from Chapter 1, $C_c^\infty(\Gamma\backslash G/K)$ is dense in $L^2(\Gamma\backslash G/K)$. Similarly, on adele groups $G_{\mathbb{A}}$ is in Chapters 2 and 3, $C_c^\infty(Z^+G_k\backslash G_{\mathbb{A}})$ is dense in $L^2(Z^+G_k\backslash G_{\mathbb{A}})$.

Proof: The argument for $\Gamma\backslash G/K$ is the same as that for $Z^+G_k\backslash G_{\mathbb{A}}$, simply dropping any reference to non-archimedean phenomena. We give the adele-group argument, to include the non-archimedean aspects. By density of continuous, compactly supported functions in either case, it suffices to L^2-approximate continuous, compactly supported functions f by *smooth*, compactly supported functions. The standard device uses a *smooth Dirac sequence*[5] $\{\varphi_i\}$ on G, where smoothness at archimedean places means indefinitely continuously differentiable, and at non-archimedean places means locally constant. Put

$$(\varphi_i \cdot f)(g) = \int_{G_{\mathbb{A}}} \varphi_i(h)\, f(gh)\, dh$$

As in [14.5] and [14.6], via Gelfand-Pettis integrals, $\varphi_i \cdot f$ is in $C_c^o(Z^+G_k\backslash G_{\mathbb{A}})$, and $\varphi_i \cdot f \to f$ in the L^2 topology, and, in fact, in the finer LF-topology on $C_c^o(Z^+G_k\backslash G_{\mathbb{A}})$. Changing variables in the integral,

$$(\varphi_i \cdot f)(g) = \int_{G_{\mathbb{A}}} \varphi_i(g^{-1}h)\, f(h)\, dh$$

That is, the function-valued function

$$h \longrightarrow (g \to \varphi_i(g^{-1}h)\, f(h))$$

is a continuous, compactly supported, $C_c^\infty(Z^+G_k\backslash G_{\mathbb{A}})$-valued function. Thus, it has a Gelfand-Pettis integral in $C_c^\infty(Z^+G_k\backslash G_{\mathbb{A}})$. That is, each $\varphi_i \cdot f$ is a smooth, compactly supported function. We saw that the sequence of these approaches f. ///

[5] A smooth *Dirac sequence* or *approximate identity* $\{\varphi_i\}$ on a unimodular group G is a sequence of smooth, compactly supported real-valued functions φ_i so that $\int_G \varphi_i(g)\, dg = 1$ and $0 \le \varphi \le 1$ for all i, and for every neighborhood N of the identity $e \in G$, there is i_o such that for all $i \ge i_o$ the support of φ_i is inside N.

[6.5.2] Corollary: The Casimir operators Ω_v on the archimedean factors G_v of $G_{\mathbb{A}}$ are *symmetric* on $C_c^\infty(Z^+G_k\backslash G_{\mathbb{A}})$, that is,

$$\langle \Omega_v f, F\rangle_{Z^+G_k\backslash G_{\mathbb{A}}} = \langle f, \Omega_v F\rangle_{Z^+G_k\backslash G_{\mathbb{A}}}$$

for $f, F \in C_c^\infty(Z^+G_k\backslash G_{\mathbb{A}})$, and *negative semidefinite* on right K_∞-invariant functions $C_c^\infty(Z^+G_k\backslash G_{\mathbb{A}})^{K_\infty}$:

$$\langle \Omega_v f, f\rangle_{Z^+G_k\backslash G_{\mathbb{A}}} \leq 0 \qquad (\text{for } f \in C_c^\infty(Z^+G_k\backslash G_{\mathbb{A}}))$$

Proof: As in [4.2], the Casimir element Ω_v on archimedean G_v, or a constant multiple of it, is described as follows. Let $\langle \alpha, \beta\rangle = \mathrm{Re}\,\mathrm{tr}(\alpha\beta)$ be the *trace pairing* on the Lie algebra $\mathfrak{g}$ of G_v, where *trace* is just matrix trace. As in [4.2], for any basis $\{x_j\}$ of $\mathfrak{g}$ and dual basis $\{x_j^*\}$, $\Omega_v = \sum_j x_j x_j^* \in U\mathfrak{g}$. In fact, the pairing is negative-definite on the Lie algebra $\mathfrak{k}$ of K_v, which is $O(n,\mathbb{R})$ or $U(n)$, and positive-definite on their complements $\mathfrak{s}$ consisting of symmetric real matrices and Hermitian-symmetric complex matrices. So we can choose an orthogonal basis $\{x_j\}$ for $\mathfrak{s}$ with $x_j^* = x_j$, and an orthogonal basis $\{\theta_j\}$ for $\mathfrak{k}$ with $\theta_j^* = -\theta_j$. Thus,

$$\Omega_v = \sum_j x_j^2 - \sum_j \theta_j^2$$

The action of $x \in \mathfrak{g}$ on $f \in C_c^\infty(Z^+G_k\backslash G_{\mathbb{A}})$ is

$$(xf)(g) = \frac{\partial}{\partial t}\bigg|_{t=0} f(ge^{tx})$$

To properly indicate the order of operations, the evaluation at $t = 0$ should come *after* the derivative, so for clarity write

$$(xf)(g) = \bigg|_{t=0} \frac{\partial}{\partial t} f(ge^{tx})$$

For $f, F \in C_c^\infty(Z^+G_k\backslash G_{\mathbb{A}})$,

$$\langle xf, F\rangle = \int_{Z^+G_k\backslash G_{\mathbb{A}}} \bigg|_{t=0} \frac{\partial}{\partial t} f(ge^{tx}) \cdot \overline{F}(g)\, dg$$

It would be needlessly reckless to claim that the integrand is a compactly supported, continuous $C_c^\infty(\mathbb{R})$-valued function

$$g \longrightarrow (t \longrightarrow f(ge^{tx}) \cdot \overline{F}(g))$$

because the compact support in t can easily fail. Instead, restrict t to a small interval $[-\varepsilon, \varepsilon]$. The integrand *is* a compactly supported, continuous $C^\infty[-\varepsilon,\varepsilon]$-valued function, where $C^\infty[-\varepsilon,\varepsilon]$ has its natural Fréchet-space structure, as in [13.8] and [13.9]. Evaluation at $t = 0$ is a continuous linear functional on $C^\infty[-\varepsilon, \varepsilon]$, so by Gelfand-Pettis the evaluation commutes with the integral:

$$\int_{Z^+G_k\backslash G_\mathbb{A}} \Big|_{t=0} \frac{\partial}{\partial t} f(ge^{tx}) \cdot \overline{F}(g)\, dg$$

$$= \Big|_{t=0} \int_{Z^+G_k\backslash G_\mathbb{A}} \frac{\partial}{\partial t} f(ge^{tx}) \cdot \overline{F}(g)\, dg$$

Similarly, $\partial/\partial t$ is a continuous map of $C^\infty[-\varepsilon, \varepsilon]$ to itself, so commutes with the integral:

$$\Big|_{t=0} \int_{Z^+G_k\backslash G_\mathbb{A}} \frac{\partial}{\partial t} f(ge^{tx}) \cdot \overline{F}(g)\, dg = \Big|_{t=0} \frac{\partial}{\partial t} \int_{Z^+G_k\backslash G_\mathbb{A}} f(ge^{tx}) \cdot \overline{F}(g)\, dg$$

This legitimizes the change of variables replacing g by $ge^{-t\gamma}$:

$$\Big|_{t=0} \frac{\partial}{\partial t} \int_{Z^+G_k\backslash G_\mathbb{A}} f(ge^{tx}) \cdot \overline{F}(g)\, dg = \Big|_{t=0} \frac{\partial}{\partial t} \int_{Z^+G_k\backslash G_\mathbb{A}} f(g) \cdot \overline{F}(ge^{-t\gamma})\, dg$$

The differentiation and evaluation can be moved back inside the integral for the same reasons. Thus, $\langle xf, F\rangle = -\langle f, xF\rangle$, and

$$\langle x^2 f, F\rangle = -\langle xf, xF\rangle = \langle f, x^2 F\rangle$$

Thus,

$$\Big\langle (\sum_j x_j^2 - \sum_j \theta_j^2) f,\ F\Big\rangle = \Big\langle f,\ (\sum_j x_j^2 - \sum_j \theta_j^2) F\Big\rangle$$

This is the symmetry. On f, F right K_v-invariant, $\theta_j \in \mathfrak{k}$ acts by 0, so

$$\Big\langle (\sum_j x_j^2 - \sum_j \theta_j^2) f,\ f\Big\rangle = \Big\langle (\sum_j x_j^2) f,\ f\Big\rangle$$

$$= \sum_j \langle x_j^2 f, f\rangle = -\sum_j \langle x_j f, x_j f\rangle \le 0$$

giving the nonpositiveness. ///

6.6 An Instance of Schur's Lemma

The general idea of *Schur's lemma* is that endomorphisms of an *irreducible* G-representation space commuting with the G-action must be *scalar*.

The group $G = GL_n(\mathbb{Q}_p)$ is abstracted to consider G *totally disconnected*, in the sense that for every $x \neq y$ in X there are open sets $x \in U$, $y \in V$ so that

$U \cap V = \phi$ and $U \cup V = X$. Since we only consider Hausdorff topological groups, the sets U, V in the definition are not only open but *closed.* The first assertions admit explicit arguments for $GL_n(\mathbb{Q}_p)$, but, in fact, use only general features of totally disconnected groups.

[6.6.1] Claim: At every point x of a locally compact totally disconnected space X there is a local basis consisting of compact *open* sets.

Proof: By local compactness, take an open set V containing x so that the closure $\overline{V}$ is compact. The boundary $\partial V = \overline{V} \cap (X - V) \subset \overline{V}$ is closed, so is compact. For $y \in \partial V$, there are open (and closed) sets $x \in U_y$ and $y \in V_y$ so that $U_y \cap V_y = \emptyset$ and $U_y \cup V_y = X$. Take a finite subcover $V_{y_1}, \ldots, V_{y_n}$ of ∂V. The set

$$V - (\bigcup_i \bar{V}_{y_i}) = \bar{V} - (\bigcup_i V_{y_i})$$

contains x, and is both open and closed. Being a closed subset of the compact set $\overline{V}$ in a Hausdorff space, it is compact. ///

[6.6.2] Claim: A locally compact totally disconnected topological group G has a local basis at $1 = 1_G$ consisting of compact open subgroups. Further, for a fixed compact open subgroup K_1, there are *normal* compact open subgroups of K_1 such that $K_1 \supset K_2 \supset K_3 \supset \cdots$ and these K_j are a local basis at 1.

Proof: Let V be a compact open subset of G containing 1, by the previous. The set

$$K = \{x \in G : xV \subset V \text{ and } x^{-1}V \subset V\}$$

is a *subgroup* of G, and

$$K = (\bigcap_{v \in V} Vv^{-1}) \cap (\bigcap_{v \in V} Vv^{-1})^{-1}$$

so K is the continuous image of a compact set, so is compact. What remains to be shown is that K is open.

To the latter end, show that the compact-open topology on G constructed from the original topology on G is the original topology on G. That is, show that, for compact C in G and for open V in G, the set

$$U = U_{C,V} = \{x \in G : xC \subset V\}$$

is *open* in G. Take U nonempty, and $x \in U$. For all points $xy \in xC$ for $y \in C$, there is a small-enough open neighborhood U_y of 1 so that the open neighborhood $xU_y y$ of xy is contained in V. By continuity of the multiplication in G, there is an open neighborhood W_y of 1 so that $W_y W_y \subset U_y$. The sets $xW_y y$ cover xC. Let $xW_{y_1}y_1, \ldots, xW_{y_n}y_n$ be a finite subcover. Put $W = \bigcap_i W_{y_i}$. Then xW is a neighborhood of x and

$$xW \cdot C \subset xW \cdot \bigcup_i W_{y_i} y_i$$

and $xWW_{y_i}y_i \subset xW_{y_i}W_{y_i}y_i \subset xU_{y_i}y_i$. Thus, U is open.

Let $H_1 \supset H_2 \supset \cdots$ be a local basis of compact open subgroups. Put $K_1 = H_1$. Of course, K_1 is a union of cosets $K_1 = \bigcup_{x \in K_1} xH_2$. By compactness, there is a finite subcover $K_1 = x_1H_2 \cup \cdots \cup x_nH_2$. Thus, H_2 is of finite index in K_1. Then

$$K_2 = \bigcap_{x \in K_1} xH_2x^{-1} = \bigcap_{x \in K_1/H_2} xH_2x^{-1} = x_1H_2x_1^{-1} \cap \cdots \cap x_nH_2x_n^{-1}$$

is a finite intersection of compact open subgroups, so is compact and open, and is normal in K_1. Replace H_3 by $H_3 \cap K_2$ and continue inductively. ///

A representation $G \times V \to V$ of G on a complex vector space V is *smooth* when, for all $v \in V$, the isotropy group $G_v = \{g \in G : g \cdot v = v\}$ is *open*. Because of the total-disconnectedness, this condition is equivalent to an expression of V as an ascending union, or strict colimit [13.8], [13.9],

$$V = \bigcup_K V^K = \mathrm{colim}_K\, V^K$$

with $V^K = K$-fixed vectors in V, where K runs over compact open subgroups of G. All our topological groups are countably based, so the colimit has a countable cofinal subsystem. Since every G-homomorphism $V \to W$ preserves K-fixed-ness, smoothness is automatically preserved by G-homomorphisms. The representation V is *admissible* when each V^K is finite-dimensional. In that case, V is a strict colimit of finite-dimensional topological vector spaces.[6] Finite-dimensional topological vectors spaces have unique topologies [13.4], and every linear map from a finite-dimensional space is continuous. Thus, admissible V has a topology so that *every* linear $T : V \to W$ is continuous. In particular, the continuous dual V^* of admissible V consists of *all* linear functions on V. Similarly, for a topological vector space V that is a strict colimit of finite-dimensional spaces, *every* group action $G \times V \to V$ is continuous.[7]

[6] Happily, but not obviously, many important representations of such groups are admissible, which reduces the topological or analytical delicacy.

[7] Some sources give the impression that a colimit of finite-dimensional spaces has *no* topology, or that topology is ignored. However, as noted here, there is a canonical topology, and every linear map *from* it to any topological vector space is continuous.

Generally, given an arbitrary (continuous) representation $G \times V \to V$ on a vectorspace V, the subspace V^∞ of *smooth vectors* is

$$V^\infty = \{v \in V : G_v \text{ is open}\}$$

This subspace is G-stable, so the restriction of $G \times V \to V$ to $G \times V^\infty \to V^\infty$ is a G-subrepresentation of V. For a smooth representation $G \times V \to V$, the *(smooth) dual* or *(smooth) contragredient* $V^\vee$ is the representation of G on the smooth vectors $(V^*)^\infty$ in the continuous linear dual V^*, where, as always, G acts on V^* by $(g \cdot \lambda)(v) = \lambda(g^{-1} \cdot v)$. A smooth representation $G \times V \to V$ is *(topologically) irreducible* when it contains no proper closed subspace W stable under G. It is *(algebraically) irreducible* when it contains no proper (not necessarily closed) subspace W stable under G.

[6.6.3] Claim: Every subspace of a strict colimit of finite-dimensional spaces is closed.

Proof: Let W be a subspace of $\bigcup_i V_i$ where $V_i \subset V_{i+1}$ and every V_i is finite-dimensional. Given $v \in V$ and $v \notin W$, we can make a linear functional λ_v on V that vanishes on W and is nonzero on v, as follows. Let j be large-enough so that $v \in V_j$. Let $\lambda_j : V_j \to \mathbb{C}$ be linear with $\lambda_j(W \cap V_j) = 0$ and $\lambda_j v = 1$. Extend λ_j to a linear functional λ_{j+1} on V_{j+1} by choosing a complementary subspace X_{j+1} to V_j inside V_{j+1}, and making $\lambda_{j+1} X_{j+1} = 0$, while agreeing with λ_j on the copy of V_j inside V_{j+1}. Continue by induction. This defines λ_v on the strict colimit, and continuity is automatic. Thus, W is the intersection of the kernels of all such λ_v over all $v \notin W$, so is closed. ///

The *(full) Hecke algebra* on G is the space $\mathcal{H} = \mathcal{H}(G) = C_c^\infty(G)$ of smooth, compactly supported complex-valued functions on G. Here *smooth* means *locally constant*, in the sense that, given $\eta \in \mathcal{H}$ and $g \in G$, there is a neighborhood U of g such that $\eta(g') = \eta(g)$ for all $g' \in U$. The Hecke algebra acts on V for any representation $G \times V \to V$ as usual by

$$\eta \cdot v = \int_G \eta(g)\, g \cdot v\, dg \qquad \text{(with right-invariant measure)}$$

and convolution $\eta * \psi$ on $\mathcal{H}$ is characterized by

$$(\eta * \psi)v = \eta \cdot (\psi \cdot v) \qquad (\text{for } v \in V \text{ and } \eta, \psi \in \mathcal{H})$$

That is,

$$\begin{aligned}(\eta * \psi)v = \eta \cdot (\psi \cdot v) &= \eta \cdot \int_G \psi(x) x \cdot v\, dx \\ &= \int_G \eta(y) y \cdot \left(\int_G \psi(x) x \cdot v\, dx \right) dy \\ &= \int_G \int_G \eta(y)\psi(x)\, y \cdot x \cdot v\, dx\, dy\end{aligned}$$

using properties of Gelfand-Pettis integrals of vector-valued functions [14.1]. Changing the order of integration and replacing y by yx^{-1}, then changing the order back, and again using Gelfand-Pettis, this is

$$\int_G \int_G \eta(yx^{-1})\psi(x)\, y\cdot v \; dy\, dx = \int_G \left(\int_G \eta(yx^{-1})\psi(x)\, dx \right) y \cdot v \; dy$$

Thus, we discover an expression for the convolution, without assuming unimodularity of G,

$$(\eta * \psi)(y) = \int_G \eta(yx^{-1})\, \psi(x)\, dx$$

Unless G is discrete, which is not of interest here, the algebra $\mathcal{H}$ with convolution has no unit element 1. Fortunately, there are *sufficiently many idempotents*: for compact open subgroup K of G, let

$$e_K = \frac{\text{ch}_K}{\text{meas}\,(K)} \in \mathcal{H}$$

with characteristic function ch_K of K. These are idempotents: $e_K * e_K = e_K$. Indeed, for any $K' \subset K$, we have $e'_K * e_K = e_K * e_{K'} = e_K$. There are *sufficiently many* of these idempotents in $\mathcal{H}$ in the sense that, given $\eta \in \mathcal{H}$ there is e_K such that $e_K * \eta = \eta = \eta * e_K$. Smoothness of a G-representation space V is that for all $v \in V$ there is a small-enough K so that $e_K \cdot v = v$.

A complex vectorspace V which is a module over the ring $\mathcal{H}$ is *smooth* when for every $v \in V$ there is a small-enough compact open subgroup K so that $e_K v = v$.

[6.6.4] Theorem: The category of smooth $\mathcal{H}$-modules is the same as the category of smooth G-representations.

Proof: We have already seen how to get smooth $\mathcal{H}$-modules from smooth G-representations. We need to recover the action of G from the $\mathcal{H}$-module structure. Slightly generalizing previous notation, for a compact open subset X of G, let

$$e_X = \frac{\text{ch}_X}{\text{meas}\,(X)}$$

For v in a smooth $\mathcal{H}$-module V with K a small-enough compact open subgroup so that $e_K v = v$, and for $g \in G$, try go define the action of g by $g \cdot v = e_{gK} v$. To see that this is well-defined, check that K may be made smaller without altering $e_{gK} \cdot v$. Since $e_K \cdot v = v$, by associativity it suffices to show that $e_{gK'} * e_K = e_{gK}$ for $K' \subset K$:

$$\begin{aligned} \text{meas}\,(gK') \cdot \text{meas}\,(K) \cdot (e_{gK'} * e_K)(x) &= \int_G \text{ch}_{gK'}(xy^{-1})\, \text{ch}_K(y)\, dy \\ &= \int_K \text{ch}_{gK'}(xy^{-1})\, dy \end{aligned}$$

The integrand is nonzero exactly for $y \in K$ and $xy^{-1} \in gK'$, that is, for $y \in K$ and $y \in K'g^{-1}x$. Since $K' \subset K$, the set $K'g^{-1}x \cap K$ is nonempty exactly for $g^{-1}x \in K$, which is $x \in gK$, in which case the intersection has measure meas (K'). That is, with modular function δ,

$$\begin{aligned} e_{gK'} * e_K(x) &= \frac{\text{meas}\,(K')}{\text{meas}\,(gK') \cdot \text{meas}\,(K)} \cdot \text{ch}_{gK}(x) \\ &= \frac{\text{ch}_{gK}(x)}{\delta(g) \cdot \text{meas}\,(K)} = \frac{\text{ch}_{gK}(x)}{\text{meas}\,(gK)} \end{aligned}$$

as claimed. Second, check that this gives a group homomorphism: given compact open K and given $h \in G$, take K' small-enough so that $e_{K'} * e_{hK} = e_{hK}$. This condition is

$$\begin{aligned} \text{ch}_{hK}(x) &= \text{meas}\,(K') \cdot \int_G \text{ch}_{K'}(xy^{-1})\, \text{ch}_{hK}(y)\, dy \\ &= \text{meas}\,(K') \cdot \int_G \text{ch}_{K'}(y^{-1})\, \text{ch}_{hK}(yx)\, dy \\ &= \text{meas}\,(K') \cdot \int_{K'} \text{ch}_{hK}(yx)\, dy \end{aligned}$$

It suffices that ch_{hK} be left K'-invariant, since then the integrand is $\text{ch}_{hK}(x)$ for all $y \in K'$. This left invariance is that $k'x \in hK$ for $x \in hK$ and $k' \in K'$. That is, $k'hK \subset hK$, or $k'h \in hK$, or $k' \in hKh^{-1}$. Thus, $K' = hKh^{-1}$ suffices. We must show that $e_{gK'} * e_{hK} = e_{ghK}$. Noting that

$$\text{meas}\,(K') = \text{meas}\,(hKh^{-1}) = \text{meas}\,(hK) = \delta(h) \cdot \text{meas}\,(K)$$

we have

$$\begin{aligned} \text{meas}\,(gK') \cdot \text{meas}\,(hK) \cdot (e_{gK'} * e_{hK})(x) &= \int_G \text{ch}_{gK'}(xy^{-1})\, \text{ch}_{hK}(y)\, dy \\ &= \delta(h) \int_G \text{ch}_{gK'}(xy^{-1}h^{-1})\, \text{ch}_K(y)\, dy \\ &= \delta(h) \int_K \text{ch}_{ghKh^{-1}}(xy^{-1}h^{-1})\, dy \\ &= \delta(h) \int_K \text{ch}_{ghK}(xy)\, dy \end{aligned}$$

The integrand is nonzero exactly when $xy \in ghK$, which is $y \in K \cap x^{-1}ghK$. This is nonzero exactly for $x^{-1}ghK \subset K$, which is equivalent to $x \in ghK$. Then y is integrated over K, giving $\delta(h)\text{meas}\,(K)\text{ch}_{ghK}(x)$. Thus,

$$\begin{aligned} e_{gK'} * e_{hK} &= \frac{\delta(h)\text{meas}\,(K) \cdot \text{ch}_{ghK}}{\text{meas}\,(gK') \cdot \text{meas}\,(hK)} \\ &= \frac{\delta(h)\text{meas}\,(K) \cdot \text{ch}_{ghK}}{\delta(g)\delta(h)\text{meas}\,(K) \cdot \delta(h)\text{meas}\,(K)} = \frac{\text{ch}_{ghK}}{\text{meas}\,(ghK)} \end{aligned}$$

This is the homomorphism property. Visibly, G-homomorphisms and $\mathcal{H}$-module homomorphisms are interchanged under this bijection. ///

In the following, recall that for $G \times V \to V$ with V a strict colimit of finite-dimensional complex vectorspaces, for example, *admissible*, algebraic irreducibility and topological irreducibility are equivalent. For larger, not necessarily admissible, representations, algebraic irreducibility is usually strictly stronger.

[6.6.5] Theorem: *(Schur's Lemma)* Let $G \times V \to V$ be an algebraically irreducible smooth representation of G. Let T be a $\mathbb{C}$-linear endomorphism of V commuting with all maps $v \to g \cdot v$ with $g \in G$. Then T is a *scalar*, that is, multiplication by an element of $\mathbb{C}$.

Proof: (Jacquet) Since G has a countable basis, $\mathcal{H}$ has countable dimension as $\mathbb{C}$-vectorspace. Algebraic irreducibility implies $\mathcal{H} \cdot v = V$ for $v \neq 0$ in V, so V is of countable $\mathbb{C}$-dimension. An $\mathcal{H}$-endomorphism T is completely determined by Tv for one $v \neq 0$, since $T(\eta v) = \eta T(v)$ for $\eta \in \mathcal{H}$. Thus, the ring D of $\mathcal{H}$-endomorphisms of V has countable $\mathbb{C}$-dimension. As V is algebraically irreducible, for all $T \in D$ both the kernel and image of T are $\mathcal{H}$-submodules, so can be only 0 or V. Thus, D is a division ring with $\mathbb{C}$ in its center.

Since $\mathbb{C}$ is algebraically closed, nonscalar $T \in D$ are *transcendental* over $\mathbb{C}$. Therefore, for $T \in D$ *not* a scalar the elements $S_\lambda = (T - \lambda)^{-1} \in D$ with λ varying over $\mathbb{C}$ are *linearly independent* over $\mathbb{C}$, by uniqueness of partial fraction expansions in $\mathbb{C}(T)$. As $\mathbb{C}$ is uncountable, this would yield an uncountable set of linearly independent elements of D, contradiction. ///

6.7 Duality of Induced Representations

The group G is still *totally disconnected*, and representations $G \times V \to V$ are *smooth*. For a closed subgroup H of G, there is the *forgetful functor* Res_H^G from smooth G-representations to smooth H-representations, by forgetting all but the action of H. We eventually want a *right adjoint*[8] Ind_H^G to this forgetful functor from representations of H to representations of G, in the sense that there should be natural bijections

$$\mathrm{Hom}_G(V, \mathrm{Ind}_H^G W) \approx \mathrm{Hom}_H(\mathrm{Res}_H^G V, W)$$

for all smooth G-representation V and smooth H-representation W. This adjunction is *Frobenius reciprocity*, proven in the next section, and is viewed

[8] It is not obvious, but there is no *left* adjoint to Res_H^G in this situation.

there as approximately analogous to the *Cartan-Eilenberg adjunction*

$$\mathrm{Hom}_{\mathbb{Z}}(A,\ \mathrm{Hom}_{\mathbb{Z}}(B, C)) \approx \mathrm{Hom}_{\mathbb{Z}}(A \otimes_{\mathbb{Z}} B,\ C)$$

for $\mathbb{Z}$-modules A, B, C, with $\varphi_\Phi \leftarrow \Phi$ by $\varphi_\Phi(a)(b) = \Phi(a \otimes b)$, and $\varphi \to \Phi_\varphi$ by $\Phi_\varphi(a \otimes b) = \varphi(a)(b)$. The relatively simple argument for this will also be recalled in the next section.

The present section constructs *induced representations* $\mathrm{Ind}_H^G W$ of G made from representations W of H, and *compactly induced* representations c-$\mathrm{Ind}_H^G W$ as spaces of W-valued functions on G, and proves a duality

$$(\text{c-}\mathrm{Ind}_H^G W,\ \mathbb{C})^\vee = \mathrm{Hom}_G(\text{c-}\mathrm{Ind}_H^G W,\ \mathbb{C}) \approx \mathrm{Ind}_H^G(\rho \cdot \mathrm{Hom}_G(W^\vee, \mathbb{C}))$$

with $\rho = \delta_H/\delta_G$ for modular functions on H and G (proof below), whose relevance to Frobenius reciprocity is suggested by the simpler Cartan-Eilenberg adjunction.

For a smooth representation W of a closed subgroup H of G, let $C_c^\infty(H\backslash G, W)$ be the space of W-valued functions f on G that are compactly supported left-modulo H (in the sense that the images in $H\backslash G$ of their supports are compact), locally constant, and so that

$$f(hg) = h \cdot f(g) \qquad (\text{for } h \in H \text{ and } g \in G)$$

The *compact-induced representation* c-$\mathrm{Ind}_H^G W$ has representation space $C_c^\infty(H\backslash G, W)$ with the right translation action of G

$$(g \cdot f)(g') = f(g'g) \qquad (\text{for } g, g' \in G)$$

The *induced representation* $\mathrm{Ind}_H^G W$ has representation space consisting of *uniformly* locally constant W-valued functions f on G satisfying

$$f(hg) = h \cdot f(g) \qquad (\text{for } h \in H \text{ and } g \in G)$$

with the right translation action

$$(g \cdot f)(g') = f(g'g) \qquad (\text{for } g, g' \in G)$$

The *uniform* locally constant condition on f in this space of functions is that there is a compact open subgroup Θ so that

$$f(g\theta) = f(g) \qquad (\text{for all } g \in G \text{ and for all } \theta \in \Theta)$$

As a variant of [5.2.1] about invariant measures on quotients, for the present discussion, we need a slightly different *unwinding*:

[6.7.1] Lemma: Let δ_H, δ_G be the modular functions of H, G, for a closed subgroup H of G, and $\rho = \delta_H/\delta_G$. There is a nontrivial right G-invariant functional u on c-$\mathrm{Ind}_H^G\, \rho$, unique up to scalar multiples.

Proof: Let

$$\alpha \; : \; C_c^\infty(G) \longrightarrow \text{c-Ind}_H^G \, \rho$$

by

$$\alpha f(g) = \int_H \rho^{-1}(h) \cdot f(hg) \, dh$$

be the appropriate averaging map, as in [5.1] and [5.2], using right Haar measure on H. For totally disconnected groups and locally constant, compactly supported functions, the surjectivity of α allows an even simpler argument than the case treated in [5.1].

Attempt to define a right G-invariant $\mathbb{C}$-valued u on c-Ind$_H^G\rho$ by $u(\alpha f) = \int_G f$ with right Haar measure on G. The telling issue is well-definedness: replacing $g \to f(g)$ by $g \to f(hg)$ gives, on one hand,

$$\int_G f(hg) \, dg = \delta_H(h^{-1}) \cdot \int_G f(g) \, dg = u(\alpha f)$$

On the other hand,

$$\begin{aligned} \alpha(h' \to f(hh')) &= \int_H \rho^{-1}(h') \, f(hh'g) \, dh' \\ &= \delta_G(h^{-1})\rho^{-1}(h^{-1}) \int_H \rho^{-1}(h') \, f(h'g) \, dh' \end{aligned}$$

Thus, ρ is the only possible choice. As in the proof of [5.2.1], it suffices to show that $\int_G f(g) \, dg = 0$ for $\alpha f = 0$. Indeed, for $\alpha f = 0$, for all $F \in C_c^\infty(G)$, the integral of F against αf is certainly 0. Rearrange

$$\begin{aligned} 0 = \int_G F(g) \, \alpha f(g) \, dg &= \int_G \int_H F(g) \, \rho(h)^{-1} f(hg) \, dh \, dg \\ &= \int_H \int_G F(g) \, \rho(h)^{-1} f(hg) \, dg \, dh \\ &= \int_H \int_G F(h^{-1}g) \, f(g) \, \delta_G(h^{-1}) \, dg \, dh \end{aligned}$$

by replacing g by $h^{-1}g$. Replacing h by h^{-1} replaces right Haar measure by $\delta_H(h) \, dh$, so

$$0 = \int_G \left(\int_H \rho(h)^{-1} \cdot F(hg) \, dh \right) f(g) \, dg = \int_G \alpha F(g) \, f(g) \, dg$$

Surjectivity of α shows that F can be chosen so that αF is identically 1 on the support of f. Then the integral of f is 0, as claimed, proving the well-definedness. ///

[6.7.2] Claim: Let W be a smooth representation of a closed subgroup H of G. Let $\rho = \delta_H/\delta_G$. The smooth dual of the compactly induced c-$\mathrm{Ind}_H^G W$ is the induced representation $\mathrm{Ind}_H^G(\rho \cdot W^\vee)$, by the map described as follows. With $\langle , \rangle$ the duality pairing on $W \times W^\vee$, for $F \in \mathrm{Ind}_H^G(\rho \cdot W^\vee)$, and u : c-$\mathrm{Ind}_H^G \rho$ as in the previous lemma, define

$$\lambda_F(f) = u\big(g \longrightarrow \langle f(g),\, F(g)\rangle\big) \qquad \big(\text{for all } f \in \text{c-Ind}_H^G W\big)$$

Proof: First, claim that the $\mathbb{C}$-valued function $\varphi(g) = \langle f(g), F(g)\rangle$ is in c-$\mathrm{Ind}_H^G \rho$. Since F is uniformly locally constant and f is locally constant and compactly supported left modulo H, φ is locally constant and compactly supported left modulo H. Further, from the definition of the action on the dual $W^\vee$,

$$\begin{aligned}\varphi(hg) &= \langle f(hg),\, F(hg)\rangle = \langle h \cdot f(g),\, \rho(h) \cdot h \cdot F(g)\rangle \\ &= \rho(h) \cdot \langle f(g),\, F(g)\rangle = \rho(h) \cdot \varphi(g)\end{aligned}$$

Therefore, these functionals constitute a $\mathbb{C}$-linear subspace of the smooth dual of c-$\mathrm{Ind}_H^G W$. That $F \to \lambda_F$ is a G-homomorphism is also apparent.

To see that $F \to \lambda_F$ is *injective*, take $F \in (\mathrm{Ind}_H^G \rho \cdot W^\vee)^K$ for some compact open K, take $x \in G$ so that $0 \neq F(x) \in W^\vee$. Let $w \in W$ with $\langle w, F(x)\rangle = 1$. Without loss of generality, $w \in W^K$: by properties [14.1] of Gelfand-Pettis integrals,

$$\begin{aligned}\left\langle \int_K k \cdot w \, dk,\, F(x)\right\rangle &= \int_K \langle k \cdot w,\, F(x)\rangle \, dk = \int_K \langle w,\, k^{-1} \cdot F(x)\rangle \, dk \\ &= \mathrm{meas}\,(K) \cdot \langle w,\, F(x)\rangle\end{aligned}$$

Define $f \in$ c-$\mathrm{Ind}_H^G W$ by

$$f(g) = \int_H \mathrm{ch}_{xK}(hg) \cdot h \cdot w \, dh$$

Again using [14.1] to move the integration outside the pairing $\langle , \rangle$,

$$\begin{aligned}\langle f(g),\, F(g)\rangle &= \int_H \mathrm{ch}_{xK}(hg) \cdot \langle h \cdot w,\, F(g)\rangle \, dh \\ &= \int_H \rho(h)^{-1} \cdot (\mathrm{ch}_{xK}(hg) \cdot \langle h \cdot w,\, F(hg)\rangle) \, dh\end{aligned}$$

This expresses $g \to \langle f(g),\, F(g)\rangle \in$ c-$\mathrm{Ind}_H^G \rho$ as an image of the averaging map of [6.7.1], so

$$\begin{aligned}\lambda_F(f) &= u(g \to \langle f(g),\, F(g)\rangle) = \int_G \mathrm{ch}_{xK}(g) \cdot \langle w,\, F(g)\rangle \, dg \\ &= \int_{xK} \langle w,\, F(g)\rangle \, dg = \left\langle w,\, \int_{xK} F(g) \, dg\right\rangle \\ &= \mathrm{meas}\,(xK) \cdot \langle w,\, F(x)\rangle = \mathrm{meas}\,(xK) \cdot 1 \neq 0\end{aligned}$$

Prove *surjectivity* by proving surjectivity to each $((\text{c-Ind}_H^G W)^*)^K$, for compact open subgroups K. The quotient $(H\backslash G)/K$ is *discrete* since K is open, and Hausdorff since K is closed. Fix a set of representatives x_i for $H\backslash G/K$, and let f_i be the characteristic function of $x_i K \subset G$. For $\lambda \in ((\text{c-Ind}_H^G W)^K)^*$, define a smooth functional $\lambda_i \in W^\vee$ by

$$\lambda_i(w) = \lambda\Big(\int_H f_i(h) \cdot h \cdot w \, dh\Big) \qquad (\text{for } w \in W)$$

Define F piecewise by $F(hx_i\theta) = \rho(h)\lambda_i$, so $\lambda_F(f) = u(g \to \langle f(g), F(g)\rangle)$ is the original λ. ///

6.8 An Instance of Frobenius Reciprocity

The following simplest instance of the fundamental *adjunction* is a precursor to the adjunction that is the assertion of Frobenius reciprocity proven just below:

[6.8.1] Claim: *(Cartan-Eilenberg adjunction)* For $\mathbb{Z}$-modules A, B, C,

$$\text{Hom}_\mathbb{Z}(A, \text{Hom}_\mathbb{Z}(B, C)) \approx \text{Hom}_\mathbb{Z}(A \otimes_\mathbb{Z} B, C)$$

with $\varphi_\Phi \leftarrow \Phi$ by $\varphi_\Phi(a)(b) = \Phi(a \otimes b)$, and $\varphi \to \Phi_\varphi$ by $\Phi_\varphi(a \otimes b) = \varphi(a)(b)$.

Proof: Given $\Phi \in \text{Hom}_\mathbb{Z}(A \otimes B, C)$, certainly $\varphi_\Phi(a)(b) = \Phi(a \otimes b)$ is immediately well defined. Oppositely, the universal property of tensor products produces a unique linear map $A \otimes B \to C$ for each bilinear $A \times B \to C$. Applied to $a \times b \to \varphi(a)(b)$ produces a well-defined $\Phi_\varphi \in \text{Hom}(A \otimes B, C)$ by $\Phi_\varphi(a \otimes b) = \varphi(a)(b)$. ///

The duality of compact-induced and induced proven in the previous section gives

$$\left(\text{c-Ind}_H^G W\right)^\vee = \text{Hom}_G\left(\text{c-Ind}_H^G W, \mathbb{C}\right) \approx \text{Ind}_H^G(\rho \cdot W^\vee)$$

However, to apply such ideas in the present context, there would be several technical complications: tensor products of smooth bi-modules over noncommutative rings without units, even with sufficiently idempotents as $\mathcal{H}$, are not as simply behaved as over $\mathbb{Z}$. These complications cannot be avoided in the following section, but we can prove Frobenius reciprocity more directly:

[6.8.2] Theorem: *(Frobenius Reciprocity)* There is a natural $\mathbb{C}$-vectorspace isomorphism

$$\text{Hom}_G\left(V, \text{Ind}_H^G W\right) \longrightarrow \text{Hom}_H\left(\text{Res}_H^G V, W\right)$$

by $\Phi \to \varphi_\Phi$ where $\varphi_\Phi(v) = \Phi(v)(1)$. The inverse is $\Phi_\varphi \leftarrow \varphi$ where $\Phi_\varphi(v)(g) = \varphi(g \cdot v)$.

Proof: Once the formula for the inverse is conceived, the several things to be checked are fairly straightforward. The H-homomorphism property of φ_Φ follows from

$$h \cdot \varphi_\Phi(v) = h \cdot \Phi(v)(1) = \Phi(v)(1 \cdot h) = \Phi(h \cdot v)(1) = \varphi_\Phi(h \cdot v)$$

The G-homomorphism property of Φ_φ follows from

$$(g \cdot \Phi_\varphi(v))(x) = \Phi_\varphi(v)(xg) = \varphi(xg \cdot v) = \varphi(x \cdot (g \cdot v)) = \Phi_\varphi(g \cdot v)(x)$$

for $g, x \in G$. That the two maps are mutual inverses is easy in one direction:

$$\varphi_{\Phi_\varphi}(v) = \Phi_\varphi(v)(1) = \varphi(1 \cdot v) = \varphi(v)$$

and in the other direction

$$\begin{aligned}\Phi_{\varphi_\Phi}(v)(x) &= \varphi_\Phi(x \cdot v) = \Phi(x \cdot v)(1) = x \cdot \Phi(v)(1) \\ &= \Phi(v)(1 \cdot x) = \Phi(v)(x)\end{aligned}$$

This proves Frobenius reciprocity in this situation. ///

6.9 Induction in Stages

The description that follows of compactly induced representations c-$\mathrm{Ind}_H^G W$ as *tensor products* $\mathcal{H}_G \otimes_{\mathcal{H}_H} W$ suggests the possibility of *inducing in stages*, from the associativity of *tensor products*.

First, we describe a purely algebraic context that encourages optimism about the more complicated situation at hand. Consider not-necessarily-commutative rings with a copy of $\mathbb{C}$ in their centers, and $1_\mathbb{C}$ is the multiplicative unit 1_R of R. All R-modules will be *unital*, in the sense that $1_R \cdot v = v$ for all $v \in V$. The tensor product $M \otimes_R N$ of a *right* R-module M and *left* R-module N is the quotient of $M \otimes_\mathbb{C} N$ by the submodule generated by all expressions

$$m \cdot r \otimes n - m \otimes r \cdot n$$

In general, this tensor product is no longer an R-module. However, when M is both a right R-module and a left S-module for another ring S, that is, is an S, R-*bimodule*, there does remain the left multiplication by S on the tensor product over R, namely, $s \cdot (m \otimes n) = (s \cdot m) \otimes n$. In particular, when $M = S$, for a left R-module N the tensor product $S \otimes_R N$ is a left S-module. This is one notion of *induced module*. However, for present purposes, we only need a comparison:

[6.9.1] Claim: Let R, S, T be $\mathbb{C}$-algebras, with S a right R-algebra, and T a right S-algebra. Make T a right R-module by $t \cdot r = t \cdot (1_S \cdot r)$. Let M be a left

R-module. Then

$$T \otimes_S (S \otimes_R M) \approx T \otimes_R M$$

by

$$t \otimes (s \otimes m) \longrightarrow t \cdot s \otimes m \qquad \text{and} \qquad t \otimes (1_S \otimes m) \longleftarrow t \otimes m$$

Proof: The two maps are well-defined as maps

$$\varphi : T \otimes_{\mathbb{C}} S \otimes_{\mathbb{C}} M \longrightarrow T \otimes_S M$$
$$\psi : T \otimes_{\mathbb{C}} M \longrightarrow T \otimes_S (S \otimes_R M)$$

Thus, it suffices to show that these maps factor through the corresponding quotients. For $t \in T$, $s, s' \in S$, $r \in R$, and $m \in M$,

$$\varphi\big(t \otimes (ss' \otimes m) - t \cdot s \otimes (s' \otimes m)\big) = t \cdot (ss') \otimes m - (t \cdot s) \cdot s' \otimes m = 0$$

by associativity of the right action of S on T. Similarly,

$$\varphi\big(t \otimes (s \otimes r \cdot m) - t \otimes (s \cdot r \otimes m)\big) = (t \cdot s) \cdot r \otimes m - t \cdot (s \cdot r) \otimes m = 0$$

by the associativity

$$(t \cdot s) \cdot r = (t \cdot s) \cdot (1_S \cdot r) = t \cdot (s \cdot (1_S \cdot r)) = t \cdot ((s \cdot 1_S) \cdot r) = t \cdot (s \cdot r)$$

In the other direction,

$$\begin{aligned} \psi\big(t \otimes r \cdot m - t \cdot r \otimes m\big) &= t \otimes 1_S \otimes r \cdot m - t \cdot (1_S \cdot r) \otimes m \\ &= (t \otimes 1_S) \cdot r \otimes m - t \cdot (1_S \cdot r) \otimes m = 0 \end{aligned}$$

again by the associativity

$$(t \cdot 1_S) \cdot r = (t \cdot 1_S) \cdot (1_S \cdot r) = t \cdot (1_S \cdot (1_S \cdot r)) = t \cdot (1_S \cdot r) = t \cdot (1_S \cdot r)$$

Thus, the two maps are mutual inverses. ///

The intention is to use the same idea in application to Hecke algebras $\mathcal{H}_B$, $\mathcal{H}_P$, $\mathcal{H}_G$ for closed subgroups $B \subset P \subset G$, with suitable right $\mathcal{H}_P$ structure on $\mathcal{H}_G$, and right $\mathcal{H}_B$ structure on $\mathcal{H}_P$. These are rings without units but with sufficiently many idempotents.

We first describe relevant right $\mathcal{H}_H$-module structure on $\mathcal{H}_G$, for closed subgroup H of G. Let δ_H be the modular function on H. Give $\mathcal{H}_G = C_c^\infty(G)$ a *right* $\mathcal{H}_H$-module structure by

$$(f \cdot \eta)(g) = \int_H f(hg)\, \eta(h)\, \delta_H(h)\, dh$$

with $\eta \in \mathcal{H}_H$ and $f \in \mathcal{H}_G$. The insertion of the modular function is a normalization choice, which becomes sensible in hindsight, in what follows. The action of $h \in H$ on the argument of f is literally left multiplication in G, but the H-structure or $\mathcal{H}_H$-module structure should be notated on the *right*, to have associativity

$$(f \cdot \eta_1) \cdot \eta_2 = f \cdot (\eta_1 * \eta_2) \qquad (\text{with } \eta_1, \eta_2 \in \mathcal{H}_H \text{ and } f \in \mathcal{H}_G)$$

To check this associativity:

$$\begin{aligned}((f \cdot \eta_1) \cdot \eta_2)(g) &= \int_H \int_H f(xyg)\, \eta_1(x)\, \delta(x)\, \eta_2(y)\, \delta_H(y)\, dx\, dy \\ &= \int_H \int_H f(xg)\, \eta_1(xy^{-1})\, \delta(xy^{-1})\, \eta_2(y)\, \delta_H(y)\, dx\, dy \\ &= \int_H f(xg)(\int_H \eta_1(xy^{-1})\, \eta_2(y)\, dy) \delta_H(x)\, dx \\ &= (f \cdot (\eta_1 * \eta_2))(g)\end{aligned}$$

A smooth representation $H \times W \to W$ is a smooth $\mathcal{H}_H$-module as in [6.6.4]. The tensor product $\mathcal{H}_G \otimes_{\mathcal{H}_H} W$ is $\mathcal{H}_G \otimes_{\mathbb{C}} W$ modulo all relations

$$(f \cdot \eta) \otimes w = f \otimes (\eta \cdot w)$$

and has left $\mathcal{H}_G$-module structure

$$\zeta \cdot (f \otimes w) = (\zeta * f) \otimes w \qquad (\text{convolution in } \mathcal{H}_G)$$

[6.9.2] Theorem: We have an $\mathcal{H}_G$-isomorphism $\mathcal{H}_G \otimes_{\mathcal{H}_H} W \approx \text{c-Ind}_H^G W$ by

$$\beta(f \otimes w)(g) = \int_H f(hg) \cdot h^{-1} w\, dh \qquad (\text{for } f \in \mathcal{H}_G \text{ and } w \in W)$$

Proof: To see that β commutes with the $\mathcal{H}_G$ action, that is, that $\beta(\zeta * f \otimes w) = \zeta * \beta(f \otimes w)$ for $\zeta, f \in \mathcal{H}_G$, is just a change of order of integration:

$$\begin{aligned}\beta(\zeta * f \otimes w)(x) &= \beta\left(\left(y \to \int_G \zeta(yz^{-1})\, f(z)\, dz\right) \otimes w\right)(x) \\ &= \int_H \left(\int_G \zeta(hxz^{-1})\, f(z)\, dz\right) \cdot h^{-1} w\, dh \\ &= \int_G \int_H \zeta(hxz^{-1}) \cdot h^{-1} w\, dh\, f(z)\, dz \\ &= \int_G \beta\left(y \to \zeta(yz^{-1}) \otimes w\right)(x)\, f(z)\, dz = \left(\zeta * \beta(f \otimes w)\right)(x)\end{aligned}$$

as claimed. A change of variables in the integral shows that the image $\beta(f \otimes w)$ lies in the indicated compact-induced representation space:

$$\beta(f \otimes w)(hg) = \int_H f(h'hg) \cdot (h')^{-1} w \, dh'$$

$$= \int_H f(h'g) \cdot (h'h^{-1})^{-1} w \, dh' = \int_H f(h'g) \cdot h \cdot (h')^{-1} w \, dh'$$

$$= h \cdot \int_H f(h'g) \cdot (h')^{-1} w \, dh' = h \cdot \beta(f \otimes w)(g)$$

To show that the map factors through the tensor product over $\mathcal{H}_H$ we must show that

$$\beta(f \cdot \eta \otimes v) = \beta(f \otimes \eta \cdot v) \qquad (\text{for } \eta \in \mathcal{H}_H)$$

Here the normalization by insertion of the modular function will play a role. Take $\eta \in \mathcal{H}_H$, $w \in W$, and $f \in \mathcal{H}_G$:

$$\beta(f \cdot \eta \otimes w)(g) = \beta\left(x \to \int_H f(hx)\,\eta(h)\,\delta_H(h)\,dh \otimes w\right)(g)$$

$$= \int_H \left(\int_H f(hyg)\eta(h)\,\delta_H(h)\,dh\right) y^{-1} w \, dy$$

$$= \int_H \eta(h) \left(\int_H f(hyg)\, y^{-1} w \, dy\right) \delta_H(h)\,dh$$

$$= \int_H \eta(h) \left(\int_H f(yg)\,\delta_H(h^{-1})\, y^{-1} hw \, dy\right) \delta_H(h)\,dh$$

by replacing y by $h^{-1}y$. The $\delta_H(h^{-1})$ and $\delta_H(h)$ cancel, giving

$$\int_H \eta(h) \left(\int_H f(yg)\; y^{-1} hw \, dy\right) dh = \int_H f(yg) y^{-1} \cdot \left(\int_H \eta(h)\, hw \, dh\right) dy$$

$$= \int_H f(yg) y^{-1} \cdot (\eta \cdot w)\, dy$$

$$= \beta(f \otimes \eta \cdot w)(g)$$

Thus, β factors through a map $\gamma : \mathcal{H}_G \otimes_{\mathcal{H}_H} W \to \text{c-Ind}_H^G W$.

To make an inverse map, make an inverse on right K-fixed elements for each compact open subgroup K of G. Given K, fix representatives $\{x_i\}$ for $H\backslash G/K$, and let ch_i be the characteristic function of $x_i K$. Let $q : \mathcal{H}_G \otimes_{\mathbb{C}} W \to \mathcal{H}_G \otimes_{\mathcal{H}_H} W$ be the quotient map. For $F \in \text{c-Ind}_H^G W$, put

$$\Phi(F) = q\left(\sum_i \text{ch}_i \otimes F(x_i)\right) \in \mathcal{H}_G \otimes_{\mathcal{H}_H} W$$

This is the inverse to γ. ///

[6.9.3] Corollary: Let $B \subset P \subset G$ be closed subgroups of a totally disconnected group G, and W a smooth representation of B. Then inducing W from B to G produces the same outcome as inducing from B to P and then from P to G:

$$\text{c-Ind}_B^G W \approx \text{c-Ind}_P^G(\text{c-Ind}_B^P W)$$

Proof: Grant for a moment that for a class of rings without units containing Hecke algebras $\mathcal{H}_G$, $\mathcal{H}_P$, and $\mathcal{H}_B$, and and containing smooth representations V of them,

$$T \otimes_S (S \otimes_R V) \approx T \otimes_R V$$

The theorem gives

$$\text{c-Ind}_B^G W \approx \mathcal{H}_G \otimes_{\mathcal{H}_B} V \approx \mathcal{H}_G \otimes_{\mathcal{H}_P} (\mathcal{H}_P \otimes_{\mathcal{H}_B} V) \approx \text{c-Ind}_P^G(\text{c-Ind}_B^P W)$$

The collapsing of tensor products does hold for *idempotented* rings R, S, T where T is a smooth right S-module, and S is a smooth right R-module. A ring R is *idempotented* when, for every finite $X \subset R$, there is an idempotent element $e \in R$ so that $ex = x = xe$ for all $x \in X$. This property holds for these Hecke algebras, using idempotents e_K as in the proof of [6.6.4]. Smoothness of a module V over an idempotent ring R is that for every finite $X \subset V$, there is an idempotent $e \in R$ such that $ex = x = xe$ for all $x \in X$.

Then T has a smooth right R-module structure $t \cdot r = t \cdot (e' \cdot t)$ where $e' \in S$ is an idempotent in S fixing t, invoking smoothness of T over S. Unlike the case of rings with units, we must check that this is well-defined. Let e_1, e_2 be idempotents in S both fixing t, let e' be an idempotent in S such that $e'e_1 = e_1e' = e_1$ and $e'e_2 = e_2e' = e_2$. Then

$$\begin{aligned} t \cdot (e_1 \cdot r) &= t \cdot (e_1e' \cdot r) = (t \cdot e_1) \cdot (e' \cdot r) = t \cdot (e' \cdot r) = (t \cdot e_2) \cdot (e' \cdot r) \\ &= t \cdot (e_2e' \cdot r) = t \cdot (e_2 \cdot r) \end{aligned}$$

which gives the well-definedness. For an idempotent $e \in R$ such that $e' \cdot e = e'$, we have smoothness $t \cdot e = t \cdot (e' \cdot e) = t \cdot e' = t$. The isomorphisms are

$$t \otimes s \otimes v \longrightarrow t \cdot s \otimes v \qquad \text{and} \qquad t \otimes e' \otimes \longleftarrow t \otimes v$$

for any idempotent $e \in S$ such that $t \cdot e = t$. Certification of the well-definedness of the second map is similar to the previous argument: let $t \cdot e_1 = t = t \cdot e_2$ for two idempotents $e_1, e_2 \in S$, and let e be another idempotent in S such that $e_1e = e_1$ and $e_2e = e_2$. It suffices to compute in $T \otimes_S S$:

$$t \otimes e_1 = t \otimes e_1e = t \cdot e_1 \otimes e = t \otimes e = t \cdot e_2 \otimes e = t \otimes e_2e = t \otimes e_2$$

This gives the well-definedness. ///

6.10 Representations of Compact G/Z

We still consider totally disconnected G and smooth representations. The general case of representations of compact groups on topological vector spaces is treated in [9.C].

Let Z be a closed subgroup of G inside the center of G, and suppose that G/Z is *compact*. Consider representations V with central character $\omega : Z \to \mathbb{C}^\times$, i.e., so that $z \cdot v = \omega(z)v$ for all $v \in V$ and $z \in Z$. The simple situation that G is compact and $Z = \{1\}$ is already useful.

[6.10.1] Proposition: Every finitely generated smooth representation V of G with central character ω is finite-dimensional.

Proof: Take a compact open subgroup K small enough so that a (finite) set X of generators for V lies inside V^K. Let Y be a choice of a set of representatives for G/ZK; since G/Z is compact, Y is finite. The set of all vectors $g \cdot v$ with $v \in X$ and $g \in G$ is contained in the span of the *finite* set of vectors $y \cdot x$ for $y \in Y$ and $x \in X$. ///

[6.10.2] Corollary: Every irreducible smooth representation of G having a central character for Z is finite-dimensional. ///

[6.10.3] Proposition: Let $f : M \to N$ be a surjective G-homomorphism of two smooth G-representation spaces, both with central character ω. Suppose there is a small-enough compact open subgroup K of G so that $M^K = M$ and $N^K = N$, as for M, N finitely generated. There is a G-homomorphism $\varphi : N \to M$ so that $f \circ \varphi$ is the identity map id_N on N.

Proof: Let n be the cardinality of G/ZK. Let $\psi : N \to M$ be any k-vectorspace map so that $f \circ \psi = \mathrm{id}_N$: take any k-vectorspace N_1 in M complementary to the kernel of f, and let ψ be the inverse of the restriction of f to N_1. Define

$$\varphi v = \frac{1}{n} \sum_{h \in G/ZK} h^{-1} \psi h v$$

The hypotheses ensure that this φ is independent of the choice of representatives for G/ZK, and it is immediate (by changing variables in the sum) that this averaged version of ψ is a G-homomorphism providing a one-sided inverse to f. ///

[6.10.4] Corollary: Let $f : M \to N$ be an injective G-homomorphism of two G-representation spaces, both with central character ω (for Z). Suppose that there is a compact open subgroup K of G so that $M^K = M$ and $N^K = N$ (as is the case if M, N are finitely generated). There is a G-homomorphism $\varphi : M \to N$

so that $\varphi \circ f$ is the identity map id_M on M. In particular, every G-submodule of N has a complementary submodule.

Proof: Let $Q = N/fM$ be the quotient, and $q : N \to Q$ the quotient map. The previous proposition yields $\psi : Q \to N$ so that $q \circ \psi = \mathrm{id}_Q$. Since $N = fM \oplus \psi Q$ and $fM \approx M$, $N/\psi Q$ is naturally isomorphic to M, and the composition

$$N \longrightarrow N/\psi \approx M$$

is the desired φ. ///

[6.10.5] Corollary: (*Complete Reducibility*) Every smooth representation of G with central character ω (for Z) is a direct sum of irreducible smooth representations, each with central character ω for Z.

Proof: This will follow from the previous and from Zorn's Lemma.

First, show that a finite-dimensional smooth representation M contains a nonzero irreducible. Since M is finite-dimensional, it is finitely generated, so has an irreducible quotient $q : M \to Q$. By the preceding discussion, there is a G-subspace M' of M so that as G-spaces $M \approx M' \oplus Q$. Thus, M contains the irreducible Q.

Let $M = \oplus_\alpha M_\alpha$ be a maximal direct sum of (necessarily finite-dimensional) irreducibles inside N, and suppose that $M \neq N$. Take $x \in N$ not lying in M, and let X be the G-subspace of N generated by x. Then X is finitely generated, so is finite dimensional, and has a nonzero irreducible quotient Q. From the foregoing, there is a copy Q' of Q inside X and $X = Q' \oplus X'$ for some X'. By the maximality of M, Q must be inside M already. Apply the same argument to X', so by induction on dimension, conclude that X was 0. ///

6.11 Gelfand-Kazhdan Criterion

Subgroups H of groups G with the property that restrictions of most or all irreducibles V of G to H are *multiplicity-free*, that is, so that $\dim_{\mathbb{C}} \mathrm{Hom}_H(W, \mathrm{Res}^G_H V) \leq 1$ for most or all irreducibles W of H, are called *(strongly) Eulerian*, or G, H is a *(strong) Gelfand pair*.

It is important to know that induced representations are *multiplicity-free*, meaning contain at most one copy of a given irreducible representation, whenever this is the case, to produce Euler factorization of global integrals, for example. The idea of the proof, useful already in the representation theory of finite groups, is that if *no* irreducible *occurs twice* inside a representation, then the endomorphism algebra should be *commutative*, and vice-versa. Unfortunately, this principle is not quite valid more generally, for infinite-dimensional representations of nonfinite groups. After sufficient adaptations are made, we have the *Gelfand-Kazhdan criterion* that follows. As in the Gelfand criterion for commutativity of the spherical Hecke algebra [2.4.5] (echoed in [3.5.4]), the criterion for multiplicity-free-ness depends on identifying an involutive anti-automorphism to interchange the order of factors but that nevertheless acts as the identity on suitable subalgebras.

Let G be a unimodular, totally disconnected group. The space $\mathcal{D} = \mathcal{D}(G)$ of *test functions* on G is the space of a compactly supported, locally constant complex-valued functions on G. As a colimit (that is, direct limit) of finite-dimensional complex vector spaces, this space has a uniquely determined topology. The space $\mathcal{D}^* = \mathcal{D}(G)^*$ of *distributions* on G is the space of continuous complex-linear functionals on $\mathcal{D}$.

Let H be a closed, unimodular subgroup of G. For a one-dimensional complex representation ψ of H,

$$\mathrm{Ind}_H^G \psi = \big\{\mathbb{C}\text{-valued functions } f \text{ on } G, \textit{uniformly} \text{ locally constant} \\ \text{so that } f(hg) = \psi(h) f(g) \text{ for all } g \in G \text{ and } h \in H\big\}$$

The case of trivial ψ is already interesting. Let $\mathbb{C}$ denote the trivial representation of G or of H. Say (G, H) is a *Gelfand pair*, or equivalently, H is an *Euler subgroup* of G if, for all irreducible *admissible* representations π of G,

$$\dim \mathrm{Hom}_G\big(\pi, \mathrm{Ind}_H^G \mathbb{C}\big) \times \dim \mathrm{Hom}_G\big(\pi^\vee, \mathrm{Ind}_H^G \mathbb{C}\big) \le 1$$

where $\pi^\vee$ is the contragredient of π. By Frobenius Reciprocity, this is equivalent to

$$\dim \mathrm{Hom}_H\big(\mathrm{Res}_H^G \pi, \mathbb{C}\big) \times \dim \mathrm{Hom}_H\big(\mathrm{Res}_H^G \pi^\vee, \mathbb{C}\big) \le 1$$

An *involutive anti-automorphism* σ on a group G is a bijection $G \to G$ so that $(gh)^\sigma = h^\sigma g^\sigma$. The action of σ on functions is by $f^\sigma(g) = f(g^\sigma)$, and on distributions u by $u^\sigma(f) = u(f^\sigma)$ for $f \in \mathcal{D}$.

[6.11.1] Theorem: Let ψ be a one-dimensional representation of a closed unimodular subgroup H of G. Suppose H is *stabilized* by an involutive anti-automorphism σ of G. Put $\psi^\sigma(h) = \psi(h^\sigma)$. Suppose that $u^\sigma = u$ for all distributions u on G possessing equivariance

$$u(L_h \eta) = \psi(h) \cdot u(\eta) \qquad \text{and} \qquad u(R_h \eta) = \psi^\sigma(h)^{-1} \cdot u(\eta)$$

for $\eta \in \mathcal{D}(G)$. Then

$$\dim \mathrm{Hom}_G\big(\pi, \mathrm{Ind}_H^G \psi\big) \times \dim \mathrm{Hom}_H\big(\mathrm{Res}_H^G \pi^\vee, \psi^\sigma\big) \le 1$$

Proof: By Frobenius reciprocity $\mathrm{Hom}_G(\pi, \mathrm{Ind}_H^G\psi) \approx \mathrm{Hom}_H(\mathrm{Res}_H^G\pi, \psi)$, existence of nontrivial G-maps $\pi \to \mathrm{Ind}_H^G\psi$ and $\pi^\vee \to \mathrm{Ind}_H^G\psi^\sigma$ is equivalent to existence of nonzero H-homomorphisms $s : \pi \longrightarrow \psi$ and $t : \pi^\vee \longrightarrow \psi^\sigma$. We obtain G-homomorphisms $S : \mathcal{D} \longrightarrow \pi^\vee$ and $T : \mathcal{D} \longrightarrow (\pi^\vee)^\vee$ by integrating:

$$(S\eta)(v) = \int_G \eta(x)\, s(x \cdot v)\, dx \qquad (T\eta)(\lambda) = \int_G \eta(x)\, t(x \cdot \lambda)\, dx$$

for $\eta \in \mathcal{D}(G)$, $v \in \pi$, $\lambda \in \check{\pi}$. The assumed admissibility of π implies that π is *reflexive*, that is, that $(\pi^\vee)^\vee \approx \pi$. By direct computation, right translation R_g by $g \in G$, and left translation L_h by $h \in H$ interact with S and T by

$$S(R_g\eta) = g \cdot (S\eta) \qquad T(R_g\eta) = g \cdot (T\eta)$$
$$S(L_h\eta) = \psi(h) \cdot S\eta \qquad T(L_H\eta) = \psi^\sigma(h) \cdot T\eta$$

The first assertion, for example, is verified as follows: for $v \in \pi$,

$$S(R_g\eta)(v) = \int_G \eta(xg)\, s(x \cdot v)\, dx = \int_G \eta(x)\, s(xg^{-1} \cdot v)\, dx$$
$$= S\eta(g^{-1}v)$$

by replacing x by xg^{-1}. The last expression is simply the contragredient action of g, that is, on $\pi^\vee$. The left H-invariance follows by

$$S(L_h\eta)(v) = \int_G \eta(h^{-1}x)\, s(x \cdot v)\, dx = \int_G \eta(x)\, s(hx \cdot v)\, dx$$
$$= \int_G \eta(x)\, \psi(h)\, s(x \cdot v)\, dx = S(\eta)(v)$$

by replacing x by hx and then invoking the H-equivariance of s. The corresponding assertions for T are proven similarly. That is, both S and T are left H-equivariant as indicated and are right G-equivariant, giving G-homomorphisms from $\mathcal{D}(G)$ (with right regular representation) to π and $\pi^\vee$.

Let $\langle , \rangle : \pi \times \pi^\vee \to \mathbb{C}$ be the canonical complex-bilinear pairing $\langle v, \lambda \rangle = \lambda(v)$. Let the induced complex-*linear* map on the tensor product be $\beta : \pi \otimes_{\mathbb{C}} \pi^\vee \longrightarrow \mathbb{C}$. Define

$$B = \beta \circ (T \otimes S) : \mathcal{D}(G) \otimes \mathcal{D}(G) \longrightarrow \pi \otimes \pi^\vee \longrightarrow \mathbb{C}$$

Note the reversal of S and T. The functional B is in the space of distributions $\mathcal{D}(G \times G)^*$, is left $(H, \psi^\sigma) \times (H, \psi)$-equivariant and right G^Δ-invariant, where G^Δ is the diagonal copy of G in $G \times G$. Reversal of ψ and ψ^σ due to the reversal of S and T.

[6.11.2] Lemma: With B, t, S as shown earlier, for η, φ in $\mathcal{D}(G)$, $B(\eta \otimes \varphi) = t(S(\varphi * \eta))$.

Proof: Apart from an issue of interchange of integration and linear operators, this is a direct computation:

$$B(\eta \otimes \varphi) = T\eta(S\varphi) = \int_G \eta(x)\, t(x \cdot S\varphi)\, dx$$
$$= \int_G \eta(x)\, t(S(R_x^{-1} \cdot \varphi))\, dx$$

by the G-equivariance of S. Moving the integral inside $t \circ S$, this becomes

$$(t \circ S)\left(\int_G \eta(x)\, R_x^{-1} \cdot \varphi\, dx\right) = (t \circ S)(\varphi * \eta)$$

Exchange of integration and application of the operator $t \circ S$ is easily justified, since the indicated integral is actually a finite sum. More generally, $\mathcal{D}(G)$ is an an LF-space [13.10], so is quasi-complete [13.12], and Gelfand-Pettis integrals of compactly supported continuous $\mathcal{D}(G)$-valued function f exist, and, as in [14.1], for any continuous linear operator L on $\mathcal{D}(G)$,

$$L\left(\int_G f(x)\, dx\right) = \int_G L(f(x))\, dx$$

The desired exchange is a special case of this. ///

[6.11.3] Corollary: The distribution u on G defined by $u(\eta) = t(S(\eta))$ is left H-equivariant by ψ and right H-equivariant by $(\psi^\sigma)^{-1}$, meaning that

$$u(L_h\eta) = \psi(h) \cdot u(\eta) \qquad \text{and} \qquad u(R_h\eta) = \psi^\sigma(h)^{-1} \cdot u(\eta)$$

Proof: Given $\eta \in \mathcal{D}(G)$ and given $h \in H$, we claim that there is $\varphi \in \mathcal{D}(G)$ so that $R_h\eta * \varphi = R_h\eta$. For example, for K a small-enough compact open subgroup of G so that $R_h\eta$ is left K-invariant, take φ to be meas $(K)^{-1}$ on K and 0 off K. Then

$$u(R_h\eta) = (t \circ S)(R_h\eta) = (t \circ S)((R_h\eta) * \varphi)$$
$$= (t \circ S)\big(\eta * L_h^{-1}\varphi\big) = B\big(L_h^{-1}\varphi \otimes \eta\big)$$

by the way that convolutions and translations interact. By the left H-equivariance of B by ψ^σ in its first argument,

$$B\big(L_h^{-1}\varphi \otimes \eta\big) = \psi^\sigma(h)^{-1} \cdot B(\varphi \otimes \eta)$$

Going back by the same procedure, $u(R_h\eta) = \psi^\sigma(h)^{-1} \cdot u(\eta)$. Even more simply, for φ so that $(L_h\eta) * \varphi = L_h\eta$, we compute that

$$u(L_h\eta) = B(\varphi \otimes L_h\eta) = \psi(h) \cdot B(\varphi \otimes \eta) = \psi(h) \cdot u(\eta)$$

giving the equivariance. ///

As usual,

[6.11.4] Lemma: For η, φ in $\mathcal{D}(G)$,

$$(\eta * \varphi)^\sigma = \varphi^\sigma * \eta^\sigma$$

Proof: For $g \in G$,

$$(\eta * \varphi)^\sigma(g) = (\eta * \varphi)(g^\sigma) = \int_G \eta(g^\sigma x^{-1})\,\varphi(x)\,dx = \int_G \eta(x^{-1})\,\varphi(xg^\sigma)\,dx$$

$$= \int_G \eta(x)\,\varphi(x^{-1}g^\sigma)\,dx = \int_G \eta(x^\sigma)\,\varphi((gx^{-1})^\sigma)\,dx$$

replacing x successively by xg, x^{-1}, and x^σ. This is

$$\int_G \eta^\sigma(x)\,\varphi^\sigma(gx^{-1})\,dx = (\varphi^\sigma * \eta^\sigma)(g)$$

as claimed. ///

[6.11.5] Corollary: $B(\eta \otimes \varphi) = B(\varphi^\sigma \otimes \eta^\sigma)$.

Proof: $u(\eta^\sigma) = u(\eta)$ for all $\eta \in \mathcal{D}(G)$, so

$$B(\eta \otimes \varphi) = u(\varphi * \eta) = u((\varphi * \eta)^\sigma) = u(\eta^\sigma * \varphi^\sigma) = B(\varphi^\sigma \otimes \eta^\sigma)$$

///

[6.11.6] Corollary: For η in $\mathcal{D}(G)$, $T\eta = 0$ implies $S(\eta^\sigma) = 0$, and similarly $S\eta = 0$ implies $T(\eta^\sigma) = 0$.

Proof: For, $T\eta = 0$, for all φ in $\mathcal{D}(G)$

$$0 = \langle T\eta, S\varphi\rangle = B(\eta \otimes \varphi) = B(\varphi^\sigma \otimes \eta^\sigma) = \langle T\varphi^\sigma, S\eta^\sigma\rangle$$

by the previous corollary. That is, $S\eta^\sigma$ gives the trivial linear functional on π, so must be 0 in $\pi^\vee$. The other assertion is similarly proven. ///

That is, $\ker T$ determines $\ker S$ and vice versa.

Since π is irreducible, by Schur's lemma the kernel of $S : \mathcal{D}(G) \to \pi$ determines S uniquely up to a constant, and the same assertion holds for T. We can recover $s : \pi \to \mathbb{C}$ unambiguously from S. Given $v \in \pi$, let η be meas $(K)^{-1}$

times the characteristic function of K, where K is any sufficiently small compact open subgroup of G. Then

$$(S\eta)(v) = \int_G \eta(x)\, s(x \cdot v)\, dx = s(v)$$

That is, from $\ker S$ we recover S uniquely up to a constant, and then recover s uniquely up to a constant. The analogous assertion holds for $\ker T$, T, and t.

Then t certainly determines T, which determines $\ker T$. From the foregoing, $\ker T$ determines $\ker S$, which (by the previous paragraph) determines s up to a constant. We could have fixed t and let s be arbitrary, which would show that *if* the space of t's were positive-dimensional, then the space of s's would be *at most* one-dimensional. The symmetrical argument reversing the role of s and t goes through in the same manner, wherein we use the assumed admissibility of π and, thus, of $\pi^\vee$. This proves the theorem. ///

6.A Appendix: Action of Compact Abelian Groups

Let A be a *compact, abelian* topological group, and countably-based. We grant that A has a translation-invariant measure, that is, a Haar measure. Let $|\cdot|$ be the corresponding norm on $L^2(A)$. The analogue of Fourier series expansion is

[6.A.1] Theorem: $L^2(A)$ is the completion of the direct sum $\bigoplus_\chi \mathbb{C} \cdot \chi$ as χ ranges over continuous homomorphisms $\chi : A \to \mathbb{C}^\times$.

Proof: Certainly every χ is in $L^2(A)$. On the other hand, $a \in A$ acts on $F \in L^2(A)$ by translation $(a \cdot F)(b) = F(a + b)$, and every continuous f on A acts correspondingly on $F \in L^2(A)$ by an integral operator

$$(f \cdot F)(b) = \int_A f(a)\, F(a + b)\, da$$

Replacing a by $a - b$, this is

$$(f \cdot F)(b) = \int_A f(a - b)\, F(a)\, da$$

expressing the map $F \to f \cdot F$ as a *Hilbert-Schmidt* operator with integral kernel $K(a, b) = f(a - b)$. Hilbert-Schmidt operators are *compact* [9.A.4]. For f real-valued and *even*, in the sense that $f(-a) = f(a)$, the corresponding integral kernel is symmetric and real-valued, so gives a *self-adjoint* operator. The spectral theorem [9.A.6] for compact self-adjoint operators gives an eigenspace decomposition of $L^2(A)$ with respect to the operator given by f, and all eigenspaces are finite-dimensional except possibly the 0-eigenspace.

As usual, composition of two such operators is by the action of their convolution, as illustrated already in [2.4] for nonabelian groups: for f, g real-valued in $C^o(A)$ and $F \in L^2(A)$,

$$f \cdot (g \cdot F) = \int_A f(a)\, a \cdot \left(\int_A g(b)\, b \cdot F\, db \right) da$$
$$= \int_A f(a) \left(\int_A g(b)\, a \cdot (b \cdot F)\, db \right) da$$

The integrand $b \to g(b)\, b \cdot F$ of the inner integral is a continuous, compactly supported, $L^2(A)$-valued function of $b \in A$, by the continuity of the action of A on $L^2(A)$ and the continuity of g on the compact A. Thus, the action of a passes inside the vector-valued integral for general reasons [14.1]. By Fubini, this is

$$\int_A \int_A f(a)\, g(b)\, (a+b) \cdot F\, da\, db = \int_A \int_A f(a-b)\, g(b)\, a \cdot F\, da\, db$$

by replacing a by $a - b$. This is

$$\int_A \left(\int_A f(a-b)\, g(b)\, db \right) a \cdot F\, da = \int_A (f * g)(a)\, a \cdot F\, da = (f * g) \cdot F$$

Since the group is abelian, the convolution product is abelian:

$$(f * g)(a) = \int_A f(a-b)\, g(b)\, db = \int_A f(-b)\, g(a+b)\, db$$
$$= \int_A f(b)\, g(a-b)\, db$$

by replacing b by $b + a$ and then replacing b by $-b$. Since the group is abelian, the Haar measure is invariant under $b \to -b$ as well. Thus, these self-adjoint compact operators *commute* with each other. Commuting operators preserve each others' eigenspaces: for v in the λ-eigenspace for T, and for $ST = TS$,

$$T(Sv) = (TS)v = (ST)v = S(Tv) = S(\lambda v) = \lambda \cdot SV$$

Thus, $L^2(A)$ decomposes into simultaneous eigenspaces for all these operators. For that matter, the action $a \times F \longrightarrow a \cdot F$ stabilizes eigenspaces:

$$a \cdot (f \cdot F) = a \cdot \int_A f(b)\, b \cdot F\, db = \int_A a \cdot f(b)\, b \cdot F\, db$$
$$= \int_A f(b)\, (a+b) \cdot F\, db = \int_A f(b)\, (b+a) \cdot F\, db$$
$$= \int_A f(b)\, b \cdot (a \cdot F)\, db = f \cdot (a \cdot F)$$

Again, the action of $a \in A$ passes inside the integral by Gelfand-Pettis theory [14.1].

The nondegeneracy result [14.1.5], that for every $0 \neq v \in V$ there is $\varphi \in C_c^o(G)$ such that $\varphi \cdot v \neq 0$, implies that the simultaneous 0-eigenspace for all the integral operators is *trivial*.

Since the integral operators are self-adjoint, distinct eigenspaces are mutually orthogonal: given two distinct eigenspaces, let real-valued $f \in C^o(A)$ be such that the two eigenvalues $\lambda, \mu \in \mathbb{R}$ are different. For v, w in the respective eigenspaces, using the self-adjointness,

$$\begin{aligned}\lambda \cdot \langle v, w\rangle &= \langle f \cdot v, w\rangle = \langle v, f \cdot w\rangle \\ &= \langle v, \mu \cdot w\rangle = \mu \cdot \langle v, w\rangle\end{aligned}$$

We claim that each of these finite-dimensional spaces V_o decomposes into simultaneous eigenspaces for A itself. Again, as earlier, each is stabilized by the (unitary) action of the abelian group A. We do a descending induction on dimension. If V_o is a simultaneous eigenspace for A, we are done. Otherwise, there is $a_1 \in A$ such that V_o decomposes properly into a_1-eigenspaces, noting that a_1 acts unitarily. Let V_1 be a proper a_1-eigenspace inside V_o. If V_1 is a simultaneous eigenspace for all A, we are done. Otherwise, take $a_2 \in A$ such that V_1 decomposes properly into a_2-eigenspaces. Continuing, by the finite-dimensionality of V_o, the process must stop, producing a simultaneous eigenspace for A. That is, $L^2(A)$ has an orthogonal decomposition into simultaneous eigenspaces for A.

Let V^λ be such a simultaneous eigenspace, where $a \cdot v = \lambda_a \cdot v$ for $\lambda_a \in \mathbb{C}$. The collection of eigenvalues λ_a is a *group homomorphism* $A \to C^\times$:

$$\begin{aligned}\lambda_a \cdot \lambda_b \cdot v &= \lambda_b \cdot \lambda_a \cdot v = \lambda_b \cdot a \cdot v = a \cdot \lambda_b \cdot v = a \cdot b \cdot v \\ &= (a+b) \cdot v = \lambda_{a+b} \cdot v\end{aligned}$$

We claim that the λ-eigenspace V^λ is simply all scalar multiples of λ itself: for $F \in V^\lambda$,

$$F(a) = F(0+a) = (a \cdot F)(0) = (\lambda_a \cdot F)(0) = \lambda_a \cdot F(0)$$

That is, $F = F(0) \cdot \lambda$. Last, we show that such λ is *continuous*: from the continuity of $A \times L^2(A) \to L^2(A)$, the restriction $A \times V^\lambda \to V^\lambda$ is continuous, with one-dimensional $V^\lambda = \mathbb{C} \cdot \lambda$. Thus, $a \to (a \cdot \lambda) = \lambda_a \cdot \lambda$ is continuous, so $a \to \lambda_a$ must be continuous. Thus, $L^2(A)$ is the completion of the orthogonal direct sum of $\mathbb{C} \cdot \chi$ with χ running over all continuous group homomorphisms $A \to \mathbb{C}^\times$. ///

Let compact, abelian A act on a Hilbert space V by *unitary* operators. For a continuous group homomorphism $\chi : A \to \mathbb{C}^\times$, the χ-*isotype* V^χ in V is

$$V^\chi = \{v \in V : a \cdot v = \chi(a) \cdot v, \ \text{for all } a \in A\}$$

[6.A.2] Claim: V decomposes as the completion of the direct sum

$$V = (\text{completion of}) \bigoplus_\chi V^\chi$$

with χ ranging over continuous homomorphisms $A \to \mathbb{C}^\times$.

Proof: First, we claim that the orthogonal projection $P_\chi : V \to V^\chi$ is given by the integral operator

$$P_\chi v = \int_A \overline{\chi}(a)\, a \cdot v \, da$$

with compact A having total measure normalized to 1. Indeed, the image is in V^χ: using properties of vector-valued integrals [14.1] to allow the action to pass inside the integral,

$$\begin{aligned} a \cdot P_\chi v &= a \cdot \int_A \overline{\chi}(b)\, b \cdot v \, db = \int_A \overline{\chi}(b)\, a \cdot (b \cdot v)\, db \\ &= \int_A \overline{\chi}(b)\, (a+b) \cdot v \, db = \int_A \overline{\chi}(b-a)\, b \cdot v \, db \\ &= \int_A \chi(a)\overline{\chi}(b)\, b \cdot v \, db = \chi(a) \cdot \int_A \overline{\chi}(b)\, b \cdot v \, db = \chi(a) \cdot P_\chi v \end{aligned}$$

Next, $P_\chi \circ P_\chi = P_\chi$: using properties of vector-valued integrals [14.1] and using the normalization that the total measure of A is 1,

$$\begin{aligned} P_\chi(P_\chi v) &= \int_A \int_A \overline{\chi}(a)\overline{\chi}(b)\, a \cdot b \cdot v \, da\, db \\ &= \int_A \int_A \overline{\chi}(a)\overline{\chi}(b)\, (a+b) \cdot v \, da\, db \\ &= \int_A \int_A \overline{\chi}(a-b)\overline{\chi}(b)\, a \cdot v \, da\, db \\ &= \int_A \int_A \overline{\chi}(a)\chi(b)\overline{\chi}(b)\, a \cdot v \, da\, db = \int_A \int_A \overline{\chi}(a)\, a \cdot v \, da\, db \\ &= \int_A \overline{\chi}(a)\, a \cdot v \, da \cdot \int_A 1 \, db = \int_A \overline{\chi}(a)\, a \cdot v \, da = P_\chi v \end{aligned}$$

Last, we show that the orthogonal complement of all the images $P_\chi V$ is just $\{0\}$. Let $\{\eta_i\}$ be an *approximate identity* in $C^o(A)$, as in [6.5] and [14.1], meaning that $0 \le \eta_i(a) \le 1$ for all i and $a \in A$, that $\int_A \eta_i = 1$ for all i, and the supports shrink to $\{1_A\}$. By [14.1.4], $\eta_i \cdot v \to v$ in V for each fixed $v \in V$. Each η_i has a Fourier expansion $\eta_i = \sum_\chi \widehat{\eta}_i(\chi) \cdot \chi$ converging in $L^2(A)$, summing over continuous homomorphisms $\chi : A \to \mathbb{C}^\times$. For a finite set X of χs, let η_i^X be the corresponding finite partial sum $\sum_{\chi \in X} \widehat{\eta}_i(\chi) \cdot \chi$ of the Fourier expansion of η_i. These are finite linear combinations of continuous functions, so are certainly continuous.

We claim that $\eta_i^X \cdot v \longrightarrow v$ in V. Since the partial sums η_i^X approach η_i in $L^2(A)$, it suffices to show that for $\eta \in C^o(A)$ with $|\eta|_{L^2(A)}$ small, $|\eta \cdot v|_V$ is small, for fixed $v \in V$. Indeed, invoking properties of vector-valued integrals [14.1] to exchange inner products and integrals,

$$|\eta \cdot v|_V^2 = \left\langle \int_A \eta(a)\, a \cdot v\, da,\ \int_A \eta(b)\, b \cdot v\, db \right\rangle$$

$$= \int_A \int_A \eta(a) \cdot \overline{\eta}(b) \cdot \langle a \cdot v,\ b \cdot v \rangle\, da\, db$$

By Cauchy-Schwarz-Bunyakowsky and the unitariness, $|\langle a \cdot v,\ b \cdot v \rangle| \le |av| \cdot |bv| = |v| \cdot |v|$, so

$$|\eta \cdot v|_V^2 \le \int_A \int_A |\eta(a)| \cdot |\eta(b)|\, da\, db$$

$$= \int_A |\eta(a)|\, da \cdot \int_A |\eta(b)|\, db \le |\eta| \cdot |\eta|$$

again by Cauchy-Schwarz-Bunyakowsky, since $|\eta(a)| = \eta(a) \cdot \mu(a)$, where $|\mu(a)| = 1$, and the total measure of A is 1. This shows that $\eta_i^X \cdot v \to v$.

The action $\chi \cdot v$ is that of the projector $P_{\overline{\chi}}$:

$$\chi \cdot v = \int_A \chi(a)\, a \cdot v\, da = \int_A \overline{\overline{\chi}}(a)\, a \cdot v\, da = P_{\overline{\chi}} v$$

Thus,

$$\eta_i^X \cdot v = \sum_{\chi \in X} \widehat{\eta}_i(\chi) \cdot \chi \cdot v = \sum_{\chi \in X} \widehat{\eta}_i(\chi) \cdot P_{\overline{\chi}} v$$

Since $\eta_i \cdot v \to v$ and $\eta_i^X \cdot v \to \eta_i \cdot v$, not all $P_\chi v$ can be 0 for nonzero v. ///

7

Discrete Decomposition of Cuspforms

The general result here is that test functions $\varphi \in C_c^\infty(G)$ act as *compact operators* on spaces of square-integrable cuspforms. Similar arguments and outcomes hold for both the four simplest archimedean examples from Chapter 1, adelic $Z^+GL_2(k)\backslash GL_2(\mathbb{A})$ from Chapter 2, and archimedean or adelic versions of GL_n from Chapter 3, with appropriate senses of *test functions* depending on context.

The most general argument refers to *noncommutative* rings of compact operators, closed under adjoints, and the irreducible representations of such rings are generally infinite-dimensional. In a larger context, this is the truth of the matter, but without knowing more about the irreducibles, it is not easy to recover more tangible information about the behavior of the Laplacian or spherical Hecke operators.

Thus, we also give more special arguments using *commutative* rings of compact operators closed under adjoints, so that a more tangible notion of *simultaneous eigenspace* takes the place of (infinite-dimensional) *irreducible representation*. This gives the decomposition of spaces of right K-invariant functions by Laplacians and spherical Hecke algebras. The result for spherical Hecke algebras obtained in this fashion is nearly optimal, but there is still some imprecision in corollaries on eigenfunctions for Laplacians. The method of Chapter 10, directly considering the spectral behavior of pseudo-Laplacians and pseudo-cuspforms, gives better results of that sort.

Beyond the spectral decomposition theorems [7.1.1], [7.2.1], and [7.3.1] and their immediate corollaries, we also conclude that there is an orthonormal basis for cuspforms consisting of *smooth* functions of *rapid decay* in Siegel sets: [7.1.20], [7.2.19], [7.2.20], and [7.3.19].

For perspective, Appendix [7.B] gives the much easier argument for discreteness of decomposition of $L^2(\Gamma\backslash G)$ for *compact* quotients $\Gamma\backslash G$, for a unimodular topological group G and discrete subgroup Γ, although we do not give explicit examples of such G, Γ. In fact, this is a variant on the discrete decomposition [9.C.2] of $L^2(K)$ for compact topological groups K.

7.1 The Four Simplest Examples

For this section, let G, Γ, K be as in any of the four examples $SL_2(\mathbb{R})$, $SL_2(\mathbb{C})$, $Sp^*_{1,1}$, $SL_2(\mathbb{H})$ from Chapter 1. As earlier, *test functions* $C_c^\infty(G)$ are compactly supported, smooth functions, and act on functions f on $\Gamma\backslash G$ by

$$(\eta \cdot f)(g) = \int_G \eta(h)\, f(gh)\, dh \qquad (\text{for } \eta \in C_c^\infty(G))$$

As earlier, functions on $\Gamma\backslash G/K$ are identified with right K-invariant functions on $\Gamma\backslash G$, and the action of the spherical convolution algebra of left-and-right K-invariant test functions $C_c^\infty(K\backslash G/K)$ stabilizes the subspace of such functions. As we will recall subsequently, the spherical convolution algebra $C_c^\infty(K\backslash G/K)$ is *commutative*. The main results of this section are as follows:

[7.1.1] Theorem: The *spherical* convolution algebra $C_c^\infty(K\backslash G/K)$ of left and right K-invariant test functions on G acts on square-integrable right K-invariant cuspforms $L^2_o(\Gamma\backslash G/K)$ by *compact* operators, the collection of such operators is closed under adjoints, and is *non-degenerate* in the sense that for every $f \in L^2_o(\Gamma\backslash G/K)$ there is $\varphi \in C_c^\infty(K\backslash G/K)$ such that $\varphi \cdot f \neq 0$. *(Proof in the sequel.)*

[7.1.2] Corollary: The space $L^2_o(\Gamma\backslash G/K)$ of right K-invariant square-integrable cuspforms decomposes into simultaneous eigenspaces for operators in the *commutative* convolution algebra $C_c^\infty(K\backslash G/K)$, each finite-dimensional. The simultaneous eigenfunctions are *smooth*. *(Proof below.)*

[7.1.3] Corollary: There is an orthonormal Hilbert-space basis for the space of K-invariant square-integrable cuspforms consisting of simultaneous eigenfunctions for the invariant Laplacian Δ. *(Proof below.)*

[7.1.4] Remark: The argument here does not immediately prove that the eigenspaces for Δ are of finite multiplicity, since it only indirectly refers to Δ.

The unbounded-operator argument in Chapter 9 gives a stronger result about Δ-eigenspaces.

Let P, M, N, A^+ be as in Chapter 1, and let η be a *height function*

$$\eta(na_t k) = t^{r_o}$$

for $n \in N$, $k \in K$, with $a_t = \begin{pmatrix} \sqrt{t} & 0 \\ 0 & 1/\sqrt{t} \end{pmatrix}$ for $t > 0$, with $r_o = 1, 2, 3, 4$ in the respective cases. Thus, $\eta(m) = \delta(m)$, the modular function for P. In an *Iwasawa decomposition* $G = N \cdot M \cdot K$ for $x \in G$, write $x = n_x \cdot m_x \cdot k_x$ with $n_x \in N$, $m_x \in M$, and $k_x \in K$. This notation n_x is in conflict with the use of that notation in the earlier discussion of the smaller examples, but those former uses will not be needed here. For $t > 0$, compact $C \subset N$, the corresponding *Siegel set* in G is

$$\mathfrak{S} = \mathfrak{S}_{C,\tau} = C \cdot \{a_t \in A^+ : t \geq \tau\} \cdot K = C \cdot \{m \in M : \eta(m) \geq \tau\} \cdot K$$

[7.1.5] Claim: A point in a Siegel set is well approximated by its M-component in an Iwasawa decomposition, in the sense that, for $x \in \mathfrak{S}_{C,\tau}$ with $\tau > 0$ and C compact in N, there is another compact subset C' of N such that $x \in m_x \cdot C' \cdot K$.

Proof: $x \in \mathfrak{S}_{C,\tau}$ gives

$$x = n_x \cdot m_x \cdot k_x \in C \cdot m_x \cdot K = m_x \cdot m_x^{-1} C m_x \cdot K$$

The lower bound $\eta(m) \geq \tau$ gives a compact set C' in N depending only upon τ and C such that

$$m^{-1}Cm \subset C' \qquad (\text{for } m \in M \text{ with } \eta(m) \geq \tau)$$

In particular, $m_x^{-1} C m_x \subset C'$. Thus, $x \in m_x \cdot C' \cdot K$. ///

For strictly upper-triangular square matrices x with entries in any field of characteristic zero, the series for the *matrix exponential* $e^x = \exp(x) = \sum_{\ell \geq 0} x^\ell / \ell!$ is *finite*. Let $\mathfrak{n}$ be the Lie algebra of N:

$$\mathfrak{n} = \{\begin{pmatrix} 0 & v \\ 0 & 0 \end{pmatrix}\}$$

where v is in $\mathbb{R}$, $\mathbb{C}$, $\{w \in \mathbb{H} : w + \overline{w} = 0\}$, or $\mathbb{H}$ in the four cases. Because N is abelian, the exponential map is a *diffeomorphism* $\exp : \mathfrak{n} \longrightarrow N$, and there is a discrete *additive* subgroup Λ in $\mathfrak{n}$ such that $\exp(\Lambda) = \Gamma \cap N$:

$$\exp \begin{pmatrix} 0 & v \\ 0 & 0 \end{pmatrix} = \begin{pmatrix} 1 & v \\ 0 & 1 \end{pmatrix}$$

namely, take v in $\mathbb{Z}$, $\mathbb{Z}[i]$, $\mathbb{Z}i + \mathbb{Z}j + \mathbb{Z}k$, or the Hurwitz quaternion integers $\mathfrak{o}$, respectively. For test function $\varphi \in C_c^\infty(G)$, *wind up* the integral for $f \to \varphi \cdot f$ along $\exp(\Lambda) = N \cap \Gamma$: for $y \in G$,

$$(\varphi \cdot f)(y) = \int_G \varphi(x)\, f(yx)\, dx = \int_G \varphi(y^{-1}x)\, f(x)\, dx$$

$$= \int_{\exp(\Lambda)\backslash G} \left(\sum_{\gamma \in \exp(\Lambda)} \varphi(y^{-1}\gamma x) \right) f(x)\, dx$$

$$= \int_{\exp(\Lambda)\backslash G} \left(\sum_{\nu \in \Lambda} \varphi(y^{-1} \cdot \exp(\nu) \cdot x) \right) f(x)\, dx$$

The *kernel function* for the $(N \cap \Gamma)$-wound-up operator is the latter inner sum:

$$K_\varphi(x, y) = \sum_{\nu \in \Lambda} \varphi(y^{-1} \cdot \exp(\nu) \cdot x)$$

[7.1.6] Claim: For a fixed Siegel set $\mathfrak{S}$ and for fixed compact $E \subset C_c^\infty(G)$, there is compact $C_M \subset M$ such that if there exist $n \in N$, $x, y \in \mathfrak{S}$, and $\varphi \in E$ with $\varphi(y^{-1} \cdot n \cdot x) \neq 0$, then $m_x \in m_y \cdot C_M$. That is, $K_\varphi(x, y) = 0$ for all $x, y \in \mathfrak{S}$ and all $\varphi \in E$ unless $m_x \in m_y \cdot C_M$.

Proof: From the previous claim, there is compact $C' \subset N$ such that $m_y^{-1} y \in C' \cdot K$. A compact set of test functions has a common compact support C_G, because a compact set is *bounded* in the topological vector space sense, and a bounded subset of an LF-space such as $C_c^\infty(G)$ lies in some Fréchet limitand, by [13.8.5]. Nonvanishing of $\varphi(y^{-1}nx)$ implies $y^{-1}nx \in C_G$, so

$$nx \in y \cdot C_G \subset m_y \cdot C' \cdot K \cdot C_G \subset m_y \cdot C'_G$$

with $C'_G = C'KC_G =$ compact. That is,

$$C'_G \ni m_y^{-1} \cdot nx = m_y^{-1} \cdot nn_x \cdot m_x \cdot k_x = (m_y^{-1} nn_x m_y) \cdot m_y^{-1} m_x \cdot k_x$$

That is,

$$(m_y^{-1} nn_x m_y) \cdot m_y^{-1} m_x \in C'_G \cdot K = \text{compact}$$

Since M normalizes N, the element $m_y^{-1} nn_x m_y$ is in N. Since $N \cap M = \{1\}$, the multiplication map $N \times M \to NM$ is a homeomorphism. Thus, for the product $(m_y^{-1} nn_x m_y) \cdot m_y^{-1} m_x$ to lie in a compact set in G requires that its N-component lies in a compact set in N and its M-component lies in a compact set in M. Thus, there is compact $C_M \subset M$ such that $m_y^{-1} m_x \in C_M$, as claimed. ///

[7.1.7] Corollary: For fixed Siegel set $\mathfrak{S}$ and fixed compact $E \subset C_c^\infty(G)$, there is a compact $C_M \subset M$ such that, if $\varphi(y^{-1}nx) \neq 0$ for some $x, y \in \mathfrak{S}$, some $n \in N$ and some $\varphi \in E$, then $m_y^{-1}x \in C_M$.

Proof: By [7.1.5], there is a compact C_G in G such that $x \in m_x \cdot C_G$. By [7.1.6], there is a compact C_M in M such that $m_x \in m_y C_M$. Thus, $x \in m_x C_G \subset (m_y C_M)C_G'$, rearranging to give the claim. ///

[7.1.8] Corollary: With $\omega_y = y^{-1}m_y$ and $\omega_{x,y} = m_y^{-1}x$, the functions $\nu \longrightarrow \varphi_{x,y}(\nu) = \varphi(\omega_y \cdot \exp(\nu) \cdot \omega_{x,y})$ for x, y in fixed Siegel set, $\varphi(y^{-1}nx) \neq 0, \varphi \in E$, constitute a *compact* subset of the Schwartz space $\mathscr{S}(\mathfrak{n})$.

Proof: The left and right translation actions of G on test functions are continuous $G \times G \times C_c^\infty(G) \to C_c^\infty(G)$ by [6.4]. With fixed Siegel set $\mathfrak{S}$, by [7.1.5] and [7.1.7], $\{\omega_y : y \in \mathfrak{S}\}$ and $\{\omega_{x,y} : x \in \mathfrak{S},\ y \in \mathfrak{S}\}$ are compact. This gives compactness of the image of

$$\{\omega_y : y \in \mathfrak{S}\} \times \{\omega_{x,y} : x \in \mathfrak{S},\ y \in \mathfrak{S}\} \times E \longrightarrow C_c^\infty(G)$$

with $\varphi(y^{-1}nx) \neq 0$. Since N is *closed* in G, the restriction map $C_c^\infty(G) \to C_c^\infty(N) \approx C_c^\infty(\mathfrak{n})$ is continuous. A continuous image of a compact set is compact. Comparing the topologies from [13.7] and [13.9], as in [13.9.3], the inclusion $C_c^\infty(\mathfrak{n}) \subset \mathscr{S}(\mathfrak{n})$ is continuous, giving compactness of the image. ///

Poisson summation for the lattice $\Lambda \subset \mathfrak{n}$ gives

$$K_\varphi(x, y) = \sum_{\nu \in \Lambda} \varphi(y^{-1} \cdot \exp(\nu) \cdot x) = \sum_{\psi \in \Lambda^*} \int_{\mathfrak{n}} \overline{\psi}(\nu)\, \varphi(y^{-1} \cdot \exp(\nu) \cdot x)\, d\nu$$

where Λ^* is the collection of $\mathbb{C}^\times$-valued characters on $\mathfrak{n}$ that are trivial on Λ. Rearrange slightly to

$$K_\varphi(x, y) = \sum_{\psi \in \Lambda^*} \int_{\mathfrak{n}} \overline{\psi}(\nu) \cdot \varphi\left(y^{-1}m_y \cdot \exp(m_y^{-1}\nu m_y) \cdot m_y^{-1}x\right) d\nu$$

Replacing ν by $m_y \nu m_y^{-1}$ and letting $\varphi_{x,y}(\nu) = \varphi\left(y^{-1}m_y \cdot \nu \cdot m_y^{-1}x\right)$,

$$K_\varphi(x, y) = \delta(m_y) \sum_{\psi \in \Lambda^*} \widehat{\varphi}_{x,y}(\psi^{m_y})$$

where $\widehat{\varphi}_{x,y}(\psi^{m_y})$ is the Fourier transform $\widehat{\varphi}_{x,y}$ of $\varphi_{x,y}$ evaluated at ψ, and $\psi^{m_y}(\nu) = \psi(m_y \nu m_y^{-1})$.

[7.1.9] Theorem: Fix compact $E \subset C_c^\infty(G)$ and Siegel set $\mathfrak{S}$. For any given $q > 0$ there is a uniform bound

$$|(\varphi \cdot f)(y)| \ll_q \eta(y)^{-q} \cdot |f|_{L^2(\Gamma\backslash G)}$$

for all $y \in \mathfrak{S}$, for all $\varphi \in E$, for all L^2 cuspforms f.

Proof: For convenient discussion of the Schwartz seminorms on $\mathfrak{n}$, give the real vector space $\mathfrak{n}$ a positive-definite inner product $\langle , \rangle$ invariant under conjugation action of $M \cap K$, allowing identification of $\mathfrak{n}$ with its dual $\mathfrak{n}^*$ when desired, for simplicity. Let $| \cdot |$ be the associated norm on either. The compactness [7.1.8] and continuity of Fourier transform on $\mathscr{S}(\mathfrak{n})$ give a *uniform* estimate on Fourier transforms: for fixed Siegel set $\mathfrak{S}$, for given $q > 0$, $|\widehat{\varphi}_{x,y}(\psi)| \ll_r (1 + |\psi|)^{-q}$ for all $x, y \in \mathfrak{S}$. Then

$$\delta(m_y)\, |\widehat{\varphi}_{x,y}(\psi^{m_y})| \ll_q \delta(m_y) \cdot (1 + |\psi^{m_y}|)^{-q} \qquad \text{(for all } x, y \in \mathfrak{S})$$

Next, toward [7.1.9] we need

[7.1.10] Claim: For fixed Siegel set $\mathfrak{S}$ and $r \gg 1$, the kernel $K_\varphi(x, y)$ with its 0th Fourier component removed satisfies

$$|K_\varphi(x, y) - \widehat{\varphi}_{x,y}(\psi_0)| \ll_q |\eta(y)|^{-q} \qquad \text{(for } x, y \in \mathfrak{S})$$

Proof: First, we claim that, given Λ and given a Siegel set $\mathfrak{S} = \mathfrak{S}_{C,\tau}$, there is an implied constant such that

$$\eta(y) \ll 1 + |\psi^{m_y}|^{r_o} \qquad \text{(for all } y \in \mathfrak{S}, \text{ for all } 0 \neq \psi \in \Lambda^*)$$

Again, $\eta(y) = t^{r_o}$ for $y = na_tk$ with $t > 0$, and $|\psi^{m_y}| = t \cdot |\psi|$. Since the norms of nonzero elements of Λ^* have a positive inf,

$$|\psi^{m_y}|^{r_o} = \Big(\eta(y)^{1/r_o} \cdot |\psi|\Big)^{r_o} \geq \eta(y) \cdot \inf_{0 \neq \psi \in \Lambda^*} |\psi|^{r_o}$$

Since $\widehat{\varphi}_{x,y}$ is a Schwartz function, $|\widehat{\varphi}_{x,y}(\psi)| \ll_\ell (1 + |\psi|)^{-\ell}$ for every $\ell > 0$. By the comparison of $\eta(y)$ to $|\psi^{m_y}|$, for $0 \neq \psi \in \Lambda^*$,

$$\begin{aligned}
\delta(m_y) \cdot (1 + |\psi^{m_y}|)^{-\ell} &= \eta(y) \cdot (1 + |\psi^{m_y}|)^{-\ell} \\
&\ll \eta(y) \cdot (1 + |\psi^{m_y}|)^{-(q+1)\cdot r_o - (\ell - (q+1)\cdot r_o)} \\
&\ll \eta(y) \cdot \eta(y)^{-(q+1)} \cdot (1 + |\psi^{m_y}|)^{-(\ell - (q+1)\cdot r_o)}
\end{aligned}$$

For ℓ sufficiently large depending on q, the latter sum over $0 \neq \psi \in \Lambda^*$ converges, giving the claim. ///

[7.1.11] Claim: Cuspforms f ignore the ψ_0^{th} Fourier component of $K_\varphi(x,y)$:

$$(\varphi \cdot f)(y) = \int_{\exp(\Lambda)\backslash G} \Big(K_\varphi(x,y) - \widehat{\varphi}_{x,y}(\psi_0)\Big) \cdot f(x)\, dx$$

Proof: For trivial character ψ_0 on $\mathfrak{n}$, the corresponding function $\widehat{\varphi}_{x,y}(\psi_0)$ is left N-invariant in x: let $n \in N$, and replace x by nx in the original integral defining $\widehat{\varphi}_{x,y}(\psi_0)$, with $n = \exp(\nu')$, obtaining

$$\int_{\mathfrak{n}} \overline{\psi}(\nu) \cdot \varphi(y^{-1}\exp(\nu) \cdot nx)\, d\nu = \int_{\mathfrak{n}} \overline{\psi}(\nu) \cdot \varphi(y^{-1}\exp(\nu + \nu') \cdot x)\, d\nu$$

using the abelian-ness of N. Replacing ν by $\nu - \nu'$ in the integral gives the left N-invariance in x:

$$\int_{\mathfrak{n}} \overline{\psi}(\nu) \cdot \varphi(y^{-1}\exp(\nu) \cdot nx)\, d\nu = \psi(\nu') \cdot \widehat{\varphi}_{x,y}(\psi) = \widehat{\varphi}_{x,y}(\psi)$$

Therefore, in

$$(\varphi \cdot f)(y) = \delta(m_y) \sum_{\psi \in \Lambda^*} \int_{(N\cap\Gamma)\backslash G} \widehat{\varphi}_{x,y}(\psi^{m_y}) \cdot f(x)\, dx$$

the integral for trivial character ψ_0 is

$$\begin{aligned}
&\int_{(N\cap\Gamma)\backslash G} \widehat{\varphi}_{x,y}(\psi_0^{m_y}) \cdot f(x)\, dx \\
&= \int_{N\backslash G} \int_{(N\cap\Gamma)\backslash N} \widehat{\varphi}_{nx,y}(\psi_0^{m_y}) \cdot f(nx)\, dn\, dx \\
&= \int_{N\backslash G} \widehat{\varphi}_{x,y}(\psi_0^{m_y}) \cdot \left(\int_{(N\cap\Gamma)\backslash N} f(nx)\, dn \right) dx \\
&= \int_{N\backslash G} \widehat{\varphi}_{x,y}(\psi_0^{m_y}) \cdot 0\, dx = 0 \qquad \text{(cuspform } f)
\end{aligned}$$

proving the claim. ///

The proof of [7.1.9] is almost complete. From the foregoing, for y in a fixed Siegel set $\mathfrak{S}$ and for fixed test function φ, there is a compact $C_M \subset A^+$ such that, for $\varphi(y^{-1}nx)$ to be nonzero, the Iwasawa component m_x of x must lie in $m_y \cdot C_M$. Thus,

$$\{x \in \mathfrak{S} : \varphi(y^{-1}nx) \neq 0 \text{ for some } n \in N\} \subset m_y C_M \cdot K$$

Combining this with the estimate just obtained, for cuspform f,

$$|(\varphi \cdot f)(y)| \ll_q |\eta(y)|^{-q} \cdot \int_{\Gamma\backslash\Gamma(C \cdot m_y C_M \cdot K)} |f(x)|\, dx$$

By Cauchy-Schwarz-Bunyakowsky,

$$\left(\int_{\Gamma\backslash\Gamma(C\cdot m_y C_M\cdot K)} |f(x)|\,dx\right)^2$$
$$\leq \int_{\Gamma\backslash\Gamma(C\cdot m_y C_M\cdot K)} 1\,dx \cdot \int_{\Gamma\backslash\Gamma(C\cdot m_y C_M\cdot K)} |f(x)|^2\,dx \ll |f|^2_{L^2}$$

This gives the desired decay, proving theorem [7.1.9]. ///

We are getting closer to the compactness of the operators $f \to \varphi \cdot f$ on cuspforms f. Recall that a collection E of continuous functions on G or $\Gamma\backslash G$ is *(uniformly) equicontinuous* when, given $\varepsilon > 0$, there is a neighborhood U of 1 in G such that

$$|f(x) - f(y)| < \varepsilon$$

for all $f \in E$, for all $x \in G$, for all $y \in x \cdot U$. We have the expected

[7.1.12] Lemma: For $X \in \mathfrak{g}$, the left-derivative map

$$C_c^\infty(G) \longrightarrow C_c^\infty(G)$$

by

$$\varphi \longrightarrow X^{\text{left}} \cdot \varphi = \left(g \to \frac{d}{dt}\bigg|_{t=0} \varphi(e^{-tX}\,g)\right)$$

is *continuous*.

Proof: $C_c^\infty(G)$ is an LF-space, a (strict) colimit of Fréchet spaces, the limit being taken over spaces $\mathcal{D}_\Omega$ of smooth functions on G supported on compact $\Omega \subset G$. The topology on each $\mathcal{D}_\Omega$ is given by seminorms taking sups of *derivatives* of all orders, but of course the notion of *derivative* is ambiguous, since there are at least two choices of global vectorfields, *left* derivatives by $\mathfrak{g}$, and *right* derivatives by $\mathfrak{g}$.

But a reasonable general assertion is true, for fairly elementary reasons. On a smooth manifold, let X^i be a tuple of (smooth) vector fields on an open U containing a given compact set B such that, for every x in U, the values X^i_x at x are a basis for the tangent space at x. Another such tuple Y^j can be expressed (smoothly, pointwise) as linear combinations of the X^i, and vice versa. Every entry of the matrix of coefficients is bounded, the determinant of the matrix of coefficients does not vanish on the compact B, so is bounded and bounded away from 0, the coefficients are smooth functions, and the inverse is smooth on B. Thus, the two sets of seminorms, corresponding to left or right first derivatives,

$$\sup_{x\in B}\sup_i (X_i\varphi)(x) \qquad \text{or} \qquad \sup_{x\in B}\sup_j (Y_j\varphi)(x)$$

are *topologically equivalent*. With this ambiguity removed, a (left or right) derivative is a continuous map $C^k(B) \to C^{k-1}(B)$, and so gives a continuous map on the limit: $C^\infty(B) \to C^\infty(B)$. The subspace $\mathcal{D}_B$ is a closed subspace of $C^\infty(B)$ described by closed conditions: all derivatives vanishing on the boundary. Thus, a derivative map is still continuous $\mathcal{D}_B \to \mathcal{D}_B$. This induces a continuous map on the colimit. ///

[7.1.13] Corollary: For a compact set E of test functions on G, for a compact $C_\mathfrak{g}$ in $\mathfrak{g}$, and for f ranging over cuspforms in the unit ball in $L^2(\Gamma\backslash G)$, there is a *uniform* implied constant such that

$$\big|(X \cdot (\varphi \cdot f))(g)\big| \ll 1$$

for all $g \in G$, for all $\varphi \in E$, for all $X \in C_\mathfrak{g}$.

Proof: The differentiation of $\varphi \cdot f$ can be rewritten as a differentiation of φ, followed by action of the resulting function on f:

$$\begin{aligned}(X \cdot \varphi \cdot f)(x) &= \frac{d}{dt}\bigg|_{t=0} \int_G \varphi(y)\, f(x\, e^{tX}\, y)\, dy \\ &= \frac{d}{dt}\bigg|_{t=0} \int_G \varphi(e^{-tX} y)\, f(x y)\, dy \\ &= \int_G \left(\frac{d}{dt}\bigg|_{t=0} \varphi(e^{-tX} y)\right) f(x y)\, dy \qquad \text{(replacing } y \text{ by } e^{-tX} y)\end{aligned}$$

justifying interchange of differentiation and integration by the continuity of differentiation on test functions φ, and Gelfand-Pettis integral properties [14.1]. That is, $X \cdot \varphi \cdot f = (X^{\text{left}} \varphi) \cdot f$ with X^{left} the *left* action. Since $\mathfrak{g}$ is a finite-dimensional real vector-space and the action is linear in X, this gives the continuity in $X \in \mathfrak{g}$. Thus, the collection of test functions $X^{\text{left}} \varphi$ with $X \in C_\mathfrak{g}$ and $\varphi \in E$ is again compact in $C_c^\infty(\mathcal{G})$. Thus, by the bound of theorem [7.1.9],

$$|(X \cdot \varphi \cdot f)(y)| = |(X^{\text{left}} \varphi) \cdot f)(y)| \ll_r \eta(y)^{-r} \cdot |f|_{L^2(\Gamma\backslash G)}$$

for all $y \in \mathfrak{S}, X \in C_\mathfrak{g}, \varphi \in E$. For large-enough Siegel set to cover the quotient, and any $r > 0$, this gives

$$\sup |X \cdot \varphi \cdot f(y)| \ll_r |f|_{L^2}$$

proving *uniform* boundedness for $|f|_{L^2} \leq 1$. ///

The smoothing property of $f \to \varphi \cdot f$ as in [14.5] ensures that each $\varphi \cdot f$ is in $C^\infty(G)$. A uniform bound on derivatives implies uniform continuity:

[7.1.14] Lemma: Let F be a smooth function on G, with a uniform pointwise bound on all $X \cdot F$ with X in a *compact* neighborhood $C_\mathfrak{g}$ of 0 in $\mathfrak{g}$, namely,

$$|(X \cdot F)(x)| \leq B \qquad \text{(for all } x \in G, \text{ all } X \in C_\mathfrak{g})$$

Then F is *uniformly continuous*: for every $\varepsilon > 0$, there is a neighborhood U of 1 in G such that $|F(x) - F(y)| < \varepsilon$ for all $x \in G$ and $y \in xU$.

Proof: Let V be a small enough open containing 1 such that V is contained in $\exp C_\mathfrak{g}$, and that the exponential map on $\exp^{-1} V$ is injective to V. Let $y = x \cdot e^{sX}$ for $X \in C_\mathfrak{g}$ and $0 \leq s \leq 1$. By hypothesis, the function $h(t) = F(x \cdot e^{sX})$ has

$$h'(s) = \frac{d}{dt}\bigg|_{t=0} h(s+t) = F(x \cdot e^{sX} \cdot e^{tX})$$

bounded by B. From the mean value theorem, $|F(x \cdot e^{tX}) - F(x)| \leq t \cdot B$. Thus, for all $|t| < B \cdot \varepsilon$ we have the desired inequality. ///

[7.1.15] Corollary: For a compact set E of test functions on G, and for f ranging over cuspforms in the unit ball in $L^2(\Gamma\backslash G)$, the family of images $\varphi \cdot f$ is *equicontinuous* on G. ///

We are almost done with the proof that $f \to \varphi \cdot f$ is compact on cuspforms.

[7.1.16] Lemma: Let E be a *equicontinuous*, uniformly bounded, set of functions on $\Gamma\backslash G$. Then E has *compact closure* in $L^2(\Gamma\backslash G)$.

[7.1.17] Remark: Superficially, this lemma is reminiscent of the Arzela-Ascoli theorem, which asserts that an equicontinuous and uniformly bounded subset of $C^o(K)$ (with sup norm) for a *compact* topological space K has compact closure in $C^o(K)$. Indeed is common to end a sketch of the discrete decomposition of cuspforms by an allusion to Arzela-Ascoli. However, we need *less*, fortunately, since adaptation of the literal Arzela-Ascoli result to the present circumstance seems awkward.

Proof: The proof is a maneuver to invoke the fact that a *totally bounded* subset of a complete metric space has compact closure. If the quotient $\Gamma\backslash G$ were *compact* then we could simply invoke Arzela-Ascoli, but this is perhaps exactly the difficulty. Without loss of generality, all the functions in E are bounded (in absolute value) by 1, and the total measure of $\Gamma\backslash G$ is 1. Take $\varepsilon > 0$. Using the equicontinuity, let U be a small enough neighborhood (with compact closure) of 1 in G such that for any $x \in G$ and $y \in xU$ we have $|F(x) - F(y)| < \varepsilon$ for all $F \in E$.

Let $\{x_i\}$ be a countable set dense in G. Let $U_i = x_i U$. Let $q : G \to \Gamma\backslash G$ be the quotient map. Let V_1 be the image of U_1 in $\Gamma\backslash G$, and recursively take

$$V_{n+1} = \{x \in \Gamma\backslash G : \ x \in qU_i \ \text{ but } \ x \notin q(U_1 \cup \ldots \cup U_n)\}$$

Since the V_i are disjoint and their union is $\Gamma\backslash G$, which has finite measure, $\sum_i \operatorname{meas}(V_i) < +\infty$. In particular, the measures $\operatorname{meas} V_i$ go to 0 as $i \to \infty$. Take m large enough such that $\sum_{i>m} \operatorname{meas}(V_i) < \varepsilon$. Let X be a *finite* set of complex numbers such that any complex number of absolute value at most 1 is within $\varepsilon/2$ of an element of X. For each m-tuple $\xi = (\xi_1, \ldots, \xi_m)$ of elements of X, define a function $F_{i,\xi}$ on $\Gamma\backslash G$ by

$$F_{i,\xi}(x) = \begin{cases} \xi_i & (\text{for } x \in V_i,\ i \le m) \\ 0 & (\text{for } x \in V_i,\ i > m) \end{cases}$$

Given $F \in E$, for each $i \le m$ choose ξ_i such that $|F(x_i) - \xi_i| < \varepsilon$. By the choice of U,

$$|F(x) - \xi_i| \le |F(x) - F(x_i)| + |F(x_i) - \xi_i| < 2\varepsilon \qquad (\text{for } x \in V_i)$$

Then

$$\int_{\Gamma\backslash G} |F - F_\xi|^2 < \int_{V_1 \cup \ldots \cup V_m} (2\varepsilon)^2 + \int_{V_{m+1} \cup \ldots} 1$$
$$\le 4\varepsilon^2 \cdot \operatorname{meas} \Gamma\backslash G + \operatorname{meas}(V_{m+1} \cup \ldots) \le 4\varepsilon^2 + \varepsilon$$

Tweaking the estimates as desired, for given $\varepsilon > 0$, we can cover E by a finite number of balls of radius $\varepsilon > 0$ in $L^2(\Gamma\backslash G)$. Total boundedness implies compact closure. ///

[7.1.18] Corollary: For $\varphi \in C_c^\infty(G)$, the operator $f \to \varphi \cdot f$ is a *compact* operator $L^2_o(\Gamma\backslash G) \longrightarrow L^2_o(\Gamma\backslash G)$.

Proof: The asymptotics of the kernels prove pointwise boundedness of the image of the unit ball B of $L^2_o(\Gamma\backslash G)$. Consideration of derivatives proves equicontinuity of the image of B. The faux Arzela-Ascoli compactness lemma proves compactness of the closure of $\varphi \cdot B$. Being integrated versions of *right* translations, these operators stabilize the subspace of cuspforms, as the latter is defined by a *left* integral condition. Thus, φ maps the unit ball to a set with compact closure, so is a compact operator. ///

In these examples, the space of right K-invariant cuspforms, $L^2_o(\Gamma\backslash G/K) = L^2_o(\Gamma\backslash G)^K$ is of main interest, and the left-and-right K-invariant test functions act there:

[7.1.19] Corollary: For $\varphi \in C_c^\infty(K\backslash G/K)$, the operator $f \to \varphi \cdot f$ is a *compact* operator $L_o^2(\Gamma\backslash G/K) \longrightarrow L_o^2(\Gamma\backslash G/K)$.

Proof: A restriction of a compact operator is still compact. It suffices to show that the K-fixed subspace is *closed* in $L_o^2(\Gamma\backslash G)$, since the spherical Hecke algebra $C_c^\infty(K\backslash G/K)$ stabilizes it, by direct computation. From [6.1], the right action of G on $L^2(\Gamma\backslash G)$ is *unitary*, so continuous. Thus, the condition of right K-invariance is a *closed* condition. ///

Adjoints of the operators $f \to \varphi \cdot f$ are easily computed: letting $\langle , \rangle$ be the inner product on $L^2(\Gamma\backslash G)$,

$$\begin{aligned}\langle \varphi \cdot f, F\rangle &= \int_{\Gamma\backslash G}\int_G \varphi(x)\, f(yx)\, \overline{F}(y)\, dx\, dy \\ &= \int_{\Gamma\backslash G}\int_G \varphi(x)\, f(y)\, \overline{F}(yx^{-1})\, dx\, dy \\ &= \int_{\Gamma\backslash G}\int_G f(y)\, \varphi(x^{-1})\, \overline{F}(yx)\, dx\, dy = \int_{\Gamma\backslash G} f(y)\, \overline{\varphi^\vee \cdot F}(y)\, dy\end{aligned}$$

where $\varphi^\vee(x) = \overline{\varphi}(x^{-1})$ as suggested by the computation. The space $C_c^\infty(K\backslash G/K)$ is stable under the operation $\varphi \to \varphi^\vee$.

The *nondegeneracy* is essentially [14.1.5], but we need a right K-averaged version. Let φ_i be an approximate identity, so that $\varphi_i \cdot f \to f$, by [14.1.4]. For f right-invariant, the K-averaged versions

$$\alpha(\varphi_i \cdot f)(x) = \int_K (\varphi_i \cdot f)(xk)\, dk \qquad \text{(giving } K \text{ total measure 1)}$$

of $\varphi_i \cdot f$ must approach f, since K-averaging is an orthogonal projector to the space of K-invariant functions:

$$\begin{aligned}\langle \alpha f, \alpha F\rangle &= \int_K\int_K\int_G f(xh)\, F(xk)\, dx\, dk\, dh \\ &= \int_K\int_K\int_G f(x)\, F(xh^{-1}k)\, dx\, dk\, dh \\ &= \int_K\int_K\int_G f(x)\, F(xk)\, dx\, dk\, dh = \langle f, \alpha F\rangle\end{aligned}$$

Thus, for $f \neq 0$, for all sufficiently large i we have $\varphi_i \cdot f \neq 0$. These averaged versions are obtained by averaging φ_i:

$$\begin{aligned}\alpha(\varphi_i \cdot f)(x) &= \int_K (\varphi_i \cdot f)(xk)\, dk = \int_K\int_G \varphi_i(y) \cdot f(xky)\, dy\, dk \\ &= \int_K\int_G \varphi_i(k^{-1}y) \cdot f(xy)\, dy\, dk = \int_G \Big(\int_K \varphi_i(k^{-1}y)\, dk\Big) \cdot f(xy)\, dy\end{aligned}$$

Again using the right K-invariance of f,

$$(\varphi \cdot f)(x) = \int_G \varphi(y)\, f(xy)\, dy = \int_G \varphi(y)\, f(xy)\, dy \cdot \int_K 1\, dk$$
$$= \int_K \int_G \varphi(y)\, f(xyk)\, dy\, dk = \int_G \Big(\int_K \varphi(yk^{-1})\, dk\Big) f(xy)\, dy$$

Thus, for K-invariant f, there is left and right K-invariant φ such that $\varphi \cdot f \neq 0$. This is the nondegeneracy of the action of $C_c^\infty(K\backslash G/K)$ on $L^2_o(\Gamma\backslash G/K)$. This proves theorem [7.1.1]. ///

Proof (of corollary [7.1.2]): Just as in [2.4.5], the Gelfand commutativity criterion for the convolution algebra $C_c^\infty(K\backslash G/K)$ is that there should be an involutive anti-automorphism σ on G, that is, $g \to g^\sigma$ such that $(gh)^\sigma = h^\sigma g^\sigma$ and $(g^\sigma)^\sigma = g$ for all $g, h \in G$, and stabilizing double cosets for K, that is, $(KgK)^\sigma = KgK$ for all $g \in G$. Then the convolution algebra $C_c^\infty(K\backslash G/K)$ is *commutative*. Again, transpose-conjugation $g^\sigma = g^*$ is such an anti-automorphism, because the Cartan decomposition $G = KA^+K$ from [1.2] shows that left and right K-invariant test functions are determined by their values on A^+, and A^+ is pointwise fixed under transpose-conjugation.

By this commutativity and by the theorem [7.1.1], we have a commutative ring of compact operators, closed under adjoints, acting on a Hilbert space $V = L^2_o(\Gamma\backslash G/K)$ nondegenerately. The closed-ness under adjoints ensures that any operator T in that commutative algebra is a complex linear combination of self-adjoint operators in that algebra:

$$T = \frac{T + T^*}{2} + i \cdot \frac{T - T^*}{2i}$$

For a nonzero self-adjoint operator T on V, by the spectral theorem for self-adjoint compact operators [9.A], V decomposes into finite-dimensional eigenspaces V_λ for T with nonzero, real eigenvalues, and an orthogonal complement V':

$$V = \Big(\text{completion of } \bigoplus_{\lambda \neq 0} V_\lambda\Big) \oplus V'$$

The eigenspaces V_λ for nonzero eigenvalues λ of a nonzero operator self-adjoint compact T in that algebra are stabilized by every operator commuting with it, by the usual argument:

$$T(Sv) = (TS)v = (ST)v = S(Tv) = S(\lambda v) = \lambda \cdot Sv$$

for $v \in V_\lambda$. Then V_λ decomposes further into S-eigenspaces. By finite-dimensionality, we can do a downward induction to decompose V_λ into simultaneous eigenspaces for all of $C_c^\infty(K\backslash G/K)$. The eigenvalue map from operators to eigenvalues on a given simultaneous eigenspace must be a ring homomorphism.

This motivates one formulation of a proof of the corollary. Let X be the collection of not-identically-zero commutative-ring homomorphisms χ of $C_c^\infty(K\backslash G/K)$ to $\mathbb{C}$ such that

$$V_\chi = \{v \in V : Sv = \chi(S) \cdot v \text{ for all } S \in C_c^\infty(K\backslash G/K)\}$$

is not $\{0\}$. Each V_χ is finite-dimensional, by the spectral theorem, since χ is not identically 0. The orthogonal complement of the sum of all the V_χ is stable under the action of $C_c^\infty(K\backslash G/K)$, and every $\varphi \in C_c^\infty(K\backslash G/K)$ acts by 0 there. But this contradicts the nondegeneracy of the action, from the theorem. Thus, indeed, V is the completion of the orthogonal direct sum of the V_χ with χ not identically 0.

For nonzero f in V_χ, invoking the nondegeneracy, let φ be a test function such that $\chi(\varphi) \neq 0$, and put $\eta = \varphi/\chi(\phi)$. Then $\eta \cdot f = f$. By smoothing, as in [14.5] for example, this entails that f is smooth. This finishes the proof of [7.1.2]. ///

Proof (of corollary [7.1.3]): As in [4.2], the invariant Laplacian Δ on $\Gamma\backslash G/K$ or G/K is the Casimir operator Ω restricted to right K-invariant functions. Since Ω commutes with both right and left translation action of G, it commutes with the integrated action of $\varphi \in C_c^\infty(G)$:

$$\Omega(\varphi \cdot f)(x) = \Omega \int_G \varphi(y)\, f(xy)\, dy = \int_G \varphi(y)\, \Omega_x f(xy)\, dy$$

$$= \int_G \varphi(y)\, (\Omega_x f)(xy)\, dy = (\varphi \cdot \Omega f)(x)$$

using the Gelfand-Pettis characterization [14.1] and the fact that $y \to (x \to \varphi(y)f(xy))$ is a continuous, compactly supported, $C^\infty(\Gamma\backslash G)$-valued function and Ω is a continuous map of $C^\infty(\Gamma\backslash G)$ to itself. Thus, for right K-invariant smooth f on $\Gamma\backslash G$ and $\varphi \in C_c^\infty(K\backslash G/K)$,

$$\Delta(\varphi \cdot f) = \Omega(\varphi \cdot f) = \varphi \cdot (\Omega f) = \varphi \cdot \Delta f$$

Thus, Δ stabilizes each of the finite-dimensional simultaneous eigenspaces V_χ, each of which consists of smooth functions. The restriction of Δ to V_χ still has the *symmetry* proven in [6.5], so V_χ has an orthonormal basis of Δ-eigenfunctions. ///

Further implications of the following corollary will be apparent in Chapter 8:

[7.1.20] Corollary: The space of L^2 cuspforms has an orthonormal basis of cuspforms f such that there is a test function $\varphi \in C_c^\infty(K\backslash G/K)$ such that $\varphi \cdot f = f$. Such a cuspform f is *smooth* and of *rapid decay* in the sense that, on a standard Siegel set $\mathfrak{S}$, for every $q > 0$,

$$|f(g)| \ll_q \eta(g)^{-q} \qquad \text{(for all } g \in \mathfrak{S})$$

Proof: By the commutativity of $\mathcal{H} = C_c^\infty(K\backslash G/K)$ from the foregoing, its irreducible representations consist of simultaneous eigenfunctions and are one-dimensional. On each such $\mathcal{H}$ necessarily acts by an algebra homomorphism $\lambda : \mathcal{H} \to \mathbb{C}$. By the spectral consequences of the compactness above, the λth simultaneous eigenspace V_λ has finite dimension.

That is, for $f \in V_\chi$ and for test function φ, $\varphi \cdot f = \lambda(\varphi) \cdot f$. From the nondegeneracy result in [7.1.1], there is φ such that $\lambda(\varphi) \neq 0$. Replace φ by $\varphi/\lambda(\varphi)$ for the result. Then [7.1.9] applies to $f = \varphi \cdot f$ to prove rapid decay. Smoothness follows as in [14.5] and [14.6]. ///

7.2 $Z^+GL_2(k)\backslash GL_2(\mathbb{A})$

The next example shows how to adapt the argument of the previous section to adele groups $G = GL_2$ over number fields k. This incorporates spherical Hecke operators at good primes and shows how to decouple bad primes. Use notation as in Chapter 2. Test functions $C_c^\infty(G_\mathbb{A})$ on $G_\mathbb{A}$ are compactly supported and smooth, where smoothness at archimedean places means indefinite continuous differentiability, and at finite places means local constancy. Test functions φ act on functions f on $Z^+G_k\backslash G_\mathbb{A}$ as usual by

$$(\varphi \cdot f)(x) = \int_{G_\mathbb{A}} \varphi(y)\, f(xy)\, dy \qquad (\text{for } x \in G_\mathbb{A})$$

The compactness of suitable operators on cuspforms is a *global* property, and the kernel function is a *global* object. Thus, we cannot expect purely local arguments to suffice. In particular, the (purely local) Hecke operators of Chapter 2 are not quite adequate.

Let $K_\infty = \prod_{v|\infty} K_v$. As in Chapter 2, for simplicity, we will eventually restrict attention to right K_∞-invariant functions on $Z^+G_k\backslash G_\mathbb{A}$ rather than track K_∞ types. Commensurately, we will eventually restrict attention to left and right K_∞-invariant test functions on $G_\mathbb{A}$.

The main results of this section can be specialized to situations involving *commutative* Hecke algebras, which admit simultaneous eigenfunctions. The noncommutative Hecke algebras entering more general assertions mostly do *not* admit simultaneous eigenfunctions and need a more complicated notion of *irreducible representation*, as follows.

[7.2.1] Theorem: $C_c^\infty(G_\mathbb{A})$ acts on square-integrable cuspforms $L_o^2(Z^+G_k\backslash G_\mathbb{A})$ by *compact* operators. The collection of such operators is closed under adjoints and is *nondegenerate* in the sense that for every $f \in L_o^2(Z^+G_k\backslash G_\mathbb{A})$, there is $\varphi \in C_c^\infty(G_\mathbb{A})$ such that $\varphi \cdot f \neq 0$. *(Proof below.)*

[7.2.2] Corollary: The space $L_o^2(Z^+G_k\backslash G_\mathbb{A})$ decomposes discretely with finite multiplicities into irreducibles for $C_c^\infty(G_\mathbb{A})$. *(Proof below.)*

The assertion deserves clarification. Thinking of $A = C_c^\infty(G_\mathbb{A})$, let A be a (not necessarily commutative) associative algebra over $\mathbb{C}$, not necessarily having a unit. In the present context, a *representation* of A is a Hilbert space V on which A acts by continuous linear operators, with the expected associativity

$$\varphi \cdot (\psi \cdot v) = (\varphi * \psi) \cdot v \qquad (\text{for } \varphi, \psi \in A,\ v \in V)$$

where $*$ is the multiplication in A (convolution in $C_c^o(G)$). The space V is *(topologically) irreducible* (with respect to A) when it has no proper *closed* subspace stable under the action of A. An A-homomorphism $T : V \to W$ of A-representation spaces V, W is a continuous linear map T commuting with A, in the sense that $T(av) = aT(v)$ for $a \in A$ and $v \in V$. The *multiplicity* of an A-irreducible V in a larger A representation (Hilbert) space H is

$$\text{multiplicity of } V \text{ in } H = \dim_\mathbb{C} \operatorname{Hom}_A(V, H)$$

A form of Schur's lemma [9.D.12] shows that $\dim_\mathbb{C} \operatorname{Hom}_A(V, V) = 1$ for irreducibles V, and its corollary [9.D.14] shows that this removes potential ambiguity or ill-definedness in the definition of multiplicity. For Hilbert spaces, also

$$\text{multiplicity of } V \text{ in } H = \dim_\mathbb{C} \operatorname{Hom}_A(H, V)$$

because closed subspaces of Hilbert spaces admit orthogonal complements.

Yet, lacking further information about the irreducible representations of these non-commutative convolution algebras, or of the groups G_v and $G_\mathbb{A}$, variant results for *commutative* Hecke algebras may be more immediately informative. Fix $K' = K_\infty \prod_{v<\infty} K_v'$, a compact subgroup of $G_\mathbb{A}$, with K_v' equal to $K_v = GL_2(\mathfrak{o}_v)$ for almost all v. The finite primes v for which K_v' is K_v are *good* primes, while the finite v for which K_v' is strictly smaller than K_v are *bad* primes. Let S be the set of bad finite primes, of course depending on K'. With notation differing from Chapter 2, a suitable *spherical* Hecke algebra, depending on K', that does not attempt to do anything with bad primes is the collection $\mathcal{H}$ of left and right K'-invariant test functions φ on $G_{\mathbb{A}}$ that vanish at $g \in G_{\mathbb{A}}$ unless the vth component g_v is in K'_v for every $v \in S$. Gelfand's criterion [2.4.5] and the p-adic and archimedean Cartan decompositions show that this Hecke algebra $\mathcal{H}$ is *commutative*, as in [2.4] and in the proof of [7.1.2].

[7.2.3] Corollary: The space $L^2_o(Z^+G_k\backslash G_{\mathbb{A}})^{K'}$ of right K'-invariant cuspforms has an orthonormal basis of simultaneous eigenfunctions for the spherical Hecke algebra $\mathcal{H}$ attached to K', with each eigenspace finite-dimensional. The simultaneous eigenfunctions are *smooth*. *(Proof follows.)*

[7.2.4] Corollary: The space $L^2_o(Z^+G_k\backslash G_{\mathbb{A}}/K_{\mathbb{A}})$ of right $K_{\mathbb{A}}$-invariant square-integrable cuspforms decomposes into simultaneous eigenspaces for operators in the maximal spherical Hecke algebra $C^\infty_c(K_{\mathbb{A}}\backslash G_{\mathbb{A}}/K_{\mathbb{A}})$, with finite multiplicities. The simultaneous eigenfunctions are *smooth*. *(Proof follows.)*

[7.2.5] Corollary: There is an orthonormal Hilbert-space basis for the space of $K_{\mathbb{A}}$-invariant square-integrable cuspforms consisting of simultaneous eigenfunctions for the invariant Laplacians on the archimedean factors G_v. *(Proof follows.)*

For strictly upper-triangular square matrices x with entries in any field of characteristic zero, the series for the *matrix exponential* $e^x = \exp(x) = \sum_{\ell \geq 0} x^\ell/\ell!$ is *finite*. Thus, such matrices give an entirely algebraic notion of Lie algebra $\mathfrak{n}$ of N. Here,

$$\mathfrak{n}_{\mathbb{A}} = \{\begin{pmatrix} 0 & u \\ 0 & 0 \end{pmatrix} : u \in \mathbb{A}\} \qquad \mathfrak{n}_v = \{\begin{pmatrix} 0 & u \\ 0 & 0 \end{pmatrix} : u \in k_v\}$$

Since N is abelian, the exponential map is an isomorphism. Locally at archimedean places, the exponential map is a *diffeomorphism* $\exp\colon \mathfrak{n}_v \longrightarrow N_v$. Locally at non-archimedean places, it is a homeomorphism and preserves local-constant-ness. The discrete subgroup $\Lambda = n_k \subset \mathfrak{n}_{\mathbb{A}}$ is mapped isomorphically to N_k.

For $x \in G_{\mathbb{A}}$ write $x = n_x \cdot m_x \cdot k_x$ with $n_x \in N_{\mathbb{A}}$, $m_x \in M_{\mathbb{A}}$, and $k_x \in K_{\mathbb{A}}$. We can further decompose $m_x = m^1_x \cdot a_x$ with $m_x \in M^1$ and a_x in the archimedean split component

$$A^+ = \{\begin{pmatrix} t^{1/r} & 0 \\ 0 & 1 \end{pmatrix} : t > 0\}$$

on the diagonal in $M_\infty = \prod_{v|\infty} M_v$. Let η be the height function as in Chapter 2: in Iwasawa decomposition,

$$\eta(n \begin{pmatrix} \alpha & 0 \\ 0 & \beta \end{pmatrix} k) = \left|\frac{\alpha}{\beta}\right|$$

with idele norm, with $n \in N_{\mathbb{A}}$, $\alpha \in \mathbb{J}$, and $k \in K_{\mathbb{A}}$. Let

$$A_\tau = \{a \in A^+ : \eta(a) \geq \tau\}$$

For this section, a slightly refined notion of *Siegel set* is convenient: for compact $C_N \subset N_{\mathbb{A}}$, compact $C_M \subset M^1$, and $\tau > 0$ the corresponding *Siegel set* is

$$\mathfrak{S} = \mathfrak{S}_{C_N, C_M, \tau} = Z^+ \cdot C_N \cdot C_M \cdot A_\tau \cdot K_{\mathbb{A}}$$

From [2.A], $M_k \subset M^1$, and $M_k\backslash M^1$ is *compact*. Thus, sufficiently large C_M surjects to $M_k\backslash M^1$, and reduction theory [2.2] showed that for sufficiently small $\tau > 0$ the Siegel set $\mathfrak{S}$ *surjects* to $Z^+G_k\backslash G_{\mathbb{A}}$.

[7.2.6] Claim: A point in a Siegel set is well approximated by its M-component in an Iwasawa decomposition, in the sense that, for $x \in \mathfrak{S}_{C_N, C_M, \tau}$ with $\tau > 0$, compact $C_N \subset N_A$, and compact $C_M \subset M^1$, there is another compact subset C' of $N_{\mathbb{A}}$ such that $x \in m_x \cdot C' \cdot C_M \cdot K_{\mathbb{A}}$.

Proof: $x \in \mathfrak{S}$ gives

$$x = n_x \cdot m_x \cdot k_x \in N_{\mathbb{A}} \cdot m_x \cdot K_{\mathbb{A}} = m_x \cdot m_x^{-1} N_{\mathbb{A}} m_x \cdot K$$

The lower bound on A_τ gives a compact set C' in $N_{\mathbb{A}}$ depending only upon τ, C_N, and C_M such that

$$m^{-1} C_N m \ \subset \ C' \qquad (\text{for } m = m_1 a \text{ with } m_1 \in C_M \text{ and } a \in A_\tau)$$

In particular, $m_x^{-1} C_N m_x \subset C'$. Thus, $x \in m_x \cdot C' \cdot K_{\mathbb{A}}$ as claimed. ///

Given test function φ, *wind up* the operator $f \to \varphi \cdot f$ along $N_k = \exp(\mathfrak{n}_k)$: for $y \in G$,

$$(\varphi \cdot f)(y) = \int_{G_{\mathbb{A}}} \varphi(x) f(yx)\, dx = \int_{G_{\mathbb{A}}} \varphi(y^{-1}x) f(x)\, dx$$

$$= \int_{\exp(\Lambda)\backslash G_{\mathbb{A}}} \left(\sum_{\nu \in \mathfrak{n}_k} \varphi(y^{-1} \exp(\nu) x) \right) f(x)\, dx$$

$$= \int_{N_k\backslash G} \left(\sum_{\nu \in \Lambda} \varphi(y^{-1} \cdot \exp(\nu) \cdot x) \right) f(x)\, dx$$

The *kernel function* for the $(N \cap \Gamma)$-wound-up operator is the latter inner sum:

$$K_\varphi(x, y) = \sum_{\nu \in \mathfrak{n}_k} \varphi(y^{-1} \cdot \exp(\nu) \cdot x)$$

[7.2.7] Claim: For x, y both in a Siegel set $\mathfrak{S}$ the N_k-wound-up kernel vanishes unless the M-components of the two are *close*. That is, for fixed compact

$E \subset C_c^\infty(G_{\mathbb{A}})$ and for $x, y \in \mathfrak{S}$, there is compact $C_M \subset M_{\mathbb{A}}$ such that if there exist $n \in N_{\mathbb{A}}$ and $\varphi \in E$ with $\varphi(y^{-1} \cdot n \cdot x) \neq 0$, then $m_x \in m_y \cdot C_M$.

Proof: From the previous claim, there is compact $C' \subset N$ such that $m_y^{-1} y \in C' \cdot K_{\mathbb{A}}$. A compact set of test functions has a common compact support C_G, because a compact set is *bounded* in the topological vector space sense, and a bounded subset of an LF-space such as $C_c^\infty(G)$ lies in some Fréchet limitand, by [13.8.5]. Nonvanishing of $\varphi(y^{-1}nx)$ implies $y^{-1}nx \in C_G$, so

$$nx \in y \cdot C_G \subset m_y \cdot C' \cdot K_{\mathbb{A}} \cdot C_G \subset m_y \cdot C'_G$$

with $C'_G = C'K_{\mathbb{A}}C_G = \text{compact}$. That is,

$$C'_G \ni m_y^{-1} \cdot nx = m_y^{-1} \cdot nn_x \cdot m_x \cdot k_x = (m_y^{-1} nn_x m_y) \cdot m_y^{-1} m_x \cdot k_x$$

That is,

$$(m_y^{-1} nn_x m_y) \cdot m_y^{-1} m_x \in C'_G \cdot K_{\mathbb{A}} = \text{compact}$$

Since $M_{\mathbb{A}}$ normalizes $N_{\mathbb{A}}$, the element $m_y^{-1} nn_x m_y$ is in $N_{\mathbb{A}}$. Since $N_{\mathbb{A}} \cap M_{\mathbb{A}} = \{1\}$, the multiplication map $N_{\mathbb{A}} \times M_{\mathbb{A}} \to N_{\mathbb{A}} M_{\mathbb{A}}$ is a homeomorphism. Thus, for the product $(m_y^{-1} nn_x m_y) \cdot m_y^{-1} m_x$ to lie in a compact set in $G_{\mathbb{A}}$ requires that n lies in a compact set in $N_{\mathbb{A}}$ and m lies in a compact set in $M_{\mathbb{A}}$. Thus, there is compact $C_M \subset M_{\mathbb{A}}$ such that $m_y^{-1} m_x \in C_M$, as claimed. ///

[7.2.8] Corollary: For x, y in a fixed Siegel set, and for fixed compact $E \subset C_c^\infty(G_{\mathbb{A}})$, there is a compact $C_M \subset M_{\mathbb{A}}$ such that, if $\varphi(y^{-1}nx) \neq 0$ for some $n \in N_{\mathbb{A}}$ and some $\varphi \in E$, then $m_y^{-1} x$ lies in C_M.

Proof: By the first of the two claims, there is a compact C_G in $G_{\mathbb{A}}$ such that $x \in m_x \cdot C_G$. By the second, there is a compact C_M in $M_{\mathbb{A}}$ such that $m_x \in m_y C_M$. Thus, $x \in m_x C_G \subset (m_y C_M) C'_G$, rearranging to give the claim. ///

The notion of Schwartz function on an archimedean vectorspace such as $\mathfrak{n}_\infty = \prod_{v|\infty} \mathfrak{n}_v$ is as in [13.7]. On non-archimedean vectorspaces $\mathfrak{n}_v$, Schwartz functions are the same as test functions, namely, locally constant, compactly supported. Similarly, on the finite-adeles part of an adelic vectorspace, Schwartz functions are simply test functions, that is, locally constant, compactly supported. Then Schwartz functions on adelic vector spaces $\mathfrak{n}_{\mathbb{A}}$ are finite sums $\sum_i f_{\infty,i} \otimes f_{o,i}$ where the functions $f_{\infty,i}$ are Schwartz functions on the archimedean part, and the functions $f_{o,i}$ are Schwartz/test functions on the non-archimedean part. Topologies on such spaces are as in [6.2], [6.3], and as simpler examples in [13.7], [13.8], and [13.9].

[7.2.9] Claim: With $\omega_y = y^{-1}m_y$ and $\omega_{x,y} = m_y^{-1}x$, the functions $\nu \longrightarrow \varphi_{x,y}(\nu) = \varphi(\omega_y \cdot \exp(\nu) \cdot \omega_{x,y})$ for x, y in fixed Siegel set, $\varphi(y^{-1}nx) \neq 0, \varphi \in E$, constitute a *compact* subset of the Schwartz space $\mathscr{S}(\mathfrak{n}_\mathbb{A})$.

Proof: The left and right translation actions $G_\mathbb{A} \times G_\mathbb{A} \times C_c^\infty(G_\mathbb{A}) \to C_c^\infty(G_\mathbb{A})$ are continuous, by [6.4]. With fixed Siegel set $\mathfrak{S}$, by [7.2.6], [7.2.7], and [7.1.7], $\{\omega_y : y \in \mathfrak{S}\}$ and $\{\omega_{x,y} : x \in \mathfrak{S},\ y \in \mathfrak{S}\}$ are compact. This gives compactness of the image of

$$\{\omega_y : y\} \times \{\omega_{x,y} : x, y\} \times E \longrightarrow C_c^\infty(G_\mathbb{A})$$

x, y in fixed Siegel set, $\varphi(y^{-1}nx) \neq 0$. Since $N_\mathbb{A}$ is *closed* in $G_\mathbb{A}$, the restriction map $C_c^\infty(G_\mathbb{A}) \to C_c^\infty(N_\mathbb{A}) \approx C_c^\infty(\mathfrak{n}_\mathbb{A})$ is continuous. From [13.9.3], the inclusion $C_c^\infty(\mathfrak{n}_\mathbb{A}) \subset \mathscr{S}(\mathfrak{n}_\mathbb{A})$ of test functions to Schwartz functions is continuous, giving compactness of the image. ///

Poisson summation for the lattice $\mathfrak{n}_k \subset \mathfrak{n}_\mathbb{A}$ gives

$$\sum_{\nu \in \mathfrak{n}_k} \varphi(y^{-1} \cdot \exp(\nu) \cdot x) = \sum_{\psi \in \mathfrak{n}_k^*} \int_{\mathfrak{n}_\mathbb{A}} \overline{\psi}(\nu)\, \varphi(y^{-1} \cdot \exp(\nu) \cdot x)\, d\nu$$

with suitably normalized measure, where $\mathfrak{n}_k^*$ is the collection of $\mathbb{C}^\times$-valued characters on $\mathfrak{n}_\mathbb{A} \approx \mathbb{A}$ trivial on the lattice $\mathfrak{n}_k \approx k$. By Appendix [7.A], given a nontrivial character ψ_1 on $\mathfrak{n}_\mathbb{A}/\mathfrak{n}_k \approx \mathbb{A}/k$, every other character is of the form $\psi_\xi(\nu) = \psi_1(\xi \cdot \nu)$ with $\xi \in k$. Thus, choice of that nontrivial character identifies $\mathbb{A}^* \approx \mathfrak{n}_\mathbb{A}^*$ with $\mathbb{A} \approx \mathfrak{n}_\mathbb{A}$ and the dual lattice of $k \approx \mathfrak{n}_k$ with k itself. Thus,

$$\sum_{\nu \in \mathfrak{n}_k} \varphi(y^{-1} \cdot \exp(\nu) \cdot x) = \sum_{\xi \in k} \int_{\mathfrak{n}_\mathbb{A}} \overline{\psi}_1(\xi \cdot \nu)\, \varphi(y^{-1} \cdot \exp(\nu) \cdot x)\, d\nu$$

We have

$$\int_{\mathfrak{n}_\mathbb{A}} \overline{\psi}_1(\xi \cdot \nu) \cdot \varphi\left(y^{-1} \cdot \exp(\nu) \cdot x\right)\, d\nu$$

$$= \int_{\mathfrak{n}_\mathbb{A}} \overline{\psi}_1(\xi \cdot \nu) \cdot \varphi\left(y^{-1}m_y \cdot \exp(m_y^{-1}\nu m_y) \cdot m_y^{-1}x\right)\, d\nu$$

Replacing ν by $m_y \nu m_y^{-1}$ and letting $\varphi_{x,y}(\nu) = \varphi\left(y^{-1}m_y \cdot \nu \cdot m_y^{-1}x\right)$,

$$K_\varphi(x, y) = \delta(m_y) \sum_{\xi \in k} \widehat{\varphi}_{x,y}(\psi_\xi^{m_y})$$

where $\widehat{\varphi}_{x,y}(\psi_\xi^{m_y})$ is the Fourier transform $\widehat{\varphi}_{x,y}$ of $\varphi_{x,y}$ evaluated at ψ_ξ, and $\psi_\xi^{m_y}(\nu) = \psi_\xi(m_y \nu m_y^{-1})$.

[7.2.10] Theorem: Fix compact $E \subset C_c^\infty(G_\mathbb{A})$ and Siegel set $\mathfrak{S}$. Given $q > 0$ there is a uniform bound

$$|(\varphi \cdot f)(y)| \ll_q \eta(y)^{-q} \cdot |f|_{L^2(\Gamma\backslash G)}$$

for all $y \in \mathfrak{S}$, for all $\varphi \in E$, for all L^2 cuspforms f.

Proof: With the self-duality identifications as already shown, Fourier transform is a continuous automorphism of $\mathscr{S}(\mathfrak{n}_\mathbb{A})$ to itself, by the archimedean case [13.15] and the p-adic case [13.17]. Thus, $\{\widehat{\varphi}_{x,y} : \varphi \in E,\ x, y \in \mathfrak{S}\}$ is a compact subset of $\mathscr{S}(\mathfrak{n}_\mathbb{A})$. The adelic Schwartz space $\mathscr{S}(\mathfrak{n}_\mathbb{A})$ is an LF-space, a strict colimit of Fréchet spaces, characterized as a countable ascending union of Fréchet subspaces described by restricting support at finite primes and by requiring *uniform* local constancy at finite primes. That is, for U (large) compact open subgroup of the finite-adele part $\mathfrak{n}_{\mathbb{A}_{\rm fin}}$ of $\mathfrak{n}_\mathbb{A}$, and for H a (small) compact open subgroup of $\mathfrak{n}_{\mathbb{A}_{\rm fin}}$, let $\mathscr{S}(\mathfrak{n}_\infty \times U)^H$ be the space of H-invariant Schwartz functions supported on $\mathfrak{n}_{\mathbb{A}_\infty} \times U$. Then

$$\mathscr{S}(\mathfrak{n}_\mathbb{A}) = \bigcup_{H,U} \mathscr{S}(\mathfrak{n}_\infty \times U)^H = \mathrm{colim}_{H,U}\ \mathscr{S}(\mathfrak{n}_\infty \times U)^H$$

as H shrinks and U grows. There is a countable cofinal collection of subgroups U and subgroups H, confirming that the adelic Schwartz space is an LF-space, [13.8] and [13.9]. In particular, a compact subset lies in some limitand $\mathscr{S}(\mathfrak{n}_\infty \times U)^H$, by [13.8.5]. Thus, the compactness [7.2.8] implies that the Schwartz functions $\psi \to \widehat{\varphi}_{x,y}(\psi)$ are inside a compact subset of some $\mathscr{S}(\mathfrak{n}_\infty \times U)^H$.

Thus, for $\widehat{\varphi}_{x,y}(\psi) \neq 0$, the finite-prime part $\xi_{\rm fin}$ of ξ is in some compact $U \subset \mathfrak{n}_{\mathbb{A}_{\rm fin}} \approx \mathbb{A}_{\rm fin}$. Thus, $\xi \in \frac{1}{h}\mathfrak{o}_k$ for some $0 < h \in \mathbb{Z}$, and the collection of infinite-prime parts ξ_∞ of such ξ is a *lattice* $\Lambda^* \subset \mathfrak{n}_\infty \approx k_\infty$. Give the real vectorspace k_∞ the real inner product

$$\langle \xi, \xi' \rangle = \sum_{v|\infty} \mathrm{Re}\left(\xi_v \cdot \overline{\xi'}_v\right)$$

with the complex conjugation to accommodate complex k_v. Let $|\cdot|$ be the associated norm. Thus, for $0 \neq \xi_\infty \in \Lambda^*$, we have $|\xi| \gg_\mathfrak{o} \frac{1}{h}$, with an implied constant depending on the ring of integers $\mathfrak{o}$. In these terms, for $F \in \mathscr{S}(\mathfrak{n}_\infty \times U)^H$, $|F(\psi)| \ll_r (1 + |\psi|)^{-\ell}$ for all $\ell > 0$. Thus, for fixed Siegel set $\mathfrak{S}$ and φ in fixed compact E, the compactness [7.2.9] gives a *uniform* implied constant depending only on ℓ so that

$$|\widehat{\varphi}_{x,y}(\psi)| \ll_\ell (1 + |\psi|)^{-\ell} \qquad \text{(for all } x, y \in \mathfrak{S}, \text{ for all } \varphi \in E)$$

For $m_y = zm'a_t$ with $z \in Z^+$, $m' \in C_M \subset M^1$, and $t > 0$, our parametrization of the archimedean split component A^+ gives $\delta(m_y) = t$, and

$$|\psi^{m_y}| \ll_{C_M} t^{1/r_o} \cdot |\psi| = \delta(m_y)^{1/r_o} \cdot |\psi|$$

where r_o is the number of archimedean completions of k modulo complex conjugation. Thus, with fixed Siegel set,

$$\delta(m_y)\,|\widehat{\varphi}_{x,y}(\psi^{m_y})| \ll_\ell \delta(m_y) \cdot (1 + \delta(m_y)^{1/r_o} \cdot |\psi|)^{-\ell}$$

for all $x, y \in \mathfrak{S}$.

[7.2.11] Claim: For fixed Siegel set $\mathfrak{S}$ and $r \gg 1$, the kernel $K_\varphi(x, y)$ with its 0th Fourier component removed satisfies

$$|K_\varphi(x, y) - \widehat{\varphi}_{x,y}(\psi_0)| \ll_q |\eta(y)|^{-q} \qquad (\text{for } x, y \in \mathfrak{S})$$

Proof: First, we claim that, with fixed lattice $\Lambda^* \subset k_\infty$ obtained by projecting $k \cap (k\infty \times U)$ to k_∞, there is an implied constant such that

$$\eta(y) = \delta(m_y) \ll |\psi^{m_y}|^{r_o}$$

for all $y \in \mathfrak{S}$, for all $0 \neq \psi \in k \cap U$. Again, $\eta(y) = t$ for $y = na_tk$ with $t > 0$, and $|\psi^{m_y}| \ll_{\mathfrak{S}} t^{1/r_o} \cdot |\psi|$. Since the norms of nonzero elements of Λ^* have a positive inf,

$$|\psi^{m_y}|^{r_o} \gg_{\mathfrak{S}} \left(\eta(y)^{1/r_o} \cdot |\psi|\right)^{r_o} \geq \eta(y) \cdot \inf_{0 \neq \lambda \in \Lambda^*} |\lambda|^{r_o}$$

Since $\widehat{\varphi}_{x,y}$ is a Schwartz function, $|\widehat{\varphi}_{x,y}(\psi)| \ll_\ell (1 + |\psi|)^{-\ell}$ for every $\ell > 0$. By the comparison of $\eta(y)$ to $|\psi^{m_y}|$, for $0 \neq \psi \in k \cap (k_\infty \times U)$,

$$\begin{aligned}
\delta(m_y) \cdot (1 + |\psi^{m_y}|)^{-\ell} &= \eta(y) \cdot (1 + |\psi^{m_y}|)^{-\ell} \\
&= \eta(y) \cdot (1 + |\psi^{m_y}|)^{-(q+1)\cdot r_o - (\ell - (q+1)\cdot r_o)} \\
&\ll \eta(y) \cdot \eta(y)^{-(q+1)} \cdot (1 + |\psi^{m_y}|)^{-(\ell - (q+1)\cdot r_o)}
\end{aligned}$$

For ℓ sufficiently large depending on q, the sum of this over $0 \neq \psi$ converges, giving the claim. ///

[7.2.12] Claim: Cuspforms f ignore the trivial Fourier component of $K_\varphi(x, y)$:

$$(\varphi \cdot f)(y) = \int_{\exp(\Lambda)\backslash G} \left(K_\varphi(x, y) - \widehat{\varphi}_{x,y}(\psi_0)\right) \cdot f(x)\, dx$$

(Direct computation, identical to [7.1.11].) ///

The proof of [7.2.9] is almost complete. From the preceding, for y in a fixed Siegel set $\mathfrak{S}$ and for fixed test function φ, there is a compact $C_M \subset A^+$ such that, for $\varphi(y^{-1}nx)$ to be nonzero, the Iwasawa A^+-component m_x of x must lie in $m_y \cdot C_M$. Thus,

$$\{x \in \mathfrak{S} : \varphi(y^{-1}nx) \neq 0 \text{ for some } n \in N\} \subset m_y C_M \cdot K$$

Combining this with the estimate just obtained, for cuspform f,

$$|(\varphi \cdot f)(y)| \ll_r |\eta(y)|^{-r} \cdot \int_{Z^+G_k\backslash G_k(C \cdot m_y C_M \cdot K_{\mathbb{A}})} |f(x)|\, dx$$

By Cauchy-Schwarz-Bunyakowsky,

$$\Big(\int_{Z^+G_k\backslash G_k(C \cdot m_y C_M \cdot K_{\mathbb{A}})} |f(x)|\, dx \Big)^2$$

$$\leq \int_{Z^+G_k\backslash G_k(C \cdot m_y C_M \cdot K_{\mathbb{A}})} 1\, dx \cdot \int_{Z^+G_k\backslash G_k(C \cdot m_y C_M \cdot K_{\mathbb{A}})} |f(x)|^2\, dx \ll |f|^2_{L^2}$$

This gives the desired decay, proving theorem [7.2.9]. ///

We are getting closer to the compactness of the operators $f \to \varphi \cdot f$ on cuspforms f. Again, a collection E of continuous functions on $G_{\mathbb{A}}$ or $Z^+G_k\backslash G_{\mathbb{A}}$ is *(uniformly) equicontinuous* when, given $\varepsilon > 0$, there is a neighborhood U of 1 in $G_{\mathbb{A}}$ such that

$$|f(x) - f(y)| < \varepsilon$$

for all $f \in E$, for all $x \in G_{\mathbb{A}}$, for all $y \in x \cdot U$. As in the previous section, for general reasons, we have

[7.2.13] Lemma: Let $\mathfrak{g}_v$ be the Lie algebra of G_v for archimedean v. For $X \in \mathfrak{g}_v$, the left-derivative map

$$C_c^\infty(G_{\mathbb{A}}) \longrightarrow C_c^\infty(G_{\mathbb{A}})$$

by

$$\varphi \longrightarrow \Big(g \to \frac{d}{dt}\Big|_{t=0} \varphi(e^{-tX} g) \Big)$$

is *continuous. (Same proof as [7.1.12].)* ///

[7.2.14] Corollary: For a compact set E of test functions on $G_{\mathbb{A}}$, for a compact $C_{\mathfrak{g}}$ in $\mathfrak{g} = \mathfrak{g}_v$, and for f ranging over cuspforms in the unit ball in $L^2(Z^+G_k\backslash G_{\mathbb{A}})$, there is a *uniform* implied constant such that

$$\left| \frac{d}{dt}\bigg|_{t=0} (\varphi \cdot f)(g e^{tX}) \right| \ll 1$$

for all $g \in G_{\mathbb{A}}$, for all $\varphi \in E$, for all $X \in C_{\mathfrak{g}}$.

Proof: As in the proof of [7.1.13], differentiation of $\varphi \cdot f$ can be rewritten as a differentiation of φ, followed by action of the resulting function on f, by changing variables:

$$\begin{aligned}(X \cdot \varphi \cdot f)(x) &= \frac{d}{dt}\bigg|_{t=0} \int_G \varphi(y)\, f(x e^{tX} y)\, dy \\ &= \frac{d}{dt}\bigg|_{t=0} \int_G \varphi(e^{-tX} y)\, f(xy)\, dy \\ &= \int_G \left(\frac{d}{dt}\bigg|_{t=0} \varphi(e^{-tX} y) \right) f(xy)\, dy \ \text{(replacing } y \text{ by } e^{-tX} y)\end{aligned}$$

justifying interchange of differentiation and integration by the continuity of differentiation on test functions φ, and Gelfand-Pettis integral properties [14.1]. That is, $X \cdot \varphi \cdot f = (X^{\text{left}} \varphi) \cdot f$ with X^{left} the *left* action. Since $\mathfrak{g}_v$ is a finite-dimensional real vector-space and the action is linear in X, this gives the continuity in $X \in \mathfrak{g}_v$. Thus, the collection of test functions $X^{\text{left}}\varphi$ with $X \in C_{\mathfrak{g}}$ and $\varphi \in E$ is again compact in $C_c^\infty(G)$. Thus, by the bound of theorem [7.2.10],

$$|(X \cdot \varphi \cdot f)(y)| = |(X^{\text{left}} \varphi) \cdot f)(y)| \ll_r \eta(y)^{-r} \cdot |f|_{L^2}$$

for all $y \in \mathfrak{S}, X \in C_{\mathfrak{g}}, \varphi \in E$. For large-enough Siegel set to cover the quotient, and any $r > 0$, this gives

$$\sup |X \cdot \varphi \cdot f(y)| \ll_r |f|_{L^2}$$

proving *uniform* boundedness for $|f|_{L^2} \le 1$. ///

The smoothing property of $f \to \varphi \cdot f$ as in [14.5] ensures that each $\varphi \cdot f$ is smooth. Smoothness of φ at finite places is *uniform*, since that of φ is: with φ left-invariant by compact open subgroup $K' \subset G_{\mathbb{A}_{\text{fin}}}$, for $h \in K'$,

$$\begin{aligned}(\varphi \cdot f)(g \cdot h) &= \int_G \varphi(x)\, f(ghx)\, dx = \int_G \varphi(h^{-1}x)\, f(gx)\, dx \\ &= \int_G \varphi(x)\, f(gx)\, dx = (\varphi \cdot f)(g)\end{aligned}$$

A uniform bound on derivatives at archimedean places will imply *uniform* continuity:

[7.2.15] Lemma: Let F be a smooth function on G, (uniformly) right K'-invariant for some compact open subgroup $K' \subset G_{\mathbb{A}_{\mathrm{fin}}}$, with a uniform pointwise bound on all $X \cdot F$ with X in a *compact* neighborhood $C_{\mathfrak{g}}$ of 0 in $\mathfrak{g}_\infty$, namely,

$$|(X \cdot F)(x)| \le B \qquad (\text{for all } x \in G_{\mathbb{A}}, \text{ all } X \in C_{\mathfrak{g}})$$

Then F is *uniformly continuous*: for every $\varepsilon > 0$ there is a neighborhood U of 1 in $G_{\mathbb{A}}$ such that $|F(x) - F(y)| < \varepsilon$ for all $x \in G_{\mathbb{A}}$ and $y \in xU$.

Proof: The only issue is at archimedean places. Let V be a small enough open containing 1 such that V is contained in $\exp(C_{\mathfrak{g}}) \cdot K'$ and that the exponential map on the archimedean part is injective to V. Let $y = x \cdot e^{sX}$ for $X \in C_{\mathfrak{g}}$ and $0 \le s \le 1$. By hypothesis, the function $h(t) = F(x \cdot e^{sX})$ has

$$h'(s) = \frac{d}{dt}\bigg|_{t=0} h(s+t) = F(x \cdot e^{sX} \cdot e^{tX})$$

which is bounded by B. From the mean value theorem, $|F(x \cdot e^{tX}) - F(x)| \le t \cdot B$. Thus, for all $|t| < B \cdot \varepsilon$ we have the desired inequality. ///

[7.2.16] Corollary: For a compact set E of test functions on $G_{\mathbb{A}}$, and for f ranging over cuspforms in the unit ball in $L^2(Z^+ G_k \backslash G_{\mathbb{A}})$, the family of images $\varphi \cdot f$ is *(uniformly) equicontinuous* on G. ///

We are almost done with the proof that $f \to \varphi \cdot f$ is compact on cuspforms. We again have a compactness lemma vaguely reminiscent of Arzela-Ascoli:

[7.2.17] Lemma: Let E be a *equicontinuous*, uniformly bounded, set of functions on $Z^+ G_k \backslash G_{\mathbb{A}}$. Then E has *compact closure* in $L^2(Z^+ G_k \backslash G_{\mathbb{A}})$. *(Same proof as [7.1.16].)*

Finally, we prove the theorem [7.2.1]. To summarize: the asymptotics of the kernels prove pointwise boundedness of the image of the unit ball B of $L^2_o(Z^+ G_k \backslash G_{\mathbb{A}})$, and consideration of derivatives proves equicontinuity of the image of B. The faux-Arzela-Ascoli compactness lemma above proves compactness of the closure of $\{\varphi \cdot B : \varphi \in E\}$. Being integrated versions of *right* translations, these operators stabilize the subspace of cuspforms, as the latter is defined by a *left* integral condition. Thus, φ maps the unit ball to a set with compact closure, so is a compact operator.

Adjoints are easily computed: letting $\langle , \rangle$ be the inner product on $L^2(Z^+G_k\backslash G_{\mathbb{A}})$,

$$\begin{aligned}\langle \varphi \cdot f, F\rangle &= \int_{Z^+G_k\backslash G_{\mathbb{A}}} \int_{G_{\mathbb{A}}} \varphi(x)\, f(yx)\, \overline{F}(y)\, dx\, dy \\ &= \int_{Z^+G_k\backslash G_{\mathbb{A}}} \int_{G_{\mathbb{A}}} \varphi(x)\, f(y)\, \overline{F}(yx^{-1})\, dx\, dy \\ &= \int_{Z^+G_k\backslash G_{\mathbb{A}}} \int_{G_{\mathbb{A}}} f(y)\, \varphi(x^{-1})\, \overline{F}(yx)\, dx\, dy \\ &= \int_{Z^+G_k\backslash G_{\mathbb{A}}} f(y)\, \overline{\varphi^{\vee} \cdot F}(y)\, dy\end{aligned}$$

where $\varphi^{\vee}(x) = \overline{\varphi}(x^{-1})$ as suggested by the computation. The space of test functions is stable under the operation $\varphi \to \varphi^{\vee}$.

The *nondegeneracy* is [14.1.5], finishing the proof of theorem [7.2.1]. ///

Proof: Now we can prove Corollary [7.2.2], that $L^2_o(Z^+G_k\backslash G_{\mathbb{A}})$ decomposes as (the closure of) a direct sum of irreducible representations of $C_c^{\infty}(G_{\mathbb{A}})$, each occurring with finite multiplicity.

[7.2.18] Claim: Let A be an *adjoint-stable* algebra of *compact* operators on a Hilbert space H, *nondegenerate* in the sense that for every nonzero $v \in H$ there is $a \in A$ with $a \cdot v \neq 0$. Then H is (the completion of) an orthogonal direct sum of *closed A-irreducible* subspaces, and each isomorphism class of A-irreducible V occurs with finite multiplicity.

Proof: To prove that H is (the completion of) a direct sum of (closed) A-irreducibles, reduce to the case that H has *no* proper irreducible A-subspaces, by replacing H by the orthogonal complement to the sum of all irreducible A-subspaces. By nondegenerateness, there is a *nonzero* self-adjoint operator T in A, since for a nonzero operator S in A, either $S + S^*$ or $S - S^*$ is nonzero (and $S + S^*$ and $(S - S^*)/i$ are self-adjoint).

From the spectral theorem for self-adjoint compact operators [9.A.6], there is a nonzero eigenvalue λ of the (nonzero) self-adjoint compact operator T on H, and the λ-eigenspace is finite-dimensional. Among all A-stable closed subspaces choose one, W, such that the λ-eigenspace W_λ is of *minimal positive* (finite) dimension. Let w be a nonzero vector in W_λ. The closure of $A \cdot w$ is a closed subspace of W, and we claim that it is irreducible. Suppose that closure$(A \cdot w) = X \oplus Y$ is a decomposition into mutually orthogonal, closed A-stable subspaces. With $w = w_X + w_Y$ the corresponding decomposition,

$$\lambda w_X + \lambda w_Y = \lambda w = Tw = T(w_X + w_Y)$$

By the orthogonality and stability,

$$\lambda w_X = T w_X \qquad \text{and} \qquad \lambda w_Y = T w_Y$$

By the minimality of the λ-eigenspace in W, either $w_X = 0$ or $w_Y = 0$. That is, $\lambda w = Tw \subset X$ or $\lambda w = Tw \subset Y$. That is, since $\lambda \neq 0$, either $w \in X$ or $w \in Y$. Thus, either $A \cdot w \subset X$ or $A \cdot w \subset Y$, and likewise for the *closures*. But this implies that one or the other of X, Y is 0. This proves the irreducibility of the closure of $A \cdot w$, contradicting the assumption that H had no irreducible A-subspaces.

For finite multiplicities: an irreducible V is nondegenerate, since otherwise the subspace annihilated by all $a \in A$ would be a proper subspace. Thus, there is an operator $T \in A$ compact and self-adjoint on H and nonzero on V. If the orthogonal direct sum $V \oplus \ldots \oplus V$ of n copies appeared inside H for arbitrarily large n, this would give T infinite multiplicities of nonzero eigenvalues on H, contradicting the spectral theorem. ///

Then the proofs of [7.2.3] and [7.2.4] are special cases, where the algebra $\mathcal{H}$ of compact operators is designed to be *commutative*, so the notion of *irreducibles* simplifies to *simultaneous eigenspace*. Gelfand's criterion [2.4.5] and the p-adic and archimedean Cartan decompositions from [2.1] and [1.2] show that this Hecke algebra $\mathcal{H}$ is *commutative*, as in [2.4] and in the proof of [7.1.2]. For nonzero f in a simultaneous eigenspace V_χ for $\mathcal{H}$, by nondegeneracy there is a test function φ such that $\varphi \cdot f = \chi(\varphi) \cdot f$ and $\chi(\varphi) \neq 0$. With $\eta = \varphi/\chi(\phi)$, then $\eta \cdot f = f$. By smoothing, as in [14.5], for example, f is smooth.

Just as in the proof of [7.1.3], the fact that the Casimir operators Ω_v on archimedean G_v commute with left and right translation implies that Ω_v commutes with the action of $C_c^\infty(G_{\mathbb{A}})$, by integrating. On right K_v-invariant functions, Ω_v is the invariant Laplacian Δ_v. Thus, each Δ_v stabilizes the simultaneous eigenspaces V_χ of $\mathcal{H}$, all of which are finite-dimensional, consisting of smooth functions. The restriction of Δ_v to V_χ is still *symmetric*, so by finite-dimensional spectral theory V_χ has a basis of Δ_v-eigenfunctions. ///

As in the previous section, the implications of the following corollaries will be apparent in Chapter 8:

[7.2.19] Corollary: The space of right $K_{\mathbb{A}}$-invariant L^2 cuspforms has an orthonormal basis of cuspforms f such that there is a test function $\varphi \in C_c^\infty(K_{\mathbb{A}}\backslash G_{\mathbb{A}}/K_{\mathbb{A}})$ such that $\varphi \cdot f = f$. Such a cuspform f is *smooth* and of *rapid decay* in the sense that, on a standard Siegel set $\mathfrak{S}$, for every $q > 0$,

$$|f(g)| \ll_q \eta(g)^{-q} \qquad (\text{for all } g \in \mathfrak{S})$$

Proof: The proof is identical to that of [7.1.20]. ///

The proof of the following more general case is subtler than the previous:

[7.2.20] Corollary: The space of L^2 cuspforms has an orthonormal basis of cuspforms f such that there is a test function $\varphi \in C_c^\infty(G_\mathbb{A})$ such that $\varphi \cdot f = f$. Such a cuspform f is of *rapid decay* in the sense that, on a standard Siegel set $\mathfrak{S}$, for every $q > 0$,

$$|f(g)| \ll_q \eta(g)^{-q} \qquad (\text{for all } g \in \mathfrak{S})$$

Proof: The irreducible modules over $\mathcal{H} = C_c^\infty(G_\mathbb{A})$ appearing in the space of L^2 cuspforms are merely finite-dimensional, each occurring with finite multiplicity. Let f be in a copy V of an irreducible module for $C_c^\infty(G_\mathbb{A})$.

Since V is irreducible, it has no proper, topologically closed $\mathcal{H}$-stable subspace. Since V is finite-dimensional, all vector subspaces are closed. Thus, $\mathcal{H} \cdot f = V$. In particular, there is a test function φ such that $\varphi \cdot f = f$. Then [7.2.20] shows that $f = \varphi \cdot f$ is of rapid decay in Siegel sets. Smoothness follows as in [14.5] and [14.6]. ///

[7.2.21] Remark: The irreducibles for adjoint-closed rings of compact operators need not be finite-dimensional, even though nonzero eigenspaces of self-adjoint compact operators are finite-dimensional. It is not so easy to give examples of such infinite-dimensional irreducibles. Perhaps the simplest examples would be the irreducibles in the decomposition of $L^2(\Gamma\backslash G)$ for compact quotients [7.B].

7.3 $Z^+GL_r(k)\backslash GL_r(\mathbb{A})$

The only new ingredient beyond the previous two sections is treatment of more complicated asymptotics for $G = GL_r$ for $r \geq 3$. The statements of results, and the proof mechanisms, are essentially identical to the previous section:

[7.3.1] Theorem: $C_c^\infty(G_\mathbb{A})$ acts on square-integrable cuspforms $L_o^2(Z^+G_k\backslash G_\mathbb{A})$ by *compact* operators. The collection of such operators is closed under adjoints, and is *nondegenerate* in the sense that for every $f \in L_o^2(Z^+G_k\backslash G_\mathbb{A})$ there is $\varphi \in C_c^\infty(G_\mathbb{A})$ such that $\varphi \cdot f \neq 0$. *(Proof follows.)*

[7.3.2] Corollary: The space $L_o^2(Z^+G_k\backslash G_\mathbb{A})$ decomposes discretely with finite multiplicities into irreducibles for $C_c^\infty(G_\mathbb{A})$. *(Proof follows.)*

As in the previous sections, without further information about the irreducible representations of these noncommutative convolution algebras, or of the groups

G_v and $G_{\mathbb{A}}$, corollaries about *commutative* Hecke algebras are more immediately informative. Fix $K' = K_\infty \prod_{v<\infty} K'_v$, a compact subgroup of $G_{\mathbb{A}}$, with K'_v equal to $K_v = GL_r(\mathfrak{o}_v)$ for almost all v. The finite primes v for which K'_v is K_v are *good* primes, while the finite v for which K'_v is strictly smaller than K_v are *bad* primes. Let S be the set of bad finite primes, of course depending on K'. With notation differing from Chapter 3, a suitable *spherical* Hecke algebra, depending on K', that does not attempt to do anything with bad primes is the collection $\mathcal{H}$ of left and right K'-invariant test functions φ on $G_{\mathbb{A}}$ which vanish at $g \in G_{\mathbb{A}}$ unless the vth component g_v is in K'_v for every $v \in S$. Gelfand's criterion [2.4.5] and the p-adic and archimedean Cartan decompositions [3.2] show that $\mathcal{H}$ is *commutative.*

[7.3.3] Corollary: The space $L^2_o(Z^+G_k\backslash G_{\mathbb{A}})^{K'}$ of right K'-invariant cuspforms has an orthonormal basis of simultaneous eigenfunctions for the spherical Hecke algebra $\mathcal{H}$ attached to K', with each eigenspace finite-dimensional. The simultaneous eigenfunctions are *smooth. (Proof follows.)*

[7.3.4] Corollary: The space $L^2_o(Z^+G_k\backslash G_{\mathbb{A}}/K_{\mathbb{A}})$ of right $K_{\mathbb{A}}$-invariant square-integrable cuspforms decomposes into simultaneous eigenspaces for operators in the maximal spherical Hecke algebra $C^\infty_c(K_{\mathbb{A}}\backslash G_{\mathbb{A}}/K_{\mathbb{A}})$, with finite multiplicities. The simultaneous eigenfunctions are *smooth. (Proof follows.)*

[7.3.5] Corollary: There is an orthonormal Hilbert-space basis for the space of $K_{\mathbb{A}}$-invariant square-integrable cuspforms consisting of simultaneous eigenfunctions for the invariant Laplacians on the archimedean factors G_v. *(Proof follows.)*

As in the previous section, the discussion needs a slightly refined version of *Siegel set.* Let Φ^o be the collection of *positive simple roots* (composed with norms), namely, the characters on diagonal matrices given by

$$\alpha_j \begin{pmatrix} m_1 & & \\ & \ddots & \\ & & m_r \end{pmatrix} = \left|\frac{m_j}{m_{j+1}}\right|$$

for $1 \le j \le r-1$, with idele norm. Let $B = P^{\min} = P^{1,1,\ldots,1}$ be the minimal parabolic. Put

$$M^1 = \{\begin{pmatrix} m_1 & & \\ & \ddots & \\ & & m_r \end{pmatrix} \in M^B_{\mathbb{A}} : |m_j| = 1, \ \text{for all } j\}$$

and

$$A^+ = \{a_t = \begin{pmatrix} t_1 & & \\ & \ddots & \\ & & t_r \end{pmatrix} : \text{all } t_j > 0,$$

$$\text{diagonally in archimedean } \prod_{v|\infty} M_v^B\}$$

Thus, for $m' \in M^1$, $\alpha_j(m'a_t) = (t_j/t_{j+1})^{r_o}$ where r_o is the number of isomorphism classes of archimedean completions of k. For $\tau > 0$, let

$$A_\tau = \{a \in A^+ : \alpha(a) \geq \tau, \text{ for all } \alpha \in \Phi^o\}$$

For $\tau > 0$, compact $C_N \subset N_{\mathbb{A}}^B$, compact $C_M \subset M^1$, and compact subgroup $K_{\mathbb{A}}$ of $G_{\mathbb{A}}$, the corresponding *Siegel set* attached to the minimal parabolic B is

$$\mathfrak{S} = \mathfrak{S}(C_N, C_M, \tau) = C_N \cdot C_M \cdot A_\tau \cdot K_{\mathbb{A}}$$

The compactness [2.A] of $\mathbb{J}^1/k^\times$ and reduction theory [3.3] showed that for sufficiently large compact C_N and C_M and sufficiently small $\tau > 0$ the corresponding Siegel set *surjects* to $Z^+G_k\backslash G_{\mathbb{A}}$.

In the following, unadorned N, M will be $N = N^B$ and $M = M^B$, and the unipotent radical and standard Levi components for other parabolics P will be N^P and M^P.

[7.3.6] Claim: Fix a Siegel set $\mathfrak{S} = \mathfrak{S}(C_N, C_M, \tau)$ with compact $C_N \subset N_{\mathbb{A}}$, compact $C_M \subset M^1$, and $\tau > 0$. Write $x \in \mathfrak{S}$ as $x = n_x m_x k_x$ with $n_x \in C_N$, $m_x = m'_x a_t$ with $m'_x \in C_M$ and $a_t \in A_\tau$, and $k_x \in K_{\mathbb{A}}$. Then there is a compact subset C' of $N_{\mathbb{A}}$ such that $x \in m_x \cdot C' \cdot K_{\mathbb{A}}$.

Proof: Rewrite the Iwasawa decomposition of $x \in \mathfrak{S}$ as

$$x = n_x \cdot m_x \cdot k_x \in C_N \cdot m_x \cdot K_{\mathbb{A}} = m_x \cdot m_x^{-1} C_N m_x \cdot K_{\mathbb{A}}$$
$$\subset m_x \cdot a_t^{-1}\Big((m'_x)^{-1} C_N m'_x\Big) a_t \cdot K_{\mathbb{A}} \subset m_x \cdot a_t^{-1}\Big(C_M^{-1} C_N C_M\Big) a_t \cdot K_{\mathbb{A}}$$

Being a continuous image of a compact set, $D = C_M^{-1} C_N C_M$ is a compact subset of $N_{\mathbb{A}}$. Now we claim that because $a_t \in A_\tau$, the union of all conjugates $a_t^{-1} D a_t$ is contained in a *compact* set. Indeed, for $u = \{u_v\} \in N_{\mathbb{A}}$, since a_t is purely archimedean, conjugation by a_t does not alter the non-archimedean components u_v of u. At archimedean places v, for $i < j$, the ij^{th} entry of $a_t^{-1}(u_v)a_t$ is

$t_i^{-1}t_j$ times the ij^{th} entry of u_υ, and

$$\frac{t_j}{t_i} = \frac{t_j}{t_{j-1}} \cdot \frac{t_{j-1}}{t_{j-2}} \cdot \ldots \cdot \frac{t_{i+1}}{t_i}$$

$$= \Big(\chi_i(a_t)\chi_{i+1}(a_t)\ldots\chi_{j-1}(a_t)\Big)^{-1/r_o} \leq (\tau^{j-i-1})^{-1/r_o}$$

where r_o is the number of isomorphism classes of archimedean completions of k. Those entries are *bounded* on D, so the entries of the conjugate are *uniformly* bounded for all $a_t \in A_\tau$. Thus, $\bigcup_{a\in A_\tau} a^{-1}Da$ is contained in a *compact* subset C' of $N_{\mathbb{A}}$, as claimed. ///

For strictly upper-triangular square matrices x with entries in any field of characteristic zero, the series for the *matrix exponential* $e^x = \exp(x) = \sum_{\ell\geq 0} x^\ell/\ell!$ is *finite*. Thus, the Lie algebra $\mathfrak{n}^P$ of the unipotent radical N^P for any standard parabolic P has a purely algebraic sense:

$$\mathfrak{n}^P = \{n\text{-by-}n\ x : \exp(x) \in N^P\}$$

For example, for the minimal parabolic $B = P^{\min}$, the Lie algebra $\mathfrak{n} = \mathfrak{n}^B$ is all upper-triangular matrices with zeros on the diagonal. For maximal proper $P = P^{i,r-i}$,

$$\mathfrak{n}^P = \{\begin{pmatrix} 0 & u \\ 0 & 0 \end{pmatrix}\} \qquad \text{(where } u \text{ is } i\text{-by-}(r-i))$$

In the latter case, because N^P is abelian, the exponential map is an *isomorphism* $\exp : \mathfrak{n}^P \longrightarrow N^P$. For all parabolics, the discrete *additive* subgroup $\mathfrak{n}_k^P \subset \mathfrak{n}_{\mathbb{A}}^P$ exponentiates to N_k^P. For test function $\varphi \in C_c^\infty(G_{\mathbb{A}})$, we can *wind up* the integral for $f \to \varphi \cdot f$ along the unipotent radical N_k of the standard minimal parabolic B: for $y \in G_{\mathbb{A}}$,

$$(\varphi \cdot f)(y) = \int_{G_{\mathbb{A}}} \varphi(x)\, f(yx)\, dx = \int_{G_{\mathbb{A}}} \varphi(y^{-1}x)\, f(x)\, dx$$

$$= \int_{N_k\backslash G_{\mathbb{A}}} \left(\sum_{\gamma\in\exp(\mathfrak{n}_k^P)} \varphi(y^{-1}\gamma x)\right) f(x)\, dx$$

$$= \int_{N_k\backslash G_{\mathbb{A}}} \left(\sum_{\nu\in\mathfrak{n}_k} \varphi(y^{-1}\cdot\exp(\nu)\cdot x)\right) f(x)\, dx$$

The *kernel function* for this wound-up form of the operator is the latter left-N_k-invariant inner sum:

$$K_\varphi(x,y) = \sum_{\nu\in\mathfrak{n}_k} \varphi(y^{-1}\cdot\exp(\nu)\cdot x)$$

As just above, write B-Iwasawa decompositions for $x \in \mathfrak{S}$ as $x = n_x m_x k_x$ with $n_x \in C_N$, $m_x = m'_x a_t$ with $m'_x \in C_M$ and $a_t \in A_\tau$, and $k_x \in K_{\mathbb{A}}$.

[7.3.7] Claim: For a fixed Siegel set $\mathfrak{S}$, fixed compact $E \subset C_c^\infty(G)$, there is compact $C'_M \subset M_{\mathbb{A}}$ such that if there exist $n \in N_{\mathbb{A}}$ and $\varphi \in E$ with $\varphi(y^{-1} \cdot n \cdot x) \neq 0$ for any $x, y \in \mathfrak{S}$, then $m_x \in m_y \cdot C'_M$. That is, $K_\varphi(x, y) = 0$ for all $x, y \in \mathfrak{S}$ and all $\varphi \in E$ *unless* $m_x \in m_y \cdot C'_M$.

Proof: From the previous claim, there is compact $D \subset N_{\mathbb{A}}$ such that $(m_y)^{-1} y \in D \cdot K_{\mathbb{A}}$. A compact set of test functions has a common compact support C_G because a compact set is *bounded* in the topological vector space sense, and a bounded subset of an LF-space such as $C_c^\infty(G)$ lies in some Fréchet limitand, by [13.8.5]. Nonvanishing of $\varphi(y^{-1}nx)$ implies $y^{-1}nx \in C_G$, so

$$nx \in y \cdot C_G \subset m_y \cdot D \cdot K_{\mathbb{A}} \cdot C_G \subset m_y \cdot C'_G$$

with $C'_G = DK_{\mathbb{A}}C_G =$ compact. That is,

$$C'_G \ni m_y^{-1} \cdot nx = m_y^{-1} \cdot nn_x \cdot m_x \cdot k_x = (m_y^{-1} nn_x m_y) \cdot m_y^{-1} m_x \cdot k_x$$

That is,

$$(m_y^{-1} nn_x m_y) \cdot m_y^{-1} m_x \in C'_G \cdot K_{\mathbb{A}} = \text{compact}$$

Since M normalizes N, the element $m_y^{-1} nn_x m_y$ is in N. Since $N_{\mathbb{A}} \cap M_{\mathbb{A}} = \{1\}$, the multiplication map $N_{\mathbb{A}} \times M_{\mathbb{A}} \to B_{\mathbb{A}}$ is a homeomorphism. Thus, for the product $(m_y^{-1})nn_x m_y) \cdot m_y^{-1} m_x$ to lie in a compact set in $G_{\mathbb{A}}$ requires that its N-component lies in a compact set in $N_{\mathbb{A}}$ and its M-component lies in a compact set in $M_{\mathbb{A}}$. Thus, there is compact $C'_M \subset M_{\mathbb{A}}$ such that $m_y^{-1} m_x \in C'_M$, as claimed. ///

[7.3.8] Corollary: For fixed Siegel set $\mathfrak{S}$ and fixed compact $E \subset C_c^\infty(G)$, there is a compact $C'_M \subset M_{\mathbb{A}}$ such that, if $\varphi(y^{-1}nx) \neq 0$ for some $x, y \in \mathfrak{S}$, some $n \in N_{\mathbb{A}}$ and some $\varphi \in E$, then $m_y^{-1} x \in C'_M$.

Proof: By [7.1.5], there is a compact C_G in G such that $x \in m_x \cdot C_G$. By [7.1.6], there is a compact C'_M in $M_{\mathbb{A}}$ such that $m_x \in m_y C'_M$. Thus, $x \in m_x C_G \subset m_y C'_M C'_G$, rearranging to give the claim. ///

As in the previous section, the notion of Schwartz function on an archimedean vectorspace such as $\mathfrak{n}_\infty = \oplus_{v|\infty} \mathfrak{n}_v$ is as in [13.7], and on non-archimedean vectorspaces $\mathfrak{n}_v$, Schwartz functions are the same as test functions, namely, locally constant, compactly supported. Similarly, on the finite-adeles part of an adelic vectorspace, Schwartz functions are simply test functions, that is, locally constant, compactly supported. Then Schwartz functions on adelic vector spaces $\mathfrak{n}_{\mathbb{A}}$ are finite sums $\sum_i f_{\infty,i} \otimes f_{o,i}$ where the functions $f_{\infty,i}$ are Schwartz functions on the archimedean part, and the functions $f_{o,i}$ are Schwartz/test functions on the non-archimedean part. Topologies on such spaces are as in [6.2] and [6.3] and as simpler examples in [13.7], [13.8], and [13.9].

[7.3.9] Claim: With $\omega_y = y^{-1} m_y$ and $\omega_{x,y} = m_y^{-1} x$, the functions $\nu \longrightarrow \varphi_{x,y}(\nu) = \varphi(\omega_y \cdot \exp(\nu) \cdot \omega_{x,y})$ for x, y in fixed Siegel set, $\varphi(y^{-1}nx) \neq 0, \varphi \in E$, constitute a *compact* subset of the Schwartz space $\mathscr{S}(\mathfrak{n}_{\mathbb{A}})$.

Proof: The left and right translation actions $G_{\mathbb{A}} \times G_{\mathbb{A}} \times C_c^\infty(G_{\mathbb{A}}) \to C_c^\infty(G_{\mathbb{A}})$ are continuous, by [6.4]. With fixed Siegel set $\mathfrak{S}$, by [7.2.6], [7.2.7], and [7.1.7], $\{\omega_y : y \in \mathfrak{S}\}$ and $\{\omega_{x,y} : x \in \mathfrak{S},\ y \in \mathfrak{S}\}$ are compact. This gives compactness of the image of

$$\{\omega_y : y\} \times \{\omega_{x,y} : x, y\} \times E \longrightarrow C_c^\infty(G_{\mathbb{A}})$$

x, y in fixed Siegel set, $\varphi(y^{-1}nx) \neq 0$. Since $N_{\mathbb{A}}$ is *closed* in $G_{\mathbb{A}}$, the restriction map $C_c^\infty(G_{\mathbb{A}}) \to C_c^\infty(N_{\mathbb{A}}) \approx C_c^\infty(\mathfrak{n}_{\mathbb{A}})$ is continuous. We are fortunate that test functions are characterized by compact support together with purely local smoothness properties, so that $C_c^\infty(N_{\mathbb{A}}) \approx C_c^\infty(\mathfrak{n}_{\mathbb{A}})$. From [13.9.3], the inclusion $C_c^\infty(\mathfrak{n}_{\mathbb{A}}) \subset \mathscr{S}(\mathfrak{n}_{\mathbb{A}})$ of test functions to Schwartz functions is continuous, giving compactness of the image. ///

Poisson summation for the lattice $\mathfrak{n}_k \subset \mathfrak{n}_{\mathbb{A}}$ gives

$$\sum_{\nu \in \mathfrak{n}_k} \varphi(y^{-1} \cdot \exp(\nu) \cdot x) = \sum_{\psi \in (\mathfrak{n}_k)^*} \int_{\mathfrak{n}_{\mathbb{A}}} \overline{\psi}(\nu)\, \varphi(y^{-1} \cdot \exp(\nu) \cdot x)\, d\nu$$

with suitably normalized measure, where $(\mathfrak{n}_k)^*$ is the collection of $\mathbb{C}^\times$-valued characters on $\mathfrak{n}_{\mathbb{A}}$ trivial on the lattice $\mathfrak{n}_k$. As in Appendix [7.A], we can identify the dual $(\mathfrak{n}_{\mathbb{A}})^*$ with $\mathfrak{n}_{\mathbb{A}}$ and the dual lattice $(\mathfrak{n}_k)^*$ with $\mathfrak{n}_k$. One reasonable identification is as follows. Make an $\mathbb{A}$-valued pairing on $\mathfrak{n}_{\mathbb{A}}$ by $\langle \nu, \xi \rangle = \sum_{i<j} \nu_{ij}\, \xi_{ij}$, and for fixed nontrivial character ψ_1 of $\mathbb{A}/k$ put

$$\psi_\xi(\nu) = \psi_1\big(\langle \nu, \xi \rangle\big) \qquad (\text{with } \xi, \nu \text{ in } \mathfrak{n}_{\mathbb{A}})$$

Thus,

$$\sum_{\nu \in \mathfrak{n}_k} \varphi(y^{-1} \cdot \exp(\nu) \cdot x) = \sum_{\xi \in k} \int_{\mathfrak{n}_{\mathbb{A}}} \overline{\psi}_\xi(\nu)\, \varphi(y^{-1} \cdot \exp(\nu) \cdot x)\, d\nu$$

We have

$$\int_{\mathfrak{n}_{\mathbb{A}}} \overline{\psi}_\xi(\nu)\cdot\varphi\left(y^{-1}\cdot\exp(\nu)\cdot x\right)\,d\nu$$

$$=\int_{\mathfrak{n}_{\mathbb{A}}} \overline{\psi}_\xi(\nu)\cdot\varphi\left(y^{-1}m_y\cdot\exp(m_y^{-1})\nu m_y)\cdot m_y^{-1}x\right)\,d\nu$$

Replacing ν by $m_y\nu m_y^{-1}$ and letting $\varphi_{x,y}(\nu)=\varphi\left(y^{-1}m_y\cdot\nu\cdot m_y^{-1}x\right)$,

$$K_\varphi(x,y)=\delta_B(m_y)\sum_{\xi\in k}\widehat{\varphi}_{x,y}(\psi_\xi^{m_y})$$

where $\widehat{\varphi}_{x,y}(\psi_\xi^{m_y})$ is the Fourier transform $\widehat{\varphi}_{x,y}$ of $\varphi_{x,y}$ along $\mathfrak{n}_{\mathbb{A}}$, evaluated at ψ_ξ, where $\psi_\xi^{m_y}(\nu)=\psi_\xi(m_y\nu(m_y)^{-1})$ and where δ_B is the modular function of $B_{\mathbb{A}}$.

[7.3.10] Theorem: For y in a fixed Siegel set attached to the minimal parabolic B, and for φ in a fixed compact $E\subset C_c^\infty(G_{\mathbb{A}})$, for every $q>0$ there is an implied constant depending only on such that, for every L^2 cuspform f,

$$|(\varphi\cdot f)(y)|\ll_q\left(\inf_{\alpha\in\Phi^o}\alpha(m_y)\right)^{-q}\cdot|f|_{L^2}$$

Proof: We need several preliminary results:

[7.3.11] Claim: Let P be a standard maximal proper parabolic. For every character ψ of $\mathfrak{n}_{\mathbb{A}}$ trivial on $\mathfrak{n}_{\mathbb{A}}^P$, the Fourier component $\widehat{\varphi}_{x,y}(\psi)$ is left $N_{\mathbb{A}}^P$-invariant in x and therefore integrates to 0 against every cuspform.

Proof: For $n\in N_{\mathbb{A}}^P$, replace x by nx in the original integral defining $\widehat{\varphi}_{x,y}(\psi)$, obtaining

$$\widehat{\varphi}_{nx,y}(\psi_0)=\int_{\mathfrak{n}_{\mathbb{A}}^P}\overline{\psi}_0(\nu)\cdot\varphi\left(y^{-1}\exp(\nu)\cdot nx\right)d\nu$$

$$=\int_{\mathfrak{n}_{\mathbb{A}}^P}\overline{\psi}_0(\nu)\cdot\varphi\left(y^{-1}\exp(\nu+\nu')\cdot x\right)d\nu$$

where ν' is a continuous function of ν determined by the obvious $\exp(\nu+\nu')=\exp(\nu)\cdot n$. That ν' is in the subalgebra $\mathfrak{n}^P$ rather than merely in $\mathfrak{n}$ follows from a computation in block decompositions of the appropriate size:

$$\exp(\nu)\cdot n=\exp\begin{pmatrix}\nu_{11}&\nu_{12}\\0&\nu_{22}\end{pmatrix}\cdot\begin{pmatrix}1&v\\0&1\end{pmatrix}=\begin{pmatrix}e^{\nu_{11}}&b\\0&e^{\nu_{22}}\end{pmatrix}\cdot\begin{pmatrix}1&v\\0&1\end{pmatrix}$$

$$=\begin{pmatrix}e^{\nu_{11}}&e^{\nu_{11}}\cdot v+b\\0&e^{\nu_{22}}\end{pmatrix}$$

for some block b. This is still of the form $\exp\begin{pmatrix} \nu_{11} & \nu'_{12} \\ 0 & \nu_{22} \end{pmatrix}$ for some ν'_{12} depending on ν and n, and we take

$$\nu' = \begin{pmatrix} 0 & \nu'_{12} \\ 0 & 0 \end{pmatrix} \in \mathfrak{n}_\alpha \qquad \text{(in suitable blocks)}$$

Replacing ν by $\nu - \nu'$ in the integral gives

$$\int_{\mathfrak{n}_{\mathbb{A}}} \overline{\psi}_0(\nu) \cdot \varphi(y^{-1} \exp(\nu) \cdot nx)\, d\nu = \psi_0(\nu') \cdot \widehat{\varphi}_{x,y}(\psi_0) = \widehat{\varphi}_{x,y}(\psi_0)$$

proving the left $N^P_{\mathbb{A}}$-invariance in x. The corresponding integral is

$$\begin{aligned}
&\int_{N_k \backslash G_{\mathbb{A}}} \widehat{\varphi}_{x,y}(\psi_0^{m_y}) \cdot f(x)\, dx \\
&= \int_{N_k N^P_{\mathbb{A}} \backslash G_{\mathbb{A}}} \int_{N^P_k \backslash N^P_{\mathbb{A}}} \widehat{\varphi}_{nx,y}(\psi_0^{m_y}) \cdot f(nx)\, dn\, dx \\
&= \int_{N_k N^P_{\mathbb{A}} \backslash G_{\mathbb{A}}} \widehat{\varphi}_{x,y}(\psi_0^{m_y}) \cdot \Big(\int_{N^P_k \backslash N^P_{\mathbb{A}}} f(nx)\, dn \Big)\, dx
\end{aligned}$$

and the inner integral is 0 because f is a cuspform. ///

Thus, for cuspforms f,

$$(\varphi \cdot f)(y) = \delta_B(m_y) \sum\nolimits^*_{\psi \in \mathfrak{n}_k} \int_{N_k \backslash G_{\mathbb{A}}} \widehat{\varphi}_{x,y}(\psi^{m_y})\, f(x)\, dx$$

where the $\sum^*$ is to mean that the sum omits $\psi \in \mathfrak{n}_k$ such that $\psi|_{\mathfrak{n}^P_{\mathbb{A}}} = 1$ identically for some maximal standard proper parabolic P.

Returning to the proof of theorem [7.3.10], Fourier transform is a continuous map of $\mathscr{S}(\mathfrak{n}_{\mathbb{A}})$ to itself, so Fourier transform maps the compact set of [7.3.9] to a compact set of Schwartz functions $\{\widehat{\varphi}_{x,y} : x, y \in \mathfrak{S},\ \varphi \in E\} \subset \mathscr{S}(\mathfrak{n}_{\mathbb{A}})$. The adelic Schwartz space $\mathscr{S}(\mathfrak{n}_{\mathbb{A}})$ is an LF-space, a strict colimit of Fréchet spaces, characterized as a countable ascending union of Fréchet subspaces described by restricting support at finite primes and by requiring *uniform* local constancy at finite primes. That is, for U (large) compact open (additive) subgroup of the finite-adele part $\mathfrak{n}_{\mathbb{A}_{\text{fin}}}$ of $\mathfrak{n}_{\mathbb{A}}$, and for H a (small) compact open (additive) subgroup of $\mathfrak{n}_{\mathbb{A}_{\text{fin}}}$, let $\mathscr{S}(\mathfrak{n}_\infty \times U)^H$ be the space of H-invariant Schwartz functions supported on $\mathfrak{n}_{\mathbb{A}_\infty} \times U$. Then

$$\mathscr{S}(\mathfrak{n}_{\mathbb{A}}) = \bigcup_{H,U} \mathscr{S}(\mathfrak{n}_\infty \times U)^H = \operatorname{colim}_{H,U} \mathscr{S}(\mathfrak{n}_\infty \times U)^H$$

as H shrinks and U grows. There is a countable cofinal collection of subgroups U and subgroups H, certifying the LF-space structure [13.8], [13.9]. In particular, a compact subset lies in some limitand $\mathscr{S}(\mathfrak{n}_\infty \times U)^H$, by [13.8.5]. Thus, the compactness [7.2.8] implies that all the Schwartz functions $\psi \to \widehat{\varphi}_{x,y}(\psi)$ lie in a compact subset of some $\mathscr{S}(\mathfrak{n}_\infty \times U)^H$.

Thus, for $\widehat{\varphi}_{x,y}(\psi_\xi) \neq 0$, the finite-prime part ξ_{fin} of ξ is in some compact $U \subset \mathfrak{n}_{\mathbb{A}_{\text{fin}}}$. Thus, $\xi \in \frac{1}{h}\mathfrak{n}_\mathfrak{o}$ for some $0 < h \in \mathbb{Z}\ell$, where $\mathfrak{n}_\mathfrak{o}$ is the collection of elements of $\mathfrak{n}_k$ with entries in the ring of algebraic integers $\mathfrak{o}$ of k. The collection of infinite-prime parts ξ_∞ of such ξ is a *lattice* $\Lambda^* \subset \mathfrak{n}_\infty$. Give the finite-dimensional $\mathbb{R}$-vectorspace $\mathfrak{n}_\infty$ a positive-definite inner product

$$\langle \xi, \xi' \rangle_\mathfrak{n} = \sum_{v|\infty} \mathrm{Re}\left((\xi_v)_{ij} \cdot \overline{(\xi'_v)_{ij}}\right)$$

with complex conjugation to accommodate complex k_v. Let $|\cdot|_\mathfrak{n}$ be the associated norm, and write $|\psi_\xi|_\mathfrak{n} = |\xi_\infty|_\mathfrak{n}$. There is the lower bound $|\xi_\infty|_\mathfrak{n} \geq h$ for $0 \neq \xi_\infty \in \Lambda^*$. In these terms, for a fixed compact subset $E' \subset \mathscr{S}(\mathfrak{n}_\infty \times U)^H$, for each $\ell > 0$, there is an *uniform* implied constant depending on ℓ, not on $\varphi' \in E'$, such that

$$|\varphi'(\psi_\xi)|_\mathfrak{n} \ll_\ell (1 + |\psi_\xi|_\mathfrak{n})^{-\ell} = (1 + |\xi_\infty|_\mathfrak{n})^{-\ell}$$

for all $\xi \in \mathfrak{n}_k$, for all $\varphi' \in E'$.

[7.3.12] Lemma: For fixed $\mathfrak{S}$ and Λ^*, there is a uniform implied constant such that, for every $\psi_\xi \in \Lambda^*$ not vanishing identically on any $\mathfrak{n}^P_\mathbb{A}$, for every $y = n_y m_y k_y \in \mathfrak{S}$, and for every $\alpha \in \Phi^o$,

$$\alpha(y) \ll |\psi_\xi^{m_y}|_\mathfrak{n}$$

Proof: As earlier, for given Λ^*, there is a lower bound $b > 0$ such that for all $\xi_\infty \in \Lambda^*$, $|\xi_{ij}| \geq b$ for all indices ij with $\xi_{ij} \neq 0$. Given $\alpha = \alpha_i \in \Phi^o$, take $P = P^{i-1,r-i+1}$. The nonzero entries of elements of $\mathfrak{n}^P$ are at $i'j'$ with $i' \leq i$ and $j' \geq i+1$. Thus, the condition that ψ_ξ restricted to $n^P_\mathbb{A}$ is not identically 1 requires that there are indices $i' \leq i$ and $j' \geq i+1$ such that the $i'j'^{\text{th}}$ component $\xi_{i'j'}$ is nonzero. With such $i'j'$,

$$|\psi_\xi^{m_y}|_\mathfrak{n} = |m_y \xi m_y^{-1}|_\mathfrak{n} \geq |(m_y \xi m_y^{-1})_{i'j'}| = |(m_y)_{i'} \cdot \xi_{i'j'} \cdot (m_y)_{j'}^{-1})|$$

where the latter two norms are on the real vector space k_∞. Since $m_y = m'_y a_t$ with m'_y in the compact C_M, there is a uniform implied constant, independent of i, i', j', ξ, and y, such that

$$|(m_y)_{i'} \cdot \xi_{i'j'} \cdot (m_y)_{j'}^{-1})| \gg |(a_t)_{i'} \cdot \xi_{i'j'} \cdot (a_t)_{j'}^{-1})| = \frac{t_{i'}}{t_{j'}} \cdot |\xi_{i'j'}| \geq \frac{t_{i'}}{t_{j'}} \cdot b$$

Every character $a_t \to t_{i'}/t_{j'}$ with $i' < j'$ is a product of nonnegative powers of the simple positive characters $a_t \to t_\ell/t_{\ell+1}$ for $1 \le \ell < r$:

$$\frac{t_{i'}}{t_{j'}} = \prod_{i' \le \ell < j'} \frac{t_\ell}{t_{\ell+1}}$$

For $i' \le i < j'$, the exponent of α_i in such an expression is 1. Thus, for $y = n_y m'_y a_t k_y \in \mathfrak{S} = C_N C_M A_\tau K_{\mathbb{A}}$,

$$\frac{t_{i'}}{t_{j'}} \ge \frac{t_i}{t_{i+1}} \cdot \prod_{i' \le \ell < j',\, \ell \ne i} \frac{t_\ell}{t_{\ell+1}} \ge \frac{t_i}{t_{i+1}} \cdot \prod_{i' \le \ell < j',\, \ell \ne i} \tau$$

$$\ge \frac{t_i}{t_{i+1}} \cdot \tau^r \gg_{\mathfrak{S}} \frac{t_i}{t_{i+1}}$$

Thus,

$$|\psi_\xi^{m_y}|_{\mathfrak{n}} \gg \frac{t_i}{t_{i+1}} = \alpha_i(m_y) \qquad (\text{for } y \in \mathfrak{S})$$

as was claimed. ///

[7.3.13] Corollary: For fixed $\mathfrak{S}$ and Λ^*, there is a uniform implied constant such that, for every $\psi_\xi \in \Lambda^*$ not vanishing identically on any $\mathfrak{n}^P_{\mathbb{A}}$, for every $y = n_y m_y k_y \in \mathfrak{S}$,

$$\delta_B(m_y) \ll |\psi_\xi^{m_y}|^{r(r-1)/2}$$

Proof: δ_B is the product of all characters $a_t \to t_i/t_j$ with $i < j$. Apply the lemma. ///

Thus, for any $q, \ell > 0$, for any $\alpha \in \Phi^o$, the part of the kernel $K_\varphi(x, y)$ that interacts with cuspforms has an estimate

$$\delta_B(m_y) \sum\nolimits^*_{\psi \in \mathfrak{n}_k} \widehat{\varphi}_{x,y}(\psi^{m_y}) \ll_\ell \sum\nolimits^*_{\psi \in \mathfrak{n}_k} |\psi_\xi^{m_y}|^{r(r-1)/2} \cdot (1 + |\psi_\xi|)^{-\ell}$$

$$\ll \sum\nolimits^*_{\psi \in \mathfrak{n}_k} \alpha(m_y)^{-q} |\psi_\xi^{m_y}|^{r(r-1)/2+q-\ell}$$

$$= \alpha(m_y)^{-q} \sum\nolimits^*_{\psi \in \mathfrak{n}_k} |\psi_\xi^{m_y}|^{r(r-1)/2+q-\ell}$$

For given q, for ℓ sufficiently large, the sum $\sum^*_\psi |\psi|^{r(r-1)/2+q-\ell}$ is convergent, so

$$\delta_B(m_y) \sum\nolimits^*_{\psi \in \mathfrak{n}_k} \widehat{\varphi}_{x,y}(\psi^{m_y}) \ll_q \alpha(m_y)^{-q} \qquad (\text{for every } \alpha \in \Phi^o)$$

From this estimate, for cuspform f,

$$|(\varphi \cdot f)(y)| \ll_q \Big(\inf_{\alpha \in \Phi^o} \alpha(m_y)\Big)^{-q} \cdot \int_{Z^+ G_k \backslash G_k \mathfrak{S}} |f(x)|\, dx$$

and by Cauchy-Schwarz-Bunyakowsky

$$\begin{aligned}
&\int_{Z^+G_k\backslash G_k\mathfrak{S}} |f(x)|\,dx \\
&\leq \left(\int_{Z^+G_k\backslash G_k\mathfrak{S}} 1\,dx\right)^{1/2} \cdot \left(\int_{Z^+G_k\backslash G_k\mathfrak{S}} |f(x)|^2\,dx\right)^{1/2} \\
&\leq \text{meas}\,(Z^+G_k\backslash G_\mathbb{A}) \cdot |f|_{L^2}
\end{aligned}$$

That is, at last, for y in a fixed Siegel set,

$$|(\varphi \cdot f)(y)| \ll_q \left(\inf_{\alpha\in\Phi^o} \alpha(m_y)\right)^{-q} \cdot |f|_{L^2}$$

This is the decay property of $\varphi \cdot f$ asserted in theorem [7.3.10]. ///

The remainder of the arguments for theorem [7.3.1] and corollaries [7.3.2]–[7.3.5] is essentially identical to that for theorem [7.2.1] and corollaries [7.2.2] through [7.2.5]. We review the points of the argument.

As earlier, for general reasons, we have

[7.3.14] Lemma: Let $\mathfrak{g}_v$ be the Lie algebra of G_v for archimedean v. For $X \in \mathfrak{g}_v$, the left-derivative map

$$C_c^\infty(G_\mathbb{A}) \longrightarrow C_c^\infty(G_\mathbb{A})$$

by

$$\varphi \longrightarrow \left(g \to \frac{d}{dt}\bigg|_{t=0} \varphi(e^{-tX}\,g)\right)$$

is *continuous. (Proof as [7.1.12].)* ///

[7.3.15] Corollary: For a compact set E of test functions on $G_\mathbb{A}$, for a compact $C_\mathfrak{g}$ in $\mathfrak{g} = \mathfrak{g}_v$, and for f ranging over cuspforms in the unit ball in $L^2(Z^+G_k\backslash G_\mathbb{A})$, there is a *uniform* implied constant such that

$$\left|\frac{d}{dt}\bigg|_{t=0} (\varphi \cdot f)(g\,e^{tX})\right| \ll 1$$

for all $g \in G_\mathbb{A}$, for all $\varphi \in E$, for all $X \in C_\mathfrak{g}$. *(Proof as [7.2.14].)* ///

The smoothing property of $f \to \varphi \cdot f$ as in [14.5] assures that each $\varphi \cdot f$ is smooth. Smoothness of φ at finite places is *uniform*, since that of φ is: with φ

left-invariant by compact open subgroup $K' \subset G_{\mathbb{A}_{\text{fin}}}$, for $h \in K'$,

$$(\varphi \cdot f)(g \cdot h) = \int_G \varphi(x)\, f(ghx)\, dx = \int_G \varphi(h^{-1}x)\, f(gx)\, dx$$
$$= \int_G \varphi(x)\, f(gx)\, dx = (\varphi \cdot f)(g)$$

A uniform bound on derivatives at archimedean places will imply *uniform* continuity:

[7.3.16] Lemma: Let F be a smooth function on $G_{\mathbb{A}}$, (uniformly) right K'-invariant for some compact open subgroup $K' \subset G_{\mathbb{A}_{\text{fin}}}$, with a uniform pointwise bound on all $X \cdot F$ with X in a *compact* neighborhood $C_{\mathfrak{g}}$ of 0 in $\mathfrak{g}_\infty$, namely,

$$|(X \cdot F)(x)| \le B \qquad (\text{for all } x \in G_{\mathbb{A}}, \text{ all } X \in C_{\mathfrak{g}})$$

Then F is *uniformly continuous*: for every $\varepsilon > 0$ there is a neighborhood U of 1 in $G_{\mathbb{A}}$ such that $|F(x) - F(y)| < \varepsilon$ for all $x \in G_{\mathbb{A}}$ and $y \in xU$. *(Proof as [7.2.15].)* ///

[7.3.17] Corollary: For a compact set E of test functions on $G_{\mathbb{A}}$, and for f ranging over cuspforms in the unit ball in $L^2(Z^+G_k\backslash G_{\mathbb{A}})$, the family of images $\varphi \cdot f$ is *(uniformly) equicontinuous* on G. ///

Again, a compactness lemma reminiscent of Arzela-Ascoli:

[7.3.18] Lemma: Let E be a *equicontinuous*, uniformly bounded, set of functions on $Z^+G_k\backslash G_{\mathbb{A}}$. Then E has *compact closure* in $L^2(Z^+G_k\backslash G_{\mathbb{A}})$. *(Proof as [7.1.16].)*

Finally, we prove the theorem [7.3.1]. To summarize: the asymptotics of the kernels prove pointwise boundedness of the image of the unit ball B of $L^2_o(Z^+G_k\backslash G_{\mathbb{A}})$, and consideration of derivatives proves equicontinuity of the image of B. The faux-Arzela-Ascoli compactness lemma proves compactness of the closure of $\{\varphi \cdot B : \varphi \in E\}$. Being integrated versions of *right* translations, these operators stabilize the subspace of cuspforms, as the latter is defined by a *left* integral condition. Thus, φ maps the unit ball to a set with compact closure, so is a compact operator.

As in earlier examples, by direct computation, the adjoint of $f \to \varphi \cdot f$ is $f \to \varphi^\vee \cdot f$, where $\varphi^\vee(x) = \overline{\varphi}(x^{-1})$. The space of test functions is stable under the operation $\varphi \to \varphi^\vee$.

Again, the general *nondegeneracy* result is [14.1.5], finishing the proof of [7.3.1]. ///

The proof of corollary [7.3.2], that $L^2_o(Z^+G_k\backslash G_{\mathbb{A}})$ decomposes as (the closure of) a direct sum of irreducible representations of $C_c^\infty(G_{\mathbb{A}})$, each occurring with finite multiplicity, is identical to the proof of [7.2.2]. ///

The proofs of [7.3.3] and [7.3.4] are special cases, where the algebra $\mathcal{H}$ of compact operators is designed to be *commutative*, so the notion of *irreducibles* simplifies to *simultaneous eigenspace*. Gelfand's criterion [2.4.5] and the p-adic and archimedean Cartan decompositions from [3.2] show that this Hecke algebra $\mathcal{H}$ is *commutative*. For nonzero f in a simultaneous eigenspace V_χ for $\mathcal{H}$, by nondegeneracy there is a test function φ such that $\varphi \cdot f = \chi(\varphi) \cdot f$ and $\chi(\varphi) \neq 0$. With $\eta = \varphi/\chi(\phi)$, then $\eta \cdot f = f$. By smoothing, as in [14.5], for example, f is smooth.

Just as in the proof of [7.1.3], the fact that the Casimir operators Ω_v on archimedean G_v commute with left and right translation implies that Ω_v commutes with the action of $C_c^\infty(G_{\mathbb{A}})$, by integrating. On right K_v-invariant functions, Ω_v is the invariant Laplacian Δ_v. Thus, each Δ_v stabilizes the simultaneous eigenspaces V_χ of $\mathcal{H}$, all of which are finite-dimensional, consisting of smooth functions. The restriction of Δ_v to V_χ is still *symmetric*, so by finite-dimensional spectral theory V_χ has a basis of Δ_v-eigenfunctions. ///

[7.3.19] Corollary: The space of L^2 cuspforms has an orthonormal basis of cuspforms f such that there is a test function $\varphi \in C_c^\infty(G_{\mathbb{A}})$ such that $\varphi \cdot f = f$. Such a cuspform f is *smooth* and of *rapid decay* in the sense that, given a standard Siegel set $\mathfrak{S}$ and $q > 0$,

$$|f(g)| \ll_q \left(\inf_{\alpha \in \Phi^o} \alpha(m_g) \right)^{-q}$$

Proof: The proof is the same as that of [7.2.19]: From above, the irreducibles modules over $\mathcal{H} = C_c^\infty(G_{\mathbb{A}})$ appearing in the space of L^2 cuspforms are finite-dimensional, each occurring with finite multiplicity. Let f be in a copy V of an irreducible module for $C_c^\infty(G_{\mathbb{A}})$.

Since V is irreducible, it has no proper, topologically closed $\mathcal{H}$-stable subspace. Since V is finite-dimensional, all vector subspaces are closed. Thus, $\mathcal{H} \cdot f = V$. In particular, there is a test function φ such that $\varphi \cdot f = f$. Then [7.3.10] applies to $f = \varphi \cdot f$. Smoothness follows as in [14.5] and [14.6]. ///

7.A Appendix: Dualities

For an *abelian* topological group G and $\mathbb{T}$ the unit circle in $\mathbb{C}$, the *unitary dual* of G is

$$\widehat{G} = \mathrm{Hom}^o(G, S^1) = \{\text{continuous group homomorphisms } G \to \mathbb{T}\}$$

Pointwise multiplication makes $\widehat{G}$ an abelian group. A reasonable topology[1] on $\widehat{G}$ is the *compact-open* topology, with a subbasis of opens

$$U = U_{C,E} = \{f \in \widehat{G} \, : \, f(C) \subset E\}$$

for compact C in G, open E in $\mathbb{T}$. From [7.A.4], the compact-open topology makes $\widehat{G}$ a abelian (locally compact, Hausdorff) topological group:

[7.A.1] Claim: The unitary dual of a *compact* abelian group is *discrete*. The unitary dual of a *discrete* abelian group is *compact*.

Proof: Let G be compact. Let E be a small-enough open in $\mathbb{T}$ so that E contains no nontrivial subgroups of G. Noting that G itself is *compact*, let $U \subset \widehat{G}$ be the open

$$U = \{f \in \widehat{G} \, : \, f(G) \subset E\}$$

Since E is *small*, $f(G) = \{1\}$. That is, f is the trivial homomorphism. This proves discreteness of $\widehat{G}$. For G discrete, *every* group homomorphism to $\mathbb{T}$ is continuous. The space of *all* functions $G \to \mathbb{T}$ is the cartesian product of copies of $\mathbb{T}$ indexed by G. By Tychonoff's theorem, with the product topology, this product is *compact*. Indeed, for *discrete* X, the compact-open topology on the space $C^o(X, Y)$ of continuous functions from $X \to Y$ *is* the product topology on copies of Y indexed by X. The subset of functions f satisfying the group homomorphism condition

$$f(gh) = f(g) \cdot f(h) \qquad (\text{for } g, h \in G)$$

is *closed*, since the group multiplication $f(g) \times f(h) \to f(g) \cdot f(h)$ in $\mathbb{T}$ is continuous. Since the product is also *Hausdorff*, $\widehat{G}$ is also compact. ///

[7.A.2] Claim: Local fields k_v are self-dual, as are the adeles of a number field k: $\mathbb{A}^\vee \approx \mathbb{A}$.

Proof: For *compact totally disconnected* G, since $\mathbb{C}^\times$ contains no *small subgroups* [2.4.3], every element of $G^\vee$ has image in roots of unity in $\mathbb{C}^\times$, which

[1] The reasonableness of the compact-open topology is in its function. First, on a compact topological space X, the space $C^o(X)$ of continuous $\mathbb{C}$-valued functions with the *sup norm* (of absolute value) is a *Banach space*. On noncompact X, the semi-norms given by sups of absolute values on compacts make $C^o(X)$ a *Fréchet space*. The compact-open topology accommodates spaces of continuous functions $C^o(X, Y)$ where the target space Y is not a subset of a normed real or complex vector space and is most interesting when Y is a *topological group*. In the latter case, when the source X is also a topological group, the subset of all continuous functions $f : X \to Y$ consisting of *group homomorphisms* is a (locally compact, Hausdorff) topological group, as proven subsequently.

can be identified with $\mathbb{Q}/\mathbb{Z}$. Thus, for compact totally disconnected G,

$$G^\vee \approx \mathrm{Hom}^o(G, \mathbb{Q}/\mathbb{Z}) \qquad \text{(continuous homomorphisms)}$$

where $\mathbb{Q}/\mathbb{Z} = \mathrm{colim}\, \frac{1}{N}\mathbb{Z}/\mathbb{Z}$ is *discrete*. As a topological group, $\mathbb{Z}_p = \lim \mathbb{Z}/p^\ell\mathbb{Z}$. It is also useful to observe that $\mathbb{Z}_p$ is a limit of the corresponding quotients of itself, namely,

$$\mathbb{Z}_p \approx \lim \mathbb{Z}_p/p^\ell\mathbb{Z}_p$$

Indeed, more generally, every abelian *totally disconnected* topological group G has the property that

$$G \approx \lim_K G/K$$

where K ranges over compact open subgroups of G. Also, as a topological group,

$$\mathbb{Q}_p = \bigcup \frac{1}{p^\ell}\mathbb{Z}_p = \mathrm{colim}\, \frac{1}{p^\ell}\mathbb{Z}_p$$

Because of the *no small subgroups* property [2.4.3] of the unit circle in $\mathbb{C}^\times$, every continuous element of $\mathbb{Z}_p^\vee$ factors through *some* limitand

$$\mathbb{Z}_p/p^\ell\mathbb{Z}_p \approx \mathbb{Z}/p^\ell\mathbb{Z}$$

Thus,

$$\mathbb{Z}_p^\vee = \mathrm{colim}\left(\mathbb{Z}_p/p^\ell\mathbb{Z}_p\right)^\vee = \mathrm{colim}\, \frac{1}{p^\ell}\mathbb{Z}_p/\mathbb{Z}_p$$

since $\frac{1}{p^\ell}\mathbb{Z}_p/\mathbb{Z}_p$ is the dual to $\mathbb{Z}_p/p^\ell\mathbb{Z}_p$ under the pairing

$$\frac{1}{p^\ell}\mathbb{Z}_p/\mathbb{Z}_p \times \mathbb{Z}_p/p^\ell\mathbb{Z}_p \approx \frac{1}{p^\ell}\mathbb{Z}/\mathbb{Z} \times \mathbb{Z}/p^\ell\mathbb{Z} \ni \left(\frac{x}{p^\ell} + \mathbb{Z}\right) \times \left(y + p^\ell\mathbb{Z}\right)$$

$$\longrightarrow xy + \mathbb{Z} \in \mathbb{Q}/\mathbb{Z}$$

The transition maps in the colimit expression for $\mathbb{Z}_p^\vee$ are inclusions, so

$$\mathbb{Z}_p^\vee = \mathrm{colim}\, \frac{1}{p^\ell}\mathbb{Z}_p/\mathbb{Z}_p \approx \left(\mathrm{colim}\, \frac{1}{p^\ell}\mathbb{Z}_p\right)/\mathbb{Z}_p \approx \mathbb{Q}_p/\mathbb{Z}_p$$

Thus,

$$\mathbb{Q}_p^\vee = \left(\mathrm{colim}\, \frac{1}{p^\ell}\mathbb{Z}_p\right)^\vee = \lim(\frac{1}{p^\ell}\mathbb{Z}_p^\vee$$

As a topological group, $\frac{1}{p^\ell}\mathbb{Z}_p \approx \mathbb{Z}_p$ by multiplying by p^ℓ, so the dual of $\frac{1}{p^\ell}\mathbb{Z}_p$ is isomorphic to $\mathbb{Z}_p^\vee \approx \mathbb{Q}_p/\mathbb{Z}_p$. However, the inclusions for varying ℓ are not the identity map, so for compatibility take

$$\left(\frac{1}{p^\ell}\mathbb{Z}_p\right)^\vee = \mathbb{Q}_p/p^\ell\mathbb{Z}_p$$

Thus,

$$\mathbb{Q}_p^\vee = \lim \mathbb{Q}_p/p^\ell\mathbb{Z}_p \approx \mathbb{Q}_p$$

because, again, any abelian totally disconnected group is the projective limit of its quotients by compact open subgroups. The same argument applies to $\widehat{\mathbb{Z}} = \lim \mathbb{Z}/N\mathbb{Z}$ and finite adeles $\mathbb{A}_{\text{fin}} = \text{colim}\, \frac{1}{N}\widehat{\mathbb{Z}}$, proving the self-duality of $\mathbb{A}_{\text{fin}}$. Fourier inversion asserts the self-duality of $\mathbb{R}$ and $\mathbb{C}$, giving the self-duality of $\mathbb{A}$. The same argument applies over an arbitrary finite extension k_v of $\mathbb{Q}_p$, but now the pairing is composed with the local *trace* from k_v to $\mathbb{Q}_p$ and the dual lattice to the local integers $\mathfrak{o}_v$ is (by definition) the *inverse different.*

[7.A.3] Claim: The unitary dual $(\mathbb{A}/k)^\frown$ of the compact quotient $\mathbb{A}/k$ is isomorphic to k. In particular, given any nontrivial character ψ on $\mathbb{A}/k$, *all* characters on $\mathbb{A}/k$ are of the form $x \to \psi(\alpha \cdot x)$ for some $\alpha \in k$.

Proof: Because $\mathbb{A}/k$ is compact, $(\mathbb{A}/k)^\frown$ is *discrete.* Since multiplication by elements of k respects cosets $x + k$ in $\mathbb{A}/k$, the unitary dual has a k-vectorspace structure given by

$$(\alpha \cdot \psi)(x) = \psi(\alpha \cdot x) \qquad (\text{for } \alpha \in k, x \in \mathbb{A}/k)$$

There is no topological issue in this k-vectorspace structure, because $(\mathbb{A}/k)^\frown$ is discrete. The quotient map $\mathbb{A} \to \mathbb{A}/k$ gives a natural *injection* $(\mathbb{A}/k)^\frown \to \widehat{\mathbb{A}}$.

Given nontrivial $\psi \in (\mathbb{A}/k)^\frown$, the k-vectorspace $k \cdot \psi$ inside $(\mathbb{A}/k)^\frown$ injects to a copy of $k \cdot \psi$ inside $\widehat{\mathbb{A}} \approx \mathbb{A}$. *Assuming* for a moment that the image in $\mathbb{A}$ is essentially the same as the diagonal copy of k, the quotient $(\mathbb{A}/k)^\frown/k$ injects to the compact $\mathbb{A}/k$. The topology of $(\mathbb{A}/k)^\frown$ is discrete, and the quotient $(\mathbb{A}/k)^\frown/k$ is still discrete. Because all these maps are continuous group homomorphisms, the image of $(\mathbb{A}/k)^\frown/k$ in $\mathbb{A}/k$ is a discrete subgroup of a compact group, so is *finite*. Since $(\mathbb{A}/k)^\frown$ is a k-vectorspace, the quotient $(\mathbb{A}/k)^\frown/k$ must be a singleton. This proves that $(\mathbb{A}/k)^\frown \approx k$, granting that the image of $k \cdot \psi$ in $\mathbb{A} \approx \widehat{\mathbb{A}}$ is the usual diagonal copy.

To see how $k \cdot \psi$ is imbedded in $\mathbb{A} \approx \widehat{\mathbb{A}}$, fix nontrivial ψ on $\mathbb{A}/k$, and let ψ be the induced character on $\mathbb{A}$. The self-duality of $\mathbb{A}$ is that the action of $\mathbb{A}$ on $\widehat{\mathbb{A}}$ by $(x \cdot \psi)(y) = \psi(xy)$ gives an *isomorphism*. The subgroup $x \cdot \psi$ with $x \in k$ is certainly the usual diagonal copy.

For completeness, we prove

[7.A.4] Claim: The unitary dual $\widehat{G}$ of an abelian (locally compact, Hausdorff) topological group is an abelian (locally compact, Hausdorff) topological group.

[7.A.5] Remark: We do not prove the *local compactness* in general. The important special cases, that the dual of discrete is compact and vice versa, give the local compactness of the duals in those cases.

Proof: That the unitary dual is *abelian* is immediate because the multiplication is pointwise by values, and the target group $\mathbb{T}$ is abelian. First, verify that the topology is *invariant*. That is, given a subbasis open

$$U(C, E) = \{f \in \widehat{G} \; : \; f(c) \in E, \text{ for all } c \in C\}$$

with C compact in G, E open in $\mathbb{T}$. and given $f_o \in \widehat{G}$, show that $f_o \cdot U(C, E)$ is open. This is not completely trivial, as $f_o \cdot U(C, E)$ is not obviously of the form $U(C', E')$:

$$f_o \cdot U(C, E) = \{f \in \widehat{G} \; : \; f(c) \in f_o(c) \cdot E, \text{ for all } c \in C\}$$

To show that $f_o \cdot U(C, E)$ is open, we show that every point is contained in a finite intersection of the basic opens, with that intersection contained in $f_o \cdot U(C, E)$.

Fix $f \in f_o \cdot U(C, E)$. Since $f_o^{-1}(c)f(c) \in E$, each $c \in C$ has a neighborhood N_c such that $f_o^{-1}(N_c) \cdot f(N_c) \subset E$. Shrink each N_c to have compact closure $\overline{N}_c$, and so that $f_o^{-1}(\overline{N}_c) \cdot f(\overline{N}_c) \subset E$. By compactness of C, it has a finite subcover $N_i = N_{c_i}$. Thus,

$$f(\overline{N}_i) \subset f_o(c') \cdot E \qquad \text{(for all } i \text{, for all } c' \in \overline{N}_i)$$

From the result of the following subsection, an intersection of a *compact* family of opens is open, so

$$E_i = \bigcap_{c' \in \overline{N}_i} f_o(c') \cdot E = \text{open}$$

This open E_i is nonempty, since it contains $f(\overline{N}_i)$. Thus,

$$f \in \bigcap_i U(\overline{N}_i, E_i) \qquad \text{(a finite intersection)}$$

On the other hand, with c_i and $\overline{N}_i$ determined by f, take

$$f' \in \bigcap_i U(\overline{N}_i, E_i)$$

Then

$$f'(\overline{N}_i) \subset f_o(c) \cdot E \qquad \text{(for all } c \in \overline{N}_i)$$

In particular,

$$f'(c) \in f_o(c) \cdot E \qquad (\text{for all } c \in \overline{N}_i)$$

Since the sets $\overline{N}_i$ cover C, we have $f' \in f_o \cdot U(C, E)$. That is,

$$\bigcap_i U(\overline{N}_i, E_i) \subset f_o \cdot U(C, E)$$

This proves that the translate $f_o \cdot U(C, E)$ is open, in the compact-open topology. That is, the compact-open topology is translation-invariant.

Now we prove the fact needed earlier, that *compact intersections of opens are open*, in the following sense. Let H be a topological group, Hausdorff, but not necessarily locally compact. We claim that

$$\bigcap_{k \in K} k \cdot U = \text{open} \qquad (\text{for } U \subset H \text{ open, and } K \subset H \text{ compact})$$

For $u \in k \cdot U$ for all k, by the continuity of inversion and the group operation, there are neighborhoods U_k of u and V_k of k such that

$$V_k^{-1} \cdot U_k \subset U$$

Let $V_i = V_{k_i}$ be a finite subcover of K, and put $U_i = U_{k_i}$. Thus, for $k \in V_i$,

$$k^{-1} \cdot U_i \subset U \qquad (\text{for } k \in V_i)$$

Thus,

$$k^{-1} \cdot \bigcap_i U_i \subset U \qquad (\text{for all } k \in K)$$

Since *finite* intersections of opens are open, the intersection of the U_i, each containing u, is an open neighborhood of u. That is, the intersection of the translates $k \cdot U$ is open. This proves the claim.

Next, show that the pointwise multiplication operation

$$(f_1 \cdot f_2)(x) = f_1(x) \cdot f_2(x) \qquad (\text{for } f_i \in \widehat{G} \text{ and } x \in G)$$

in $\widehat{G}$ is continuous in the compact-open topology. Given a subbasis neighborhood $U(C, E)$ of $f_1 \cdot f_2$, the already-demonstrated invariance of the topology implies that $(f_1 f_2)^{-1} U(C, E)$ is open and is a neighborhood of the trivial character. Thus, without loss of generality, take $f_1 = f$ and $f_2 = f^{-1}$. Given a subbasic neighborhood $U(C, E)$ of the trivial character in $\widehat{G}$, show that there are neighborhoods U_1 of f and U_2 of f^{-1} such that $U_1 \cdot U_2 \subset U(C, E)$. For $U(C, E)$ to be a neighborhood of the trivial character means exactly that $1 \in E$ (and $C \neq \phi$).

Let E' be an open neighborhood of 1 such that $E' \cdot E' \subset E$. For

$$f' \in \big(f \cdot U(C,E')\big) \cdot \big(f^{-1} \cdot U(C,E')\big)$$

we have

$$f'(c) \in \big(f(c) \cdot E'\big) \cdot \big(f^{-1}(c) \cdot E'\big) = E' \cdot E' \subset E \qquad \text{(for all } c \in C\text{)}$$

That is,

$$\big(f \cdot U(C,E')\big) \cdot \big(f^{-1} \cdot U(C,E')\big) \subset U(C,E)$$

This proves continuity of multiplication. Continuity of inversion is similar.

Finally, Hausdorffness: take $f_1 \neq f_2$ in $\widehat{G}$. For some $g \in G$, $f_1(g) \neq f_2(g)$. Since the target $\mathbb{T}$ is Hausdorff, there are opens $E_1 \ni f_1(g)$ and $E_2 \ni f_2(g)$ with $E_1 \cap E_2 = \phi$. Since the source G is Hausdorff, the singleton $\{g\}$ is compact. Thus, $f_i \in U(\{g\}, E_i)$, and these opens are disjoint. This completes the discussion of the unitary dual. ///

7.B Appendix: Compact Quotients $\Gamma\backslash G$

Here we see the enormous simplification in the argument for discrete decomposition when $\Gamma\backslash G$ is *compact*, so that Gelfand's cuspform condition is vacuously met. Let G be any unimodular topological group, and Γ a discrete subgroup so that $\Gamma\backslash G$ is *compact*. As in [6.1] and [6.2], G acts continuously on $L^2(\Gamma\backslash G)$ by right translation, in fact by unitary operators, and $\eta \in C_c^o(G)$ acts continuously by the integral operators

$$(\eta \cdot f)(x) = \int_G \eta(g)\, f(xg)\, dg$$

[7.B.1] Theorem: $C_c^o(G)$ acts on $L^2(\Gamma\backslash G)$ by Hilbert-Schmidt (hence, compact) operators. The collection of such operators is closed under adjoints and is *nondegenerate* in the sense that for every $f \in L^2(\Gamma\backslash G)$, there is $\eta \in C_c^\infty(K\backslash G/K)$ such that $\eta \cdot f \neq 0$.

Proof: Just as in the more complicated arguments concerning cuspforms, first rearrange:

$$(\eta \cdot f)(x) = \int_G \eta(x^{-1}g)\, f(g)\, dg = \int_{\Gamma\backslash G} \sum_{\gamma\in\Gamma} \eta(x^{-1}\gamma g)\, f(\gamma g)\, dg$$

$$= \int_{\Gamma\backslash G} \sum_{\gamma\in\Gamma} \eta(x^{-1}\gamma g)\, f(g)\, dg = \int_{\Gamma\backslash G} \Big(\sum_{\gamma\in\Gamma} \eta(x^{-1}\gamma g)\Big) \cdot f(g)\, dg$$

Thus, with Schwartz kernel $K_\eta(x, y) = \sum_{\gamma\in\Gamma} \eta(x^{-1}\gamma y)$ (see [Schwartz 1950]),

$$(\eta \cdot f)(x) = \int_{\Gamma\backslash G} K_\eta(x, y)\, f(y)\, dy$$

Since Γ is discrete in G, for x, y in a fixed compact subset of G, the sum for $K_\eta(x, y)$ is finite, so K_η is continuous on $\Gamma\backslash G \times \Gamma\backslash G$. Unlike the more general situation, since $\Gamma\backslash G$ is *compact*, $K_\eta \in L^2(\Gamma\backslash G \times \Gamma\backslash G)$. Thus, $f \to \eta \cdot f$ is *Hilbert-Schmidt* (see [9.A]), and therefore a compact operator.

The adjoint of $f \to \eta \cdot f$ is easily expressed by changing variables:

$$\begin{aligned}\langle \eta f, F\rangle_{L^2(\Gamma\backslash G)} &= \int_{\Gamma\backslash G}\int_G \eta(g)\, f(xg)\,\overline{F(x)}\, dg\, dx \\ &= \int_G\int_{\Gamma\backslash G} \eta(g)\, f(xg)\,\overline{F(x)}\, dx\, dg \\ &= \int_G\int_{\Gamma\backslash G} f(x)\,\overline{\overline{\eta}(g)\, F(xg^{-1})}\, dx\, dg \\ &= \langle f, \check{\eta} \cdot F\rangle_{L^2(\Gamma\backslash G)} \qquad (\text{where } \check{\eta}(g) = \overline{\eta(g^{-1})})\end{aligned}$$

The nondegeneracy is [14.1.5]. ///

A *representation* of G on a topological vector space V is a continuous map $G \times V \to V$ that sends $g \in G$ to (continuous) linear maps on V. The representation V is *irreducible* when there are no (topologically) closed G-stable subspaces except $\{0\}$ and V itself. Homomorphisms $\varphi : V \to W$ of G-representations are continuous linear maps commuting with the action of G: $\varphi(g \cdot v) = g \cdot \varphi(v)$ for all $g \in G$ and $v \in V$. The *multiplicity* of an irreducible V in another representation W of G is $\dim_{\mathbb{C}} \mathrm{Hom}_G(V, W)$, as elaborated in [9.D.14]. The same terminology applies to the integral-operator action of $C_c^o(G)$ on a G-representation.

[7.B.2] Corollary: $L^2(\Gamma\backslash G)$ is (the completion of) an orthogonal direct sum of irreducible $C_c^o(G)$-subrepresentations, each occurring with finite multiplicity.

Proof: Given the theorem, this is mostly just [7.2.18]. ///

[7.B.3] Corollary: $L^2(\Gamma\backslash G)$ is (the completion of) an orthogonal direct sum of irreducible unitary representations of G, each occurring with finite multiplicity.

Proof: By [14.1.6] and [14.1.7], $C_c^o(G)$-subrepresentations of the G-representation $L^2(\Gamma\backslash G)$ are G-subrepresentations, and $C_c^o(G)$ irreducibility implies G-irreducibility. ///

[7.B.4] Remark: Discrete subgroups Γ of $G = SL_2(\mathbb{R})$ with compact quotient $\Gamma\backslash G$ can be obtained in several ways. A purely analytical device is the *uniformization theorem*, asserting that every compact, connected Riemann surface of genus ≥ 2 is such a quotient. More number-theoretic examples are obtained by taking quaternion division algebras B over $\mathbb{Q}$ (that is, $\mathbb{Q}$-four-dimensional simple division algebras with $\mathbb{Q}$ in the center) *split* over $\mathbb{R}$, that is, so that $B \otimes_{\mathbb{Q}} \mathbb{R}$ is isomorphic to the 2-by-2 matrix algebra $M_2(\mathbb{R})$. For a maximal finite-$\mathbb{Z}$-module-rank subring $\mathfrak{o}$ of B, imbed $\mathfrak{o}^\times$ into $GL_2(\mathbb{R})$, and let $\Gamma = \mathfrak{o}^\times \cap SL_2(\mathbb{R})$. Then $\Gamma\backslash G$ is compact, by an argument resembling that in [2.A] for the compactness of $\mathbb{J}^1/k^\times$. Similarly, for a number field k and quaternion division algebra B over k, with $B \otimes_k k_v \approx M_2(\mathbb{R})$ for exactly one archimedean place v of k, and $B \otimes_k k_{v'} \approx \mathbb{H}$ for all other archimedean places, for maximal subring $\mathfrak{o}$, let Γ be the intersection of $SL_2(\mathbb{R})$ with the projection of $\mathfrak{o}^\times$ to $GL_2(k_v) \approx GL_2(\mathbb{R})$. Then $\Gamma\backslash SL_2(\mathbb{R})$ is compact. The corresponding compact quotients $\Gamma\backslash\mathfrak{H}$ are *Shimura curves*.

8

Moderate Growth Functions, Theory of the Constant Term

From [7.1.20], [7.2.20], and [7.3.19], there is an orthonormal basis for cuspforms f such that there are test functions φ with $\varphi \cdot f = f$, and these cuspforms are *of rapid decay* in Siegel sets. This is a special case of the idea that the asymptotic behavior of *moderate growth* automorphic forms f in Siegel sets is dominated by their constant terms, under the hypothesis that there is a test function φ such that $\varphi \cdot f = f$. Eisenstein series are of moderate growth, even after meromorphic continuation, and under reasonable hypotheses on the data used to form them, meet the condition $\varphi \cdot f = f$ for suitable test function. Thus, although Eisenstein series are not in L^2, they admit good asymptotic approximations by their constant terms.

The underlying mechanism for the results of this chapter is essentially the fundamental theorem of calculus. Thus, these results are essentially archimedean. Thus, the second section indicates how to reduce the $GL_2(\mathbb{A})$ example to the four simple examples, and the third section treats only the simplest purely archimedean version of GL_r.

8.1 The Four Small Examples

First, consider the four small examples from Chapter 1, with $G, \Gamma, P, M, N, A^+, K$ as there. For this section, write Iwasawa decompositions as $x = n_x a_x k_x$ with $a_x \in A^+$. The *height* function η is

$$\eta(nak) = \eta(a) = \delta_P(a) = t^{2r}$$

for $n \in N$, $a = \begin{pmatrix} t & 0 \\ 0 & 1/t \end{pmatrix} \in A^+$, and $k \in K$, where $r = 1, 2, 3, 4$ in the respective examples, and δ_P is the modular function. A left $N \cap \Gamma$-invariant function $\mathbb{C}$-valued f on G or on G/K is *of moderate growth of exponent* $\lambda \in \mathbb{R}$ on a fixed standard Siegel set $\mathfrak{S}$ when

$$|f(x)| \ll_{\mathfrak{S}} \eta(x)^\lambda \qquad (\text{for } x \in \mathfrak{S})$$

A left $N \cap \Gamma$-invariant function $\mathbb{C}$-valued f on G or G/K is *of moderate growth of exponent* λ (on standard Siegel sets) when it is of moderate growth of exponent λ on *every* standard Siegel set. Say f is *of moderate growth* if it is of moderate growth of exponent λ for *some* λ. Since constant terms

$$c_P f(x) = \int_{(\Gamma \cap N)\backslash N} f(nx)\, dn$$

of automorphic forms f are merely $N \cap \Gamma$-invariant, the notion of moderate growth needs to be more broadly applicable than just to Γ-invariant functions. A left $N \cap \Gamma$-invariant function $\mathbb{C}$-valued f on G or G/K is *of rapid decay* on standard Siegel sets when

$$|f(x)| \ll_{\lambda,\mathfrak{S}} \eta(x)^\lambda \qquad (\text{for } x \in \mathfrak{S}, \text{ for every } \mathfrak{S}, \text{ for every } \lambda \in \mathbb{R})$$

We can see directly that, for f of moderate growth of exponent λ, the constant term $c_P f$ is also of moderate growth of exponent λ, and for f of rapid decay, the constant term $c_P f$ is also of rapid decay: on a fixed standard Siegel set $\mathfrak{S}$,

$$|c_P f(x)| \le \int_{(N\cap\Gamma)\backslash N} |f(nx)|\, dn \ll_{\mathfrak{S}} \int_{(N\cap\Gamma)\backslash N} \eta(nx)^\lambda\, dn$$

$$= \int_{(N\cap\Gamma)\backslash N} \eta(x)^\lambda\, dn = \eta(x)^\lambda$$

For $f \in C^o(G)$ and for $\varphi \in \mathcal{D}(G)$, as usual

$$(\varphi \cdot f)(x) = \int_G \varphi(g)\, f(xg)\, dg \qquad (\text{for } \varphi \in \mathcal{D}(G))$$

This converges at least as a $\mathbb{C}$-valued integral, for each $x \in G$.

[8.1.1] Theorem: For f on $(N \cap \Gamma)\backslash G/K$ of moderate growth, if there is $\varphi \in \mathcal{D}(G)$ with $\varphi \cdot f = f$, then $f - c_P f$ is of rapid decay on Siegel sets.

Proof: The proof of the theorem is in several stages. First, the action of test functions does preserve moderate growth of a given exponent:

[8.1.2] Claim: Let f be a function on $(\Gamma \cap N)\backslash G$ of moderate growth of exponent λ in standard Siegel sets. Then, for every test function $\varphi \in \mathcal{D}(G)$, the function $\varphi \cdot f$ is also of moderate growth of exponent λ in standard Siegel sets.

Proof: Let compact $C \subset G$ contain the support of φ. Without loss of generality, replace C by $C \cdot K$ to make C right K-stable. Fix a Siegel set $\mathfrak{S} = C_N A_\tau K$ with compact $C_N \subset N$ and $A_\tau = \{a \in A^+ : \eta(a) \geq \tau\}$. For $x \in \mathfrak{S}$,

$$(\varphi \cdot f)(x) = \int_G f(xy)\,\varphi(y)\,dy = \int_G f(y)\,\varphi(x^{-1}y)\,dy$$
$$= \int_{xC} f(y)\,\varphi(x^{-1}y)\,dy$$

[8.1.3] Lemma: Given compact $C \subset G$, there is compact $C_A \subset A^+$ such that $a_y \in a_x C_A$ for $y \in xC$.

Proof: (of lemma) For right K-stable C, $C \subset (NA^+ \cap C) \cdot K$, since in Iwasawa coordinates $pk \in C$ with $p \in NA^+$ and $k \in K$ implies that $p = (pk) \cdot k^{-1}$ is also in C. Since $N \cdot A^+ \approx N \times A^+$ is a topological product, there are compacts $C'_N \subset N$, $C_A \subset A^+$ so that $K \cdot C \subset C'_N \cdot C_A \cdot K$. Then

$$xC \subset Na_xK \cdot C \subset Na_x \cdot C'_N C_A K \subset Na_x \cdot NC_A K$$
$$\subset N \cdot (a_x N a_x^{-1}) \cdot (a_x C_A) \cdot K \subset N \cdot (a_x C_A) \cdot K$$

That is, $a_y \in a_x C_A$ for $y \in xC$, as claimed. ///

Returning to the proof of the claim, for $x \in \mathfrak{S}$, for y in the support xC of the integral,

$$\eta(y) = \eta(a_y) \in \eta(a_x C_A) = \eta(a_x) \cdot \eta(C_A)$$

Let

$$\mu = \inf_{a \in C_A} \eta(a) \qquad\qquad \sigma = \sup_{a \in C_A} \eta(a)^\lambda$$

Take compact $C''_N \subset N$ large enough to surject to $(N \cap \Gamma)\backslash N$. Then, up to adjustment by $N \cap \Gamma$, for $x \in \mathfrak{S}$, $y \in xC$ implies that y is in $\mathfrak{S}' = C_{N''} A_{\mu \cdot \tau} K$. Invoking the moderate growth of f on $\mathfrak{S}'$,

$$|\varphi \cdot f(x)| \leq \sup|\varphi| \int_{xC} |f(y)|\,dy \ll_{f,\mathfrak{S}} \sup|\varphi| \cdot \int_{xC} \eta(y)^\lambda\,dy$$
$$\ll \sup|\varphi| \cdot \sigma \cdot \int_{xC} \eta(x)^\lambda\,dy$$
$$= \sup|\varphi| \cdot \sigma \cdot \eta(x)^\lambda \cdot \text{meas}\,(C) \ll_\varphi \eta(x)^\lambda \qquad (\text{for } x \in \mathfrak{S})$$

This argument applies to every standard Siegel set $\mathfrak{S}$, giving the moderate growth of f. ///

[8.1.4] Claim: For f of moderate growth of exponent λ, if $\varphi \cdot f = f$ for some $\varphi \in \mathcal{D}(G)$, then f is *smooth*, and is of *uniform* moderate growth of exponent λ, in the sense that for any L in the universal enveloping algebra $U\mathfrak{g}$ of the Lie algebra $\mathfrak{g}$ of G, the derivative Lf is of moderate growth with exponent λ on standard Siegel sets.

Proof: The image $\varphi \cdot f$ is smooth, by [14.5]. The point is that the left-G-invariant differential operators attached to X in the Lie algebra of G by

$$Xf(x) = \frac{\partial}{\partial s}\Big|_{s=0} f(x \cdot e^{sX})$$

can be absorbed into the action of varying $\varphi \in \mathcal{D}(G)$ on f:

$$\begin{aligned} X(\varphi \cdot f)(x) &= \frac{\partial}{\partial s}\Big|_{s=0} \int_G f(x \cdot e^{sX} g)\, \varphi(g)\, dg \\ &= \frac{\partial}{\partial s}\Big|_{s=0} \int_G f(g)\, \varphi(e^{-sX} x^{-1} g)\, dg \end{aligned}$$

by replacing g by $e^{-sX}x^{-1}g$. This is

$$\int_G f(x) \frac{\partial}{\partial s}\Big|_{s=0} \varphi(e^{-sX} x^{-1} g)\, dg = \int_G f(x) X^{\text{left}} \varphi(x^{-1} g)\, dg$$

where X^{left} is the (*right-G-invariant*) differential operator on the left naturally attached to X via the *left* translation action. The interchange of differentiation and integration is justified by Gelfand-Pettis [14.1], observing that the integral is compactly supported, continuous, and takes values in a quasi-complete locally convex topological vector space on which differentiation is a continuous linear map. Using $\varphi \cdot f = f$,

$$Xf(x) = X(\varphi \cdot f)(x) = ((X^{\text{left}} \varphi) \cdot f)(x)$$

which is of moderate growth of exponent λ, by the previous claim. By induction on the degree of the differential operator L, Lf is of moderate growth of exponent λ. ///

The key bootstrapping property is the following:

[8.1.5] Claim: For f smooth and left $(N \cap \Gamma)$-invariant, of uniform moderate growth of exponent λ on standard Siegel sets, on a standard Siegel set $\mathfrak{S}$

$$|(f - c_P f)(x)| \ll_{\mathfrak{S}} \eta(x)^{\lambda - 1}$$

Proof: For notational simplicity, we first carry out the argument for the example with $G = SL_2(\mathbb{R})$. Normalizing the measure of $(\Gamma \cap N)\backslash N$ to be 1,

$$(c_P f - f)(x) = \int_{(\Gamma\cap N)\backslash N} f(nx) - f(x)\, dn$$

$$= \int_{0\le t\le 1} f(e^{tX}\cdot x) - f(x)\, dt$$

where $X = \begin{pmatrix} 0 & 1 \\ 0 & 0 \end{pmatrix}$ in the Lie algebra of N. By the fundamental theorem of calculus,

$$f(e^{tX}\cdot x) - f(x) = \int_0^t \frac{\partial}{\partial u}\Big|_{u=0} f(e^{(u+s)X}\cdot x)\, ds$$

$$= \int_0^t f(e^{sX}\cdot x\cdot e^{ux^{-1}Xx})\, ds = \int_0^t (x^{-1}Xx\cdot f)(e^{sX}\cdot x)\, ds$$

Let $x = n_x a_x k_x$ with $n_x \in N$, $a_x \in A^+$, $k_x \in K$. Then

$$x^{-1}Xx = (k_x^{-1}a_x^{-1}n_x^{-1})X(n_x a_x k_x)$$

$$= (k_x^{-1}a_x^{-1})X(a_x k_x)$$

Further,

$$a_x^{-1}Xa_x = \eta(a_x)^{-1}\cdot X$$

Then

$$x^{-1}Xx = (k_x^{-1}\, a_x^{-1})X(k_x^{-1}\, a_x^{-1})$$

$$= \eta(a_x)^{-1}k_x^{-1}Xk_x = \eta(a_x)^{-1}\cdot \sum_i c_i(k_x)X_i$$

where the c_i are continuous functions (depending upon X) on K and $\{X_i\}$ is a basis for the Lie algebra of G. Since the c_i are continuous on the compact set K, they have a uniform bound c in absolute value. Altogether,

$$(c_P f - f)(x) = \int_{0\le t\le 1}\int_{0\le s\le t} \eta(a_x)^{-1}\cdot\left(-\sum_i c_i(k_x)X_i\right) f(e^{sX}\cdot x)\, ds\, dt$$

$$= \eta(a_x)^{-1}\cdot\sum_i c_i(k_x)\int_{0\le t\le 1}\int_{0\le s\le t} (-X_i f)(e^{sX}\cdot x)\, ds\, dt$$

$$= \eta(a_x)^{-1}\cdot\sum_i c_i(\theta_x)\int_{0\le t\le 1} (-X_i f)(e^{tX}\cdot x)\, dt$$

$$= \eta(a_x)^{-1}\cdot\sum_i c_i(\theta_x)\cdot c_P(-X_i f)(x)$$

so

$$\begin{aligned}|(c_P f - f)(x)| &\leq \eta(a_x)^{-1} \cdot \sum_i |c_i(\theta_x)| \cdot c_P|(-X_i f)(x)| \\ &\leq \eta(a_x)^{-1} \cdot \sum_i c \cdot c_P|(-X_i f)(x)| \\ &\ll_{\mathfrak{S}} \eta(a_x)^{-1} \cdot \sum_i c \cdot \eta(x)^{\lambda} \ll \eta(a_x)^{\lambda-1}\end{aligned}$$

as claimed, for $G = SL_2(\mathbb{R})$. For the other three small examples, since N is $r = 2, 3, 4$-dimensional, respectively, commensurately more differentiations are needed, but the pattern is the same:

$$\begin{aligned}(c_P f - f)(x) &= \int_{(\Gamma \cap N)\backslash N} f(nx) - f(x)\, dn \\ &= \int_{0 \leq t_1 \leq 1} \cdots \int_{0 \leq t_r \leq 1} f(e^{t_1 X_1 + \ldots + t_r X_r} \cdot x) - f(x)\, dt_1 \ldots dt_r\end{aligned}$$

where $X_1, \ldots, X_r$ is a suitable basis for the Lie algebra of N. By the fundamental theorem of calculus,

$$\begin{aligned}&f(e^{t_1 X_1 + \ldots + t_r X_r} \cdot x) - f(x) \\ &\quad = \int_0^{t_1} \cdots \int_0^{t_r} \frac{\partial}{\partial u_1}\bigg|_{u_1=0} \cdots \frac{\partial}{\partial u_r}\bigg|_{u_r=0} \\ &\qquad \times f(e^{(u_1+s_1)X_1 + \ldots + (u_r+s_r)X_r} \cdot x) - f(x)\, ds_1 \ldots ds_r \\ &\quad = \int_0^{t_1} \cdots \int_0^{t_r} f(e^{s_1 X_1 + \ldots + s_r X_r} \cdot x \cdot e^{u_1 x^{-1} X_1 x + \ldots + u_r x^{-1} X_r x})\, ds_1 \ldots ds_r \\ &\quad = \int_0^{t_1} \cdots \int_0^{t_r} (x^{-1} X_1 x \cdots x^{-1} X_r x \cdot f)(e^{s_1 X_1 + \ldots + s_r X_r} \cdot x)\, ds_1 \ldots ds_r\end{aligned}$$

As in the simpler version above,

$$\begin{aligned}x^{-1} X_j x &= (k_x^{-1}\, a_x^{-1}) X_j (k_x^{-1}\, a_x^{-1}) = \eta(a_x)^{-1/r} k_x^{-1} X_j k_x \\ &= \eta(a_x)^{-1/r} \sum_i c_{ij}(k_x) X_i \qquad \left(\text{with } \eta \begin{pmatrix} t & 0 \\ 0 & 1/t \end{pmatrix} = t^{2r}\right)\end{aligned}$$

for some continuous functions c_{ij} on K. Thus, again, these functions have a bound c, and

$$\left|\int_0^{t_1} \cdots \int_0^{t_r} (x^{-1}X_1x \cdots x^{-1}X_rx \cdot f)(e^{s_1X_1+\ldots+s_rX_r} \cdot x)\, ds_1 \ldots ds_r\right|$$

$$\ll \eta(x)^{-1} \sum_{i_1,\ldots,i_r} \int_0^{t_1} \cdots \int_0^{t_r} (X_{i_1} \cdots X_{i_r} f)(e^{s_1X_1+\ldots+s_rX_r} \cdot x)\, ds_1 \ldots ds_r$$

$$\ll \eta(x)^{-1} \sum_{i_1,\ldots,i_r} c_P \left|X_{i_1} \cdots X_{i_r} f(x)\right| \ll \eta(x)^{-1} \cdot \eta(x)^{\lambda}$$

giving the claim. ///

Now finish the proof of the theorem. Take $\varphi \cdot f = f$ on $(\Gamma \cap N)\backslash G$ of moderate growth of exponent λ in Siegel sets. Then αf is of moderate growth of exponent λ for all $\alpha \in U\mathfrak{g}$, and $f - c_P f$ is of moderate growth of exponent $\lambda - 1$. Then $\alpha(f - c_P f)$ is of moderate growth of exponent $\lambda - 1$ for all $\alpha \in U\mathfrak{g}$, and $(f - c_P f) - c_P(f - c_P f)$ is of moderate growth of exponent $(\lambda - 1) - 1$. But

$$(f - c_P f) - c_P(f - c_P f) = f - c_P f - c_P f + c_P c_P f$$
$$= f - c_P f - c_P f + c_P f = f - c_P f$$

By induction, $f - c_P f$ is of moderate growth of exponent $\lambda - \ell$ for every $\ell \in \mathbb{Z}$. ///

[8.1.6] Remark: In fact, the preceding arguments apply to f *eventually* having the property $f = \varphi \cdot f$, in the sense that this property holds in a region where $\eta(a_x)$ is sufficiently large.

8.2 $GL_2(\mathbb{A})$

The purely archimedean argument of the previous section applies to the archimedean local factors G_v of groups $GL_2(\mathbb{A})$. Again, the fundamental device is the fundamental theorem of calculus. For simplicity, we consider only trivial central characters.

A function f on $Z_{\mathbb{A}} P_k \backslash G_{\mathbb{A}}/K$ is *of moderate growth of exponent* λ on standard Siegel sets $\mathfrak{S}$ when

$$|f(nmk)| \ll_{f,\mathfrak{S}} |m_1/m_2|^{\lambda}$$

where $n \in N_{\mathbb{A}}$, $m = \begin{pmatrix} m_1 & 0 \\ 0 & m_2 \end{pmatrix} \in M_{\mathbb{A}}$, $k \in K_{\mathbb{A}}$. Such a function is *of rapid decay* (on Siegel sets) when it is of moderate growth of exponent λ for all $\lambda \in \mathbb{R}$, with implied constant allowed to depend on λ. We may suppress the phrase *on Siegel sets*, but this is implied throughout.

[8.2.1] Theorem: For f on $P_k\backslash G_\mathbb{A}/K$ of moderate growth, $K_\mathbb{A}$-finite, if there is $\varphi \in \mathcal{D}(G_\mathbb{A})$ with $\varphi \cdot f = f$, then $f - c_P f$ is of rapid decay.

Proof: The proof is completely parallel to that of the previous sections, so we merely outline it, highlighting differences and adaptations. First, the action of test functions preserves moderate growth of a given exponent:

[8.2.2] Claim: Let f be a function on $Z_\mathbb{A} P_k\backslash G$ of moderate growth of exponent λ in standard Siegel sets. Then, for every test function $\varphi \in \mathcal{D}(G)$, the function $\varphi \cdot f$ is also of moderate growth of exponent λ in standard Siegel sets. *(Same proof as [8.1.2].)* ///

[8.2.3] Claim: For f of moderate growth of exponent λ, if $\varphi \cdot f = f$ for some $\varphi \in \mathcal{D}(G)$, then f is *smooth*, and is of *uniform* moderate growth of exponent λ, in the sense that for any L in the universal enveloping algebra $U\mathfrak{g}$ of the Lie algebra $\mathfrak{g}$ of G, the derivative Lf is of moderate growth with exponent λ on standard Siegel sets. *(Same proof as [8.1.4].)* ///

Again, we have the bootstrapping property

[8.2.4] Claim: For f smooth and left $(N \cap \Gamma)$-invariant, of uniform moderate growth of exponent λ on standard Siegel sets, on a standard Siegel set $\mathfrak{S}$

$$|(f - c_P f)(nmk)| \ll_\mathfrak{S} |m_1/m_2|^{\lambda-1} \qquad \left(\text{with } m = \begin{pmatrix} m_1 & 0 \\ 0 & m_2 \end{pmatrix}\right)$$

(Same proof as [8.1.5], using the fundamental theorem of calculus.) ///

The proof of the theorem is finished up as follows. Take $\varphi \cdot f = f$ on $Z_\mathbb{A} G_k\backslash G_\mathbb{A}$ and of moderate growth of exponent λ in Siegel sets. Then αf is of moderate growth of exponent λ for all $\alpha \in U\mathfrak{g}$, and $f - c_P f$ is of moderate growth of exponent $\lambda - 1$. Then $\alpha(f - c_P f)$ is of moderate growth of exponent $\lambda - 1$ for all $\alpha \in U\mathfrak{g}$, and $(f - c_P f) - c_P(f - c_P f)$ is of moderate growth of exponent $(\lambda - 1) - 1$. But

$$\begin{aligned}(f - c_P f) - c_P(f - c_P f) &= f - c_P f - c_P f + c_P c_P f \\ &= f - c_P f - c_P f + c_P f = f - c_P f\end{aligned}$$

By induction, $f - c_P f$ is of moderate growth of exponent $\lambda - \ell$ for every $\ell \in \mathbb{Z}$. ///

8.3 $SL_3(\mathbb{Z}), SL_4(\mathbb{Z}), SL_5(\mathbb{Z}), \ldots$

For these larger examples, for simplicity the archimedean aspects are emphasized. The general case is a superposition of copies of the archimedean, as in the previous section. The proofs for $SL_r(\mathbb{Z})$ are in essence mild extensions of those of [6.1], repeated for intelligibility, with appropriate modifications.

Let $G = SL_r(\mathbb{R})$, $\Gamma = SL_r(\mathbb{Z})$, and

$$A = \{\begin{pmatrix} * & & \\ & \ddots & \\ & & * \end{pmatrix} \in SL_r(\mathbb{R})\}$$

and let A^+ be the connected component of the identity in A, namely, diagonal matrices with positive entries. Let B be the standard minimal parabolic (Borel subgroup) of upper triangular matrices, so A is its standard Levi component. Let N^B be the unipotent radical of B, namely, the upper-triangular unipotent matrices. With $K = SO_n(\mathbb{R})$, an Iwasawa decomposition of G is $G = N^B \cdot A^+ \cdot K$. The function $g \to a_g$ defined by expressing $g = na_gk$ with $n \in N^B$, $a_g \in A^+$, $k \in K$ is well-defined.

Let $\log : A^+ \to \mathfrak{a}$ be the inverse of the Lie exponential map $\alpha \to e^\alpha$ from the Lie algebra $\mathfrak{a}$ of A^+ to A^+ itself. For λ in the space of characters $\mathfrak{a}^*$ of $\mathfrak{a}$, write

$$a^\lambda = e^{\lambda(\log a)}$$

The *roots* of A^+ or $\mathfrak{a}$ on the Lie algebra $\mathfrak{g}$ of G are the characters λ such that the λ-eigenspace

$$\mathfrak{g}_\lambda = \{x \in \mathfrak{g} : axa^{-1} = a^\lambda \cdot x, \quad \text{for all } a \in A^+\}$$

is nonzero, and then the eigenspace is called the λ-*rootspace*. The nontrivial roots are

$$\chi_{ij}\begin{pmatrix} m_1 & & & \\ & m_2 & & \\ & & \ddots & \\ & & & m_n \end{pmatrix} = \frac{m_i}{m_j} \qquad (\text{for } i \neq j)$$

As in [3.3], [3.10], [3,12], for $G = SL_r$ the standard *simple* roots are $\chi_{i,i+1}$. A left $N^B \cap \Gamma$-invariant $\mathbb{C}$-valued function f on G is *of moderate growth of exponent* λ on a fixed standard Siegel set

$$\mathfrak{S} = \mathfrak{S}_t = \{x \in G : a_x^\alpha \geq t \ \text{ for all simple roots } \alpha\}$$

when $|f(g)| \ll_{\mathfrak{S}} a_g^\lambda$ for $g \in \mathfrak{S}$. Such f is *of rapid decay* on a standard Siegel set $\mathfrak{S}$ when $|f(g)| \ll_{\mathfrak{S},\lambda} a_g^\lambda$ for *all* characters $\lambda \in \mathfrak{a}^*$.

Recall that, for standard parabolic P with unipotent radical N^P, for left $(\Gamma \cap N^P)$-invariant f on G, the constant term of f along P is

$$c_P f(x) = \int_{(N^P \cap \Gamma)\backslash N^P} f(nx)\, dn$$

Left-invariance under a larger subgroup of P than just $\Gamma \cap N^P$ is inherited by the constant term, since N^P is *normal* in P. In particular, left $(\Gamma \cap N^B)$-invariance is inherited. For f left $\Gamma \cap N^B$-invariant, of moderate growth of exponent λ in standard Siegel sets, for maximal proper standard parabolic P, the constant term $c_P f$ is also of moderate growth of exponent λ in standard Siegel sets: normalizing the measure of $(N^P \cap \Gamma)\backslash N^P$ to be 1,

$$|c_P f(\boldsymbol{x})| \le \int_{(N^P \cap \Gamma)\backslash N^P} |f(nx)|\, dn \ll_{f,\mathfrak{S}} \int_{(N^P \cap \Gamma)\backslash N^P} a_{nx}^{\lambda}\, dn$$
$$= \int_{(N^P \cap \Gamma)\backslash N^P} a_x^{\lambda}\, dn = a_x^{\lambda}$$

Similarly, for f of rapid decay in standard Siegel sets, the constant term $c_P f$ is also of rapid decay.

Standard *maximal* proper parabolics P have the convenient feature that their unipotent radicals N^P are *abelian*. Further, the standard maximal proper parabolics are in bijection with simple roots, as follows. For two roots λ, μ, write $\lambda \ge \mu$ if $\lambda - \mu$ is a linear combination of simple roots with *nonnegative* coefficients. In $G = SL_r$, we have $\chi_{ij} \ge \chi_{i'j'}$ if and only if $i \le i'$ and $j \ge j'$. Then the maximal proper parabolic associated to a simple root α is specified by saying that the Lie algebra $\mathfrak{n}$ of its unipotent radical $N = N^P$ is the sum of root-spaces $\mathfrak{g}_\beta$ for $\beta \ge \alpha$. In terms of matrix entries, for $\alpha = \chi_{i,i+1}$, the parabolic has diagonal blocks of sizes $i \times i$ and $(r-i) \times (r-i)$, and the unipotent radical N consists of an $i \times (r-i)$ block in the upper right.

For $f \in C^o(G)$, write

$$(\varphi \cdot f)(x) = \int_G \varphi(g)\, f(xg)\, dg \qquad (\text{for } \varphi \in \mathcal{D}(G))$$

This converges at least as a $\mathbb{C}$-valued integral, for each $x \in G$.

[8.3.1] Theorem: Let f be a left $N^B \cap \Gamma$-invariant $\mathbb{C}$-valued function on G, of moderate growth on every standard Siegel set. Suppose that there is $\varphi \in \mathcal{D}(G)$ such that $\varphi \cdot f = f$. Then for each simple root α and associated standard maximal parabolic P, $f - c_P f$ is of rapid decay in the direction α, in the sense that $f - c_P f \ll_{\mathfrak{S}} a^{-\ell \cdot \alpha}$ for all $\ell \in \mathbb{Z}$.

[8.3.2] Corollary: For a left $N^B \cap \Gamma$-invariant $\mathbb{C}$-valued function f on G, of moderate growth on every standard Siegel set, with $\varphi \in \mathcal{D}(G)$ such that $\varphi \cdot f = f$, if $c_P f = 0$ for every standard maximal compact P, then f is of rapid decay in standard Siegel sets. ///

Proof: The proof of the theorem is in stages. First, show that the action of $\mathcal{D}(G)$ preserves moderate growth on standard Siegel sets:

[8.3.3] Claim: For any $\varphi \in \mathcal{D}(G)$, if f on $(N^B \cap \Gamma)\backslash G$ is of moderate growth of exponent λ on standard Siegel sets, then $\varphi \cdot f$ is of moderate growth of exponent λ on standard Siegel sets.

Proof (of claim): For a compact set C containing the support of

$$\varphi, \varphi \cdot f(x) = \int_G f(xg)\, \varphi(g)\, dg = \int_G f(g)\, \varphi(x^{-1}g)\, dg$$

$$= \int_{xC} f(g)\, \varphi(x^{-1}g)\, dg$$

The proof of the following is identical to the proof of [8.1.3]:

[8.3.4] Lemma: Let C be a compact set in G, $x \in G$. Then there is a compact subset C_A of A^+ such that $y \in xC$ implies $a_y \in a_x \cdot C_A$. ///

Take x in a standard Siegel set

$$\mathfrak{S} = \mathfrak{S}_t = \{x \in G : a_x^\alpha \geq t \text{ for all simple roots } \alpha\}$$

For y in the support xC of the integral, for simple root α,

$$a_y^\alpha =\in \{(a_x \cdot a)^\alpha : a \in C_A\} = a_x^\alpha \cdot \{a^\alpha : a \in C_A\}$$

Let

$$\mu = \inf_\alpha \inf_{a \in C_A} a^\alpha \qquad\qquad \sigma = \sup_{a \in C_A} a^\lambda$$

Take compact $C_N'' \subset N^B$ large enough to surject to $(N^B \cap \Gamma)\backslash N^B$. Then, up to adjustment by $N^B \cap \Gamma$, $x \in \mathfrak{S}$ and $y \in xC$ implies that y is in the Siegel set $\mathfrak{S}' = C_{N''} A_{\mu \cdot t} K$. Invoking the moderate growth of f on $\mathfrak{S}'$,

$$|\varphi \cdot f(x)| \leq \sup |\varphi| \int_{xC} |f(y)|\, dy \ll_{f,\mathfrak{S}} \sup |\varphi| \cdot \int_{xC} \eta(y)^\lambda\, dy$$

$$\ll \sup |\varphi| \cdot \sigma \cdot \int_{xC} \eta(x)^\lambda\, dy$$

$$= \sup |\varphi| \cdot \sigma \cdot \eta(x)^\lambda \cdot \text{meas}\,(C) \ll_\varphi a_x^\lambda \qquad (\text{for } x \in \mathfrak{S})$$

This argument applies for every standard Siegel set $\mathfrak{S}$, giving the moderate growth of f. ///

[8.3.5] Claim: For f of moderate growth of exponent λ, if $\varphi \cdot f = f$ for some $\varphi \in \mathcal{D}(G)$, then f is *smooth*, and is of *uniform* moderate growth of exponent λ, in the sense that for any L in the universal enveloping algebra $U\mathfrak{g}$ of the Lie algebra $\mathfrak{g}$ of G, the derivative Lf is of moderate growth with exponent λ on standard Siegel sets.

Proof: The image $\varphi \cdot f$ is smooth, by [14.5]. The key mechanism is that the left-G-invariant differential operators X acting on the right attached to the right regular representation of G, arising from X in the Lie algebra of G by

$$Xf(x) = \frac{\partial}{\partial s}\Big|_{s=0} f(x \cdot e^{sX})$$

interact intelligibly with the action of $\varphi \in \mathcal{D}(G)$ on f, as follows.

$$\begin{aligned} X(\varphi \cdot f)(x) &= \frac{\partial}{\partial s}\Big|_{s=0} \int_G f(x \cdot e^{sX} g)\, \varphi(g)\, dg \\ &= \frac{\partial}{\partial s}\Big|_{s=0} \int_G f(g)\, \varphi(e^{-sX} x^{-1} g)\, dg \end{aligned}$$

by replacing g by $e^{-sX}x^{-1}g$. This is

$$\int_G f(x) \frac{\partial}{\partial s}\Big|_{s=0} \varphi(e^{-sX} x^{-1} g)\, dg = \int_G f(x)\, X^{\text{left}} \varphi(x^{-1} g)\, dg$$

where X^{left} is the (*right-G-invariant*) differential operator attached to X via the *left* regular representation.[1] Thus, since $\varphi \cdot f = f$,

$$Xf(x) = X(\varphi \cdot f)(x) = ((X^{\text{left}} \varphi) \cdot f)(x)$$

which is of moderate growth of exponent λ, by the previous. By induction on the degree of the differential operator L, Lf is of moderate growth of exponent λ. ///

[8.3.6] Claim: Let P be a *maximal* (proper) parabolic attached to simple root α. Let f be smooth and left $(N \cap \Gamma)$-invariant. Suppose that for all $Y \in \mathfrak{g}$ the (right) Lie derivative Yf is of moderate growth of exponent λ in Siegel sets. Then

$$|(f - c_P f)(x)| \ll a_x^{\lambda - \alpha}$$

[1] As usual, the interchange of differentiation and integration is justified by observing that the integral is compactly supported, continuous, and takes values in a quasi-complete locally convex topological vector space on which differentiation is a continuous linear map. See [14.1] and [14.2].

Proof: Normalizing the measure of $(N^P \cap \Gamma)\backslash N^P$ to be 1,

$$(f - c_P f)(x) = \int_{(N^P \cap \Gamma)\backslash N^P} f(nx) - f(x)\, dn$$

$$= \int_{[0,1]^k} f(e^{t_1 X_1 + \cdots + t_k X_k} \cdot x) - f(x)\, dt_1 \ldots dt_k$$

where $X_1, \ldots, X_k$ is a basis for the Lie algebra of N so that

$$\{t_1 X_1 + \cdots + t_k X_k \,:\, 0 \le t_i \le 1,\ 1 \le i \le k\}$$

maps bijectively to $(N^P \cap \Gamma)\backslash N^P$, using the abelian-ness to see that this is easily possible. By the fundamental theorem of calculus, for X in the Lie algebra,

$$f(e^{tX} \cdot x) - f(x) = \int_0^t \frac{\partial}{\partial r}\Big|_{r=0} f(e^{(r+s)X} \cdot x)\, ds$$

$$= \int_0^t -X^{\text{left}} f(e^{sX} \cdot x)\, ds$$

where X^{left} is the natural *right*-G-invariant operator attached to X. For X in the β rootspace $\mathfrak{g}_\beta$ in the Lie algebra $\mathfrak{n}^P$ of N^P, writing $\mathrm{Ad}(g)(X)$ for gXg^{-1},

$$\mathrm{Ad}\,(a_x^{-1})(X) = a_x^{-\beta} \cdot X$$

and

$$\mathrm{Ad}\,(\theta_x^{-1}\, a_x^{-1})(X) = a_x^{-\beta} \cdot \mathrm{Ad}\,(\theta_x^{-1})(X) = a_x^{-\beta} \cdot \sum_{1 \le i \le k} c_i(\theta_x) Y_i$$

where the c_i are continuous functions (depending upon X) on K and $\{Y_i\}$ is a basis for the Lie algebra of G. Since the c_i are continuous on the compact K, they have a *uniform* bound c (depending on X). Altogether,

$$\int_{0 \le t \le 1} f(e^{Y+tX} \cdot x) - f(e^Y \cdot x)\, dt$$

$$= \beta(a_x)^{-1} \cdot \sum_{1 \le i \le k} c_i(\theta_x) \int_{0 \le t \le 1} \int_{0 \le s \le t} (-Y_i f)(e^{Y+sX} \cdot x)\, ds\, dt$$

We need

[8.3.7] Lemma: For x in a fixed standard Siegel set $\mathfrak{S} = \mathfrak{S}_t$,

$$a_x^{-\beta} \ll_{\mathfrak{S}} a_x^{-\alpha} \qquad \text{(for roots } \beta \text{ with } \mathfrak{g}_\beta \subset \mathfrak{n}^P)$$

Proof: Since $\mathfrak{g}_\beta \subset \mathfrak{n}^P$, in an expression $\beta = \sum_j c_j \alpha_j$ for β in terms of the simple roots $\alpha_1, \ldots, \alpha_{n-1}$, all coefficients are non-negative, and the coefficient of

α is 1. Thus, letting $\alpha = \alpha_{j_o}$,

$$a_x^{-\beta} = a_x^{-\alpha} \cdot \prod_{j \neq j_o} a_x^{-c_j \alpha_j} \le a_x^{-\alpha} \cdot \prod_{j \neq j_o} t^{-c_j} \ll_t a_x^{-\alpha}$$

because of the inequalities $\alpha_j(a_x) \ge t$ characterizing the Siegel set $\mathfrak{S} = \mathfrak{S}_t$. ///

Continuing the proof of the claim, using the exponent λ of moderate growth of all of the functions $Y_i f$,

$$\int_{0 \le t \le 1} f(e^{Y+tX} \cdot x) - f(e^Y \cdot x)\, dt = O(a_x^{\lambda - \alpha})$$

or,

$$\int_{0 \le t \le 1} f(e^{t_1 X_1 + \cdots + t_i X_i} \cdot x) - f(e^{t_1 X_1 + \cdots + t_{i-1} X_{i-1}} \cdot x)\, dt_i = O(a_x^{\lambda - \alpha})$$

Integrating in $dt_1, \ldots, dt_{i-1}$ and in $dt_{i+1}, \ldots, dt_k$ over copies of [0, 1] gives the same estimate for the k-fold integral:

$$\int_{[0,1]^k} f(e^{t_1 X_1 + \cdots + t_i X_i} \cdot x) - f(e^{t_1 X_1 + \cdots + t_{i-1} X_{i-1}} \cdot x)\, dt_1 \ldots dt_k = O(a_x^{\lambda - \alpha})$$

This is the assertion. ///

Continuing in this context, returning to the proof of the theorem, the previous claim shows that since every Xf is of exponent λ, $f - c_P f$ is of exponent $\lambda - \alpha$. The uniform moderate growth ensures that every $X(f - c_f)$ is of exponent $\lambda - \alpha$, as well. Applying the last claim again,

$$(Xf - Xc_P f) - c_P(Xf - Xc_P f) = Xf - Xc_P f = X(f - c_P f)$$

is of exponent $\lambda - 2 \cdot \alpha$, beginning an induction that proves the theorem. ///

[8.3.8] Corollary: For $f = \varphi \cdot f$ of moderate growth, if $c_P f = 0$ for all maximal proper parabolics P, then f is of rapid decay in standard Siegel sets $\mathfrak{S}$. ///

8.4 Moderate Growth of Convergent Eisenstein Series

Now that the importance of the moderate growth property is clearer, we use an even simpler version of the approximation methods of [3.11] to prove moderate growth of convergent Eisenstein series, at least for parameters sufficiently far into the region of convergence. Chapter 11 will use this partial result to prove that Eisenstein series meromorphically continue as functions of moderate growth and therefore are of moderate growth everywhere.

We carry out the argument in detail for the maximal proper parabolic cuspidal-data Eisenstein series $E^P_{s,f}$ on $SL_r(\mathbb{Z})\backslash SL_r(\mathbb{R})/SO_n(\mathbb{R})$. After that proof, we indicate how to obtain a corresponding result more generally.

[8.4.1] Theorem: With maximal proper parabolic P in SL_r and cuspidal data $f = f_1 \otimes f_2$ on M^P with f_j cuspforms in a strong sense on the factors of M^P, the Eisenstein series $E^P_{s,f}$ is of moderate growth on Siegel sets, and uniformly so for s in compacts.

Proof: From Chapter 7, strong-sense cuspforms are *bounded*, so as in [3.11], it suffices to treat potentially degenerate Eisenstein series $E^P_s = \sum_{\gamma \in P_k \backslash GL_r(k)} \varphi^P_\sigma \circ \gamma$ formed from

$$\varphi^P_\sigma(nmk) = |\det m_1|^{\sigma_1} \cdot |\det m_2|^{\sigma_2}$$

with $n \in N^P$, $k \in K$, and $m = \begin{pmatrix} m_1 & 0 \\ 0 & m_2 \end{pmatrix} \in M^P$, where $P = P^{r_1,r_2}$ with $r_1 + r_2 = r$, $\sigma = (sig_1, \sigma_2)$. For convergence, it suffices to take $\sigma_1 \gg \sigma_2$. Allowing nontrivial central character in the notation and computations helps avoid some needless concern over artifactual details.

We reduce to the case of parabolics of the form $P^{r-1,1}$. Given $P = P^{q,r} \subset GL_{p+q}$, let $\rho : GL_{p+q} \to GL_N$ where $N = \binom{q+r}{r}$, by acting on $\wedge^r(k^{q+r})$. The parabolic $P^{q,r}$ maps to the stabilizer P' in GL_N of the line generated by $v_o = e_{q+1} \wedge \ldots \wedge e_{q+r}$ in $\wedge^r(k^{q+r})$. Certainly

$$(e_{q+1} \wedge \ldots \wedge e_{q+r}) \cdot \wedge^r \begin{pmatrix} 1_q & 0 \\ 0 & m_2 \end{pmatrix} = (e_{q+1} \wedge \ldots \wedge e_{q+r}) \cdot \det m_2$$

for $m_2 \in GL_r$, and $\det(\wedge^r g) = (\det g)^r$ for $g \in GL_{q+r}$. Thus, for *diagonal* $m_1 \in GL_q$ and $m_2 \in GL_r$,

$$\varphi^{P^{q,r}}_{\sigma_1,\sigma_2}\begin{pmatrix} m_1 & 0 \\ 0 & m_2 \end{pmatrix}$$

$$= |\det m_1|^{\sigma_1} \cdot |\det m_2|^{\sigma_2} = \left|\det \begin{pmatrix} m_1 & 0 \\ 0 & m_2 \end{pmatrix}\right|^{\sigma_1} \cdot |\det m_2|^{\sigma_2 - \sigma_1}$$

$$= \left|\det(\wedge^r \begin{pmatrix} m_1 & 0 \\ 0 & m_2 \end{pmatrix})\right|^{\sigma_1 \cdot r/\binom{q+r}{r}} \cdot |\wedge^r m_2|^{\sigma_2 - \sigma_1}$$

which in turn can be written in the corresponding form $\varphi^{P'}_{\tau_1,\tau_2}$ on GL_N for suitable τ_1, τ_2, whose precise form is inessential. Since $\rho(P^{q,r}) \subset P'$, there is the immediate domination for $g \in GL_{q+r}$:

$$\sum_{\gamma \in P^{q,r}_k \backslash GL_{q+r}(k)} \varphi^{P^{q,r}}_{\sigma_1,\sigma_2}(\gamma \cdot g) \le \sum_{\gamma' \in P'_k \backslash GL_N(k)} \varphi^{P'}_{\tau_1,\tau_2}(\gamma' \cdot \rho(g))$$

Thus, it suffices to prove

[8.4.2] Claim: Let $|\cdot|$ be the usual norm on $\mathbb{R}^r$.

$$\Phi_\sigma(g) = \sum_{0\neq v\in\mathbb{Z}^r} \frac{1}{|v\cdot g|^{2\sigma}} \qquad (\text{with } 1 \ll \sigma \in \mathbb{R})$$

is *bounded* on Siegel sets.

Proof: Indeed, up to powers of absolute value of the determinant, the indicated sum dominates the sum for a degenerate Eisenstein series attached to the $P^{r-1,1}$ parabolic in GL_r. We can take Siegel sets to be of the form

$$\mathfrak{S}_{t,C_N} = \{nmk : n \in C_N \subset N^B,\ m = \begin{pmatrix} m_1 & & \\ & \ddots & \\ & & m_r \end{pmatrix} \in M^B,$$
$$k \in K,\ |m_j/m_{j+1}| \geq t\}$$

where C_N is *compact* and $t > 0$. With operator norm $\|\cdot\|$, letting $g = nmk$ with $n \in C_N$, we have

$$|v\cdot g| = |v\cdot nmk| = |v\cdot nm| = |v\cdot m\cdot m^{-1}nm|$$

From

$$|v\cdot m| = |v\cdot m\cdot m^{-1}nm\cdot m^{-1}n^{-1}m| \leq |v\cdot m\cdot m^{-1}nm|\cdot\|m^{-1}n^{-1}m\|$$
$$= |v\cdot m\cdot m^{-1}nm|\cdot\|n^{-1}\|$$

for $\sigma > 0$ we have

$$\frac{1}{|v\cdot nmk|^{2\sigma}} \leq \frac{1}{|v\cdot m|^{2\sigma}}\cdot\|n^{-1}\|^{2\sigma}$$
$$\leq \frac{1}{|v\cdot m|^{2\sigma}}\cdot\sup_{n\in C_N}\|n^{-1}\|^{2\sigma} \ll_{C_N} \frac{1}{|v\cdot m|^{2\sigma}}$$

since the operator norm is continuous and C_N is compact. Thus,

$$\Phi_\sigma(nmk) \ll_{C_N} \sum_{0\neq v\in\mathbb{Z}^r} \frac{1}{|v\cdot m|^{2\sigma}}$$
$$= \sum_{0\neq v\in\mathbb{Z}^r} \frac{1}{\big((v_1m_1)^2 + \ldots + (v_rm_r)^2\big)^\sigma}$$

Without loss of generality, take $m_r = 1$. With $\alpha_j = |m_j/m_{j+1}| \geq t > 0$, this is

$$\sum_{0\neq v\in\mathbb{Z}^r} \frac{1}{\big((v_1\alpha_1\alpha_2\cdots\alpha_{r-1}m_1)^2 + \ldots + (v_{r-2}\alpha_{r-2}\alpha_{r-1})^2 + v_r^2\big)^\sigma}$$
$$\leq \frac{1}{\min(t,1)^{2\sigma}} \sum_{0\neq v\in\mathbb{Z}^r} \frac{1}{|v|^{2\sigma}}$$

This is finite for $2\sigma > r$, and proves that Φ_σ is bounded on Siegel sets. ///

Powers of determinants are certainly of uniformly moderate growth on Siegel sets, so the theorem is proven. ///

The same type of argument gives a more general result, as follows.

From Chapter 7, strong-sense cuspforms are *bounded*, so as in [3.11], it suffices to treat potentially degenerate Eisenstein series $E_s^P = \sum_{\gamma \in P_k \backslash GL_r(k)} \varphi_\sigma^P \circ \gamma$ formed from

$$\varphi_\sigma^P(nmk) = |\det m_1|^{\sigma_1} \cdot |\det m_2|^{\sigma_2} \cdot \ldots$$

with $n \in N^P$, $k \in K$, and $m = \begin{pmatrix} m_1 & & \\ & m_2 & \\ & & \ddots \end{pmatrix} \in M^P$, where $P = P^{r_1, r_2, \ldots}$ with $r_1 + r_2 + \cdots = r$, $\sigma = (sig_1, \sigma_2, \ldots)$, and $m_j \in GL_{r_j}$. For easy convergence, it suffices to take $\sigma_1 \gg \sigma_2 \gg \cdots$.

In Iwasawa coordinates for the minimal parabolic B, such φ_σ^P can readily be expressed as a product of analogous functions attached to *maximal proper* parabolics. For example, given $\sigma_1 \gg \sigma_2 \gg \sigma_3$,

$$\begin{aligned} &|\det m_1|^{\sigma_1} \cdot |\det m_2|^{\sigma_2} \cdot |\det m_3|^{\sigma_3} \\ &= \left(|\det m_1 \cdot \det m_2|^a \cdot |\det m_3|^b\right) \cdot \left(|\det m_1|^{a'} \cdot |\det m_2 \cdot \det m_3|^{b'}\right) \end{aligned}$$

is the requirement

$$a + b = \sigma_1 \qquad a + b' = \sigma_2 \qquad a' + b' = \sigma_3$$

This is readily satisfied by taking b' large negative, so that the a' determined from the last equation satisfies $a' \gg b'$, then a determined from the second equation and b determined from the first satisfy $a \gg b$. Thus, we write

$$\varphi_\sigma^P = \prod_j \varphi_{\tau_1^j, \tau_2^j}^{P^{j,r-j}}$$

where j indexes maximal proper parabolics and $\tau_1^j \gg \tau_2^j$ for all j. There is an immediate domination

$$\sum_{\gamma \in P_k \backslash GL_r(k)} \varphi_\sigma^P \circ \gamma \le \prod_j \sum_{\gamma \in P_k^{j,r-j} \backslash GL_r(k)} \varphi_{\tau_1^j, \tau_2^j}^{P^{j,r-j}} \circ \gamma$$

Certainly moderate growth is preserved by products. Thus, it suffices to prove moderate growth for degenerate Eisenstein series attached to maximal proper parabolics.

A similar, complementary device reduces to the case of groundfield $k = \mathbb{Q}$, thus essentially reducing to the case explicitly treated.

8.5 Integral Operators on Cuspidal-Data Eisenstein Series

Having seen the significance of the property $\varphi \cdot f = f$ for some test function φ, the analogous property [7.2.20] for cuspforms f_1, f_2 on $GL_{r_1} \times GL_{r_2} \subset GL_{r_1+r_2}$ can be invoked to prove a similar property for Eisenstein series $E^P_{s,f}$ with $f = f_1 \otimes f_2$ and $P = P^{r_1,r_2}$.

As in the computation [3.11.9] of the general form of constant terms, we need to assume a *multiplicity-one* property of f_1 and f_2, namely, that each is the unique cuspform on respective $SL_{r_j}(\mathbb{Z})\backslash SL_{r_j}(R)/SO(r_j, \mathbb{R})$ with given Laplacian eigenvalue, up to constant multiples.

Let $M^1 = SL_{r_1} \times SL_{r_2} \subset M^P$, and

$$\varphi_{s,f}(nm'z_y k) = y^s \cdot f(m') \qquad (\text{with } n \in N^P, m' \in M^1, k \in K)$$

and with

$$z_y = \begin{pmatrix} y^{\frac{1}{r_1 r_2}} \cdot 1_{r_1} & 0 \\ 0 & 1_{r_2} \end{pmatrix}$$

Let $E_{s,f} = \sum_{\gamma \in (\Gamma \cap P)\backslash\Gamma} \varphi_{s,f} \circ \gamma$, for $\mathrm{Re}(s) \gg 1$ for convergence.

[8.5.1] Claim: For every $\eta \in C_c^\infty(K\backslash G/K)$, there is an entire $\mathbb{C}$-valued function $s \to \mu_{s,f}(\eta)$ such that $\eta \cdot \varphi_{s,f} = \mu(\eta) \cdot \varphi_{s,f}$. At least in $\mathrm{Re}(s) > 1$, similarly, $\eta \cdot E_{s,f} = \mu_{s,f} \cdot E_{s,f}$. Given s, f, there is η such that $\mu_{s,f}(\eta)$ is not identically 0.

Proof: In the region of convergence, $E_{s,f}$ is a sum of left translates of $\varphi_{s,f}$, so it suffices to prove the eigenfunction property of $\varphi_{s,f}$. The eigenfunction property for meromorphically continued $E_{s,f}$ will follow by the identity principle from complex analysis.

Certainly the right action of such η preserves left N-invariance. Since η is K-bi-invariant, it preserves right K-invariant functions, as well. Computing directly, using an Iwasawa decomposition $G = P \cdot K = N^P \cdot M^P \cdot K$, noting that $P \cap K$ is compact, for $m_o \in M^P$,

$$\begin{aligned}(\eta \cdot \varphi_{s,f})(m_o) &= \int_G \eta(h)\, \varphi_{s,f}(m_o h)\, dh \\ &= \int_{N^P} \int_{M^P} \int K_\eta(nmk)\, \varphi_{s,f}(m_o nmk)\, \delta(m)^{-1}\, dn\, dm\, dk\end{aligned}$$

with modular function δ on P. Continuing, this is

$$\begin{aligned}&\int_{N^P} \int_{M^P} \int_K \eta(nmk)\, \varphi_{s,f}(m_o n m_o^{-1} \cdot m_o m)\, \delta(m)^{-1}\, dn\, dm\, dk \\ &= \int_{M^P} \left(\delta(m)^{-1} \int_{N^P} \eta(nm)\, dn\right) \cdot \varphi_{s,f}(m_o \cdot m)\, dm\end{aligned}$$

As a function of $m \in M^P$, the inner integral $\eta'(m)$ is left and right $K \cap M^P$-invariant, smooth, and compactly supported. Action of $\eta' \in C_c^\infty((K \cap M^P)\backslash M^P/(K \cap M^P))$ on functions on M^P commutes with Casimir operators on the factors of M^P and preserves central equivariance $u(mz_t) = |t|^s \cdot u(m)$. The right action on functions $m'a_y \to y^s \cdot f(m')$ with $f = f_1 \otimes f_1$ preserves cuspidality of f. Since $f = f_1 \otimes f_1$ is assumed to be the unique cuspform with its eigenvalue, the action of η' sends $y^s \cdot f(m')$ to a multiple of itself, by scalar $\mu_{s,f}(\eta)$. Since $s \to \varphi_{s,f}$ is a holomorphic $C^o(G)$-valued function (for example), and since the integral giving the action of η or η' exists as a Gelfand-Pettis integral, the function $s \to \mu_{s,f}(\eta)$ is holomorphic in s.

For a sequence $\{\eta_j\}$ of functions forming an approximate identity, $\mu_{s,f}(\eta) \cdot f = \eta_j \cdot f \to f$. Thus, for f not the zero vector, $\mu_{s,f}(\eta_j) \neq 0$ for sufficiently large j. ///

8.A Appendix: Joint Continuity of Bilinear Functionals

This is a corollary of Banach-Steinhaus [13.12.3], useful in removing ambiguities in considering *averaged* actions, for example, the action $C_c^\infty(\mathbb{R}) \times V \to V$ of test functions.

[8.A.1] Corollary: Let $\beta : X \times Y \to Z$ be a bilinear map on Fréchet spaces X, Y, Z, continuous in each variable separately. Then β is *jointly* continuous.

Proof: Fix a neighborhood N of 0 in Z Let $x_n \to x_o$ in X and $y_n \to y_o$ in Y. For each $x \in X$, by continuity in Y, $\beta(x, y_n) \to \beta(x, y_o)$. Thus, for each $x \in X$, the set of values $\beta(x, y_n)$ is *bounded* in Z. The linear functionals $x \to \beta(x, y_n)$ are *equicontinuous*, by Banach-Steinhaus, so there is a neighborhood U of 0 in X so that $b_n(U) \subset N$ for all n. In the identity

$$\beta(x_n, y_n) - \beta(x_o, y_o) = \beta(x_n - x_o, y_n) + \beta(x_o, y_n - y_o)$$

we have $x_n - x_o \in U$ for large n, and $\beta(x_n - x_o, y_o) \in N$. Also, by continuity in Y, $\beta(x_o, y_n - y_o) \in N$ for large n. Thus, $\beta(x_n, y_n) - \beta(x_o, y_o) \in N + N$, proving *sequential* continuity. Since $X \times Y$ is metrizable, sequential continuity implies continuity. ///

[8.A.2] Corollary: The same conclusion holds when X is an LF-space.

Proof: Continuous linear functionals from an LF-space are exactly given by compatible families of continuous maps from the limitands. ///

Bibliography

[Arthur 1978] J. Arthur, *A trace formula for reductive groups, I. Terms associated to classes in $G(\mathbb{Q})$*, Duke. Math. J. **45** (1978), 911–952.

[Arthur 1980] J. Arthur, *A trace formula for reductive groups, II. Application of a truncation operator*, Comp. Math. J. **40** (1980), 87–121.

[Avakumović 1956] V.G. Avakumović, *Über die Eigenfunktionen auf geschlossenen Riemannschen Mannigfaltigkeiten*, Math. Z. **65** (1956), 327–344.

[Bargmann 1947] V. Bargmann, *Irreducible unitary representations of the Lorentz group*, Ann. Math. **48** (1947), 568–640.

[Berezin 1956] F.A. Berezin, *Laplace operators on semisimple Lie groups*, Dokl. Akad. Nauk SSSR **107** (1956), 9–12.

[Berezin-Faddeev 1961] F.A. Berezin, L.D. Faddeev, *Remarks on Schrödinger's equation with a singular potential*, Soviet Math. Dokl. **2** (1961), 372–375.

[Bethe-Peierls 1935] H. Bethe, R. Peierls, *Quantum theory of the diplon*, Proc. Royal Soc. London **148a** (1935), 146–156.

[Bianchi 1892] L. Bianchi, *Sui gruppi di sostituzioni lineari con coefficienti appartenenti a corpi quadratici immaginari*, Math. Ann. **40** (1892), 332–412.

[Birkhoff 1908] G.D. Birkhoff, *On the asymptotic character of the solutions of certain linear differential equations containing a parameter*, Trans. Amer. Math. Soc. **9** (1908), 219–231.

[Birkhoff 1909] G.D. Birkhoff, *Singular points of ordinary linear differential equations*, Trans. Amer. Math. Soc. **10** (1909), 436–470.

[Birkhoff 1913] G.D. Birkhoff, *On a simple type of irregular singular point*, Trans. Amer. Math. Soc. **14** (1913), 462–476.

[Birkhoff 1935] G. Birkhoff, *Integration of functions with values in a Banach space*, Trans. AMS **38** (1935), 357–378.

[Blaustein-Handelsman 1975] N. Blaustein, R.A. Handelsman, *Asymptotic Expansions of Integrals*, Holt, Rinehart, Winston, 1975, reprinted 1986, Dover.

[Blumenthal 1903/1904] O. Blumenthal, *Über Modulfunktionen von mehreren Veränderlichen*, Math. Ann. Bd. **56** (1903), 509–548; **58** (1904), 497–527.

[Bôcher 1898/1899] M. Bôcher, *The theorems of oscillation of Sturm and Klein*, Bull. AMS **4** (1898), 295–313; **5** (1899), 22–43.

[Bochner 1932] S. Bochner, *Vorlesungen über Fouriersche Integrale*, Akademie-Verlag, 1932.

[Bochner 1935] S. Bochner, *Integration von Funktionen deren Werte die Elemente eines Vektorraumes sind*, Fund. Math. **20** (1935), 262–276.

[Bochner-Martin 1948] S. Bochner, W.T. Martin, *Several Complex Variables*, Princeton University Press, Princeton, 1948.

[Borel 1962] A. Borel, *Ensembles fondamentaux pour les groupes arithmétiques*, Colloque sur la théorie des groupes algébriques, Bruxelles, 1962, 23–40.

[Borel 1963] A. Borel, *Some finiteness properties of adele groups over number fields*, IHES Sci. Publ. Math. **16** (1963), 5–30.

[Borel 1965/1966a] A. Borel, *Introduction to automorphic forms*, in *Algebraic Groups and Discontinuous Subgroups, Boulder, 1965*, Proc. Symp. Pure Math. **9**, AMS, New York, 1966, 199–210.

[Borel 1965/1966b] A. Borel, *Reduction theory for arithmetic groups*, in *Algebraic Groups and Discontinuous Subgroups, Boulder, 1965*, Proc. Symp. Pure Math. **9**, AMS, New York 1966, 20–25.

[Borel 1969] A. Borel, *Introductions aux groupes arithmeétiques*, Publ. l'Inst. Math. Univ. Strasbourg, XV, Actualités Sci. et Industrielles, no. 1341, Hermann, Paris, 1969.

[Borel 1976] A. Borel, *Admissible representations of a semi-simple group over a local field with vectors fixed under an Iwahori subgroup*, Inv. Math. **35** (1976), 233–259.

[Borel 1997] A. Borel, *Automorphic Forms on* $SL_2(\mathbb{R})$, Cambridge Tracts in Math. **130**, Cambridge University Press, Cambridge, 1997.

[Borel 2007] A. Borel, *Automorphic forms on reductive groups*, in *Automorphic Forms and Applications*, eds. P. Sarnak and F. Shahidi, IAS/ParkCity Math Series **12**, AMS, 2007.

[Borel-HarishChandra 1962] A. Borel, Harish-Chandra, *Arithmetic subgroups of algebraic groups*, Ann. Math. **75** (1962), 485–535.

[Bourbaki 1987] N. Bourbaki, *Topological Vector Spaces*, ch. 1–5, Springer-Verlag, Berlin-Heidelberg 1987.

[Braun 1939] H. Braun, *Konvergenz verallgemeinerter Eisensteinscher Reihen*, Math. Z. **44** (1939), 387–397.

[Brooks 1969] J.K. Brooks, *Representations of weak and strong integrals in Banach spaces*, Proc. Nat. Acad. Sci. U.S.A., 1969, 266–270.

[Casselman 1978/1980] W. Casselman, *Jacquet modules for real reductive groups*, Proc. Int. Cong. Math. (1978), 557–563, Acad. Sci. Fennica, Helskinki, 1980.

[Casselman 1980] W. Casselman, *The unramified principal series of p-adic groups. I. The spherical function*, Comp. Math. **40** (1980), no. 3, 387–406.

[Casselman 2005] W. Casselman, *A conjecture about the analytical behavior of Eisenstein series*, Pure and Applied Math. Q. **1** (2005) no. 4, part 3, 867–888.

[Casselman-Miličić 1982] W. Casselman, D. Miličić, *Asymptotic behavior of matrix coefficients of admissible representations*, Duke J. Math. **49** (1982), 869–930.

[Casselman-Osborne 1975] W. Casselman, M.S. Osborne, *The* $\mathfrak{n}$*-cohomology of representations with an infinitesimal character*, Comp. Math **31** (1975), 219–227.

[Casselman-Osborne 1978] W. Casselman, M.S. Osborne, *The restriction of admissible representations to* $\mathfrak{n}$, Math. Ann. **233** (1978), 193–198.

[Cogdell–Li-PiatetskiShapiro-Sarnak 1991] J. Cogdell, J.-S. Li, I.I. Piatetski-Shapiro, P. Sarnak, *Poincaré series for* $SO(n, 1)$, Acta Math. **167** (1991), 229–285.

[Cogdell-PiatetskiShapiro 1990] J. Cogdell, I. Piatetski-Shapiro, *The Arithmetic and Spectral Analysis of Poincaré Series*, Perspectives in Mathematics, Academic Press, San Diego, 1990.

[Cohen-Sarnak 1980] P. Cohen, P. Sarnak, *Selberg Trace Formula*, ch. 6 and 7, *Eisenstein series for hyperbolic manifolds*, www.math.princeton.edu/sarnak/

[Colin de Verdière 1981] Y. Colin de Verdière, *Une nouvelle démonstration du prolongement méromorphe des séries d'Eisenstein*, C. R. Acad. Sci. Paris Sér. I Math. **293** (1981), no. 7, 361–363.

[Colin de Verdière 1982/1983] Y. Colin de Verdière, *Pseudo-laplaciens, I, II*, Ann. Inst. Fourier (Grenoble) **32** (1982) no. 3, xiii, 275–286; ibid, **33** (1983) no. 2, 87–113.

[Conway-Smith 2003] J. Conway, D. Smith, *On Quaternions and Octonians*, A. K. Peters, Natick, MA, 2003.

[DeCelles 2012] A. DeCelles, *An exact formula relating lattice points in symmetric spaces to the automorphic spectrum*, Illinois J. Math. **56** (2012), 805–823.

[DeCelles 2016] A. DeCelles, *Constructing Poincaré series for number-theoretic applications*, New York J. Math. **22** (2016) 1221–1247.

[Dirac 1928a/1928b] P.A.M. Dirac, *The quantum theory of the electron*, Proc. R. Soc. Lond. A **117** (1928), 610–624; *II*, ibid, **118** (1928), 351–361.

[Dirac 1930] P.A.M. Dirac, *Principles of Quantum Mechanics*, Clarendon Press, Oxford, 1930.

[Dirichlet 1829] P.G.L. Dirichlet, *Sur la convergence des séries trigonométriques qui servent à représenter une fonction arbitraire entre des limites données*, J. Reine Angew. Math **4** (1829), 157–169 (Werke I, 117–132).

[Douady 1963] A. Douady, *Parties compactes d'un espace de fonctions continues a support compact*, C. R. Acad. Sci. Paris **257** (1963), 2788–2791.

[Elstrodt 1973] J. Elstrodt, *Die Resolvente zum Eigenwertproblem der automorphen Formen in der hyperbolischen Ebene, I*, Math. Ann. **203** (1973), 295–330, *II*, Math. Z. **132** (1973), 99–134.

[Elstrodt-Grunewald-Mennicke 1985] J. Elstrodt, E. Grunewald, J. Mennicke, *Eisenstein series on three-dimensional hyperbolic spaces and imaginary quadratic fields*, J. reine und angew. Math. **360** (1985), 160–213.

[Elstrodt-Grunewald-Mennicke 1987] J. Elstrodt, E. Grunewald, J. Mennicke, *Zeta functions of binary Hermitian forms and special values of Eisenstein series on three-dimensional hyperbolic space*, Math. Ann. **277** (1987), 655–708.

[Epstein 1903/1907] P. Epstein, *Zur Theorie allgemeiner Zetafunktionen*, Math. Ann. **56** (1903), 615–644; ibid, **65** (1907), 205–216.

[Erdélyi 1956] A. Erdélyi, *Asymptotic Expansions*, Technical Report 3, Office of Naval Research Reference No. NR-043-121, reprinted by Dover, 1956.

[Estermann 1928] T. Estermann, *On certain functions represented by Dirichlet series*, Proc. London Math. Soc. **27** (1928), 435–448.

[Faddeev 1967] L. Faddeev, *Expansion in eigenfunctions of the Laplace operator on the fundamental domain of a discrete group on the Lobacevskii plane*, AMS Transl. Trudy (1967), 357–386.

[Faddeev-Pavlov 1972] L. Faddeev, B.S. Pavlov, *Scattering theory and automorphic functions*, Seminar Steklov Math. Inst. **27** (1972), 161–193.

[Fay 1977] J.D. Fay, *Fourier coefficients of the resolvent for a Fuchsian group*, J. für reine und angewandte Math. (Crelle) **293–294** (1977), 143–203.

[Fourier 1822] J. Fourier, *Théorie analytique de la chaleur*, Firmin Didot Père et Fils, Paris, 1822.

[Friedrichs 1934/1935] K.O. Friedrichs, *Spektraltheorie halbbeschränkter Operatoren*, Math. Ann. **109** (1934), 465–487, 685–713; *II*, Math. Ann. **110** (1935), 777–779.

[Gårding 1947] L. Gårding, *Note on continuous representations of Lie groups*, Proc. Nat. Acad. Sci. USA **33** (1947), 331–332.

[Garrett vignettes] P. Garrett, *Vignettes*, www.math.umn.edu/~garrett/m/v/

[Garrett mfms-notes] P. Garrett, *Modular forms notes*, www.math.umn.edu/~garrett/m/mfms/

[Garrett fun-notes] P. Garrett, *Functional analysis notes*, www.math.umn.edu/~garrett/m/fun/

[Garrett alg-noth-notes] P. Garrett, *Algebraic number theory notes*, www.math.umn.edu/~garrett/m/number_theory/

[Gelfand 1936] I.M. Gelfand, *Sur un lemme de la theorie des espaces lineaires*, Comm. Inst. Sci. Math. de Kharkoff, no. 4, **13** (1936), 35–40.

[Gelfand 1950] I.M. Gelfand, *Spherical functions in Riemannian symmetric spaces*, Dokl. Akad. Nauk. SSSR **70** (1950), 5–8.

[Gelfand-Fomin 1952] I.M. Gelfand, S.V. Fomin, *Geodesic flows on manifolds of constant negative curvature*, Uspekh. Mat. Nauk. **7** (1952), no. 1, 118–137. English translation, AMS ser. 2, **1** (1965), 49–65.

[Gelfand-Graev 1959] I.M. Gelfand, M.I. Graev, *Geometry of homogeneous spaces, representations of groups in homogeneous spaces, and related problems of integral geometry*, Trudy Moskov. Obshch. **8** (1962), 321–390.

[Gelfand-Graev-PiatetskiShapiro 1969] I. Gelfand, M. Graev, I. Piatetski-Shapiro, *Representation Theory and Automorphic Functions*, W.B. Saunders Co., Philadelphia, 1969.

[Gelfand-Kazhdan 1975] I.M. Gelfand, D. Kazhdan, *Representations of the group $GL(n, k)$ where k is a local field*, in *Lie Groups and Their Representations*, Halsted, New York, 1975, pp. 95–118.

[Gelfand-PiatetskiShapiro 1963] I.M. Gelfand, I.I. Piatetski-Shapiro, *Automorphic functions and representation theory*, Trudy Moskov. Obshch. **8** (1963), 389–412. [Trans.: Trans. Moscow Math. Soc. **12** (1963), 438–464.]

[Gelfand-Shilov 1964] I.M. Gelfand, G.E. Shilov, *Generalized Functions, I: Properties and Operators*, Academic Press, New York, 1964.

[Gelfand-Vilenkin 1964] I.M. Gelfand, N.Ya. Vilenkin, *Generalized Functions, IV: Applications of Harmonic Analysis*, Academic Press, NY, 1964.

[Godement 1962–1964] R. Godement, *Domaines fondamentaux des groupes arithmetiques*, Sem. Bourb. **257** (1962–3).

[Godement 1966a] R. Godement, *Decomposition of $L^2(\Gamma\backslash G)$ for $\Gamma = SL(2, Z)$*, in Proc. Symp. Pure Math. **9** (1966), 211–224.

[Godement 1966b] R. Godement, *The spectral decomposition of cuspforms*, in Proc. Symp. Pure Math. **9** (1966), AMS 225–234.

[Green 1828] G. Green, *An essay on the application of mathematical analysis to the theories of electricity and magnetism*, arXiv:0807.0088 [physics.hist-ph]. Re-typeset 2008 from Crelle's J. reprint 1850–1854.

[Green 1837] G. Green, *On the Laws of Reexion and Refraction of Light at the Common Surface of Two Non-crystallized Media* Trans. Camb. Phil. Soc. **68** (1837), 457–462.

[Grothendieck 1950] A. Grothendieck, *Sur la complétion du dual d'un espace vectoriel localement convexe*, C. R. Acad. Sci. Paris **230** (1950), 605–606.

[Grothendieck 1953a,1953b] A. Grothendieck, *Sur certaines espaces de fonctions holomorphes, I*, J. Reine Angew. Math. **192** (1953), 35–64; *II*, **192** (1953), 77–95.

[Haas 1977] H. Haas, *Numerische Berechnung der Eigenwerte der Differentialgleichung $y^2\Delta u + \lambda u = 0$ für ein unendliches Gebiet im $\mathbb{R}^2$*, Diplomarbeit, Universität Heidelberg (1977), 155 pp.

[Harish-Chandra 1954] Harish-Chandra, *Representations of semisimple Lie groups, III*, Trans. AMS **76** (1954), 234–253.

[Harish-Chandra 1959] Harish-Chandra, *Automorphic forms on a semi-simple Lie group*, Proc. Nat. Acad. Sci. **45** (1959), 570–573.

[Harish-Chandra 1968] Harish-Chandra, *Automorphic forms on semi-simple Lie groups*, notes G.J.M. Mars., SLN **62**, Springer-Verlag, 1968.

[Hartogs 1906] F. Hartogs, *Zur Theorie der analytischen Funktionene mehrerer unabhängiger Veränderlichen, insbesondere über die Darstellung derselben durch Reihen, welche nach Potenzen einer Veränderlichen fortschreiten*, Math. Ann. **62** (1906), 1–88.

[Hejhal 1976/1983] D. Hejhal, *The Selberg trace formula for $PSL_2(\mathbb{R})$, I*, SLN **548**, Springer-Verlag, 1976; *II*, ibid, **1001**, Springer-Verlag, 1983.

[Hejhal 1981] D. Hejhal, *Some observations concerning eigenvalues of the Laplacian and Dirichlet L-series*, in *Recent Progress in Analytic Number Theory*, ed. H. Halberstam and C. Hooley, vol. 2, Academic Press, New York, 1981, pp. 95–110.

[Hilbert 1909] D. Hilbert, *Wesen und Ziele einer Analysis der unendlich vielen unabhängigen Variablen*, Rendiconti Circolo Mat. Palermo **27** (1909), 59–74.

[Hilbert 1912] D. Hilbert, *Grundzüge einer allgemeinen Theorie der linearen Integralgleichungen*, Teubner, Leipzig-Berlin, 1912.

[Hildebrandt 1953] T.H. Hildebrandt, *Integration in abstract spaces*, Bull. AMS, **59** (1953), 111–139.

[Hille-Phillips 1957] E. Hille with R. Phillips, *Functional Analysis and Semigroups*, AMS Coll. Pub., 2nd edition, Providence, RI, 1957.

[Hörmander 1973] L. Hörmander, *An Introduction to Complex Analysis in Several Variables*, 2nd edition, North-Holland, 1973.

[Horvath 1966] J. Horvath, *Topological Vector Spaces and Distributions*, Addison-Wesley, Boston, 1966.

[Huber 1955] H. Huber, *Über eine neue Klasse automorpher Funktionen und ein Gitterpunkt Problem in der hypbolischen Ebene I*, Comm. Math. Helv. **30** (1955), 20–62.

[Hurwitz 1898] A. Hurwitz, *Über die Komposition der quadratische Formen von beliebig vielen Variabeln*, Nachr. König. Gesellschaft der Wiss. zu Göttingen (1898), 309–316.

[Hurwitz 1919] A. Hurwitz, *Vorlesungen úber die Zahlentheorie der Quaternionen*, Springer, Berlin, 1919.

[Iwaniec 2002] H. Iwaniec, *Spectral Methods of Automorphic Forms*, 2nd edition, AMS, Providence, 2002. [First edition, Revisto Mathematica Iberoamericana, 1995.]

[Jacquet 1982/1983] H. Jacquet, *On the residual spectrum of $GL(n)$*, in *Lie group representations, II, College Park, MD*, 185–208, SLN 1041, Springer, Berlin, 1984.

[Kodaira 1949] K. Kodaira, *The eigenvalue problem for ordinary differential equations of the second order and Heisenberg's theory of S-matrices*, Amer. J. Math. **71** (1949), 921–945.

[Krein 1945] M.G. Krein, *On self-adjoint extension of bounded and semi-bounded Hermitian transformations*, Dokl. Akad. Nauk. SSSR **48** (1945), 303–306 [Russian].

[Krein 1947] M.G. Krein, *The theory of self-adjoint extension of semi-bounded Hermitian transformations and its applications*, I. Mat. Sbornik **20** (1947), 431–495 [Russian].

[Kurokawa 1985a,b] N. Kurokawa, *On the meromorphy of Euler products, I*, Proc. London Math. Soc. **53** (1985) 1–49; *II*, ibid **53** (1985) 209–236.

[Lang 1975] S. Lang, $SL_2(\mathbb{R})$, Addison-Wesley, Boston, 1975.

[Langlands 1971] R. Langlands, *Euler Products*, Yale University Press, New Haven, 1971.

[Langlands 1967/1976] R.P. Langlands, *On the functional equations satisfied by Eisenstein series*. Lecture Notes in Mathematics, vol. 544, Springer-Verlag, Berlin and New York, 1976.

[Laplace 1774] P.S. Laplace, *Mémoir on the probability of causes of events*, Mémoires de Mathématique et de Physique, Tome Sixième. (English trans. S.M. Stigler, 1986. Statist. Sci., 1 **19**, 364–378).

[Lax-Phillips 1976] P. Lax, R. Phillips, *Scattering theory for automorphic functions*, Ann. Math. Studies, Princeton, 1976.

[Levi 1906] B. Levi, *Sul Principio di Dirichlet*, Rend. del Circolo Mat. di Palermo **22** (1906), 293–300.

[Lindelöf 1908] E. Lindelöf, *Quelques remarques sur la croissance de la fonction $\zeta(s)$*, Bull. Sci. Math. **32** (1908), 341–356.

[Liouville 1837] J. Liouville, J. Math. Pures et Appl. **2** (1837), 16–35.

[Lützen 1984] J. Lützen, *Sturm and Liouville's work on ordinary differential equations. The emergence of Sturm-Liouville theory*, Arch. Hist. Exact Sci **29** (1984) no. 4, 309–376. Retrieved July 16, 2013, from www.math.ku.dk/~lutzen/

[Maaß 1949] H. Maaß, *Über eine neue Art von nicht analytischen automorphen Funktionen und die Bestimmung Dirichletscher Reihen durch Funktionalgleichungen*, Math. Ann. **121** (1949), 141–183.

[MSE 2017] Math Stack Exchange, *Residual spectrum of a Hermitian operator*, retrieved July 19, 2017, from https://math.stackexchange.com/questions/2363904/

[Matsumoto 1977] H. Matsumoto, *Analyse harmonique dans les systèmes de Tits bornologiques de type affine*, SLN 590, Springer, Berlin, 1977.

[Minakshisundaram-Pleijel 1949] S. Minakshisundaram, Å. Pleijel, *Some properties of the eigenfunctions of the Laplace-operator on Riemannian manifolds*, Canadian J. Math. **1** (1949), 242–256.

[Moeglin-Waldspurger 1989] C. Moeglin, J.-L. Waldspurger, *Le spectre résiduel de GL_n*, with appendix *Poles des fonctions L de pairs pour GL_n*, Ann. Sci. École Norm. Sup. **22** (1989), 605–674.

[Moeglin-Waldspurger 1995] C. Moeglin, J.-L. Waldspurger, *Spectral Decompositions and Eisenstein series*, Cambridge University Press, Cambridge, 1995.

[Müller 1996] W. Müller, *On the analytic continuation of rank one Eisenstein series*, Geom. Fun. Ann. **6** (1996), 572–586.

[Myller-Lebedev 1907] Wera Myller-Lebedev, *Die Theorie der Integralgleichungen in Anwendung auf einige Reihenentwicklungen*, Math. Ann. **64** (1907), 388–416.

[Neunhöffer 1973] H. Neunhöffer, *Über die analytische Fortsetzung von Poincaréreihen*, Sitzungsberichte Heidelberg Akad. Wiss. (1973), no. 2.

[Niebur 1973] D. Niebur, *A class of nonanalytic automorphic functions*, Nagoya Math. J. **52** (1973), 133–145.

[Olver 1954] F.W.J. Olver, *The asymptotic solution of linear differential equations of the second order for large values of a parameter*, Phil. Trans. **247** (1954), 307–327.

[Paley-Wiener 1934] R. Paley, N. Wiener, *Fourier transforms in the complex domain*, AMS Coll. Publ. XIX, New York, 1934.

[Pettis 1938] B.J. Pettis, *On integration in vector spaces*, Trans. AMS **44** (1938), 277–304.

[Phragmén-Lindelöf 1908] L.E. Phragmén, E. Lindelöf, *Sur une extension d'un principe classique de l'analyse et sur quelques propriétés des fonctions monogènes dans le voisinage d'un point singuliere*, Acta Math. **31** (1908), 381–406.

[Piatetski-Shapiro 1979] I.I. Piatetski-Shapiro, Multiplicity-one theorems, in *Automorphic Forms, Representations, and L-functions*, Proc. Symp. Pure Math. XXXIII, part 1, 315–322, AMS, 1979.

[Picard 1882] E. Picard, *Sur une classe de groupes discontinus de substitutions linéaires et sur les fonctions de deux variables indépendantes restant invariables par ces substitutions*, Acta Math. **1** no. 1 (1882), 297–320.

[Picard 1883] E. Picard, *Sur des fonctions de deux variables indépendantes analogues aux fonctions modulaires* Acta Math. no. 1 **2** (1883), 114–135.

[Picard 1884] E. Picard, *Sur un groupe de transformations des points de l'espace situés du même coîé d'un plan*, Bull. Soc. Math. France **12** (1884), 43–47.

[Povzner 1953] A. Povzner, *On the expansion of arbitrary functions in characteristic functions of the operator $-\Delta u + cu$*, Math. Sb. **32** no. 74 (1953), 109–156.

[Rankin 1939] R. Rankin, *Contributions to the theory of Ramanujan's function $\tau(n)$ and similar arithmetic functions, I*, Proc. Cam. Phil. Soc. **35** (1939), 351–372.

[Riesz 1907] F. Riesz, *Sur les systèmes orthogonaux de fonctions*, C.R. de l'Acad. des Sci. **143** (1907), 615–619.

[Riesz 1910] F. Riesz, *Untersuchungen über Systeme integrierbarer Funktionen*, Math. Ann. **69** (1910), 449–497.

[Roelcke 1956a] W. Roelcke, *Über die Wellengleichung bei Grenzkreisgruppen erster Art*, S.-B. Heidelberger Akad. Wiss. Math. Nat. Kl. 1953/1955 (1956), 159–267.

[Roelcke 1956b] W. Roelcke, *Analytische Fortsetzung der Eisensteinreihen zu den parabolischen Spitzen von Grenzkreisgruppen erster Art*, Math. Ann. **132** (1956), 121–129.

[Rudin 1991] W. Rudin, *Functional Analysis*, second edition, McGraw-Hill, New York, 1991.

[Schaefer-Wolff 1999] H.H. Schaefer, with M.P. Wolff, *Topological Vector Spaces*, 2nd edition, Springer, Tübingen, 1999.

[Schmidt 1907] E. Schmidt, *Zur Theorie der linearen und nichtlinearen Integralgleichungen. Teil I: Entwicklung wilkürlicher Funktionen nach Systemen vorgeschriebener*, Math. Ann. **63** (1907), 433–476.

[Schwartz 1950] L. Schwartz, *Théorie des noyaux*, Proc. Int. Cong. Math. Cambridge 1950, I, 220–230.

[Schwartz 1950/1951] L. Schwartz, *Théorie des Distributions*, I, II Hermann, Paris, 1950/1951, 3rd edition, 1965.

[Schwartz 1952] L. Schwartz, *Transformation de Laplace des distributions*, Comm. Sém. Math. Univ. Lund (1952), Tome Supplémentaire, 196–206.

[Schwartz 1953/1954] L. Schwartz, *Espaces de fonctions différentiables à valeurs vectorielles* J. d'Analyse Math. **4** (1953/1954), 88–148.

[Selberg 1940] A. Selberg, *Bemerkungen über eine Dirichletsche Reihe, die mit der Theorie der Modulformen nahe verbunden ist*, Arch. Math. Naturvid **43** (1940), 47–50.

[Selberg 1954] A. Selberg, *Harmonic Analysis, 2. Teil*, Vorlesung Niederschrift, Göttingen, 1954; *Collected Papers I*, Springer, Heidelberg, 1988, 626–674.

[Selberg 1956] A. Selberg, *Harmonic analysis and discontinuous groups in weakly symmetric spaces, with applications to Dirichlet series*, J. Indian Math. Soc. **20** (1956), 47–87.

[Shahidi 1978] F. Shahidi, *Functional equation satisfied by certain L-functions*, Comp. Math. **37** (1978), 171–208.

[Shahidi 1985] F. Shahidi, *Local coefficients as Artin factors for real groups*, Duke Math. J. **52** (1985), 973–1007.

[Shahidi 2010] F. Shahidi, *Eisenstein series and automorphic L-functions*, AMS Coll. Publ. **58**, AMS, 2010.

[Shalika 1974] J.A. Shalika, *The multiplicity one theorem for GLn*, Ann. Math. 100 (1974), 171–193.

[Sobolev 1937] S.L. Sobolev, *On a boundary value problem for polyharmonic equations (Russian)*, Mat. Sb. (NS) **2** (44) (1937), 465–499.

[Sobolev 1938] S.L. Sobolev, *On a theorem of functional analysis (Russian)*, Mat. Sb. (NS) **4** (1938), 471–497.

[Sobolev 1950] S.L. Sobolev, *Some Applications of Functional Analysis to Mathematical Physics* [Russian], Paul: Leningrad, 1950.

[Speh 1981/1982] B. Speh, *The unitary dual of $GL_3(\mathbb{R})$ and $GL_4(\mathbb{R})$*, Math. Ann. **258** (1981/1982), 113–133.

[Stone 1929] M.H. Stone, *I, II: Linear transformations in Hilbert space*, Proc. Nat. Acad. Sci. **16** (1929), 198–200, 423–425; *III: operational methods and group theory*, ibid, **16** (1930), 172–175.

[Stone 1932] M.H. Stone, *Linear transformations in Hilbert space*, AMS, New York, 1932.

[Steklov 1898] W. Steklov, *Sur le problème de refroidissement d'une barre hétérogène*, C. R. Acad. Sci. Paris **126** (1898), 215–218.

[Sturm 1836] C. Sturm, *Mémoire sur les équations différentielles du second ordre*, J. de Maths. Pure et Appl. **1** (1836), 106–186.

[Sturm 1833a/1836a] C. Sturm, *Analyse d'un mémoire sur les propriétés générales des fonctions qui dépendent d'équations différentielles linéares du second ordre*, L'Inst. J. Acad. et Soc. Sci. Nov. 9 (1833) 219–223, summary of *Mémoire sur les Équations diffŕentielles linéaires du second ordre*, J. Math. Pures Appl. **1** (1836), 106–186 [Sept. 28, 1833].

[Sturm 1833b/1836b] C. Sturm [unnamed note], L'Inst. J. Acad. et Soc. Sci. Nov. 9 (1833) 219–223, summary of *Mémoire sur une classe d'Équations à différences partielles*, J. Math. Pures Appl. **1** (1836), 373–444.

[Thomas 1935] L.H. Thomas, *The interaction between a neutron and a proton and the structure of H^3*, Phys. Rev. **47** (1935), 903–909.

[Varadarajan 1989] V.S. Varadarajan, *An Introduction to Harmonic Analysis on Semisimple Lie Groups*, Cambridge University Press, Cambridge, 1989.

[Venkov 1971] A. Venkov, *Expansion in automorphic eigenfunctions of the Laplace operator and the Selberg Trace Formula in the space $SO(n,1)/SO(n)$*, Dokl. Akad. Nauk. SSSR **200** (1971); Soviet Math. Dokl. **12** (1971), 1363–1366.

[Venkov 1979] A. Venkov, *Spectral theory of automorphic functions, the Selberg zeta-function, and some problems of analytic number theory and mathematical physics*, Russian Math. Surveys **34** no. 3 (1979), 79–153.

[vonNeumann 1929] J. von Neumann, *Allgemeine eigenwerttheorie Hermitescher Funktionaloperatoren*, Math. Ann. **102** (1929), 49–131.

[vonNeumann 1931] J. von Neumann, *Die Eindeutigkeit der Schrödingersche Operatoren*, Math. Ann. **104** (1931), 570–578.

[Watson 1918] G.N. Watson, *Asymptotic expansions of hypergeometric functions*, Trans. Cambridge Phil. Soc. **22** (1918), 277–308.

[Weyl 1910] H. Weyl, *Über gewöhnliche Differentialgleichungen mit Singularitäten and die zugehörigen Entwicklungen wilkürlicher Funktionen*, Math. Ann. **68** (1910), 220–269.

[Weyl 1925/1926] H. Weyl, *Theorie der Darstellung kontinuierlicher half-einfacher Gruppen durch lineare Transformationen, I*, Math. Z. **23** (1925), 271–309; *II*, ibid, **24** (1926), 328–376; *III (und Nachtrag)*, ibid, 377–395, 789–791.

[Wiener 1933] N. Wiener, *The Fourier Integral and Certain of Its Applications*, Cambridge University Press, Cambridge, 1933.

[Wigner 1939] E. Wigner, *On unitary representations of the inhomogeneous Lorentz group*, Ann. Math. **40** (1939), 149–204.

[Wong 1990] S.-T. Wong, *The meromorphic continuation and functional equations of cuspidal Eisenstein series for maximal cuspidal groups*, Memoirs of AMS, **83** (1990), no. 423.

Index

Printed in the United States
By Bookmasters